ENVIRONMENTAL ENGINEERING
PE LICENSE REVIEW

Philip J. Parker, PhD, PE, & Ben J. Stuart, PhD, PE

List of Chemical Compound
— names, formulas, constants

 AEC EDUCATION

This publication is designed to provide accurate and authoritative information in regard to the subject matter covered. It is sold with the understanding that the publisher is not engaged in rendering legal, accounting, or other professional service. If legal advice or other expert assistance is required, the services of a competent professional should be sought.

President: Roy Lipner
Vice President of Product Development and Publishing: Evan M. Butterfield
Editorial Project Manager: Laurie McGuire
Director of Production: Daniel Frey
Senior Managing Editor, Production: Jack Kiburz
Creative Director: Lucy Jenkins
Production Artist: Virginia Byrne

Published by Kaplan AEC Education

   30 South Wacker Drive, Suite 2500
   Chicago, IL 60606-7481
   (312) 836-4400
   www.kaplanAECengineering.com

Printed in the United States of America
07 08 09 10 9 8 7 6 5 4 3 2 1

ISBN-13: 978-1-4195-9597-4
ISBN-10:   1-4195-9597-0

PART TWO: APPLICATIONS

OUTLINE

HOW TO USE THIS BOOK

Environmental Engineering: PE License Review contains a conceptual review of topics for the Principles and Practice of Engineering (PE) Exam in environmental engineering, including key terms, equations, analytical methods, and reference data. Because it does not contain practice problems and solutions, the book can be brought into the open-book PE exam as one of your references.

Although it does not contain practice problems, the review contains many solved examples that illustrate how to apply the equations and analytical techniques presented. Some of the example contexts are more complicated than the problems you are likely to find on the exam; this is by design. It is better to be overprepared than underprepared the day of the test.

You will notice quite a few nomographs in the review, including examples of how to use them in solving practical problems. Often, you can derive a sufficiently precise answer using a nomograph faster than by using a calculator. For this reason, they can be very useful in the time-sensitive environment of the PE exam. Remember, the exam is a multiple-choice format; the precision afforded by a nomograph will often be sufficient to enable you to choose the best possible answer from among four choices.

The chapters in *Environmental Engineering: PE License Review* have been developed to relate closely to the exam topics and subtopics identified by the National Council of Examiners for Engineering and Surveying (NCEES). Table I.1 maps exam topics to the book chapters in which they are substantially covered. In addition, Part I of the book covers fundamental topics of problem solving, biology, ecology, fluid mechanics, and hydraulics that are at issue throughout the subsequent chapters. Working through the whole book sequentially is the ideal way to review for the exam. But if you have time constraints and know that there are particular topics that you want to focus on, use Table I.1 to navigate to the relevant discussions.

The exam may include some problems that require an economic analysis—for this reason, Appendix B provides a review of engineering economics. Regulatory issues are summarized in Chapter 4, but you also will find discussion of regulations specific to water, air, and solid waste in the respective chapters. For a more detailed listing of topics that may appear on the exam, visit the NCEES Web site: *www.ncees.org/exams/professional/pe_environmental_exam_specs.pdf*

Table I.1 Environmental engineering PE exam topics

NCEES Defined Topics and Subtopics (% of exam)	Review Section(s) in This Text	Page Numbers in This Text
Water (34%)		
Wastewater (11%)	Chapter 8	321–65
Stormwater (6%)	Chapter 6	183–226
Potable water (11%)	Chapter 7	229–318
Water resources (6%)	Chapter 5	113–82
Air (20%)		
Ambient air (8%)	Chapter 9	367–72
Emissions sources (4%)	Chapter 9	372–77
Control strategies (8%)	Chapter 9	377–88
Solid, Hazardous, and Special Waste (20%)		
Municipal solid waste, commercial, and industrial wastes (10%)	Chapter 10	389–423
Hazardous waste, special, and radioactive waste (10%)	Chapter 11	425–70
Environmental Assessments, Remediation, and Emergency Response (26%)		
Environmental assessments (8%)	Chapter 4	104–9
Remediation (8%)	Chapter 12	471–93
Public health and safety (10%)	Chapter 4	98–104

BECOMING A PROFESSIONAL ENGINEER

To achieve registration as a professional engineer there are four distinct steps: (1) education, (2) the Fundamentals of Engineering/Engineer-In-Training (FE/EIT) exam, (3) professional experience, and (4) the professional engineer (PE) exam, more formally known as the Principles and Practice of Engineering Exam. These steps are described in the following sections.

Education

The obvious appropriate education is a B.S. degree in environmental engineering from an accredited college or university. This is not an absolute requirement. Alternative, but less acceptable, education is a B.S. degree in something other than environmental engineering, or a degree from a non-accredited institution, or four years of education but no degree.

Fundamentals of Engineering (FE/EIT) Exam

Most people are required to take and pass this eight-hour multiple-choice examination. Different states call it by different names (Fundamentals of Engineering, E.I.T., or Intern Engineer), but the exam is the same in all states. It is prepared and graded by the National Council of Examiners for Engineering and Surveying (NCEES). Review materials for this exam are found in other Kaplan AEC books such as *Fundamentals of Engineering: FE/EIT Exam Preparation.*

Experience

Typically one must have four years of acceptable experience before being permitted to take the Professional Engineer exam. Both the length and character of the experience will be examined. It may, of course, take more than four years to acquire four years of acceptable experience.

Professional Engineer Exam

The second national exam is called Principles and Practice of Engineering by NCEES, but just about everyone else calls it the Professional Engineer or PE exam. All states, plus Guam, the District of Columbia, and Puerto Rico, use the same NCEES exam.

ENVIRONMENTAL ENGINEERING PROFESSIONAL ENGINEER EXAM

The reason for passing laws regulating the practice of environmental engineering is to protect the public from incompetent practitioners. Beginning about 1907 the individual states began passing *title* acts regulating who could call themselves environmental engineers. As the laws were strengthened, the *practice* of environmental engineering was limited to those who were registered environmental engineers, or working under the supervision of a registered environmental engineer. There is no national registration law; registration is based on individual state laws and is administered by boards of registration in each of the states. A listing of the state boards is in Table I.2.

Table I.2 State boards of registration for engineers

State	Web site	Telephone
AL	www.bels.alabama.gov	334-242-5568
AK	www.commerce.state.ak.us/occ/pael.cfm	907-465-1676
AZ	www.btr.state.az.us	602-364-4930
AR	www.arkansas.gov/pels	501-682-2824
CA	Dca.ca.gov/pels/contacts/htm	916-263-2230
CO	www.dora.state.co.us/aes	303-894-7788
CT	State.ct.us/dcp	860-713-6145
DE	www.dape.org	302-368-6708

(continued)

State	Web site	Telephone
DC	www.asisvcs.com/indhome _fs.asp?cpcat=en09statereg	202-442-4320
FL	www.fbpe.org	850-521-0500
GA	www.sos.state.ga.us/plb/pels/	478-207-1450
GU	www.guam-peals.org	671-646-3138 or 3115
HI	www.Hawaii.gov/dcca/pbl	808-586-2702
ID	www.ipels.idaho.gov	208-373-7210
IL	www.idfpr.com	217-524-3211
IN	www.in.gov/pla/bandc/engineers	317-234-3022
IA	www.state.ia.us/engls	515-281-7360
KS	www.kansas.gov/ksbtp	785-296-3054
KY	www.kyboels.ky.gov	502-573-2680
LA	www.lapels.com	225-925-6291
ME	www.maine.gov/professionalengineers/	207-287-3236
MD	www.dllr.state.md/us	410-230-6322
MA	www.mass.gov/dpl/boards/en/	617-727-9957
MI	www.michigan.gov/engineers	517-241-9253
MN	www.aelslagid.state.mn.us	651-296-2388
MS	www.pepls.state.ms.us	601-359-6160
MO	www.pr.mo.gov/apelsla.asp	573-751-0047
MP		(011) 670-664-4809
MT	www.engineer.mt.gov	406-841-2367
NE	www.ea.state.ne.us	402-471-2021
NV	www.boe.state.nv.us	775-688-1231
NH	www.state.nh.us/jtboard/home.htm	603-271-2219
NJ	www.state.nj.us/lps/ca/nonmedical/pels.htm	973-504-6460
NM	www.state.nm.us pepsboard	505-827-7561
NY	www.op.nysed.gov	518-474-3817 x140
NC	www.ncbels.org	919-791-2000
ND	www.ndpelsboard.org/	701-258-0786
OH	www.ohiopeps.org	614-466-3651
OK	www.pels.state.ok.us/	405-521-2874
OR	www.osbeels.org	503-362-2666
PA	www.dos.state.pa.us/eng	717-783-7049
PR	www.estado.gobierno.pr/ingenieros.htm	787-722-2122 x232

(continued)

State	Web site	Telephone
RI	www.bdp.state.ri.us	401-222-2565
SC	www.llr.state.sc.us/POL/Engineers	803-896-4422
SD	www.state.sd.us/dol/boards/engineer	605-394-2510
TN	www.state.tn.us/commerce/boards/ae/	615-741-3221
TX	www.tbpe.state.tx.us	512-440-7723
UT	www.dopl.utah.gov	801-530-6396
VT	vtprofessionals.org	802-828-2191
VI	www.dlca.gov.vi/pro-aels.html	340-773-2226
VA	www.dpor.virginia.gov	804-367-8512
WA	www.dol.wa.gov/engineers/engfront.htm	360-664-1575
WV	www.wvpebd.org	304-558-3554
WI	www.drl.state.wi.us	608-266-2112
WY	engineersandsurveyors.wy.gov	307-777-6155

Examination Development

Initially the states wrote their own examinations, but beginning in 1966 the NCEES took over the task for some of the states. Now the NCEES exams are used by all states. This greatly eases the ability of an environmental engineer to move from one state to another and achieve registration in the new state.

The development of the environmental engineering exam is the responsibility of the NCEES Committee on Examinations for Professional Engineers. The committee is composed of people from industry, consulting, and education, plus consultants and subject matter experts. The starting point for the exam is an environmental engineering task analysis survey, which NCEES does at roughly 5- to 10-year intervals. People in industry, consulting, and education are surveyed to determine what environmental engineers do and what knowledge is needed. From this NCEES develops what it calls a "matrix of knowledge" that forms the basis for the environmental engineering exam structure.

The actual exam questions are prepared by the NCEES committee members, subject matter experts, and other volunteers. All people participating must hold professional registration. Using workshop meetings and correspondence by mail, the questions are written and circulated for review. The problems relate to current professional situations. They are structured to quickly orient one to the requirements, so that the examinee can judge whether he or she can successfully solve it. Although based on an understanding of engineering fundamentals, the problems require the application of practical professional judgment and insight.

Examination Structure

The exam is organized into morning and afternoon four-hour sessions. Each session contains 50 multiple-choice questions with four answer choices each (A, B, C,

D). The exam is objectively scored by computer. See Table 1 for a list and percentages of topics covered. By looking at the percentages, you can calculate how many questions to expect on each topic.

Exam Dates

The National Council of Examiners for Engineering and Surveying (NCEES) prepares Environmental Engineering Professional Engineer exams for use on a Friday in April and October of each year. Some state boards administer the exam twice a year in their state, whereas others offer the exam once a year. The scheduled exam dates are:

	April	October
2007	20	26
2008	11	24
2009	24	23
2010	16	29

People seeking to take a particular exam must apply to the state board several months in advance.

Exam Procedure

Before the morning four-hour session begins, proctors will pass out an exam booklet, answer sheet, and mechanical pencil to each examinee. The provided pencil is the only writing instrument you are permitted to use during the exam. If you need an additional pencil during the exam, a proctor will supply one.

Fill in the answer bubbles neatly and completely. Questions with two or more bubbles filled in will be marked as incorrect, so if you decide to change an answer, be sure to erase your original answer completely.

The afternoon session will begin following a one-hour lunch break.

In both the morning and afternoon sessions, if you finish more than 15 minutes early you may turn in your booklet and answer sheet and leave. In the last 15 minutes, however, you must remain to the end of the exam in order to ensure a quiet environment for those still working and an orderly collection of materials.

Preparing For and Taking the Exam

Give yourself time to prepare for the exam in a calm and unhurried way. Many candidates like to begin several months before the actual exam. Target a number of hours per day or week that you will study, and reserve blocks of time for doing so. Creating a review schedule on a topic-by-topic basis is a good idea. Remember to allow time for both reviewing concepts and solving practice problems. You may want to prioritize the time you spend reviewing specific topics according to their relative weight on the exam, as identified by NCEES, or by your areas of strength and weakness.

In addition to reviewing material on your own, you may want to join a study group or take a review course. A group study environment might help you stay committed to a study plan and schedule. Group members can create additional practice problems for one another and share tips and tricks.

People familiar with the psychology of exam taking have several suggestions for people as they prepare to take an exam.

1. Exam taking really involves two skills. One is the skill of illustrating knowledge that you know. The other is the skill of exam taking. The first may be enhanced by a systematic review of the technical material. Exam-taking skills, on the other hand, may be improved by practice with similar problems presented in the exam format.

2. Since there is no deduction for guessing on the multiple choice problems, answers should be given for all of them. Even when one is going to guess, a logical approach is to attempt to first eliminate one or two of the four alternatives. If this can be done, the chance of selecting a correct answer obviously improves from 1 in 4 to 1 in 3 or 1 in 2.

3. Plan ahead with a strategy. Which is your strongest area? Can you expect to see several problems in this area? What about your second strongest area? What is your weakest area?

4. Plan ahead with a time allocation. You might allocate a little less time per problem for those areas in which you are most proficient, leaving a little more time in subjects that are difficult for you. Your time plan should include a reserve block for especially difficult problems, for checking your scoring sheet, and to make last-minute guesses on problems you did not work. Your strategy might also include time allotments for two passes through the exam—the first to work all problems for which answers are obvious to you, and the second to return to the more complex, time-consuming problems and the ones at which you might need to guess. A time plan gives you the confidence of being in control and keeps you from making the serious mistake of misallocating time in the exam.

5. Read all four multiple-choice answers before making a selection. An answer in a multiple-choice question is sometimes a plausible decoy—not the best answer.

6. Do not change an answer unless you are absolutely certain you have made a mistake. Your first reaction is likely to be correct.

7. Do not sit next to a friend, a window, or other potential distractions.

Exam Day Preparations

The exam day will be a stressful and tiring one. This will be no day to have unpleasant surprises. For this reason we suggest that an advance visit be made to the examination site. Try to determine such items as

1. How much time should I allow for travel to the exam on that day? Plan to arrive about 15 minutes early. That way you will have ample time, but not too much time. Arriving too early, and mingling with others who also are anxious, will increase your anxiety and nervousness.

2. Where will I park?

3. How does the exam site look? Will I have ample workspace? Where will I stack my reference materials? Will it be overly bright (sunglasses), cold (sweater), or noisy (earplugs)? Would a cushion make the chair more comfortable?

4. Where are the drinking fountains and lavatory facilities?

5. What about food? Should I take something along for energy in the exam? A bag lunch during the break probably makes sense.

What to Take to the Exam

The NCEES guidelines say you may bring only the following reference materials and aids into the examination room for your personal use:

1. Handbooks and textbooks, including the applicable design standards.

2. Bound reference materials, provided the materials remain bound during the entire examination. The NCEES defines "bound" as books or materials fastened securely in their covers by fasteners that penetrate all papers. Examples are ring binders, spiral binders and notebooks, plastic snap binders, brads, screw posts, and so on.

3. Battery-operated, silent, nonprinting, noncommunicating calculators. Beginning with the April 2004 exam, NCEES has implemented a more stringent policy regarding permitted calculators. For more details, see the NCEES website (*www.ncees.org*), which includes the updated policy and a list of permitted calculators. You also need to determine whether or not your state permits preprogrammed calculators. Bring extra batteries for your calculator just in case; many people feel that bringing a second calculator is also a very good idea.

At one time NCEES had a rule barring "review publications directed principally toward sample questions and their solutions" in the exam room. This set the stage for restricting some kinds of publications from the exam. *State boards may adopt the NCEES guidelines, or adopt either more or less restrictive rules.* Thus an important step in preparing for the exam is to know what will—and will not—be permitted. We suggest that if possible you obtain a written copy of your state's policy for the specific exam you will be taking. Occasionally there has been confusion at individual examination sites, so a copy of the exact applicable policy will not only allow you to carefully and correctly prepare your materials, but will also ensure that the exam proctors will allow all proper materials that you bring to the exam.

As a general rule we recommend that you plan well in advance what books and materials you want to take to the exam. Then they should be obtained promptly so you use the same materials in your review that you will have in the exam.

License Review Books

The review books you use to prepare for the exam are good choices to bring to the exam itself. After weeks or months of studying, you will be very familiar with their organization and content, so you'll be able to quickly locate the material you want to reference during the exam. Keep in mind the caveat just discussed—some state boards will not permit you to bring in review books that consist largely of sample questions and answers.

Textbooks

If you still have your university textbooks, they are the ones you should use in the exam, unless they are too out of date. To a great extent the books will be like old friends with familiar notation.

Bound Reference Materials

The NCEES guidelines suggest that you can take any reference materials you wish, so long as you prepare them properly. You could, for example, prepare several volumes of bound reference materials, with each volume intended to cover a particular category of problem. Maybe the most efficient way to use this book would be to cut it up and insert portions of it in your individually prepared bound materials. Use tabs so that specific material can be located quickly. If you do a careful and systematic review of environmental engineering, and prepare a lot of well-organized materials, you just may find that you are so well prepared that you will not have left anything of value at home.

Other Items

In addition to the reference materials just mentioned, you should consider bringing the following to the exam:

- *Clock*—You must have a time plan and a clock or wristwatch.

- *Exam assignment paperwork*—Take along the letter assigning you to the exam at the specified location. To prove you are the correct person, also bring something with your name and picture.

- *Items suggested by advance visit*—If you visit the exam site, you probably will discover an item or two that you need to add to your list.

- *Clothes*—Plan to wear comfortable clothes. You probably will do better if you are slightly cool.

- *Box for everything*—You need to be able to carry all your materials to the exam and have them conveniently organized at your side. Probably a cardboard box is the answer.

ACKNOWLEDGMENTS

Several reviewers provided valuable insight and suggestions during the development of this book. The authors and publisher are grateful to the following:

Janet L. Baldwin, PhD
Roger Williams University

Mackenzie L. Davis, PhD, PE
Michigan State University (emeritus)

Bruce DeVantier, PhD, PE
Southern Illinois University

Raghava R. Kommalapati, PhD, PE
Prairie View A & M University

Dennis D. Truax, PhD, PE
Mississippi State University

PART ONE

Fundamentals

Solving Environmental Engineering Problems

This chapter provides a review of fundamental information in order to help you avoid some common errors when solving problems on the Environmental Engineering PE exam. The use of units in solving environmental engineering problems is emphasized because the ability to obtain a correct answer on a timed multiple-choice exam requires you to be able to efficiently convert units in addition to having a sound understanding of the concepts. This chapter also includes a review of chemistry fundamentals required to understand various environmental engineering applications. Finally, this chapter presents the mass balance approach to solving environmental engineering problems and demonstrates this approach in solving problems involving reactors.

1.1 PRINCIPLES OF MATTER

The *atomic number* (*Z*) is defined as the number of protons present in the nucleus of an element. The *atomic weight* (*A*) is defined as the weight of one *mole* (6.022 × 10²³ atoms per mole) of a substance. While this weight is generally assumed to be contributed by the protons and neutrons in the nucleus, and therefore should be equal to twice the atomic number, most elements may have several isotopes. *Isotopes* are atoms with the same atomic number but that possess different numbers of neutrons, thus yielding a single element with several different atomic weights. In order to report a single atomic weight, a weighted average of all naturally occurring isotopes is used, based upon the mass of a carbon-12 atom. Most chemical species of interest to the environmental engineer contain multiple elements that are bound to form stable molecules. In these cases, the *molecular weight* (*MW*), also called the *formula weight* (*FW*), of that compound can be determined by summing the atomic weights of each of the elements that make up the molecule. A copy of the Periodic Table of the Elements is provided in Appendix C.

In acid/base reactions and the treatment of aqueous solutions that contain high quantities of dissolved inorganic species (for example, the removal of hardness in water treatment plants), it is useful to simplify the stoichiometric calculations of reactions occurring in those solutions. In these cases, it is convenient to define the *equivalent weight* (*EW*) of a compound (often called the *equivalent*) as *the formula weight divided by its equivalency*, where the equivalency can be defined as (1) *the absolute value of the ion charge*, (2) *the number of [H⁺] or [OH⁻] ions that the species may react with*, or (3) *the absolute value in the change in valency during a redox reaction*. Table 1.1 (elements) and Table 1.2 (molecules) contain several compounds that are commonly encountered in environmental engineering with their corresponding atomic or molecular weight and equivalent weights.

Table 1.1   Common elements in environmental engineering

Name	Symbol	Atomic Weight	EW
Aluminum	Al	27.0	9.0
Calcium	Ca	40.1	20.0
Carbon	C	12.0	
Chlorine	Cl	35.5	35.5
Fluorine	F	19.0	19.0
Hydrogen	H	1.0	1.0
Iodine	I	127	127
Iron	Fe	55.8	27.9
Magnesium	Mg	24.3	12.2
Manganese	Mn	54.9	27.5
Nitrogen	N	14.0	
Oxygen	O	16.0	
Phosphorus	P	31.0	
Potassium	K	39.1	39.1
Sodium	Na	23.0	23.0
Sulfur	S	32.0	16.0

Table 1.2   Common compounds in water treatment processes

Name	Formula	MW	EW
Aluminum hydroxide	$Al(OH)_3$	78.0	26.0
Aluminum sulfate	$Al_2(SO_4)_3 \cdot 14.3H_2O$	600	100
Ammonia	NH_3	17.0	
Ammonium	NH_4^+	18.0	18.0
Ammonium fluorosilicate	$(NH_4)_2SiF_6$	178	
Bicarbonate	HCO_3^-	61.0	61.0
Calcium bicarbonate	$Ca(HCO_3)_2$	162	81.0
Calcium carbonate	$CaCO_3$	100	50.0
Calcium fluoride	CaF_2	78.1	
Calcium hydroxide	$Ca(OH)_2$	74.1	37.0
Calcium hypochlorite	$Ca(ClO)_2 \cdot 2H_2O$	179	
Calcium oxide	CaO	56.1	28.0
Calcium sulfate	$CaSO_4$	136	68.0
Carbonate	CO_3^{2-}	60.0	30.0
Carbon dioxide	CO_2	44.0	22.0
Chlorine	Cl_2	71.0	35.5
Chlorine dioxide	ClO_2	67.0	
Ferric chloride	$FeCl_3$	162	54.1
Ferric hydroxide	$Fe(OH)_3$	107	35.6
Fluorosilicic acid	H_2SiF_6	144	
Hydroxyl ion	OH^-	17.0	17.0
Hypochlorite	OCl^-	51.5	51.5
Magnesium carbonate	$MgCO_3$	84.3	42.1
Magnesium hydroxide	$Mg(OH)_2$	58.3	29.1
Magnesium sulfate	$MgSO_4$	120	60.1
Nitrate	NO_3^-	62.0	62.0
Orthophosphate	PO_4^{3-}	95.0	31.7
Oxygen	O_2	32.0	16.0
Sodium bicarbonate	$NaHCO_3$	84.0	84.0
Sodium carbonate	Na_2CO_3	106	53.0
Sodium hydroxide	$NaOH$	40.0	40.0
Sodium hypochlorite	$NaClO$	74.4	
Sodium fluorosilicate	Na_2SiF_6	188	
Sodium sulfate	Na_2SO_4	142	71.0
Sulfate	SO_4^{2-}	96.0	48.0

$$EW = \frac{MW}{[H^+] \text{ or } [OH^-] \text{ equivalents}} = \frac{eq}{g} \quad UNITS$$

| Example **1.1** | Molecular and equivalent weights |

Molecular and equivalent weights

Determine the molecular and equivalent weight of sulfuric acid, H_2SO_4.

Solution

The MW is determined by summing the atomic weights of each of the contributing atoms as

$$MW_{H_2SO_4} = 2\,MW_H + MW_S + 4\,MW_O = (2)(1.0079) + 32.064 + (4)(15.9994)$$
$$= 98.0774$$

Often, solution accuracy does not require six significant figures, and a value of 98.1, or even 98, is acceptable for most calculations. Also note that the units are generally accepted to be g/mol.

Calculation of the equivalent weight requires the determination of the equivalency of sulfuric acid. Using the first definition of equivalency given above, the absolute value of the ion charge of the sulfate ion is 2, and thus 2 is the equivalency of sulfuric acid. However, if the ion charge of hydrogen is used, it might be assumed that the equivalency is equal to 1. The second definition given above helps determine the correct equivalency as it is seen that 2 hydrogen ions are available to react, yielding the true equivalency of 2. The EW may now be calculated as

$$EW_{H_2SO_4} = \frac{MW_{H_2SO_4}}{equivalency} = \frac{98.0774}{2} = 49.0387 \approx 49.0$$

In its most simple characterization, matter is said to exist in three possible phases: solid, liquid, and gas. While other states are possible, these three phases (along with all of the possible combinations of multiphase systems) are usually sufficient to describe nearly all situations an engineer will encounter in natural environments. The determination of which phase a substance will take is dependent on the specific compound, along with the prevailing temperature and pressure of the environment where the substance is located.

Solids are generally considered the lowest energy state, although all matter at a temperature above absolute zero has molecular energies. Solids have *mass densities* (ρ) [mass per unit volume], or equivalently *specific weights* (γ) [weight per unit volume], that are greater than liquids or gases. The most notable exception to this rule is water, which is slightly less dense as a solid (ice) than it is as a liquid. Solids are usually further defined on a microscopic level by their crystalline structure, or on a macroscopic level by their particle size distribution. Examples of solids that are important to environmental engineers include soils, sand beds for filtration or drying, precipitation reactions and floc management of colloidal or settleable solids in water treatment, and particulate matter in gases from combustion processes or fugitive dusting environments.

Fluids are defined as *substances that deform continuously when acted upon by any shear stress*, or more simply as *substances that take the shape of their container*. Both liquids and gases fall into this category, but material properties are quite different for each type. While liquids possess densities that are similar to that of their solid forms, the gas phase of a substance has a density that is approximately 1000 times smaller. Further, although both types of fluids are able to be poured from their container, the degree of flowability of gases is approximately

1000 greater than the liquid phase. The measure of flowability is called *dynamic* (or *absolute*) *viscosity* (μ) and is most commonly expressed in units of centipoises (*cP*). For liquids, increasing temperature usually reduces viscosity, while reducing temperature increases viscosity until such a point at which the material turns into a solid and no longer flows.

Gases at or near atmospheric pressure are said to behave ideally, referring to the fact that there exists a direct relationship between gas density and the system temperature and pressure. The *ideal gas law* can be expressed as

$$PV = nRT \qquad \text{or} \qquad P = \rho' RT$$

IDEAL GAS LAW

where P is absolute pressure, V is the gas volume, n is the number of moles of gas, R is the universal gas constant, T is the absolute temperature, and ρ' is the molar density. Several values of the universal gas constant can be found for a variety of convenient units in Table 1.3. Also, note the molar density is reported as moles per unit volume. This must be converted into a mass density (mass per unit volume) by multiplying by the species molecular weight.

$\rho' = \dfrac{mol}{Vol}$

Table 1.3   Values for the Universal Gas Constant in various units

Value	Units
8.314	m³•Pa/mol•K
0.08314	liter•bar/mol•K
0.08206	liter•atm/mol•K
62.36	liter•mm Hg/mol•K
0.7302	ft³•atm/lbmole•°R
10.73	ft³•psia/lbmole•°R
8.314	J/mol•K
1.987	cal/mol•K
1.987	Btu/lbmole•°R

1.2   UNITS IN ENVIRONMENTAL ENGINEERING

If you are very familiar with unit conversions, you may want to skip Section 1.2. However, if you feel a bit "rusty," this section will provide you with a useful and focused review.

1.2.1 Dimensionality

In this review text, variables in equations will be defined, and the units for those variables will be presented in dimensional format. In other words, each variable will be presented in terms of mass (M), length (L), and time (T). For example, velocity will be represented in terms of M/T (or $M \cdot T^{-1}$); volume will be represented as L^3; and kinematic viscosity will expressed as (L^2/T). Thus, the use of metric or English units can be readily used with the equations.

One exception to the use of the dimensional format is when empirical models are used. In such case, the units for each variable must be used exactly as specified in the equation definition. One commonly used empirical equation,

Manning's equation, is presented in Chapter 3, "Fluid Mechanics and Hydraulics," as follows:

$$V = \frac{1.00}{n} R_H^{2/3} S^{1/2} \quad \text{(SI units)}$$

$$V = \frac{1.49}{n} R_H^{2/3} S^{1/2} \quad \text{(English units)}$$

For the use of Manning's equation in SI units, the hydraulic radius, R_H, must be expressed in meters to yield a velocity (V) of m/s; in English units, R_H must be expressed in feet to yield a velocity in ft/s. All unit conversions must be completed prior to substituting those variables into the Manning's equation. In other words, for the English units, if R_H is calculated to be 6 inches, a value of 0.5 feet must be substituted into the equation; substituting in 6 inches for R_H and then expecting an answer in inches/second would be incorrect.

Example 1.2

Dimensionality as a problem-solving strategy

Show how the equation for power gained by a fluid due to a pump is dimensionally homogeneous, and calculate the power (in watts; $1W = \frac{1 \text{ kg} \cdot \text{m}^2}{\text{s}^3}$) given a flow rate of water of 5 gpm and an increase in head (h_p) equal to 10 meters. The pertinent equation follows:

$$\dot{P} = Q\gamma h_p$$

where $\dot{P}$ = power gained by fluid ($M \cdot L^2 \cdot T^{-3}$)

$\quad Q$ = flow ($L^3 \cdot T^{-1}$)

$\quad \gamma$ = specific weight of the fluid ($M \cdot T^{-2} \cdot L^{-2}$)

$\quad h_p$ = gain in head of fluid (L)

Solution

Note that this example is poorly presented (it has a mix of SI and English units) on purpose. The equation is verified to be dimensionally homogeneous as follows:

$$\frac{M \cdot L^2}{T^3} = \frac{L^3}{T} \cdot \frac{M}{T^2 L^2} L$$

$$= \frac{L^4 \cdot M}{L^2 T^3}$$

$$= \frac{L^2 \cdot M}{T^3}$$

Next, the problem will be solved incorrectly on purpose, to show how knowledge of dimensional homogeneity can prevent common mistakes from occurring.

$$\dot{P} = Q\gamma h_p$$

$$= \left(\frac{5 \text{ gal}}{\text{min}} \cdot \frac{1 \text{ m}^3}{264.172 \text{ gal}} \cdot \frac{1 \text{ min}}{60 \text{ sec}} \right) \left(\frac{1000 \text{ kg}}{\text{m}^3} \right) (10 \text{ m})$$

$$= 0.32 \frac{\text{kg} \cdot \text{m}}{\text{sec}}$$

INCORRECT SOLUTION

density NOT specific weight

At this point, it is clear that an error has occurred, given that the units (kg · m/sec) are not dimensionally similar to W. The problem occurred because the density of water was used instead of the specific weight of water, a very common mistake. Redoing the problem correctly, the final answer is found to be:

$$\dot{P} = \left(\frac{5 \text{ gal}}{\text{min}} \cdot \frac{1 \text{ m}^3}{264.172 \text{ gal}} \cdot \frac{1 \text{ min}}{60 \text{ sec}} \right) \left(\frac{9.81 \cdot 10^3 N}{\text{m}^3} \right) (10 \text{ m})$$

$$= 31 \text{ W}$$

Table 1.4 through Table 1.13 can be very useful while taking the PE exam, and you are encouraged to flag these pages.

Table 1.4   Unit conversions: area

1 acre = 4.047×10^3 m²
1 acre = 1.563×10^{-3} mi²
1 acre = 4.356×10^4 ft²
1 acre = 0.405 hectare
1 cm² = 1.076×10^{-3} ft²
1 ft² = 0.093 m²
1 ft² = 2.296×10^{-5} acres
1 ft² = 929.03 cm²
1 hectare = 1×10^4 m²
1 in² = 6.452 cm²
1 m² = 10.764 ft²
1 mi² = 640 acres

Table 1.5   Unit conversions: volume

1 acre · ft = 1.233×10^3 m³
1 acre · ft = 4.356×10^4 ft³
1 acre · ft = 3.259×10^5 gal
1 cm³ = 3.531×10^{-5} ft³
1 cm³ = 1 mL
1 ft³ = 0.028 m³
1 ft³ = 7.481 gal
1 ft³ = 28.317 L
1 gal = 3.785×10^{-3} m³
1 gal = 0.134 ft³
1 gal = 3.785 L
1 L = 0.035 ft³
1 L = 0.264 gal
1 m³ = 35.315 ft³
1 m³ = 1×10^3 L
1 m³ = 264.172 gal
1 yd³ = 27 ft³
1 yd³ = 0.765 m³
1 yd³ = 201.974 gal

Table 1.6   Unit conversions: pressure

1 atm = 1.013×10^5 Pa
1 ft H_2O = 0.433 psi
1 Pa = 9.869×10^{-6} atm
1 Pa = 1.45×10^{-4} psi
1 Pa = 2.961×10^{-4} in Hg
1 psi = 6.895×10^3 Pa
1 psi = 144 lbf/ft^2
1 Pa = 1 N/m^2
1 Pa = 3.347×10^{-4} ft H_2O

Table 1.7   Unit conversions: mass

1 g = 2.205×10^{-3} lb
1 grain = 0.065 g
1 g = 15.432 grain
1 kg = 2.205 lb
1 lb = 0.454 kg
1 lb = 7×10^3 grain
1 ton = 2×10^3 lb
1 ton = 907.185 kg
1 tonne = 1×10^3 kg
1 tonne = 1.102 ton
1 slug = 14.594 kg

Table 1.8   Unit conversions: velocity

1 ft/s = 0.305 m/s
1 ft/s = 0.682 mi/h
1 km/h = 0.278 m/s
1 m/s = 3.281 ft/s
1 mi/h = 0.447 m/s
1 mi/h = 1.467 ft/s

Table 1.9   Unit conversions: flow rate (or discharge)

1 cfs = 0.028 m^3/s
1 cfs = 0.992 acre · in/h
1 cfs = 448.831 gpm
1 gpm = 6.309×10^{-5} m^3/s
1 gpm = 2.228×10^{-3} cfs
1 gpm = 1.44×10^{-3} MGD
1 MGD = 0.044 m^3/s
1 MGD = 1.547 cfs
1 MGD = 1.547 ft^3/s
1 m^3/s = 1.585×10^4 gpm
1 m^3/h = 9.81×10^{-3} cfs

Table 1.10   Unit conversions: power

1 ft · lbf/s = 1.356 W
1 ft · lbf/s = 1.285×10^{-3} BTU/s
1 hp = 745.7 W
1 hp = 550 ft · lbf/s
1 J = 1 N · m
1 kW = 1.341 hp
1 W = 3.412 BTU/h
1 W = 1 J/s

Table 1.11   Unit conversions: force

1 dyne = 1×10^{-5} N
1 lbf = 4.448 N
1 N = 1 kg · m/s^2
1 N = 0.225 lbf

Table 1.12   Unit conversions: length

1 cm = 0.394 in
1 cm = 0.033 ft
1 ft = 0.305 m
1 in = 0.025 m
1 km = 3.281×10^3 ft
1 km = 0.621 mi
1 m = 3.281 ft
1 m = 39.37 in
1 m = 1×10^6 μm
1 m = 1×10^9 nm
1 mi = 1.609×10^3 m
1 mi = 5.28×10^3 ft
1 mi = 1.609 km

Table 1.13   Unit conversions: energy

1 BTU = 1.055×10^3 J
1 BTU = 251.996 cal
1 BTU = 2.931×10^{-4} kW · hr
1 BTU = 1×10^{-5} therm
1 cal = 1.163×10^{-6} kW · hr
1 cal = 4.187 J
1 cal = 3.968×10^{-3} BTU
1 erg = 1×10^{-7} J
1 J = 9.478×10^{-4} BTU
1 kW · hr = 3.6×10^6 J
1 kW · hr = 8.598×10^5 cal
1 therm = 1.055×10^8 J
1 therm = 1×10^5 BTU

Example **1.3**

Units manipulation

A 1 MGD (million gallons per day) water treatment facility uses alum as a coagulant at a concentration of 40 mg/L. How many pounds of alum does the facility use each day?

Solution

$$10^6 \frac{\text{gal}}{\text{day}} \cdot 40 \frac{\text{mg}}{\text{L}} \cdot \frac{1 \text{ g}}{1000 \text{ mg}} \cdot \frac{3.785 \text{L}}{1 \text{ gal}} \cdot \frac{1 \text{ lb}}{454 \text{ g}} = 333 \frac{\text{lb}}{\text{day}}$$

There are three lessons to be learned from this simple problem that are pertinent to taking the PE exam:

1. The problem could be solved without knowing anything about drinking water treatment! Indeed, a solid understanding of units manipulation is all that is necessary to solve this problem. Although such problems may be rare on the PE exam, this problem-solving strategy could still be very helpful.

2. Note how the unit conversion calculation is set up. Although time pressures are very real when taking the PE exam, test takers are strongly recommended to set up unit conversions as shown. Many mistakes occur when test takers flip a conversion (that is multiply instead of divide). Making a mistake like this will generate an answer that is every bit as wrong (on a multiple-choice test) as an answer that you might obtain by using the wrong equation and/or theory. Such is the nature of a multiple-choice exam.

3. Note that the solution could also be written as 1 MGD × 40 mg/L × 8.34, where the latter factor (8.34) takes care of all of the unit conversions. This conversion factor is very commonly used in environmental engineering problem solving.

1.2.1 Concentration Units

Probably the most common means of expressing the amount of solute in solution is the use of mass concentrations. Many environmental solutions of concern deal with water (aqueous) samples, and concentrations are expressed as mass per unit volume, such as milligrams of solute per liter of solution (mg/L) or micrograms of solute per liter of solution (μg/L). However, it is important to be familiar with the many systems employed to express concentrations of solutions in environmental systems. Molar concentrations are generally employed when dealing with equilibrium chemistry and are expressed in terms of *molarity (M)* as *moles of solute per liter of solution*. Another related means of expressing concentration is *molality (m)*, or molal concentrations, which may be defined as *moles of solute per kilogram of solvent*. In solution reactions where it is convenient to use species equivalents (like acid/base and water-softening reactions), it is most common to use the *normality (N)* of the solution, which is expressed as *equivalents of solute per liter of solution*. The use of equivalents is convenient in that various species may be compared, and even combined, using the second definition for equivalency given previously in terms of reactivity, without the need for specific balanced chemical equations.

To convert from units of mg/L (or g/L) to M (moles/L), the molecular weight of the compound must be used. A periodic table listing atomic masses is required to complete these calculations. The conversion from mg/L to moles per liter, or mg/L to mmol/L is given by

$$X \frac{mg}{L} \div \left(MW \frac{g}{mol} \times 1000 \frac{mg}{g} \right) \rightarrow Y \frac{mol}{L}$$

$$X \frac{mg}{L} \div MW \frac{mg}{mmol} \rightarrow Y \frac{mmol}{L}$$

Note again that MW has units of g/mol, which is also equivalent to mg/mmol. It is also useful to note that molarity is related to normality by the equivalency and can be expressed as

$$N = (M) \, (equivalency)$$

Air pollutants are also usually expressed as mass per unit volume, such as micrograms of contaminant per cubic meter of air ($\mu g/m^3$). If the gas is emitted at temperatures or pressures that are not standard (STP are 0°C and 101.325 kPa) or with high moisture content, concentrations are often corrected to micrograms of contaminant per dry standard cubic meter ($\mu g/dscm$). This requires calculations to remove the water mass from the air and use of the ideal gas law for temperature and pressure correction.

Concentrations of species in solids (e.g., soil contamination) and semisolids (such as sludges) are expressed as mass per unit mass, such as milligrams of solute per kilogram of solid (mg/kg) or micrograms of solute per kilogram of solid ($\mu g/kg$). Because the ratio of mg/kg is 10^{-6} and $\mu g/kg$ is 10^{-9}, mass per unit mass concentrations are often expressed as *parts per million (ppm)* or *parts per billion (ppb)*.

In air pollution, when the contaminant species is also present in the gas phase, it is common to see the 10^{-6} or 10^{-9} ratio expressed as ppmv or ppbv. However, the "v" indicates that these are volume ratios, such as milliliters of contaminant per cubic meter of gas, or microliters of contaminant per cubic meter gas. Engineers should be able to convert air concentrations between $\mu g/m^3$ and ppmv using the ideal gas law and molar weights. Further, because the mass of one liter of water is approximately one kilogram, it is not uncommon to see aqueous solution concentrations expressed in ppm and ppb units, although the mass per unit volume units are more appropriate.

Example **1.4**

Concentration units I

The secondary water standard for SO_4^{2-} is 250 mg/L. Express this concentration in units of normality.

Solution

Normality has units of equivalents per liter. First, we need to find the equivalent weight of sulfate, which is 48.031 eq/g, which may also be expressed as 48.031 meq/mg. We can now simply convert units as follows:

$$250 \frac{mg}{L} \times \frac{1 \, meq}{48.031 \, mg} = 5.2 \frac{meq}{L} = 5.2 \, mN$$

Example **1.5**

Concentration units II

The National Ambient Air Quality Standard (NAAQS) for the criteria pollutant carbon monoxide is 40,000 µg/m³. Express this concentration in units of ppmv if the temperature is 25°C.

Solution

First, determine the number of moles of CO:

$$\text{MW}_{CO} = 28 \text{ g therefore } 40 \text{ mg CO} = \frac{0.04 \text{ g CO}}{28 \dfrac{\text{g CO}}{\text{mol CO}}} = 0.00143 \text{ mol CO}$$

Next, determine the molar volume of a gas at 1 atm and 25°C:

$$\frac{V}{n} = \frac{RT}{P} = \frac{\left(0.08206 \dfrac{\text{L} \cdot \text{atm}}{\text{mol} \cdot \text{K}}\right)(298.15 \text{ K})}{1 \text{ atm}} \Rightarrow 24.47 \frac{\text{L}}{\text{mol}} = 0.0245 \frac{\text{m}^3}{\text{mol}}$$

Now, calculate the volume of gas occupied by 0.00143 moles of gas:

$$0.0245 \text{ m}^3/\text{mol} \times 0.00143 \text{ mol} = 0.000035 \text{ m}^3$$

Finally, determine the volume ratio using 1 m³ of air as a basis:

$$[CO] = \frac{0.000035 \text{ m}^3}{1 \text{ m}^3} = 35 \times 10^{-6} \frac{\text{m}^3}{\text{m}^3} \text{ or equivalently } [CO] = 35 \text{ ppmv}$$

1.3   CHEMISTRY FUNDAMENTALS

Environmental engineers are often tasked with the design of treatment units or systems that address the negative impacts of waste materials that are discharged to the environment. These tasks require a solid understanding of the fundamentals that describe the materials of concern, not only their composition but also how those compounds interact with other substances. It is primarily the ability to manipulate the laws of basic and life sciences that allows the environmental engineer to design effective and efficient treatment units. To that end, examinees should be familiar with reaction order and kinetics, equilibrium chemistry, and redox reactions. This section also provides an introduction to the basic water quality indicators that form the basis for the treatment of water and wastewaters.

1.3.1 Reaction Order and Kinetics

Elements or molecules that combine or dissociate in the presence of specific compounds are said to have reacted to form new products. Chemical reactions occur in specific combinations and ratios, dependent on the energies of the specific compounds participating in a given reaction. The accounting procedure that is used to ensure all reactants are present in the products (adherence to the laws of conservation of mass) is called *stoichiometry*. It is critical to ensure that all reacting and forming species are included in the chemical reactions written, and that all reactions are balanced prior to any further assessment of a reactive system.

Some reactions may be assumed to be permanent such that the products cannot be further changed to reform the original reactants. These reactions are defined as irreversible, or one-way, and may be represented as

$$aA + bB \rightarrow cC + dD$$

Reversible reactions involve species that may reform the original reactants from the products, depending on changes in the environmental conditions that influence the reaction (such as temperature, pressure, and specific species concentrations) and may be expressed as

$$aA + bB \leftrightarrow cC + dD$$

Notice the two-headed arrow suggests that the reaction may proceed from left to right (forward direction) or from right to left (backwards or reverse direction).

Reaction order is a function of the mechanism of the reaction and is usually determined through experimentation. Most often, reaction rates are written explicitly as a function of reactant consumption and, as such, are always negative, although convention usually reports reaction rates in their positive form. The most simple reaction is not dependent upon the concentration of the reactant; rather, the rate is controlled by other, often physical or environmental, parameters. Such reactions are known as "zero" order reactions and are usually expressed as

$$r_A = \frac{dC_A}{dt} = -k_0$$

where r_A is the rate of reaction for species A, C_A is the concentration of A expressed in molar units, t is time, and k_0 is the zero-order reaction coefficient. Integration of this expression yields

$$dC_A = -k_0 \, dt \implies (C_A)_f - (C_A)_i = -k_0 \, (t_f - t_i)$$

where the subscripts f and i indicate the final and initial conditions, respectively. If it is further assumed that the experiment started at $t = 0$, we may express the concentration at any time t as

$$(C_A)_f = (C_A)_i - k_0 t$$

[handwritten note: ZERO ORDER RXN]

Note that the linearized equation would allow for the plotting of experimental data $(C_A)_f$ vs. time to yield slope of $-k_0$ and a y-intercept of $(C_A)_i$.

First-order reactions are dependent upon the concentration of species A and may be written as

$$r_A = \frac{dC_A}{dt} = -k_1 C_A$$

where k_1 is the first-order reaction rate coefficient. Again, assuming $t_i = 0$, this may be integrated to yield

$$\frac{dC_A}{C_A} = -k_1 \, dt \implies \ln\left[\frac{(C_A)_f}{(C_A)_i}\right] = -k_1 t \implies (C_A)_f = (C_A)_i \, e^{[-k_1 t]}$$

[handwritten note: 1st ORDER RXN]

The linearized form of this expression requires plotting the natural log of the concentration vs. time to yield a y-intercept of $\ln[(C_A)_i]$ and a slope of $-k_1$.

Another form of the first-order rate expression found in environmental engineering is used when the concentration of the reacting species has a solubility limit, such as oxygen in aqueous solutions. In these cases, the rate of reaction of species A (oxygen) is dependent upon a concentration gradient that is calculated as the difference between the measured value and the value that would occur if the solution was saturated with oxygen under equivalent environmental conditions (same temperature, ion concentration, etc.). In the specific case of oxygen reaeration of water, the reaction rate expression may take the following form

$$r = \frac{dC}{dt} = k_1\left(C - C^S\right) = -k_1\left(C^S - C\right)$$

where C^S is the saturation concentration of oxygen specified at the given environmental conditions. Integration of this equation can be represented as

$$\frac{C^S - C_f}{C^S - C_i} = e^{-k_1 t}$$

The linearized form of this expression requires plotting the natural log of (C^S-C) vs. time to yield a y-intercept of C^S-C_i and a slope of $-k_1$. Additional discussion on the use of the oxygen transfer expression for wastewater treatment systems can be found in Chapter 8.

Second-order reactions may take many different forms, depending on the species dependence. All second-order reactions may be expressed as

$$r = \frac{dC}{dt} = -k_2 C^2$$

where k_2 is the second-order reaction rate coefficient. However, with the possibility of multiple reactants, this expression may take several forms, such as

$$r = -k_2 C_A^2 \quad \text{or} \quad r = -k_2 C_A C_B \quad \text{or} \quad r = -k_2 C_A^{1.5} C_B^{0.5}$$

2nd ORDER RXN

As indicated above, in reality the second-order expression may take any number of forms, as long as the total order of the concentration exponents sums to the value 2. For simplicity, if it is assumed that the reaction rate is only dependent upon the concentration of one reactant species, the expression will take the following form

$$r_A = \frac{dC_A}{dt} = -k_2 C_A^2$$

Again assuming $t_i = 0$, this may be integrated to yield

2nd ORDER RXN

$$\frac{dC_A}{C_A^2} = -k_2\, dt \quad \Rightarrow \quad \frac{1}{\left(C_A\right)_f} = \frac{1}{\left(C_A\right)_i} + k_2 t$$

The linearized form of this expression requires plotting $1/(C_A)_f$ vs. time to yield a y-intercept of $1/(C_A)_i$ and a slope of k_2.

Table 1.14 and Table 1.15 provide a summary comparing zero-, first-, and second-order reactions.

Table 1.14 Reaction rate comparison

Order of Reaction	$n =$	Rate Law	Concentration vs. Time	Dimensions of k	Sample Units of k
Zero	0	$\dfrac{dC_A}{dt} = -k_0$		$[M \cdot T^{-1} \cdot L^{-3}]$	mg/L·sec mol/L·min
First	1	$\dfrac{dC_A}{dt} = -k_1 \cdot C$		$[T^{-1}]$	sec^{-1} min^{-1} hour^{-1}
Second	2	$\dfrac{dC_A}{dt} = -k_2 \cdot C^2$		$[L^3 \cdot M^{-1} \cdot T^{-1}]$	L/mg·sec L/mol·min

Table 1.15 Linearization process for zero-, first-, and second-order reactions

Order of Reaction	Linearized Plot	Value of $k =$
Zero		Negative slope of best-fit line
First		Negative slope of best-fit line
Second		Slope of best-fit line

The individual rate law equations are more effectively used in their integrated form (that is, $C = f((C_A)_i, t)$). Moreover, a value often very useful when solving environmental engineering problems is the *half-life,* which is the time required to reduce the initial concentration by 50%. The half-life can be readily deduced from the integrated forms of the rate laws by substituting in $(C_A)_f/(C_A)_i =$

0.5 and solving for *t*. However, in the interest of ease of use while taking the PE exam, the integrated forms and the equations for half-life have been summarized in Table 1.16.

Table 1.16 Useful equations for zero-, first-, and second-order reactions

Order of Reaction	Integrated Form	Half-Life Equation
Zero	$(C_A)_f = (C_A)_i - k_0 \cdot t$	$(C_A)_i / 2 \cdot k_0$
First	$(C_A)_f = (C_A)_i \cdot e^{-k \cdot t}$	$\ln(2)/k_1$
Second	$(C_A)_f = \left[k_2 \cdot t + \dfrac{1}{(C_A)_i} \right]^{-1}$	$1/k_2 \cdot (C_A)_i$

Some reactions in environmental engineering are dependent upon a reacting species; however, the reaction rate expression may change as that species approaches a certain value. An example of such a reaction includes biological reactions occurring under certain environmental conditions, where the reaction rate is first order until a certain concentration limit is reached, at which point the reaction reverts to a near-zero order rate. Often, these reactions are referred to as saturation reactions and may be expressed in their general form as

$$ r_A = \frac{dC_A}{dt} = -\frac{k\,C_A}{K + C_A} $$

where *k* is the rate constant and *K* is the saturation constant. This form of the kinetics expression will be covered in more detail in the discussion of Monod growth kinetics for bacteria in wastewater treatment systems in sections 2.1.2 and 8.5.1.

While it may be easier to assume that the target species will have only one possible reaction pathway, it is often the case that several reactions are possible with the same set of reactants. One type of complex pathway is called consecutive (or series) reactions and is usually expressed as

$$ A + B \xrightarrow{k_C} C \qquad \text{and} \qquad A + C \xrightarrow{k_D} D $$

where k_C and k_D are the rates of formation of the identified species. The overall reaction rate may be controlled by either k_C or k_D, depending on the relative magnitudes of the two rates. The reaction that possesses the smaller rate coefficient of the two is often called the "rate limiting step" and would be used as the basis for designing the specific treatment unit. However, care must be taken to evaluate the rate coefficients under all anticipated conditions, as changes in the temperature or other environmental conditions may cause the rate limiting reaction to be assigned to different reactions under different conditions.

Another type of complex reaction occurs when there is a competition between two or more reactions that share a common reactant. These reactions are usually called competitive (or parallel) reactions and may be expressed as

$$ A + B \xrightarrow{k_C} C \qquad \text{and} \qquad A + D \xrightarrow{k_E} E $$

The competition for reactive pathway is generally decided by reaction kinetics with the most energetic pathway favored. Further, if species A in the reactions above is in excess, it may be the case that both reactions will proceed to completion, albeit at different rates, upon the exhaustion of either species B or D.

Example **1.6**

Zero-order and first-order reaction rates

The data in Exhibit 1 shows how the concentration of BTEX (benzene, toluene, ethylbenzene, and xylene) varies as a function of time.

(a) Estimate a rate constant, assuming zero-order decay.

(b) Estimate a rate constant, assuming first-order decay.

Time (min)	C (mg/L)
0	400
5	320
10	820 ?
20	200
30	190
40	110
50	50
60	40

Exhibit 1

Solution

(a) The data has been linearized and a best fit line is used, as shown in Exhibit 2. Given the form of the best fit line for a zero-order reaction is $(C_A)_f = (C_A)_i - k_0 \cdot t$, the first-order rate constant is 7 mg/L·min.

Zero-Order Reaction

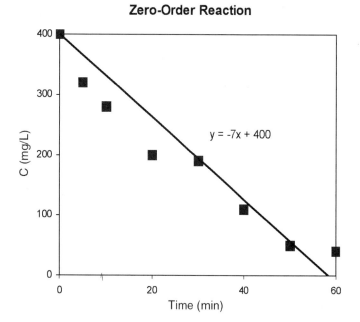

$y = -7x + 400$

Exhibit 2   Best fit line for zero-order decay

(b) As shown in Table 1.5, $\ln(C)$ vs. time should produce a nearly linear plot for a first-order equation. This plot is shown in Exhibit 3. From this plot, the reaction rate constant k_1 can be estimated to be 0.04 min^{-1}.

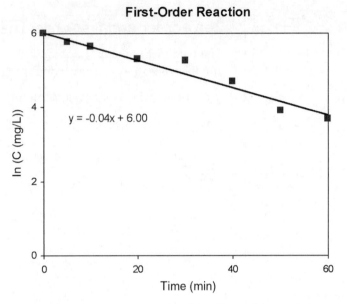

Exhibit 3   Linearization of data for first-order reaction

It is very important to note that the units differ depending on the order of the reaction. Also, in an exam situation, you might not have the luxury of plotting the data and fitting a trendline; however, a reaction rate constant of suitable accuracy can still be obtained by using a straightedge.

Example 1.7

Reaction kinetics

Given the following set of data, determine the initial concentration of species A and predict the concentration of species A after two hours.

Time [min]	15	30	45	60	90
C_A [mg/L]	256	164	102	67	29

Solution

First, the order of the reaction is needed to establish the appropriate rate expression. This can be done by plotting the linearized forms of the rate expressions. This can be immediately done for zero-order reactions (plot C vs. t), but transformation of the data is required for first($\ln[C]$ vs. t)-order and second($1/C$ vs. t)-order reactions. The transformed data is shown below.

	Zero Order	1st Order	2nd Order
t	C	$\ln(C)$	$1/C$
15	256	5.545	0.0039
30	164	5.100	0.0061
45	102	4.625	0.0098
60	67	4.205	0.0149
90	29	3.367	0.0345

Now plot the data on rectilinear graph paper and find the best linear fit.

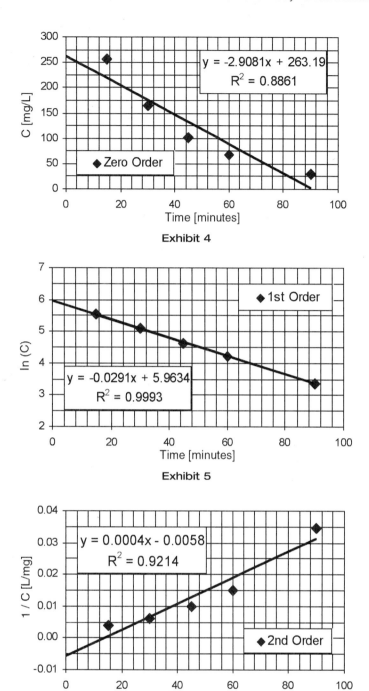

Exhibit 4

Exhibit 5

Exhibit 6

From these plots, the best linear fit is obviously the first-order expression with an R^2 of 0.9993. From the y-intercept (at $t = 0$ minutes), we can determine the initial concentration of species A as

$$\ln[(C_A)_i] = 5.9634 \qquad \text{or} \qquad (C_A)_i = \exp[5.9634] = 389 \frac{\text{mg}}{\text{L}}$$

The concentration at 120 minutes (2 hours) can now be estimated as

$$\left(C_A\right)_f = \left(C_A\right)_i e^{[-k_1 t]} = \left(389 \frac{\text{mg}}{\text{L}}\right) \exp\left[\left(-0.0291 \text{ min}^{-1}\right)\left(120 \text{ min}\right)\right] = 11.84 \approx 12 \frac{\text{mg}}{\text{L}}$$

While spreadsheet software makes this problem easier to solve, it cannot be used on the PE exam. Examinees, therefore, need to be able to recognize the most appropriate linear fit and determine the slope and intercept by taking the difference between two points that fall on the line that was fit to the data. A reasonable estimate would be that the y-intercept was at a value of 6, therefore $(C_A)_i$ would be determined as

$$\ln[(C_A)_i] = 6 \quad \text{or} \quad (C_A)_i = \exp[6] = 403 \, \frac{\text{mg}}{\text{L}}$$

Now, look for another point on the linear fit that crosses clear axis values, such as $\ln(C)$ equal to 4 at a time of 68 minutes. Now determine the rate constant as

$$\ln\left[\frac{(C_A)_f}{(C_A)_i}\right] = -k_1 t \quad \text{or} \quad \frac{\ln(C_A)_f - \ln(C_A)_i}{t_f - t_i} = -k_1 = -\frac{4-6}{68-0} = 0.0294 \, \text{min}^{-1}$$

Plugging this into the rate expression as before yields

$$(C_A)_f = (C_A)_i \, e^{[-k_1 t]} = \left(403 \, \frac{\text{mg}}{\text{L}}\right) \exp\left[\left(-0.0294 \, \text{min}^{-1}\right)\left(120 \, \text{min}\right)\right] = 11.85 \approx 12 \, \frac{\text{mg}}{\text{L}}$$

1.3.2 pH and Equilibrium Chemistry

Hydrogen ion activity (pH), often the most reported water quality indicator (WQI), is generally used to describe an aqueous environment as alkaline (basic) or acidic. The determination of pH is defined as

$$\text{pH} = -\log_{10} [\text{H}^+]$$

where the quantity $[\text{H}^+]$ is the molar hydrogen ion concentration. An increase in $[\text{H}^+]$ decreases pH (more acidic), while a decrease in $[\text{H}^+]$ increases pH (more alkaline).

Although rarely used in practice, the hydroxide $[\text{OH}^-]$ activity can be defined similarly as

$$\text{pOH} = -\log_{10} [\text{OH}^-]$$

where the quantity $[\text{OH}^-]$ is the molar hydroxide ion concentration. Further, by definition

$$[\text{H}^+][\text{OH}^-] = 10^{-14}$$

or, equivalently

$$\text{pH} + \text{pOH} = 14$$

Solution neutrality dictates that hydrogen ion and hydroxide activity (and therefore concentration) must be equal, and this occurs at

$$[\text{H}^+] = [\text{OH}^-] = 10^{-7}$$

or equivalently

$$\text{pH} = \text{pOH} = 7$$

When hydrogen ion and hydroxide are present in solution at concentrations that yield a concentration product above 10^{-14}, it is most likely that they will combine to form water (H_2O) through the following neutralization reaction

$$H^+ + OH^- \rightarrow H_2O$$

For reversible reactions, the calculation of an equilibrium constant (K_{eq}) from the molar concentrations of reactive species and their products in dilute solutions, like in natural waters, may be expressed as follows:

$$K_{eq} = \frac{k_f}{k_b} = \frac{[C]^c [D]^d}{[A]^a [B]^b}$$

where k_f is the rate constant for the forward reaction, k_b is the rate constant for the backward reaction, and the bracketed quantities are molar concentrations of the target species. This equilibrium relationship is invaluable to environmental engineers, as will be further explored in the following section.

Example **1.8**

Equilibrium

A solution has a Ca^{2+} concentration of 50 mg/L. Determine the concentration of $Ca(OH)_2$ in mg/L if the solution pH is 8.5. You may assume a K_{eq} for $Ca(OH)_2$ of 5×10^{-9}.

Solution

We start with the equilibrium expression for dissociation of calcium hydroxide as follows:

$$Ca(OH)_2 \leftrightarrow Ca^{2+} + 2\left[OH^-\right]$$

From this we may express the equilibrium constant as

$$K_{eq} = \frac{\left[Ca^{2+}\right]\left[OH^-\right]^2}{\left[Ca(OH)_2\right]} = 5 \times 10^{-9}$$

We can determine the hydroxide ion concentration from the pH data as follows:

$$pH + pOH = 14 \quad \Rightarrow \quad pOH = 14 - pH = 14 - 8.5 = 5.5 = -\log\left[OH^-\right]$$

$$\text{therefore} \quad \left[OH^-\right] = 10^{-5.5} = 3.16 \times 10^{-6} \frac{mol}{L}$$

We can convert the concentration units for Ca^{2+} from mg/L to mol/L as follows:

$$\left[Ca^{2+}\right] = 50 \frac{mg}{L} \times \frac{1 \ mmol}{40.078 \ mg} = 1.25 \frac{mmol}{L} = 1.25 \times 10^{-3} \frac{mol}{L}$$

Substituting these values into the equilibrium constant expression yields

$$5 \times 10^{-9} = \frac{\left(1.25 \times 10^{-3}\right)\left(3.16 \times 10^{-6}\right)^2}{X} \Rightarrow X = \left[Ca(OH)_2\right] = 2.5 \times 10^{-6} \frac{mol}{L}$$

Since our answer must have units of mg/L, we can convert mol/L as follows:

$$2.5 \times 10^{-6} \frac{mol}{L} = 2.5 \times 10^{-3} \frac{mmol}{L} \times 74.092 \frac{mg}{mmol} = 0.185 \frac{mg}{L}$$

Several solution equilibrium relationships based on fundamental chemical principles may be important to environmental engineers. One example is in the determination of species solubility and evaluating the potential for using precipitation reactions to remove unwanted compounds. This tool for removing undesired constituents in water is used extensively in water treatment as coagulants or in the removal of hardness, Fe/Mn, heavy metals, and phosphate. The solubility product (K_{sp}) of a compound allows for the estimation of the equilibrium concentration of the target species after using reaction chemistry to produce the new species. It may be expressed as the equilibrium constant defined previously; however, the solid reactive species is assumed to have a value of 1. This yields an expression as follows:

$$aA \leftrightarrow bB + cC$$

$$K_{sp} = \frac{[B]^b[C]^c}{[A]^a} = \frac{[B]^b[C]^c}{1} = [B]^b[C]^c$$

Values of K_{sp} for typical water treatment reactions are given in Table 1.17.

Table 1.17   Solubility products for select reactions common in environmental engineering

Equilibrium Equation	K_{sp} @ 25°C	Significance
$Al(OH)_3(s) \leftrightarrow Al^{3+} + 3OH^-$	1.26×10^{-33}	Coagulation
$CaCO_3(s) \leftrightarrow Ca^{2+} + CO_3^{2-}$	4.95×10^{-9}	Hardness removal
$Ca(OH)_2(s) \leftrightarrow Ca^{2+} + 2OH^-$	7.88×10^{-6}	Hardness removal
$CaSO_4(s) \leftrightarrow Ca^{2+} + SO_4^{2-}$	4.93×10^{-5}	Flue gas desulfurization
$Ca_3(PO_4)_2(s) \leftrightarrow 3Ca^{2+} + 2PO_4^{3-}$	2.02×10^{-33}	Phosphate removal
$Cu(OH)_2(s) \leftrightarrow Cu^{2+} + 2OH^-$	2.0×10^{-19}	Heavy metal removal
$Fe(OH)_3(s) \leftrightarrow Fe^{3+} + 3OH^-$	2.67×10^{-39}	Coagulation, iron removal
$Fe(OH)_2(s) \leftrightarrow Fe^{2+} + 2OH^-$	4.79×10^{-17}	Coagulation, iron removal
$Ni(OH)_2(s) \leftrightarrow Ni^{2+} + 2OH^-$	5.54×10^{-16}	Heavy metal removal

Another important use of equilibrium constants is in solving acid/base neutralization reactions. While strong acids and bases are assumed to completely ionize in solution, several important acids and bases in the environment are considered weak, as they only partially ionize in aqueous solution. Ionization constants for common weak acids (K_A) and weak bases (K_B) are provided in Table 1.18 and allow for the calculation of solution pH for known molar concentrations of the target species.

Table 1.18   Ionization constants for acids and bases common in environmental engineering

Acid or Base	Equilibrium Equation	K_A or K_B	Significance
Acetate	$CH_3COO^- + H_2O \leftrightarrow CH_3COOH + OH^-$	5.56×10^{-10}	Organic wastes
Acetic acid	$CH_3COOH \leftrightarrow H^+ + CH_3COO^-$	1.8×10^{-5}	Organic wastes
Ammonia	$NH_3 + H_2O \leftrightarrow NH_4^+ + OH^-$	1.8×10^{-5}	Nutrient
Ammonium	$NH_4^+ \leftrightarrow H^+ + NH_3$	5.56×10^{-10}	Nitrification

(continued)

Acid or Base	Equilibrium Equation	K_A or K_B	Significance
Calcium hydroxide	$CaOH^+ \leftrightarrow Ca^{2+} + OH^-$	3.5×10^{-2}	Softening
Carbonic acid	$H_2CO_3 \leftrightarrow H^+ + HCO_3^-$	4.5×10^{-7}	Corrosion, coagulation
	$HCO_3^- \leftrightarrow H^+ + CO_3^{2-}$	4.7×10^{-11}	
Hypochlorous acid	$HOCl \leftrightarrow H^+ + OCl^-$	2.9×10^{-8}	Disinfection
Phosphoric acid	$H_3PO_4 \leftrightarrow H^+ + H_2PO_4^-$	7.6×10^{-3}	Nutrient, phosphate
	$H_2PO_4^- \leftrightarrow H^+ + HPO_4^{2-}$	6.3×10^{-8}	removal
	$HPO_4^{2-} \leftrightarrow H^+ + PO_4^{3-}$	4.8×10^{-13}	
Magnesium hydroxide	$MgOH^+ \leftrightarrow Mg^{2+} + OH^-$	2.6×10^{-3}	Softening

✳ Need to re-review/ add L examples

Example 1.9

Ionization constants and pH

The ionization constant (K_A) for acetic acid is 1.8×10^{-5}. Estimate the pH of a solution containing 0.6 g of acetic acid in 1.0 L of water.

CH₃COOH

Solution

First, the concentration is needed in molar units. Since the formula for acetic acid is CH_3COOH, it has a MW of 60 g/mol, and therefore 0.6 g = 10 mmol. In 1.0 L of water, it would have a concentration of 10 mM = 0.01 M, and we can write the acetate (Ac) balance as

$$0.01 = HAc + Ac^- \text{ or equivalently } HAc = 0.01 - Ac^-$$

Since the equilibrium expression for the dissociation is

$$HAc \leftrightarrow H^+ + Ac^-$$

we can write the ionization constant as

$$K_A = 1.8 \times 10^{-5} = \frac{[H^+][Ac^-]}{[HAc]}$$

In order to satisfy solution electroneutrality, we know that

$$[H^+] = [OH^-] + [Ac^-]$$

If we assume a pH < 6, we know that $[H^+] \gg [OH^-]$, and we can neglect $[OH^-]$, and may assume

$$[H^+] \approx [Ac^-]$$

Setting an unknown value x for $[H^+]$ and $[Ac^-]$, our equilibrium expression can be written as

$$1.8 \times 10^{-5} = \frac{x^2}{0.01 - x}$$

This expression may be solved for x as

$$1.8 \times 10^{-7} - 1.8 \times 10^{-5}\, x = x^2 \text{ or equivalently } x = 4.15 \times 10^{-4} = [H^+]$$

This is the molar hydrogen ion concentration and can be converted to pH using the relationship $pH = -\log[H^+]$ as follows

$$pH = -\log[4.15 \times 10^{-4}] = 3.38$$

Since the pH is so low, our assumption that $[H^+] \gg [OH^-]$ is valid.

1.3.3 Redox Reactions

Chemical reactions that involve the transfer of electrons from a donor to an acceptor are called redox reactions. Oxidation is a process that is characterized by the *loss of an electron*, which causes the substance to become more positively charged, while reduction is a process that is characterized by the *gain of an electron*, causing the substance to become more negatively charged. Redox reactions require that one species is oxidized as another is simultaneously reduced, such that oxidizing agents are reduced, and reducing agents are oxidized in the reaction. Generally, the simplest way to write a balance redox reaction is the use of half-reactions, which are usually written as reduction reactions as shown in Table 1.19. Oxidation reactions are written as the reverse of the reduction reaction found in the table, and complete reactions are formed by adding the two.

Table 1.19   Half-reactions (reductions) common in environmental engineering

Element	Half-Reaction
C	$\frac{1}{4} CO_2(g) + \frac{7}{8} H^+ + e^- = \frac{1}{8} CH_3COO^- + \frac{1}{4} H_2O$
Cl	$\frac{1}{2} Cl_2(aq) + e^- = Cl^-$
Fe	$Fe^{3+} + e^- = Fe^{2+}$
Fe	$\frac{1}{2} Fe^{2+} + e^- = \frac{1}{2} Fe(s)$
Mn	$\frac{1}{3} MnO_4^- + \frac{4}{3} H^+ + e^- = \frac{1}{3} MnO_2 + \frac{2}{3} H_2O$
Mn	$\frac{1}{2} MnO_2(s) + 2 H^+ + e^- = \frac{1}{2} Mn^{2+} + H_2O$
N	$\frac{1}{5} NO_3^- + \frac{6}{5} H^+ + e^- = \frac{1}{10} N_2(g) + \frac{3}{5} H_2O$
O	$\frac{1}{4} O_2(g) + H^+ + e^- = \frac{1}{2} H_2O$
S	$\frac{1}{8} SO_4^{2-} + \frac{5}{4} H^+ + e^- = \frac{1}{8} H_2S(aq) + \frac{1}{2} H_2O$

1.3.4 Ion Balance

Water quality data is most often reported to the user as mg/L in a tabular format. In order to satisfy electroneutrality requirements for a solution, the sum of all cations must equal the sum of all anions when expressed in equivalence units, most often *milliequivalents of solute per liter solution* (meq/L or mN). Often, having a visual representation of possible ionic species combinations is useful. These ion balances (sometimes referred to as chemical bar graphs) are useful in the evaluation of softening requirements as seen in Figure 1.1. Cations are listed in the order of Ca^{2+}, Mg^{2+}, Na^+, and K^+, while anions are shown below the cations and are listed in the order of OH^- (if present), CO_3^{2-} (if present), HCO_3^-, SO_4^{2-}, Cl^-. Additional divalent or trivalent cations may be represented before Na^+ if they contribute significantly to sample hardness. Further, if CO_2 is present, it is often included in the bar graph by adding it as both a cation and anion to the left of the zero line. Again, solution electroneutrality requires both bars to be the same length, and hypothetical speciation can easily be determined by inspection as shown in the bottom portion of Figure 1.1.

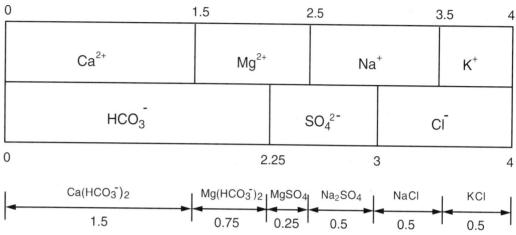

Figure 1.1   Example of ion bar graph; all units are meq/L

Example **1.10**

Ion balance

Given the following concentrations of ions, determine the probable concentration of Na_2SO_4.

Ca^{2+} = 150 mg/L as $CaCO_3$; Mg^{2+} = 100 mg/L as $CaCO_3$; Na^+ = 46 mg/L; K^+ = 39 mg/L; HCO_3^- = 225 mg/L as $CaCO_3$; SO_4^{2-} = 72 mg/L; Cl^- = 71 mg/L

Solution

To generate a chemical bar graph, first determine the normality of each chemical species as

$$Ca^{2+} = \frac{150 \text{ mg/L as } CaCO_3}{50 \, \frac{\text{mg } CaCO_3}{\text{meq}}} = 3 \text{ meq/L}$$

$$Mg^{2+} = \frac{100 \text{ mg/L as } CaCO_3}{50 \, \frac{\text{mg } CaCO_3}{\text{meq}}} = 2 \text{ meq/L}$$

$$Na^+ = \frac{46 \text{ mg/L}}{23 \, \frac{\text{mg } Na^+}{\text{meq}}} = 2 \text{ meq/L}$$

$$K^+ = \frac{39 \text{ mg/L}}{39 \, \frac{\text{mg } K^+}{\text{meq}}} = 1 \text{ meq/L}$$

$Ca\,CO_3$

40
+ 12
  32
+ 16
 100 = 50 $\frac{mg}{meq}$
  2

$$HCO_3^- = \frac{225 \text{ mg/L as CaCO}_3}{50 \dfrac{\text{mg CaCO}_3}{\text{meq}}} = 4.5 \text{ meq/L}$$

$$SO_4^{2-} = \frac{72 \text{ mg/L}}{48 \dfrac{\text{mg SO}_4^{2-}}{\text{meq}}} = 1.5 \text{ meq/L}$$

$$Cl^- = \frac{71 \text{ mg/L}}{35.5 \dfrac{\text{mg Cl}^-}{\text{meq}}} = 2 \text{ meq/L}$$

Next, draw the ion graph in units of meq/L, as shown in Exhibit 7.

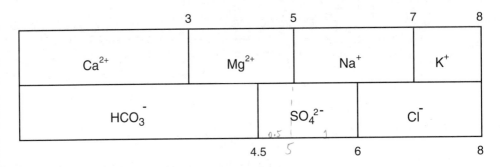

Exhibit 7

From the bar graph, the overlap for Na^+ and SO_4^{2-} occurs from 6 to 5 meq/L; therefore

$$Na_2SO_4 = 6 - 5 \text{ meq/L} = 1 \text{ meq/L}$$

Finally, this can be converted into mass concentration units by multiplying by the EW as

$$(1 \text{ meq/L}) \times (71 \text{ mg/meq}) = 71 \text{ mg/L } [Na_2SO_4]$$

1.3.5 Water Quality Indicators

The U.S. EPA establishes quality standards for several water quality indicators (WQI) for both drinking water and surface waters. Maximum contaminant levels (MCLs) are enforceable limits, established as a result of comprehensive risk assessments (covered in section 4.5), set to provide an adequate measure of safety to protect human health. Primary drinking water standards are MCLs that have been established for dozens of specific organic and inorganic chemicals, as well as biological contamination, disinfection by-products, radionuclides, and turbidity. As new information is forthcoming, the primary standards are continuously updated, both with respect to specific chemicals, as well as their MCLs, and should be researched prior to any remediation activity. Secondary contaminant standards are presented in Table 1.20 and are WQIs that are based more on aesthetic quality, rather than health hazard, and are not enforceable. However, they are generally accepted as appropriate targets and are used as an indicator as to general watershed health.

Table 1.20 Secondary contaminant standards for aesthetics of drinking water

Contaminant	Standard	Contaminant	Standard
Al	0.2 mg/L	Mn	0.05 mg/L
Cl^-	250 mg/L	odor	3 threshold #s
color	15 color units	pH	6.5–8.5
Cu	1.0 mg/L	Ag	0.1 mg/L
corrosivity	none	SO_4^{2-}	250 mg/L
F^-	2 mg/L	TDS	500 mg/L
foam	0.5 mg/L	Zn	5 mg/L
Fe	0.3 mg/L		

Acidity, Alkalinity, and Hardness

Although the most widely known of the WQIs, pH only describes hydrogen acidity (or equivalently, hydroxide alkalinity) and neglects the probability that other chemical species may be present, providing neutralization (buffer) capacity to the system. These other chemical species possess the ability to neutralize added acid or alkali, with minimal change in system pH. Alkalinity can be expressed as *ability of a solution to neutralize [H+]*, while acidity can be expressed as *ability of a solution to neutralize [OH−]*.

Natural acidity generally arises from the release of CO_2 during biological activity that is not released to the atmosphere or is not neutralized by natural elements. However, some natural environments (such as the drainage from abandoned mine sites) may contain substantial amounts of mineral acidity, dissolved species that react with oxygen in aerobic environments or are transformed in biological processes. Compounds such as pyrite (FeS_2) may oxidize to produce sulfuric acid, which in turn increases the hydrogen acidity (i.e., lowers the pH). For example, most acid mine drainage treatment systems are designed to address this acidity through alkaline dosing and removal of the metal precipitates (iron salts) formed during the reaction.

Natural waters generally have a pH between 6 and 8.5, and, therefore, the primary source of alkalinity is bicarbonate (HCO_3^-), while higher pH waters may have a significant carbonate (CO_3^{2-}) and/or hydroxide (OH^-) contribution to alkalinity, as seen in Figure 1.2. Because many compounds may contribute to alkalinity, it is necessary to employ a standard unit for expressing this quantity. This is usually done by converting all species to equivalents of $CaCO_3$ and expressing the concentration of the sum of all species as mg/L $CaCO_3$. This is accomplished mathematically by dividing the species concentration in mg/L by the species EW and multiplying by the EW of $CaCO_3$, 50 mg/meq (see Example 1.10).

Hardness can be expressed as *the sum of all multivalent cations*, although the primary source of hardness is Ca^{2+} and Mg^{2+}. Hardness is a result of water contact with minerals in subsurface environments and is an issue due to the potential for scaling in high temperature process equipment for industrial customers and, to a lesser degree, soap consumption for residential and commercial customers. As was the case with alkalinity, many compounds contribute to hardness, and it is convenient to express hardness as mg/L $CaCO_3$. Waters containing up to 300 mg/L of hardness as $CaCO_3$ are usually considered "hard" and require treatment, although many customers object to "moderately hard" water (up to 150 mg/L as $CaCO_3$). Removal of hardness from municipal water supplies is called softening, and the process is detailed in section 7.2.1.

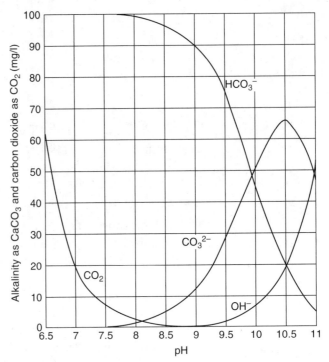

Figure 1.2   Alkalinity species dependence as a function of pH

Example **1.11**

Hardness

Determine the hardness, in units of mg/L $CaCO_3$, of a solution with a Ca^{2+} concentration of 75 mg/L and a Mg^{2+} concentration of 20 mg/L.

Solution

Hardness is defined as the sum of the multivalent cations. In this case, we need to convert the Ca^{2+} and Mg^{2+} concentrations into meq/L, add the two together, and then convert those units back to mg/L of $CaCO_3$. This may be completed as follows:

$$\left[Ca^{2+} \right] = 75 \frac{mg}{L} \times \frac{1 \, meq}{20.0 \, mg} = 3.75 \frac{meq}{L} \quad \text{and} \quad \left[Mg^{2+} \right] = 20 \frac{mg}{L} \times \frac{1 \, meq}{12.2 \, mg} = 1.64 \frac{meq}{L}$$

$$hardness = 3.75 + 1.64 = 5.39 \frac{meq}{L} \times \frac{50.0 \, mg \, CaCO_3}{1 \, meq} = 270 \frac{mg}{L} \, as \, CaCO_3$$

Solids, Turbidity, and Color

As a WQI, total solids (TS) can be expressed as *residue on evaporation* or *residue upon drying at 105°C*. TS values are determined for an aqueous sample by placing a known volume of solution into a drying vessel and drying in a 105°C oven. The value for TS is calculated as

Total Solids

$$TS = \frac{\text{weight of residue}}{\text{volume of sample dried}}$$

and expressed in units of mg/L. Total solids are comprised of many compounds, and it is often necessary to know the forms of the solid to select appropriate treatment processes. Total suspended solids (TSS) are generally particles that are not in solution but rather are discrete particles contained in a sample. TSS are often subclassified into settleable or nonsettleable categories, based on the potential to use gravity as a means of solids removal. TSS are determined by filtering a sample through a 0.45 µm filter, which was previously dried at 105°C and weighed. The filter with the solids is then dried at 105°C and reweighed. The value for TSS is calculated as

Total Suspended Solids

$$TSS = \frac{\text{weight of residue on filter}}{\text{volume of sample filtered}} \left(mg/L \right)$$

and expressed in units of mg/L. Total dissolved solids (TDS) are the sum of all dissolved species and is determined in the same manner as TS, only using the filtrate from the TSS sample. The value for TDS is calculated as

Total Dissolved Solids

$$TDS = \frac{\text{weight of residue from filtrate}}{\text{volume of sample dried}}$$

and expressed in units of mg/L. It should be noted that often, TSS or TDS can be determined by difference calculation using one of the two values and a known TDS concentration as follows:

$$TS = TSS + TDS$$

However, due to the simplicity of the method, validation by measurement is strongly suggested.

Two other important WQIs used to identify solids in water samples are total volatile solids (TVS) and total fixed solids (TFS). Fixed solids can be expressed as *residue upon igniting at 550°C*, and the volatile solids are often referred to as loss on ignition (LOI). TFS are determined by placing the TS vessel (after drying and weighing) into an ashing oven at 550°C for two hours. After ignition, the vessel is allowed to cool in a desiccator to room temperature prior to being weighed. The value for TFS is calculated as

$$TFS = \frac{\text{weight of residue after ignition}}{\text{original volume of TS sample}}$$

and expressed in units of mg/L. TVS is also expressed in units of mg/L and is determined by difference calculation using the following relationship:

$$TS = TFS + TVS$$

In the case of wastewater treatment, it is convenient to refer to the portion of suspended solids that are primarily biological organisms. This is determined by igniting the suspended solids, after determining TSS using a glass-fiber filter, and reporting the total volatile suspended solids (TVSS), which would be calculated in the same manner as TVS above.

Example **1.12**

TDS

Determine the concentration of total dissolved solids from the data given below.

Volume of water sample = 100 mL

Empty dish = 56.345 g

Dish + dried sample = 56.612 g

Filter paper = 1.629 g

Filter paper + dried solids = 1.653 g

Dish + dried sample after filtering = 56.589 g

Solution

TDS is defined as

$$TDS = \frac{\text{weight of dry residue after filtering}}{\text{volume dried}}$$

From the data given, TDS is calculated as

$$TDS = \frac{56.589 - 56.345 \text{ g}}{100 \text{ mL}} = \frac{0.244 \text{ g}}{0.1 \text{ L}} = \frac{2.44 \text{ g}}{\text{L}} = \frac{2400 \text{ mg}}{\text{L}}$$

Example **1.13**

Fixed solids

Seventy percent of the suspended solids in a wastewater sample are volatile. Determine the fixed solids of the sample if total solids are 1400 mg/L and total dissolved solids are 200 mg/L. You may assume that none of the dissolved solids are volatile.   *total = suspended + dissolved*

Solution

We first need to determine the concentration of total suspended solids as follows:

$$TS = TSS + TDS \Rightarrow TSS = TS - TDS = 1400 - 200 = 1200 \frac{mg}{L}$$

Now we can find the fraction of TSS that are volatile as follows:

$$TVSS = (0.7)\left(1200 \frac{mg}{L}\right) = 840 \frac{mg}{L}$$

Further, if we assume that none of the dissolved solids is volatile, we may write:

$$TVS = TVSS + TVDS = TVSS + 0 = 840 \frac{mg}{L}$$

Finally, fixed solids may be determined as follows:

$$TS = TVS + TFS \Rightarrow TFS = TS - TVS = 1400 - 840 = 560 \frac{mg}{L}$$

One specific WQI that arises from nonsettleable solids is termed turbidity and generally describes waters that have suspended solids that scatter light as light is passed through the water. The standard unit of turbidity is the NTU, which is measured using a set of standard samples and a procedure called nephelometry (a measure of light scattering at 90° from the incident light). These solids may be inorganic (such as clays), organic, or biological in nature and are generally comprised of colloidal suspensions. These colloids are aesthetically unpleasing and may interfere with the disinfection process in water treatment plants. Turbidity is removed through coagulation, clarification (sedimentation), and filtration processes as described in sections 7.2.3, 7.2.4, and 7.2.5, respectively.

A WQI related to turbidity is color. Color is also caused by colloidal particles, most of which are organic in origin (for example, tannins or humic matter), although some dissolved inorganic chemicals may also contribute to color (such as Fe). Care should also be employed not to confuse true color with apparent color, the latter of which may arise from suspension of settleable solids in waters with high turbulence. As colloidal particles are the cause of color, treatment is the same as was the case for turbidity.

Dissolved Oxygen, BOD, and COD

An important WQI in wastewater treatment or the assessment of the quality of habitat for biological organisms is the dissolved oxygen (DO) concentration. Oxygen solubility in water is limited and is highly dependent on temperature, with maximum DO concentrations for a given temperature listed in saturation tables, as shown in Table 1.21. As with most gas-liquid systems, oxygen solubility is inversely proportional to temperature, with a high DO concentration of approximately 14.5 mg/L at 0°C, reducing to approximately 7 mg/L at 35°C for aqueous systems in contact with air at atmosphere pressure. In some processes, oxygen solubility can be increased by increasing total system pressure or by increasing the oxygen concentration in the gas phase, although both methods are generally expensive to implement. However, oxygen solubility in water decreases with increasing solute concentrations, such as the case of salt water or wastewaters, and must be accounted for.

Table 1.21 DO saturation values for clean water exposed to air at 760 mmHg

Temp. [°C]	DO [mg/L]	Temp. [°C]	DO [mg/L]
0	14.6	16	10.0
2	13.8	18	9.5
4	13.1	20	9.2
6	12.5	22	8.8
8	11.9	24	8.5
10	11.3	26	8.2
12	10.8	28	7.9
14	10.4	30	7.6

Due to the activity of biological agents, natural waters often do not possess the maximum oxygen concentration anticipated for the system temperature and DO concentrations are reported as percent of saturation. For these cases, DO concentrations would be calculated by multiplying the saturation value from the tables by the percent saturation. Further, the discharge of organic material from municipal or industrial wastewaters can have a strong impact on DO levels in the receiving stream, because aerobic biological activity consumes DO as organisms degrade the organic constituents. Modeling of the impact of wastewater discharges on receiving streams is covered in section 2.2.

The biological transformation of organic constituents forms the basis for secondary treatment of municipal and industrial wastewaters. One analytical technique employed to determine the strength of a wastewater sample is biochemical oxygen demand (BOD). While BOD actually measures the amount of DO consumed in the biological utilization of substrate, it is related to the amount of biodegradable organic matter (OM) present. In short, the method places 300 mL of sample (including dilution water when necessary) into a sealed vessel and records the DO levels at time zero and at the end of five days of incubation at 20°C in a dark chamber. BOD_5 indicates a five-day incubation period, and the value measured in units of mg/L is calculated as

$$BOD_5 = \left(DO_i - DO_f \right) \times \left(DF \right)$$

where subscripts i and f are initial and final DO concentrations, and DF is the dilution factor, which is calculated as

$$DF = \frac{\text{volume of test bottle in mL}}{\text{volume of wastewater added in mL}}$$

Typical volumes for BOD test bottles are 300 mL. The dilution of a sample may be necessary to keep DO_f above 2 mg/L, which will ensure that any reduction in biological activity is due to substrate (OM) depletion and not depletion of the oxygen. Further, some samples require seeding, which is a process that introduces an active microbial population for testing water samples where adequate bacteria are not readily present. Presentation of the additional equations necessary to calculate the BOD in seeded bottles is beyond the scope of this review.

The five-day incubation period is chosen as a compromise between sufficient time to discern a significant change in DO and ending the test before organisms utilize some of the oxygen for nitrification, as seen in Figure 1.3. If a first-order reaction rate is assumed and DO concentrations from the BOD bottle are available for each day of the incubation period, a plot of BOD versus time can be regressed to determine the ultimate BOD (often represented as L) and reaction rate constant from the following equation:

$$BOD_t = L(1 - e^{-kt})$$

where k is the reaction rate in days^{-1}, and t is the time in days. The value of L is important in DO-sag modeling in streams as demonstrated in section 2.2.

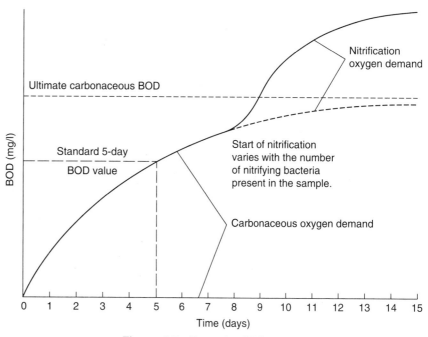

Figure 1.3   Example of BOD curve

Source: Warren Viessman, Jr., and Mark J. Hammer, *Pollution Control*, 7th ed., © 2005. Reprinted by permission of Pearson Education, Inc., Upper Saddle River, N.J.

Example **1.14**

Ultimate BOD

Determine the ultimate BOD concentration of a wastewater sample that has a BOD_5 of 185 mg/L and a biological reaction rate of 0.37 day^{-1}.

Solution

Using the ultimate BOD equation and substituting a value of five days for t, we can write

$$BOD_5 = L(1 - e^{-k5}) = 185 \text{ mg/L}$$

Since $k = 0.37$ day^{-1}, then the product $kt = (0.37 \text{ day}^{-1})(5 \text{ days}) = 1.85$, or

$$L(1 - e^{-1.85}) = 185 \text{ mg/L}$$

Rearranging, this may be expressed as

$$\frac{185}{L} = 1 - e^{-1.85} \Rightarrow 1 - \frac{185}{L} = e^{-1.85} \Rightarrow 1 - \frac{185}{L} = 0.157$$

Solving for L gives

$$0.843 = \frac{185}{L} \quad \text{or equivalently} \quad L = 219.5 \text{ mg/L}$$

Example **1.15**

$DO_f > 2 \text{ mg/L}$

Estimating BOD sample volume

You anticipate the five-day BOD concentration of a wastewater sample will be approximately 220 mg/L. Determine the maximum amount of sample that may be placed into a 300 mL BOD bottle if the initial dissolved oxygen concentration is 9.2 mg/L.

Solution

Using the expression for BOD and initial DO value given, and the fact that the final DO value must be at least 2 mg/L in order for the test to be valid, we may write

$$\text{BOD} = \left(\text{DO}_i - \text{DO}_f\right) \times \text{DF} = (9.2 - 2) \times \text{DF} = 7.2\,\text{DF} = 220\,\frac{\text{mg}}{\text{L}} \Rightarrow \text{DF} = 30.56$$

Using this value for the dilution factor and the equation given for DF, we may determine the volume of sample placed into the bottle as follows:

$$\text{DF} = \frac{300\text{ mL}}{X} = 30.56 \Rightarrow X = \frac{300\text{ mL}}{30.56} = 9.8\text{ mL}$$

Example **1.16**

BOD final DO

A wastewater with an ultimate BOD of 180 mg/L has a BOD_5 of 130 mg/L. Estimate the dissolved oxygen (DO) concentration in the test bottle after ten days if the initial DO was 8.9 mg/L and 10 mL of sample was placed in a 300 mL bottle.

Solution

From the data given in the problem statement, and the equation for BOD determination from ultimate BOD, the biological constant can be determined as

$$\text{BOD} = L\left(1 - e^{-kt}\right) \text{ for BOD}_5 \quad 130\,\frac{\text{mg}}{\text{L}} = \left(180\,\frac{\text{mg}}{\text{L}}\right)\left(1 - e^{-5k}\right) \text{ or } k = 0.256\text{ day}^{-1}$$

Now the value for BOD_{10} can be determined as

$$\text{BOD}_{10} = \left(180\,\frac{\text{mg}}{\text{L}}\right)\left\{1 - \exp\left[-(10\text{ days})(0.256\text{ day}^{-1})\right]\right\} = 166.1\,\frac{\text{mg}}{\text{L}}$$

Plugging this into the BOD equation based on DO, DO_f can be estimated as

$$\text{BOD} = \left(DO_i - DO_f\right)\left(\frac{V_{bottle}}{V_{sample}}\right) \quad \text{or} \quad 166.1\,\frac{\text{mg}}{\text{L}} = \left(8.9 - DO_f\right)\left(\frac{300\text{ mL}}{10\text{ mL}}\right)$$

$$\frac{166.1\,\dfrac{\text{mg}}{\text{L}}}{30} = 8.9 - DO_f \quad \text{or} \quad DO_f = 8.9 - 5.54 = 3.36\,\frac{\text{mg}}{\text{L}}$$

Often, in municipal and industrial wastewater discharges, five days is too long to wait for analytical results to be finalized, as flows generally are continuous and high contaminant levels could be discharged while waiting for test results. A rapid test related to BOD is chemical oxygen demand (COD), where a strong chemical oxidant (potassium dichromate, $K_2Cr_2O_7$) replaces the bacteria and DO in the oxidation of OM. After two hours in a heated mantle, samples are analyzed in a spectrometer for Cr^{3+}, which is generated from the reduction of the original Cr^{6+} as oxygen is consumed in the reaction with the OM. While the time required to obtain results is significantly reduced, the strength of the oxidant may overestimate the biodegradable fraction of the OM. This is due to the complete oxidation of all OM, including the fraction that is not biodegradable, and possibly some inorganic compounds as well.

This fact also makes the COD test especially appropriate for industrial wastewaters, where toxic compounds would otherwise inhibit biological activity and reduce or eliminate the usefulness of BOD results. However, for most domestic wastewater operations, relationships between BOD and COD can be developed from historical samples, and a weighting factor applied to the COD value can usually provide a reasonable estimate for BOD. Typical values for BOD_5 and COD are 200 mg/L and 300 mg/L, respectively, for domestic wastewaters.

Additional WQIs

Additional WQIs that may impact the quality of aquatic habitat include nutrients (generally various forms of N and P), chlorine residual (present due to disinfection processes), heavy metals (originating from geologic formations or industrial discharges), and trace organics (from runoff, landfill leachate, industrial discharges, or sanitary sewers). Each varies in its impact on aquatic systems and treatment processes necessary to reduce these constituents to levels that meet ambient water quality standards. Two additional WQIs that may arise from water contact with organic matter and biological activity are taste and odor. While it is difficult to identify a specific cause, it is generally accepted that municipal water supplies must be treated for objectionable taste and odor (see section 7.2.12). This is usually accomplished through oxidation methods (see section 7.2.8) or carbon adsorption (see section 11.4.2).

1.4 MASS BALANCE APPROACH TO PROBLEM SOLVING

This section reviews the mass balance approach to problem solving. It also addresses the use of reaction rates (section 1.3.1) and how they are used in the mass balance approach.

Many environmental engineering problems, such as those found on the PE exam, can be solved if the engineer has a good command of units and the mass balance approach. Mass balances are applied to a control volume. A *control volume* is an imaginary volume used to identify the system of study. The control volume is bounded by the *control surface*. In environmental engineering problems, the control volume may be a pond, lake, reactor, aquifer, or watershed.

Environmental engineers are often concerned with the concentration of a compound entering or exiting a control volume or the concentration of a compound within the control volume. To solve such problems, it is imperative that all mass flows of the compound *into* and *out of* the control volume be identified, and that any sources or sinks of the compound of interest *within* the control volume are identified.

In simplified terms, the mass balance approach states that all mass of a compound in a control volume must be accounted for. At *steady state* (that is, when the mass of the compound within the control volume is not changing with time), the sum of the mass of compound entering the control volume and the mass of compound being created in the control volume must equal the sum of the mass exiting the control volume and the mass being destroyed within the control volume. In mathematical terms, this can be written as

$$0 = \dot{m}_{in} - \dot{m}_{out} \pm \dot{m}_{rxn} \qquad (1.1)$$

where $\dot{m}_{in}$ is the mass flux of compound into the control volume (M/T)

$\dot{m}_{out}$ is the mass flux of compound out of the control volume (M/T)

$\dot{m}_{rxn}$ is the *reaction rate term*, the mass rate at which the compound is either created or destroyed within the control volume (M/T)

The reaction rate term in Equation 1.1 refers to the creation or destruction of mass of the compound of interest. Often, in environmental systems, the compound decays (for example, the decay of biochemical oxygen demand in a biological reactor), and the sign is negative. Alternatively, the compound is formed (for example, the creation of a compound in a reactor from two reactants) in which case the sign on this term is positive. Reaction rates are more thoroughly discussed in section 1.3.1.

It is very important to note that the mass balance equation only applies to steady state problems. The equation for non–steady state problems (in which case the mass of compound in the control volume changes with time) is given in Equation 1.2.

$$\left(\frac{dm}{dt} \right)_{cv} = \dot{m}_{in} - \dot{m}_{out} \pm \dot{m}_{rxn} \qquad (1.2)$$

where $\left(\frac{dm}{dt} \right)_{cv}$ is the change in mass in the control volume over time.

Note that the terms on the right-hand side are unchanged as compared to the steady state equation (Equation 1.1).

Steady-state problems are more desirable to solve from the standpoint that, unlike non–steady state problems, steady-state problems do not necessitate the use of differential equations. Steady state problems can be identified by such key phrases in the problem statement as "after a long time"; for example, "What is the concentration of mercury exiting the lake after the paper mill has been illicitly discharging mercury at the given concentrations for a long time?"

When solving environmental engineering problems that involve a mass balance approach, the mass fluxes entering and exiting the control volume may be given directly in the problem statement (mass/time). Alternatively, flow rates and associated concentrations may be provided, in which case the mass flux is equal to the product of the flow and concentration. Note that in Example 1.3 the product of a flow rate and a concentration yielded a mass flux.

1.5 REACTORS

The mass balance approach is typically applied to reactors. Many environmental engineering systems can be modeled as one of (or a combination of) three basic reactor types: batch reactors, continuous-flow stirred tank reactors (CSTR), or plug flow reactors (PFR).

1.5.1 Batch Reactors

A batch reactor is a reactor with no inputs and no outputs. In other words, when a control volume is defined as a batch reactor, there are no mass flux terms in the mass balance equation. Thus, the non–steady state mass balance equation for a batch reactor is simplified from Equation 1.2 as follows:

$$\left(\frac{dm}{dt}\right)_{cv} = \pm \dot{m}_{rxn} \tag{1.3}$$

The steady state solution to a batch reactor is meaningless, which can be seen from the original mass balance equation or from common sense. For such a case, Equation 1.2 would simplify to $0 = \pm \dot{m}_{rxn}$. Also, common sense tells us that nothing is happening in a reactor with no inputs and no outputs and in which there is no reaction!

One important characteristic of a batch reactor is that it is *completely stirred* (or *completely mixed*). In other words, the concentration of the compound of interest within the reactor is the same at any point within the reactor. In practice, complete mixing can be accomplished by various means, including mechanical mixers or by bubbling large quantities of air through a system. In natural systems, completely stirred conditions can be achieved due to the addition of wind energy.

Equation 1.3 can accommodate the three most common reaction rates (zero-, first-, and second-order) by substitution of the appropriate term from Table 1.22. In this table, the variable V represents the volume of the reactor. The use of Table 1.22 is demonstrated in Example 1.17.

Table 1.22 Reaction rate terms for zero-, first-, and second-order reactions

Order of Reaction	$\dot{m}_{rxn}$
Zero	$V \cdot k_0$
First	$V \cdot k_1 \cdot C_A$
Second	$V \cdot k_2 \cdot C_A{}^2$

Example **1.17**

Decay in a batch reactor

A graduate student conducts a research project using a lab-scale, well-mixed, cylindrical tank (diameter = 8 cm, height = 18 cm). The contaminant (cyanide) is degraded using UV light according to first-order kinetics ($k = 3.2$ day^{-1}). If the concentration of cyanide in the tank is 30 mg/L at the beginning of the test, at what time will 99.9% removal be obtained?

Solution

$$\frac{dm}{dt} = V k_1 C_A$$

Exhibit 8 shows a schematic of the problem.

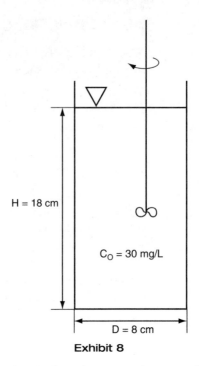

H = 18 cm

C_O = 30 mg/L

D = 8 cm

Exhibit 8

The solution begins by stating the governing equation for a batch reactor (Equation 1.3):

$$\left(\frac{dm}{dt}\right)_{cv} = \pm \dot{m}_{rxn}$$

Knowing that this is a first-order *decay* reaction and referring to Table 1.22, the right-hand side can be simplified to

$$\left(\frac{dm}{dt}\right)_{cv} = -V \cdot k_1 \cdot C$$

The rate of change of mass in the control volume can be represented as $V\frac{dC}{dt}$, which simplifies the equation to

$$V\frac{dC}{dt} =- V \cdot k_1 \cdot C$$

The volume terms cancel out from both sides and the equation can be solved by separation of variables:

$$\frac{dC}{C} = -k_1 \cdot dt$$

Integrating both sides yields $C_t = C_o \cdot exp(-k_1 \cdot t)$, where

C_t = concentration of cyanide in the batch reactor at any time t

C_o = concentration of cyanide in the batch reactor at time 0

k_1 = first-order decay coefficient

Alternatively, this equation could be solved for t:

$$t = \ln\left(\frac{C_t}{C_o}\right) \cdot (-k_1^{-1})$$

Given that 99.9% reduction is desired, the quantity C/C_o is equal to 1/1000, and the time required is calculated as

99.9 %
removal

$$t = \ln\left(\frac{1}{1000}\right) \cdot \left(\frac{-1}{3.2 \text{ day}^{-1}}\right) = 2.2 \text{ day}$$

Note that neither the volume nor the starting concentration is needed to solve this problem. The volume could be calculated and the starting concentration could be used in the solution, but these steps would have required additional time, which most people do not have when taking the PE exam.

1.5.2 Continuous-Flow Stirred Tank Reactor (CSTR)

A CSTR is sometimes called a CMFR (completely mixed flow reactor). In both cases, the names suggest that these reactors are similar to a batch reactor in that they are completely mixed and that they differ from a batch reactor as there is a flux of compound into and out of the reactor. A CSTR can be evaluated either at non-steady state or at steady state.

A generic CSTR is diagrammed in Figure 1.4. Note that the control volume is identical to the reactor's physical boundaries. The mass flux into the reactor may be the combination of many different influent sources (for example, two streams and a factory discharge into a lake), and likewise the mass fluxes leaving the reactor may be a combination of more than one "effluent" streams.

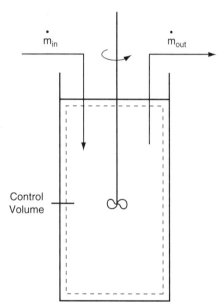

Figure 1.4 Control volume for CSTR

The average amount of time that a "parcel" of water spends in a reactor is known as the *hydraulic residence time* or *detention time*, θ. θ is calculated by the following equation:

$$\theta = \frac{V}{Q} \tag{1.4}$$

where V = the volume of the reactor [L³]

Q = the flow rate through the reactor [L³/T]

Thus, θ has units of time.

When solving CSTR problems, remember to sketch the control volume and to consider all mass fluxes across the control surface.

Example **1.18**

First-order decay in a CSTR

A 10-acre lake receives flow from an industrial facility. The lake is drained by a single stream. The industrial plant is legally discharging a 0.5 MGD effluent with a mercury concentration of 1.5 µg/L. The industrial facility is the only influent flow to the lake. The average depth of the lake is 7 feet, and the mercury decays with a rate constant of 1 day^{-1}. Find the steady state concentration in the lake.

Solution

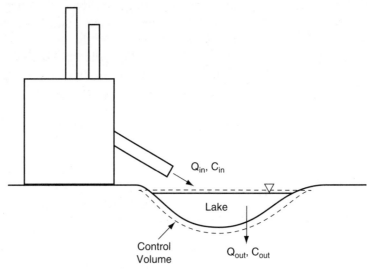

Exhibit 9   Schematic for Example 18

This problem is solved by beginning with the generic mass balance equation (Equation 1.2):

$$\left(\frac{dm}{dt}\right)_{cv} = \dot{m}_{in} - \dot{m}_{out} \pm \dot{m}_{rxn}$$

Given that the system is at steady state and that the Hg decays according to first-order kinetics, the generic mass balance equation can be simplified to

$$0 = \dot{m}_{in} - \dot{m}_{out} - V \cdot k_1 \cdot C_{out}$$

As in Example 1.17, Table 1.22 is used to make the appropriate substitution for the $\dot{m}_{rxn}$ term.

Note, that by the definition of a CSTR, the concentration anywhere in the lake is the same. Thus, the concentration of Hg in the lake is the same as the concentration of Hg exiting the lake (C_{out}).

The mass flux into the control volume is equal to $Q_{in} \cdot C_{in}$, while the mass flux exiting the control volume is equal to $Q_{out} C_{out}$. Thus, the equation simplifies to this:

$$0 = Q_{in} C_{in} - Q_{out} C_{out} - V \cdot k_1 \cdot C_{out}$$

Recognizing that $Q_{out} = Q_{in} = Q$, and rearranging this equation yields

Steady
State
CSTR

Effluent
Concentration
$$C_{out} = \frac{-Q \cdot C_{in}}{-Q - V \cdot k_1} = C_{in}\frac{1}{1 + \frac{V}{Q}k_1} = C_{in}\frac{1}{1 + \theta \cdot k_1}$$
$\theta = \frac{V}{Q}$

Values can now be substituted into this equation, being careful of units.

$$V = \text{area} \cdot \text{depth} = 10 \text{ acre } 7 \text{ ft }\frac{43,560 \text{ ft}^2}{\text{acre}}$$

Or, the volume equals $3.05 \cdot 10^6 \text{ ft}^3$.
The hydraulic residence time, θ, equals V/Q, or $\dfrac{3.05 \cdot 10^6 \text{ ft}^3}{0.5 \cdot 10^6 \frac{\text{gal}}{\text{day}}} \cdot \dfrac{7.48\text{gal}}{\text{ft}^3} = $
45.6 day.

Finally, the effluent concentration is found by substitution:

$$C_{out} = \left(1.5 \cdot 10^{-6}\frac{\text{g}}{\text{L}}\right)\frac{1}{\left(1 + 45.6 \text{ day} \cdot 1 \text{ day}^{-1}\right)} = 3.2 \cdot 10^{-8}\frac{\text{g}}{\text{L}} = 3.2 \cdot 10^{-2}\frac{\mu\text{g}}{\text{L}}$$

Discussion

1 The flow into the lake from the industrial discharge is the only flow given. Thus, its flow rate must equal the flow rate in the stream that empties the lake (neglecting evaporation, seepage, etc.).

2 The derived equation for this example *only* applies to a steady state application to a lake with a single input and a single outflow. Any other set of circumstances (for example non-steady state, multiple inputs, a PFR rather than a CMFR) requires a different equation to be derived from the governing mass balance equation.

1.5.3 Plug Flow Reactor (PFR)

A PFR differs from a CSTR in that a PFR does not have any mixing along its length. Examples of PFRs in environmental engineering include rivers and a baffled tank (shown in plan view in Figure 1.5).

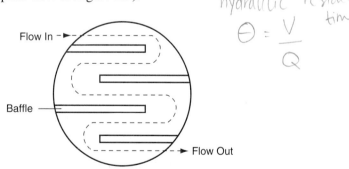

hydraulic residence
time
$\theta = \frac{V}{Q}$

Figure 1.5 A baffled tank as an example of a PFR

Since there is no mixing longitudinally within a PFR, the PFR can be envisioned as a series of individual batch reactors traveling along an imaginary conveyer belt. In a PFR, there is no mixing between the imaginary individual batch reactors.

Hydraulic residence time for a PFR is determined with the same equation used to calculate θ for a CMFR (Equation 1.4). Thus, each of the individual batch reactors on the conveyer belt is on the conveyer belt for a time equal to the hydraulic residence time of the PFR. Given this understanding of a PFR as a large number of discrete batch reactors in series, each with a residence time equal to the residence time of the PFR, we can infer that the concentration exiting a PFR with first order reaction is given by

$$C_{effluent} = C_o e^{-k \cdot \theta}$$

where C_o is the concentration entering the PFR [M/L³]

  k is the decay coefficient [T⁻¹]

  θ is the hydraulic residence time [T]

Note that this equation only applies to a first-order reaction in a PFR.

1.5.4 Summary Tables for Reactors

Example 1.18 is a relatively simple example; however, a fair amount of time is still involved in setting up and solving it. Table 1.23 provides a series of equations, derived from the generic mass balance equation, that can be used for non–steady state applications in batch reactors and steady state reactions in PFRs and CSTRs. Note that the equation for a first-order ideal CMFR is the same equation derived in Example 1.18. Thus, Table 1.23 can be used to save valuable time when solving mass balance problems.

Table 1.23 Comparison of performance* for decay reactions

Reaction Order	Ideal Batch	PFR	CMFR
0	$C_t = C_o - k_o t$	$C_{eff} = C_o - k_o \theta$	$C_{eff} = C_o - k_o \theta$
1	$C_t = C_o[\exp(-k_1 \cdot t)]$	$C_{eff} = C_o[\exp(-k_1 \theta)]$	$C_{eff} = C_o/(1 + k_1 \cdot \theta)$
2	$C_t = \dfrac{C_o}{1 + k_2 \cdot t \cdot C_0}$	$C_{eff} = \dfrac{C_o}{1 + k_2 \cdot \theta \cdot C_0}$	$C_{eff} = \dfrac{(4k_2 \cdot \theta \cdot C_o + 1)^{0.5} - 1}{2k_2 \cdot \theta}$

*C_t = Concentration at time t for batch reactor (non–steady state).
C_{eff} = Concentration in effluent of PFR or CMFR (steady state).

Source: Adapted from M. L. Davis and D. A. Cornwell, *Introduction to Environmental Engineering*, 4th ed. (McGraw-Hill, 2006).

Alternatively, rather than solving for the effluent concentration, the mass balance equation could be manipulated to solve for detention time. A series of useful equations for the most common cases is shown in Table 1.24.

Table 1.24 Comparison of steady state mean retention times* for decay reactions

Reaction Order	Ideal Batch	PFR	CMFR
0	$\theta = (C_o - C_t)/k_0$	$\theta = (C_o - C_{eff})/k_0$	$\theta = (C_o - C_{eff})/k_0$
1	$\theta = \dfrac{\ln(C_o / C_t)}{k_1}$	$\theta = \dfrac{\ln(C_o / C_{eff})}{k_1}$	$\theta = \dfrac{(C_o / C_{eff}) - 1}{k_1}$
2	$\theta = \dfrac{(C_o / C_t) - 1}{k_2 \cdot C_o}$	$\theta = \dfrac{(C_o / C_{eff}) - 1}{k_2 \cdot C_o}$	$\theta = \dfrac{(C_o / C_{eff}) - 1}{k_2 \cdot C_{eff}}$

*C_t = Concentration at time t for batch reactor (non–steady state).
C_{eff} = Concentration in effluent of PFR or CMFR (steady state).

Source: Adapted from M. L. Davis and D. A. Cornwell, *Introduction to Environmental Engineering*, 4th ed. (McGraw-Hill, 2006).

ADDITIONAL RESOURCES

1   Davis, M. L., and S. J. Masten. *Principles of Environmental Engineering and Science*. McGraw-Hill, 2003.

2   Masters, G. M. *Environmental Engineering and Science,* 2d ed. Prentice Hall, 1997.

3   Mihelcic, J. R. *Fundamentals of Environmental Engineering.* Wiley, 1998.

Biology and Ecology Review

OUTLINE

The discussion of WQI in Chapter 1 focused on chemical contaminants or, in the case of suspended or colloidal solids, physical contamination. However, in a global perspective, these are often trivial when compared to the potential negative health effects that are a result of biological contamination. While the United States has not seen large outbreaks of waterborne diseases due to biological contamination, worldwide pathogenic bacteria and other organisms in water can have devastating effects on human health. The ability to identify and quantify the presence of a single strain of organism exists; unfortunately, countless different microorganisms may be present in any ecosystem. It is necessary, therefore, to select an indicator organism, or class of organisms, to serve the role of identifying the potential for impact to human health.

By definition, microbiology is the *study of biological organisms that are generally not visible without magnification.* For environmental engineers, primary application of microbiological activity is in the areas of secondary wastewater treatment and the remediation of contaminated soils and groundwater systems. Microbial populations may also contribute to the depletion of oxygen in WWTP-receiving streams, and the contamination of surface and groundwater and the ability to provide safe drinking water through disinfection are a high priority for all populations. Additional roles for microbes include solid waste decomposition in landfills and composting systems, as well as the novel use of microbes in engineered environments (such as the reduction of the greenhouse gas CO_2, using microalgae in photobioreactors). This chapter reviews basic microbiological concepts and issues related to microbial ecology, as well as their specific relevance to environmental engineering applications.

2.1 MICROBIOLOGICAL CLASSIFICATIONS

In the microbiological world, a simple distinction can be made between organisms based on cellular complexity. Higher order microbes belong to the group known as eukaryotes and include familiar types such as fungi, algae, and protozoa (outside of the microbial world, higher plants and animals also belong to this group). Along with other specialized cell functional subunits, eukaryotes are considered the simplest life forms to possess a membrane-delimited nucleus.

The more physiologically simple form of microbes, prokaryotes, lack a membrane-delimited nucleus but instead possess a single circle of double-stranded DNA that is free to move within the cell. Of primary interest in this group are the bacteria, due primarily to their ubiquitous nature in natural ecosystems, their variation in respiratory and metabolic functionality, and their role in water and wastewater systems. The prokaryotes also include the most primitive life forms, called archaea. While archaea and bacteria share the cellular similarities that classify both as prokaryotes, the universal phylogenetic tree indicates that archaea are more closely related to the eukaryotes than they are to bacteria. A summary of microorganism types and their role in environmental engineering is provided in Table 2.1.

Table 2.1   Select microorganism classifications and their role in environmental engineering

Cell Structure	Organism	Role
Eukaryotes	Fungi	Wastewater treatment, composting, bioremediation of contaminated soils
	Algae	Wastewater treatment (stabilization ponds)
	Protozoa	Enhanced BOD removal and wastewater settling characteristics in WWTP; drinking water pathogen
Prokaryotes	Bacteria	Primary feeders in WWTP; bioremediation of contaminated soils and groundwater; drinking water pathogen
	Archaea	Thrive in extreme environments; extreme halophiles; hyperthermophiles; acidophiles

Due to the many diseases that arise from the consumption of water contaminated with pathogenic microbes (bacteria, viruses, and protozoa), special attention is given to these organisms. Rather than attempt to identify all potential pathogens in a water sample, it is convenient to perform an assay to determine if any members of a group of specific strains belonging to the same class of organisms are present. The most widely used set of indicator organisms are the coliform group of bacteria (which includes *E. coli*). They are defined as aerobic or facultative, gram-negative rods that possess the ability of fermenting lactose within 48 hours at 37°C, producing a gas by-product.

Differentiation is sometimes made between total coliforms and fecal coliforms, although the majority of total coliforms arise from the feces of warm-blooded animals. Since all waterborne pathogens are transmitted through feces, water that is coliform free is likely to be pathogen free. Therefore, total coliform tests are generally performed by preparing tubes of a lauryl tryptose–based culture media, which contain an inverted vial that is filled with the same media. Positive results are indicated when growth occurs with gas production, as indicated by some of the gas being trapped in the inverted vial. Since some coliforms may not originate in the intestines of warm-blooded animals, positive tests for total coliforms can be checked for fecal coliforms by transferring a sample of the broth to specific culture medium. Another group of indicator organisms useful in brackish or salt waters are the fecal streptococci. Further, the ratio of fecal coliforms to fecal streptococci is a strong indicator as to whether the origin of the fecal contamination is from humans or other animals.

2.1.1 Bacterial Morphology and Quantification

Bacteria shape varies from spherical (coccus) to rod (bacillus) to curved (vibrio), ranging in size from 0.5 to 2 μm in diameter and up to 5 μm in length. Individual cells may freely roam in the aqueous environment, or they may form clusters (staphylo) or chains (strepto). Some strains may form spores (protective outer covering) when subjected to harsh environmental conditions as a survival tactic, which may affect the selection of the appropriate disinfection technology or parameters employed.

While the fermentation tubes mentioned previously may indicate the presence or absence of coliforms, it is difficult to quantify organisms present and thereby determine cell concentrations. One method to quantify bacterial coliforms is the most probable number (MPN) approximation, which uses five replicates of several 1:10 dilutions to statistically estimate bacteria concentrations in the original sample. Standard Methods for the Examination of Water and Wastewater dictates that the lowest dilution that results in five positive tubes and the next two dilutions in sequence be used for estimating cell numbers. If no dilution produces five positive tubes, the first three dilutions are used. The number of positive tubes for each dilution are reported as A-B-C and compared to the MPN Index chart shown in Table 2.2. Unfortunately, initial concentrations are usually low, so dilution generates many tubes without growth, and at five samples per dilution the error is quite large and results have little quantitative meaning.

Plate counts have been used, but low concentrations also yield many plates with no colony forming units (CFUs), which can be defined as *one or more organisms that generate a colony*. To overcome the low concentrations, 0.2 or 0.45 μm filters are often employed to strain a known volume of sample (typically 100 mL), and the filters are subsequently incubated in a specific culture medium for 24 hours at 37°C. Filters are then examined for characteristic colonies (reddish or greenish tint), and results are reported as CFU/mL or CFU/100 mL.

Table 2.2 MPN Index and 95% confidence limits for five-tube 1:10 serial dilutions (Standard Methods, APHA, AWWA, WEF)

Combination of Positives	MPN Index per 100 mL	95% Confidence Limits		Combination of Positives	MPN Index per 100 mL	95% Confidence Limits	
		Lower	Upper			Lower	Upper
0-0-0	< 2	—	—	4-3-0	27	12	67
0-0-1	2	1.0	10	4-3-1	33	15	77
0-1-0	2	1.0	10	4-4-0	34	16	80
0-2-0	4	1.0	13	5-0-0	23	9.0	86
0-0-0	< 2	—	—	5-0-1	30	10	110
1-0-0	2	1.0	11	5-0-2	40	20	140
1-0-1	4	1.0	15	5-1-0	30	10	120

(continued)

Combination of Positives	MPN Index per 100 mL	95% Confidence Limits		Combination of Positives	MPN Index per 100 mL	95% Confidence Limits	
		Lower	Upper			Lower	Upper
1-1-0	4	1.0	15	5-1-1	50	20	150
1-1-1	6	2.0	18	5-1-2	60	30	180
1-2-0	6	2.0	18	5-2-0	50	20	170
2-0-0	4	1.0	17	5-2-1	70	30	210
2-0-1	7	2.0	20	5-2-2	90	40	250
2-1-0	7	2.0	21	5-3-0	80	30	250
2-1-1	9	3.0	24	5-3-1	110	40	300
2-2-0	9	3.0	25	5-3-2	140	60	360
2-3-0	12	5.0	29	5-3-3	170	80	410
3-0-0	8	3.0	24	5-4-0	130	50	390
3-0-1	11	4.0	29	5-4-1	170	70	480
3-1-0	11	4.0	29	5-4-2	220	100	580
3-1-1	14	6.0	35	5-4-3	280	120	690
3-2-0	14	6.0	35	5-4-4	350	160	820
3-2-1	17	7.0	40	5-5-0	240	100	940
4-0-0	13	5.0	38	5-5-1	300	100	1300
4-0-1	17	7.0	45	5-5-2	500	200	2000
4-1-0	17	7.0	46	5-5-3	900	300	2900
4-1-1	21	9.0	55	5-5-4	1600	600	5300
4-1-2	26	12	63	5-5-5	> 1600	—	—
4-2-0	22	9.0	56				
4-2-1	26	12	65				

Example **2.1**

MPN quantification

Given the following MPN tube results, what is the approximate number of coliforms in the original sample?

Dilution	Results	
10^0	+ + + + +	5
10^{-1}	+ + + + +	5
10^{-2}	+ + + + +	5
10^{-3}	+ + + + +	5 ← 10^{-3}
10^{-4}	+ − + + −	3
10^{-5}	− − + − +	2
10^{-6}	− + − − −	1
10^{-7}	− − − − −	0

140×10^{-3}

Solution

MPN tables and Standard Methods indicate that the correct selection of tubes is to use the lowest dilution that results in five positive tubes and the following two dilutions. From the data given, the result would be 5-3-2, with a dilution factor of 10^3. The MPN tables give a result for 5-3-2 as 140 coliforms per 100 mL. Multiplying by the dilution factor yields 140,000 coliforms/100 mL.

2.1.2 Bacterial Activity and Growth

Bacteria are also classified by their primary metabolic process, specifically the substrate that serves as energy provider (food source). Metabolism refers to two processes: catabolism, which describes the breaking down of chemical compounds for energy extraction, and anabolism, which is the assimilation and synthesis of new compounds. Autotrophs are characterized as organisms that utilize the carbon from CO_2 as their sole carbon source, while heterotrophs are characterized by the use of organic matter as a supply of carbon.

The process of converting the substrate into energy requires a terminal electron acceptor, which for aerobic processes would be oxygen. The most efficient means of respiration is to extract dissolved oxygen from water, a process that is called aerobic respiration. Aerobic biological processes require organisms, organic matter, oxygen, and water; and the complete biological transformation can be approximated by the following chemical reaction:

$$C_aH_bO_cN_d + \left(\frac{4a+b-2c+3d}{4}\right)O_2 \rightarrow aCO_2 + \left(\frac{b-3d}{2}\right)H_2O + dNH_3$$

This reaction assumes that the chemical composition given for the organic matter is comprised of the biodegradable fraction (i.e., plastics and other nonbiodegradable compounds should not be included in the chemical formulation). In addition, if sufficient time and oxygen are supplied, ammonia oxidizes to nitrate as follows:

$$NH_3 + 2O_2 \rightarrow H_2O + HNO_3$$

Under anaerobic conditions, bacteria can extract oxygen from oxygen-containing chemicals or use other electron donor compounds in the aqueous environment. Conversion of NO_2^- or NO_3^- to N_2 is called denitrification; reduction of SO_4^{2-} to H_2S is called sulfidogenesis; and CO_2 can be reduced to CH_4 in a process called methanogenesis, meaning *methane creation*. Anaerobic transformation may be expressed chemically as

$$C_aH_bO_cN_d + \left(\frac{4a-b-2c+3d}{4}\right)H_2O \rightarrow$$

$$\left(\frac{4a+b-2c-3d}{8}\right)CH_4 + \left(\frac{4a-b+2c+3d}{8}\right)CO_2 + dNH_3$$

For unknown organic compositions, rapidly biodegradable material (e.g., food wastes, paper, cardboard, and yard wastes that do not include large stumps or branches) can be estimated to have the composition $C_{70}H_{110}O_{50}N$, while other slower biodegradable matter can be estimated to have the composition $C_{20}H_{30}O_{10}N$.

Organisms that are only capable of aerobic or anaerobic respiration are termed obligate (or strict), while those that may use the most energetic pathway available to them are called facultative. Methanogenesis and sulfidogenesis are important anaerobic respiration pathways involved in the generation of landfill gases and anaerobic digestion of wastewater sludges.

There are several factors that impact biological growth, including substrate, electron acceptor, presence of nutrients, temperature, presence of toxins, biofilm formation, and synergism with other microbiological species. Many of these can be controlled by the environmental engineer when designing biological treatment systems; however, they become very difficult to control in natural environments. For example, bacteria can be classified as psychrophiles, mesophiles, or thermophiles, depending on the temperature range under which they can optimally metabolize carbon. Synergistic relationships can be competitive, cooperative, or predatory, although each has its place in the microbial community. For example, bacteria are the dominant feeders on organic matter in municipal wastewater, but they are consumed by protozoa that scavenge the weak cells and help maintain a healthy, kinetically active bacteria population, with improved settling characteristics. Another example is the consumption of organic matter in facultative ponds by bacteria, which release CO_2 that can be used by algae photosynthetically to produce O_2, which the bacteria then use to process more organic matter.

A four-phase bacterial growth curve is presented in Figure 2.1, identifying the (1) *lag* or *acclimatization*, (2) *exponential* or *log growth*, (3) *stationary*, and (4) *death* or *decay* phases. Note the use of a logarithmic scale on the vertical axis. Often, the transitions into and out of exponential growth are of particular importance and are identified as *accelerating growth* and *declining growth*, respectively. These transitions are sometimes represented as additional phases on the growth curve and have been added between the appropriate stages in Figure 2.1.

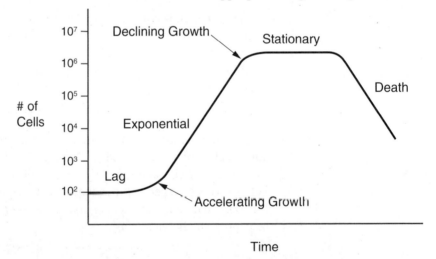

Figure 2.1 Four-phase microbial growth curve

Bacteria reproduce by binary fission, each cell dividing every 20–30 minutes under optimum conditions into two cells (often called the doubling time), which can be expressed as

$$\frac{dX}{dt} = \mu X$$

which is integrated to yield the cell mass growth curve expressed as

$$X_t = X_0 e^{\mu t}$$

where X_t is the concentration of cells at time t, X_0 is the initial cell concentration, and μ is the specific growth rate (time^{-1}). Cell concentrations are expressed in mg/L, and in municipal wastewater processes are generally equal to TVSS. The form of the equation identifies the second phase due to the exponential dependence of the growth curve.

Example 2.2

Exponential growth

If 100 bacterial cells become 10,000,000 in eight hours, find the doubling period assuming a first-order reaction rate applies.

Solution

Using a first-order reaction rate and the data given in the problem statement we may write

$$N = N_0 e^{\mu t} \quad \Rightarrow \quad \frac{N}{N_0} = \frac{10^7}{10^2} = 10^5 = e^{8\mu} \quad \Rightarrow \quad \frac{\ln\left(10^5\right)}{8} = \mu = 1.44 \text{ hr}^{-1}$$

Using this rate constant and the first-order expression, we can now determine the doubling period by selecting a value for N that is twice the value of N_0 as follows:

$$N = N_0 e^{\mu t} = 200 = 100\, e^{1.44\tau} \Rightarrow \ln\left(\frac{200}{100}\right) = 1.44\,\tau \Rightarrow$$

$$\tau = 0.48 \text{ hr} \times 60\, \frac{\min}{\text{hr}} = 28.9 \text{ min}$$

While the first-order kinetics model may work well for ideal growth conditions, growth (substrate) limiting conditions may be a more appropriate description for certain environmental systems. In these cases, it may be more appropriate to describe the growth of bacteria by the Monod kinetic model, expressed as

$$\mu = \mu_{max}\left(\frac{S}{K_s + S}\right)$$

where μ_{max} is the maximum specific growth rate (time^{-1}), S is the substrate concentration (mg/L), and K_s is the half-saturation constant (mg/L). K_s may be defined as the value of S that corresponds to one-half of μ_{max} as shown in Figure 2.2.

$$K_s \rightarrow \frac{1}{2}\,\mu_{max}$$

Monod Kinetic Model

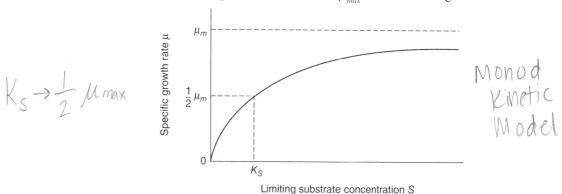

Figure 2.2 Graphical representation of Monod growth equation

Substitution of the Monod kinetic expression into the cell mass growth equation yields

$$\frac{dX}{dt} = \frac{\mu_{max}\, S\, X}{K_s + S}$$

Finally, in order to account for the death of cells during the endogenous phase, we may write

$$\left(\frac{dX}{dt}\right)_{net} = \left[\frac{\mu_{max}\, S\, X}{K_s + S}\right] - k_d X$$

where k_d is the endogenous decay constant (time^{-1}).

If all of the food in the system was converted to biomass, the rate of substrate utilization would equal the rate of biomass generation, due to the conservation of mass. However, we know that the conversion process is less than perfect, and as such a ratio of the amount of biomass generated per unit substrate utilized can be evaluated. This quantity is termed the *yield* (*Y*), and can be expressed as

$$Y = \frac{dX\,/\,dt}{dS\,/\,dt} = \frac{dX}{dS}$$

where *Y* has units of mg biomass per mg substrate. Finally, substrate utilization can be expressed in terms of the yield as

$$\frac{dS}{dt} = -\frac{\mu_{max}\, S\, X}{Y\left(K_s + S\right)}$$

where the endogenous decay constant has been dropped due to the fact that substrate is not being consumed because of cell death.

Example 2.3

Microbial kinetics I

Determine the decay (death) rate constant in the expression below, given the following biological data measured in the laboratory.

$$\left(\frac{dX}{dt}\right)_{net} = \left(\frac{dX}{dt}\right) - k_d X = 75\,\frac{mg}{L \cdot min}$$

$$\left(\frac{dS}{dt}\right) = 144\,\frac{mg}{L \cdot min}$$

$$Y = 0.7$$

$$X = 2500\ mg/L$$

Solution

Using the expression given, we may write

$$\left(\frac{dX}{dt}\right)_{net} = \left(\frac{dX}{dt}\right) - k_d X \quad\Rightarrow\quad k_d = \frac{\left(\frac{dX}{dt}\right) - \left(\frac{dX}{dt}\right)_{net}}{X} = \frac{\left(\frac{dX}{dt}\right) - 75}{2500}$$

From this, we see that we need a value for (*dX/dt*), which is the rate of biomass generated. Noticing that the yield is given and can be defined as

$$Y = \frac{\left(dX/dt\right)}{\left(dS/dt\right)}$$

we may rearrange this expression to solve for (dX/dt) as follows:

$$\left(\frac{dX}{dt}\right) = Y\left(\frac{dS}{dt}\right) = (0.7)\left(144 \ \frac{\text{mg}}{\text{L} \cdot \text{min}}\right) = 100.8 \ \frac{\text{mg}}{\text{L} \cdot \text{min}}$$

Now we may substitute this into the first expression to get k_d as follows:

$$k_d = \frac{\left(\dfrac{dX}{dt}\right) - 75}{2500} = \frac{100.8 - 75}{2500} \quad \Rightarrow \quad k_d = 0.0103 \ \text{min}^{-1} \times \frac{60 \ \text{min}}{1 \ \text{hr}} = 0.62 \ \text{hr}^{-1}$$

Example **2.4**

Microbial kinetics II

A specific growth rate curve is provided for a laboratory test on a domestic waste-water (Exhibit 1). Determine the substrate utilization rate (dS/dt) if the biomass concentration is 2500 mg/L, the microbial decay coefficient (k_d) is 0.33 hr^{-1}, and the laboratory measured growth yield is 0.58 in an activated sludge tank with a substrate concentration of 165 mg/L.

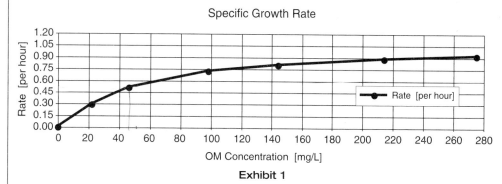

Exhibit 1

Solution

Substrate utilization can be determined as

$$\frac{dS}{dt} = \frac{\mu_{max} \ S \ X}{Y \ (K_s + S)}$$

Values for S, X, and Y are given. Therefore, only μ_{max} and K_s need to be determined. Extrapolating μ_{max} from the plot provided,

$$\mu_{max} = 1.0 \ \text{hr}^{-1}$$

Now, K_s may be evaluated as the substrate concentration at $\frac{1}{2} \mu_{max}$, which is a rate equal to 0.5 hr^{-1}. From the plot provided, $K_s = 44$ mg/L at a rate of 0.5 hr^{-1}. Now, utilization may be calculated as

$$\frac{dS}{dt} = \frac{(1 \ \text{hr}^{-1})(165 \ \text{mg/L})(2500 \ \text{mg/L})}{(0.58 \ \text{mg/mg})(44 + 165 \ \text{mg/L})} = 3403 \ \frac{\text{mg}}{\text{L}} \text{hr}^{-1} \times \frac{1 \ \text{hr}}{60 \ \text{min}} = 56.7 \ \frac{\text{mg}}{\text{L}} \text{min}^{-1}$$

2.2 DO SAG MODELING

As discussed in section 1.3.5, DO is a basic WQI that has a tremendous impact on a surface water's ability to support higher life forms. Typical minimum values for DO in streams are 5 mg/L to support most species of fish, and may be as high as 6 mg/L for more sensitive fish species, such as trout. DO may be consumed by bacteria if a wastewater effluent has a high BOD concentration, or is discharged to receiving water that does not possess sufficient dilution ability. While regulatory limits require average BOD concentrations in a discharge to be less than 30 mg/L, even at this concentration there may be a negative impact to the DO concentration in smaller receiving waters, or in receiving water where several WWTP effluents are discharged in close proximity to each other.

The decrease of DO in a stream receiving an oxygen consumptive contaminant is called DO sag. In order to model the DO sag in a stream, all sources of DO consumption and generation must be identified and quantified. This often proves to be a difficult task, so simplified models can be employed to model select reactions and examine the contributory effect of a target contaminant. For this review, we will assume that a particular stream receives BOD from a single WWTP effluent, the BOD is consumed by bacteria in the stream at a constant rate, and reaeration of the stream occurs only by diffusive flux from the atmosphere at a constant rate.

The classic model developed by Streeter and Phelps is used to predict the DO *deficit* (*D*), defined as the *difference between the DO saturation value and the actual DO concentration*, and can be expressed as

$$D = \frac{k_d L_0}{k_a - k_d}\left[\exp\left(-k_d\, t\right) - \exp\left(-k_a\, t\right)\right] + D_0 \exp\left(-k_a\, t\right)$$

where *D* is the deficit (mg/L) at time *t* (days), k_d is the deoxygenation rate constant (day^{-1}), k_a is the reaeration rate constant (day^{-1}), L_0 is the initial ultimate BOD (mg/L), and D_0 is the initial deficit (mg/L).

In this model, initial conditions are determined as the weighted averages of the stream and WWTP discharge after mixing; therefore, initial BOD would be calculated as

$$L_0 = \frac{Q_{ww} L_{ww} + Q_s L_s}{Q_{ww} + Q_s}$$

where Q_{ww} and Q_s are the volumetric flow rates of the WWTP discharge and stream, and L_{ww} and L_s are the ultimate BOD concentrations of the WWTP discharge and stream (mg/L), respectively. The units on volumetric flowrate are not important, as long as both flows have the same units.

The initial deficit is calculated as

$$D_0 = DO_T^{sat} - DO_0$$

where DO_T^{sat} is the saturation DO concentration evaluated at the temperature of the stream after receiving the WWTP discharge (mg/L), and DO_0 is the initial DO in the stream (mg/L). The weighted average temperature of the combined flows may be calculated as

$$T = \frac{Q_{ww} T_{ww} + Q_s T_s}{Q_{ww} + Q_s}$$

where T_{ww} and T_s are the temperatures (°C) of the WWTP discharge and the stream, respectively. This temperature would be located on the DO saturation table to determine DO_T^{sat}, interpolating between temperatures as necessary. Finally, DO_0 is calculated as a weighted average of the DO values in stream and WWTP discharge before mixing, and may be calculated as follows:

$$DO_0 = \frac{Q_{ww} DO_{ww} + Q_s DO_s}{Q_{ww} + Q_s}$$

where DO_{ww} and DO_s are the DO concentrations of the WWTP discharge and the stream (mg/L), respectively.

Values for k_d, the deoxygenation rate constant, are usually given or assumed based on values that are determined in the laboratory under controlled conditions. For example, often k_d is determined in the lab at a set temperature, usually 20°C. Typical values of k_d for untreated wastewater range from 0.1 to 0.5 day^{-1}, with an average value of 0.25 day^{-1} used in the absence of system-specific data. Values of k_d for treated effluents are approximately half of the untreated values.

Values for k_a, the reaeration rate constant, are dependent upon the mixing of DO in the stream as oxygen is extracted from the air, and are therefore functions of stream velocity and depth. They may be estimated using the following relationship:

$$\left(k_a\right)_{20°C} = \frac{\left(D_L V_s\right)^{1/2}}{H^{3/2}}$$

where k_a is evaluated at 20°C (s^{-1}), D_L is the oxygen diffusivity in water at 20°C (m²/s), V_s is the velocity of the stream (m/s), and H is the depth of the stream (m). At 20°C, the value of D_L is approximately equal to 2.1×10^{-9} m²/s. However, since most calculations require units for k_a of day^{-1}, the value of k_a as calculated above should be multiplied by 86,400. Combining these factors, a common expression for k_a could be written as

$$\left(k_a\right)_{20°C} = \frac{3.95 \, V_s^{1/2}}{H^{3/2}}$$

where V_s has units of m/s, H has units of m, k_a has units of day^{-1}, and the constant 3.95 accounts for the pr oper unit conversions. If insufficient data exists to calculate the reaeration constant, typical values may be used as presented in Table 2.3.

Table 2.3 Typical reaeration coefficients for various water bodies

Water Body	k_a [day^{-1}]
Small ponds	0.15
Sluggish streams, lake	0.3
Large streams, low velocity	0.4
Large streams, high velocity	0.5
Swift streams	0.8

Because both rate constants have assumed or calculated values determined at 20°C, both rate constants need to be corrected to account for reaction at the weighted average temperature using the following relationship:

$$k_T = k_{20} \, \theta^{(T-20)}$$

where T is the weighted average temperature (°C), k_T is the value of the constant at temperature T (day^{-1}), k_{20} is the given or calculated value of the constant at 20°C (day^{-1}), and θ is the temperature coefficient. The temperature coefficient for k_d may range from a value of 1.056 for temperatures between 20°C and 30°C, up to a value of 1.135 for temperatures between 4°C and 20°C. The temperature coefficient for k_a is often assumed to be 1.024.

Example 2.5

Reaeration coefficient

Calculate the reaeration coefficient of a stream 5 ft deep and 10 ft wide if it is flowing at 20 cfs and a temperature of 12°C. You may assume that $D_L = 0.002$ ft²/day and $\theta = 1.024$.

Solution

For the expression given, we will need to determine stream velocity in units of feet per day. This may be accomplished as follows:

$$V_s = \frac{Q}{A} = \frac{20 \, \frac{\text{ft}^3}{\text{s}}}{(5 \text{ ft})(10 \text{ ft})} = 0.4 \, \frac{\text{ft}}{\text{s}} \times \frac{86{,}400 \text{ s}}{\text{day}} = 34{,}560 \, \frac{\text{ft}}{\text{day}}$$

We may now estimate the reaeration coefficient at 20°C as follows:

$$\left(k_a\right)_{20°C} = \frac{\left(0.002 \, V_s\right)^{1/2}}{H^{3/2}} = \frac{\left[(0.002)(34{,}560)\right]^{1/2}}{5^{3/2}} = 0.744 \text{ day}^{-1}$$

This reaeration constant must be corrected to the stated temperature of 12°C. Using the value 1.024 given in the problem statement, we get

$$\left(k_a\right)_{12} = \left(k_a\right)_{20} (1.024)^{12-20} = (0.744)(1.024)^{-8} = 0.615 \text{ day}^{-1}$$

The Streeter-Phelps model can be solved repeatedly for small increments of t to yield the DO sag curve as shown in Figure 2.3. However, it is often desired to identify the value of the minimum DO concentration (DO_{min}), which corresponds to the maximum (or critical) DO deficit (D_c). This is accomplished by solving the deficit equation using a value of time that corresponds to the critical point (t_c), which can be calculated as

$$t_c = \frac{1}{k_a - k_d} \ln \left[\frac{k_a}{k_d} \left(1 - D_0 \frac{k_a - k_d}{k_d L_0} \right) \right]$$

where t_c is expressed in units of days. Finally, the location (X_c) downstream of the WWTP discharge where the critical deficit occurs can be found using t_c as follows:

$$X_c = t_c V_s$$

where X_c is the critical distance. It is left to the examinee to ensure dimensional homogeneity in the above expression; however, X_c is often reported in units of miles.

$DO_{sat} = DO_{min} + D_{crit}$

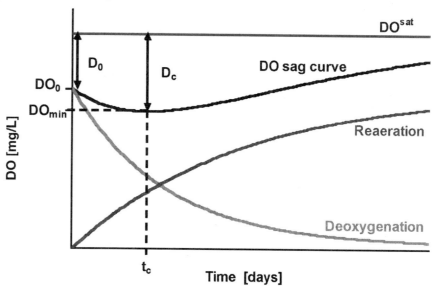

Figure 2.3 Graphical representation of DO sag curve

In practice, the Streeter-Phelps equation is used to evaluate if a known discharge will have a negative impact on the aquatic life in the receiving stream due to reduced DO levels. Because of the form of the deficit equation, one method to determine if a critical deficit will occur due to a particular discharge is to compare the $k_d L_0$ product with the $k_a D_0$ product. If sufficient BOD exists at a high enough rate of degradation, $k_d L_0$ will exceed $k_a D_0$, and a critical deficit will occur at some time and location downstream. Conversely, when a low initial BOD or slow rate of utilization exists, $k_a D_0$ will be greater than $k_d L_0$, and no DO sag will exist downstream.

The system under consideration should be reevaluated in the extreme seasons, because temperature and flow differences can have a substantial impact on the critical deficit for the same discharge. Further, if the critical deficit corresponds to a value of DO that is less than a predetermined, acceptable DO standard, then corrective measures could be attempted. Unfortunately, the only actions available to the engineer that would improve DO in the stream would be to lower the BOD in the discharge or increase the DO of the discharge. Increased DO_{ww} could be obtained through a final aeration step, while lowering BOD_{ww} may require upgrades or additions to the WWTP that would increase the overall plant BOD removal efficiency. In extreme cases, aeration of the receiving stream could be conducted as a last resort.

Example **2.6**

Streeter-Phelps modeling

A small river receives a wastewater treatment plant discharge of 5 mgd with a BOD of 20 mg/L and a DO concentration of 1.0 mg/L. The river is 40 feet wide and flows at 1 ft/sec, but depth and temperature varies by season. You may assume the BOD in the river is 0.9 mg/L and the DO concentration is at 90% of the saturation value, which also varies with temperature. You may also assume the biological activity coefficient $[(k_d)_{20}]$ is 0.3 day^{-1}, the stream reaeration coefficient $[(k_a)_{20}]$ is 0.25 day^{-1}, and the temperature correction coefficient for k_d is 1.135, while the value for k_a is 1.024.

(a) Determine the minimum DO concentration in the stream in the summer when the river temperature is 22°C and flows at a depth of 3 feet. You may assume a WWTP discharge temperature of 16°C.

(b) Determine the minimum DO concentration in the stream in the winter when the river temperature is 4°C and flows at a depth of 4 feet. You may assume a WWTP discharge temperature of 13°C.

Solution

(a) The critical (maximum) deficit occurs at the minimum DO concentration as follows:

$$D_c = DO^{sat} - DO_{min}$$

The determination of DO^{sat} requires knowledge of the temperature of the mixed flows, and D_c could be calculated using the Streeter-Phelps equation as follows:

$$D_c = \frac{k_d L_0}{k_a - k_d}\left[\exp\left(-k_d t_c\right) - \exp\left(-k_a t_c\right)\right] + D_0 \exp\left(-k_a t_c\right)$$

where

$$t_c = \frac{1}{k_a - k_d} \ln\left[\frac{k_a}{k_d}\left(1 - D_0 \frac{k_a - k_d}{k_d L_0}\right)\right]$$

First, calculate the biological use coefficient (k_d) and the reaeration coefficient (k_a) at the temperature of the mixed flows (T_m). In order to calculate the flow-weighted temperature, the flow rates of the river and wastewater must be determined in the same units as follows:

$$Q_r = \left(40\text{ ft}\right)\left(3\text{ ft}\right)\left(1\frac{\text{ft}}{\text{s}}\right) = 120\ \frac{\text{ft}^3}{\text{s}} \quad \text{and} \quad Q_{ww} = \left(5\text{ Mgd}\right)\left(1.547\frac{\text{cfs}}{\text{Mgd}}\right) = 7.735\ \frac{\text{ft}^3}{\text{s}}$$

T_m may now be calculated as:

$$T_m = \frac{Q_{ww}T_{ww} + Q_r T_r}{Q_{ww} + Q_r} = \frac{\left(7.735\text{ cfs}\right)\left(16°C\right) + \left(120\text{ cfs}\right)\left(22°C\right)}{7.735\text{ cfs} + 120\text{ cfs}} = 21.64°C$$

Correction can now be made to k_d and k_a through the temperature correction expression as follows:

$$k_T = k_{20}\ \theta^{(T-20)}$$

$$\left(k_d\right)_{21.64} = \left(k_d\right)_{20} 1.135^{(21.64-20)} = \left(0.3\right)1.135^{1.64} = 0.37\text{ day}^{-1}$$

$$\left(k_a\right)_{21.64} = \left(k_a\right)_{20} 1.024^{(21.64-20)} = \left(0.25\right)1.024^{1.64} = 0.26\text{ day}^{-1}$$

To calculate the deficit, the BOD concentration of the mixed flows is required and may be determined as:

$$L_0 = \frac{Q_{ww}L_{ww} + Q_r L_r}{Q_{ww} + Q_r} = \frac{\left(7.735\text{ cfs}\right)\left(20\ \frac{\text{mg}}{\text{L}}\right) + \left(120\text{ cfs}\right)\left(0.9\ \frac{\text{mg}}{\text{L}}\right)}{7.735\text{ cfs} + 120\text{ cfs}} = 2.057\ \frac{\text{mg}}{\text{L}}$$

The initial deficit (D_0) is also needed, which requires the initial DO in the river. This may be determined by finding 90% of the saturation value at the river temperature as follows:

$$\left(DO_i\right)_r = \left(0.9\right)\left(DO^{sat}\right)_{22°C} = \left(0.9\right)\left(8.8 \frac{mg}{L}\right) = 7.92 \frac{mg}{L}$$

This value can be used to determine the flow-weighted DO as follows:

$$DO_0 = \frac{Q_{ww}DO_{ww} + Q_rDO_r}{Q_{ww} + Q_r} = \frac{\left(7.735 \text{ cfs}\right)\left(1 \frac{mg}{L}\right) + \left(120 \text{ cfs}\right)\left(7.92 \frac{mg}{L}\right)}{7.735 \text{ cfs} + 120 \text{ cfs}} = 7.501 \frac{mg}{L}$$

From this and the value of DO^{sat} at T_m, the initial deficit may be determined as:

$$D_0 = \left(DO^{sat}\right)_{T_m} - DO_0 = \left(DO^{sat}\right)_{21.64°C} - 7.50 \frac{mg}{L} = 8.865 - 7.501 = 1.364 \frac{mg}{L}$$

Because $k_d > k_a$ and $L_0 > D_0$, we know $k_dL_0 > k_aD_0$, and therefore a deficit will occur.

Plugging the above values in the equation for critical time (t_c) above yields:

$$t_c = \frac{1}{0.26 - 0.37} \ln\left[\frac{0.26}{0.37}\left(1 - \left(1.364\right)\frac{0.26 - 0.37}{\left(0.37\right)\left(2.057\right)}\right)\right] = 1.57 \text{ days}$$

The units were left off of the previous expression for simplicity and it is left to the examinee to verify the homogeneity of units. This value must be put into the deficit equation as follows:

$$D_c = \frac{\left(0.37\right)\left(2.057\right)}{0.26 - 0.37}\left[\overset{-0.1055}{\exp\left(-0.37 \times 1.57\right)} - \overset{+0.9068}{\exp\left(-0.26 \times 1.57\right)}\right] + 1.364 \overset{0.5329}{\exp\left(-0.26 \times 1.57\right)}$$

$$\underset{\left(-6.919\right)}{} \qquad \left(-0.1282\right) \qquad$$

$$D_c = 1.64 \frac{mg}{L} \qquad D_c = 1.41$$

Again, the homogeneity of the units is left to the examinee. Finally, DO_{min} may be calculated as follows:

$$DO_{min} = \left(DO^{sat}\right)_{T_m} - D_c = 8.865 - 1.64 = 7.225 \frac{mg}{L}$$

(b) Start by calculating k_d, k_a, L_0, and D_0 as before using the cold-weather data:

$$Q_r = \left(40 \text{ ft}\right)\left(4 \text{ ft}\right)\left(1 \frac{ft}{s}\right) = 160 \frac{ft^3}{s} \quad \text{and} \quad Q_{ww} = \left(5 \text{ Mgd}\right)\left(1.547 \frac{cfs}{Mgd}\right) = 7.735 \frac{ft^3}{s}$$

$$T_m = \frac{Q_{ww}T_{ww} + Q_rT_r}{Q_{ww} + Q_r} = \frac{\left(7.735 \text{ cfs}\right)\left(13°C\right) + \left(160 \text{ cfs}\right)\left(4°C\right)}{7.735 \text{ cfs} + 160 \text{ cfs}} = 4.415°C$$

$$\left(k_d\right)_{4.415} = \left(k_d\right)_{20} 1.135^{\left(4.415-20\right)} = \left(0.3\right)1.135^{-15.585} = 0.042 \text{ day}^{-1}$$

$$\left(k_a\right)_{4.415} = \left(k_a\right)_{20} 1.024^{\left(4.415-20\right)} = \left(0.25\right)1.024^{-15.585} = 0.173 \text{ day}^{-1}$$

$$L_0 = \frac{Q_{ww}L_{ww} + Q_rL_r}{Q_{ww} + Q_r} = \frac{\left(7.735 \text{ cfs}\right)\left(20 \frac{mg}{L}\right) + \left(160 \text{ cfs}\right)\left(0.9 \frac{mg}{L}\right)}{7.735 \text{ cfs} + 160 \text{ cfs}} = 1.781 \frac{mg}{L}$$

$$\left(DO_i\right)_r = \left(0.9\right)\left(DO^{sat}\right)_{4°C} = \left(0.9\right)\left(13.1 \frac{mg}{L}\right) = 11.79 \frac{mg}{L}$$

$$DO_0 = \frac{Q_{ww}DO_{ww} + Q_r DO_r}{Q_{ww} + Q_r} = \frac{\left(7.735 \text{ cfs}\right)\left(1 \dfrac{mg}{L}\right) + \left(160 \text{ cfs}\right)\left(11.79 \dfrac{mg}{L}\right)}{7.735 \text{ cfs} + 160 \text{ cfs}} = 11.29 \frac{mg}{L}$$

$$D_0 = \left(DO^{sat}\right)_{T_m} - DO_0 = \left(DO^{sat}\right)_{4.4°C} - 11.29 \frac{mg}{L} = 12.98 - 11.29 = 1.69 \frac{mg}{L}$$

Since it is not obvious by inspection, we should now calculate the $k_d L_0$ and $k_a D_0$ products to see if a deficit occurs:

$$k_d L_0 = \left(0.042 \text{ day}^{-1}\right)\left(1.781 \frac{mg}{L}\right) = 0.075 \frac{mg}{L \cdot day}$$

$$k_a DO = \left(0.173 \text{ day}^{-1}\right)\left(1.69 \frac{mg}{L}\right) = 0.292 \frac{mg}{L \cdot day}$$

Since $k_a D_0 > k_d L_0$, reaeration is greater than demand, and the critical deficit is equal to the initial deficit. This means DO_{min} is equal to DO_0, which was determined to be 11.29 mg/L.

2.3 MICROBIAL ECOLOGY

Ecology is generally considered as the study of the interrelationship between living things and the environment in which they exist. The population and its environment together are known as an *ecosystem*. Due to their extreme importance in the practice of environmental engineering, it is important to look at certain microbiological communities and the conditions under which they interact with other biological and abiotic species. Subspecialties that exist in specific fields of study include aquatic microbiology, soil microbiology, medical sciences, nutrient cycling or food chains, and bioaccumulation (also known as biomagnification).

With a global human population of approximately 6.5 billion, the Earth may seem like it is becoming a crowded place to live. However, when looking at the microbiological world, organism counts on the order of 10^4 to 10^7 are not uncommon in a single gram of soil or water in natural environments. In these cases, a single liter of water or one kilogram (2.2 lb) of soil will have microbial populations that exceed the human population of this planet. Further, under favorable growth conditions, some microbial species can double their populations in under 30 minutes. This ubiquitous presence requires an understanding of our interactions with these populations as we attempt to address the negative impacts of material discharges to the environment in which we coexist with our biological neighbors. Understanding which populations pose a threat and which species can be utilized in the appropriate design of treatment systems offers additional options that abiotic treatment methodologies alone cannot accomplish.

Within a defined population, *cooperative* or *competitive* influences may have an impact on the health and propagation of the colony. When these influences are extended toward other populations (that is, two different species), it is usually termed *symbiosis*. An example of cooperative symbiosis includes situations where one population benefits while another is neither harmed nor benefited (*commensalism*). In another case, both populations benefit from the interaction, and this relationship may be required (*mutualism*) or optional (*synergism*). Competitive symbiosis situations may include two populations vying for the same limited resources or the extreme case where one population produces an extracellular toxin that inhibits another population (*antagonism*). Finally, many environments exist where one species primarily derives its nutritional needs at the expense of another. This may be accomplished through extracting benefit slowly over relatively long periods of time (*parasitism*), or in a single feeding where the harmed population is used as a primary food source (*predation*).

One important concept in the interactions between living organisms and the world in which they exist is the cycling of all matter. Often called *nutrient cycling* (or more appropriately termed *biogeochemical cycling*), this is the idea that essentially all mass currently providing the sustenance for the populations of today has been in existence from the genesis of the planet. The reuse of carbon, hydrogen, oxygen, nitrogen, sulfur, phosphorus, and all biologically important metals is a fact that impacts every population on a daily basis. The transfer of elements between the different ordered populations is known as the *food web*, as presented by the simplistic schematic of an idealized food web in Figure 2.4. A typical path would identify plant matter as *primary producers* that provide food for grazing animals, which in turn are food for higher predators (that is, humans), possibly along with additional plant matter. Both the *grazers* and *predators* respire CO_2 and return organic matter to the soil in their waste streams, which are used once again by the plants that also produce the O_2 necessary for the animals to live.

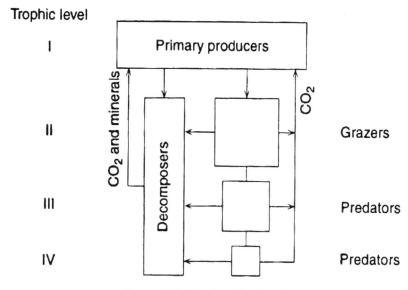

Figure 2.4 Idealized food web

Source: Ronald M. Atlas and Richard Bartha, *Microbial Ecology: Fundamentals and Applications*, 4th ed. © 1998 Benjamin/Cummings Publishing Company, Inc. Reprinted by permission of Pearson Education, Inc.

One role that microbes play in the food chain is the *mineralization* of most organic chemicals, which may be defined as complete biodegradation of large organic molecules and the subsequent release of their elemental or basic ionic species back to the ecosystem. However, several organic compounds are known to be *recalcitrant*, which means they are persistent in the environment due to their resistance to biotransformation. Many of these compounds are man-made (or *xenobiotic*) and inhibit microbial attack or have a toxic effect on the biological community. *Bioconcentration* refers to the process whereby contaminants are directly absorbed into an organism from the water surrounding it. *Bioaccumulation* occurs when the contaminants are absorbed from ingested food in addition to the water. The uptake of these chemicals by lower organisms impacts all higher species that incorporate the lower biota in their food cycles. Often, the lower organisms may adapt to the presence of the toxin, or there may be a threshold concentration that triggers a negative response in the higher organism after a period of accumulation. As it travels up the food chain, the concentration of contaminants within the higher trophic organisms becomes progressively higher, a process known as *biomagnification*.

Fluid Mechanics and Hydraulics

Individuals preparing for the Environmental Engineering PE exam are expected to be familiar with fundamental definitions and concepts in fluid mechanics and hydraulics. These topics include, but are not limited to, fluid properties, viscosity principles, manometry, forces on submerged surfaces, buoyancy, and stability. This chapter reviews the conservation laws that describe fluid flow and hydraulic equations, describes centrifugal pumps and the criteria for their selection, and discusses the basic principles of open channel flow. Many of these topics will be expanded in subsequent chapters where the basics of fluid flow are applied to areas, including stormwater collection systems, water distribution networks, and wastewater collection networks. Because the primary fluid of concern is water, values for several fluid properties are given for water in both English and SI units in Tables 3.1 and 3.2.

Table 3.1 Properties of water in English units[a]

Temperature ($°F$)	Density, ρ (slugs/ft³)	Specific Weight[b], γ (lb/ft³)	Dynamic Viscosity, μ (lb·s/ft²)	Kinematic Viscosity, v (ft²/s)	Surface Tension[c], σ (lb/ft)	Vapor Pressure, p_v [lb/in²(abs)]	Speed of Sound[d], c (ft/s)
32	1.940	62.42	3.732 E − 5	1.924 E − 5	5.18 E − 3	8.854 E − 2	4603
40	1.940	62.43	3.228 E − 5	1.664 E − 5	5.13 E − 3	1.217 E − 1	4672
50	1.940	62.41	2.730 E − 5	1.407 E − 5	5.09 E − 3	1.781 E − 1	4748

(continued)

Temperature (°F)	Density, ρ (slugs/ft³)	Specific Weight[b], γ (lb/ft³)	Dynamic Viscosity, μ (lb·s/ft²)	Kinematic Viscosity, ν (ft²/s)	Surface Tension[c], σ (lb/ft)	Vapor Pressure, p_v [lb/in²(abs)]	Speed of Sound[d], c (ft/s)
60	1.938	62.37	2.344 E − 5	1.210 E − 5	5.03 E − 3	2.563 E − 1	4814
70	1.936	62.30	2.037 E − 5	1.052 E − 5	4.97 E − 3	3.631 E − 1	4871
80	1.934	62.22	1.791 E − 5	9.262 E − 6	4.91 E − 3	5.069 E − 1	4819
90	1.931	62.11	1.500 E − 5	8.233 E − 6	4.86 E − 3	6.979 E − 1	4960
100	1.927	62.00	1.423 E − 5	7.383 E − 6	4.79 E − 3	9.493 E − 1	4995
120	1.918	61.71	1.164 E − 5	6.067 E − 6	4.67 E − 3	1.692 E + 0	5049
140	1.908	61.38	9.743 E − 6	5.106 E − 6	4.53 E − 3	2.888 E + 0	5091
160	1.896	61.00	8.315 E − 6	4.385 E − 6	4.40 E − 3	4.736 E + 0	5101
180	1.883	60.58	7.207 E − 6	3.827 E − 6	4.26 E − 3	7.507 E + 0	5195
200	1.869	60.12	6.342 E − 6	3.393 E − 6	4.12 E − 3	1.152 E + 1	5089
212	1.860	59.83	5.886 E − 6	3.165 E − 6	4.04 E − 3	1.469 E + 1	5062

[a] Based on data from *Handbook of Chemistry and Physics*, 69th ed. (CRC Press, 1988). Where necessary, values obtained by interpolation.

[b] Density and specific weight are related through the equation $\gamma = \rho g$. For this table, $g = 32.174$ ft/s².

[c] In contrast with air.

[d] From R. D. Blevins, *Applied Fluid Dynamics Handbook* (Van Nostrand Reinhold, 1984).

Source: Munson, Young, and Okiishi, *Fundamentals of Fluid Mechanics*, 5th ed., © 2006, John Wiley & Sons, Inc. Reprinted by permission.

Table 3.2 Properties of water in SI units[a]

Temperature (°C)	Density, ρ (kg/m³)	Specific Weight[b], γ (kN/m³)	Dynamic Viscosity, μ (N·s/m²)	Kinematic Viscosity, ν (m²/s)	Surface Tension[c], σ (N/m)	Vapor Pressure, p_v [N/m²(abs)]	Speed of Sound[d], c (m/s)
0	999.9	9.806	1.787 E − 3	1.787 E − 6	7.56 E − 2	6.105 E + 2	1403
5	1000.0	9.807	1.519 E − 3	1.519 E − 6	7.49 E − 2	8.722 E + 2	1427
10	999.7	9.804	1.307 E − 3	1.307 E − 6	7.42 E − 2	1.228 E + 3	1447
20	998.2	9.789	1.002 E − 3	1.004 E − 6	7.28 E − 2	2.338 E + 3	1481
30	995.7	9.765	7.975 E − 4	8.009 E − 7	7.12 E − 2	4.243 E + 3	1507
40	992.2	9.731	6.529 E − 4	6.580 E − 7	6.96 E − 2	7.376 E + 3	1526
50	988.1	9.690	5.468 E − 4	5.534 E − 7	6.79 E − 2	1.233 E + 4	1541
60	983.2	9.642	4.665 E − 4	4.745 E − 7	6.62 E − 2	1.992 E + 4	1552
70	977.8	9.589	4.042 E − 4	4.134 E − 7	6.44 E − 2	3.116 E + 4	1555
80	971.8	9.530	3.547 E − 4	3.650 E − 7	6.26 E − 2	4.734 E + 4	1555
90	965.3	9.467	3.147 E − 4	3.260 E − 7	6.08 E − 2	7.010 E + 4	1550
100	958.4	9.399	2.818 E − 4	2.940 E − 7	5.89 E − 2	1.013 E + 5	1543

[a] Based on data from *Handbook of Chemistry and Physics*, 69th ed. (CRC Press, 1988).

[b] Density and specific weight are related through the equation $\gamma = \rho g$. For this table, $g = 9.807$ m/s².

[c] In contrast with air.

[d] From R. D. Blevins, *Applied Fluid Dynamics Handbook* (Van Nostrand Reinhold, 1984).

Source: Munson, Young, and Okiishi, *Fundamentals of Fluid Mechanics*, 5th ed., © 2006, John Wiley & Sons, Inc. Reprinted by permission.

3.1 CONSERVATION LAWS

All steady state, incompressible, one-dimensional (1-D) flow systems must satisfy the conservation laws for mass and energy. Before looking at the laws in more detail, it will be useful to review some basic descriptions of fluid flow.

The vast majority of all fluid flow problems can be described as *steady*, where fluid or system properties do not change with time, and *uniform*, where fluid or system properties do not change with location. A further simplification that applies in most systems is the assumption of *one-dimensional* flow, where the velocity vector is zero in all dimensions of the coordinate system except one. Most often, the coordinate system is either rectilinear (that is, *x*, *y*, *z*), and 1-D flow is generally expressed in the *x*-direction, or cylindrical (that is, *r*, *z*, *θ*), where 1-D flow is usually expressed in the *z*-dimension and termed *axial* flow.

3.1.1 Mass

The most basic of the conservation laws is applied to the *conservation of mass* and is generally referred to as the *Equation of Continuity* (*EOC*). Because mass cannot be created or destroyed in natural environments, the EOC simply states that the mass flow rate of any material into and out of a system at steady state must be equal. In its most general form, the EOC can be expressed as

$$\dot{m}_{in} = \dot{m}_{out}$$

where $\dot{m}$ is the system mass flow rate (kg/s). Often, this is written as

$$\rho_1 Q_1 = \rho_2 Q_2$$

or

$$\rho_1 V_1 A_1 = \rho_2 V_2 A_2$$

where

ρ = the fluid density (kg/m³)

Q = the volumetric flow rate (m³/s)

V = the average fluid velocity through the conduit (m/s)

A = the conduit cross-sectional area (perpendicular to the flow path) (m²)

The subscripts 1 and 2 are used to evaluate mass flow at any two locations in the system, usually selected where flow crosses the system boundary. Multiple inlets or outlets may be summed to complete a balance on the entire system, or several subsystems can be defined and evaluated with only one inlet and one outlet. For an incompressible fluid, as in the case of water flow problems, density is assumed constant, and the EOC simplifies to

$$Q_1 = Q_2$$

or

$$V_1 A_1 = V_2 A_2$$

Water is incompressible ∴ no density term required

3.1.2 Energy

In addition to the EOC, all steady state, incompressible, 1-D flow systems must also satisfy the *energy equation* (sometimes called the *field equation*), which is an expression that ensures conservation of mechanical energy between two specified points within a system. This equation is most often expressed in units of length (also called *head*) and can be represented as

$$z_1 + \frac{V_1^2}{2g} + \frac{P_1}{\gamma} + h_P = z_2 + \frac{V_2^2}{2g} + \frac{P_2}{\gamma} + h_L + h_T$$

where

z = the elevation (m)

V = the average velocity (m/s)

g = the acceleration due to gravity (equal to 9.81 m/s²)

P = the pressure (Pa, or equivalently N/m²)

γ = the specific weight of the fluid (N/m³)

h_P = the pump head (m)

h_L = the loss head (m)

h_T = the turbine head (m)

Again, subscripts 1 and 2 are used to evaluate energy at any two locations in the system, often selected where flow crosses the system boundary, in which case subscript 1 usually specifies energy entering the system and subscript 2 usually specifies energy exiting the system. The pump and turbine head terms are often used to determine power required (for pumps) or delivered (for turbines) from a system, when the remaining terms are known or can be estimated.

Many flow systems do not contain a pump or turbine, and often these terms are dropped from the equation. If the flow is also assumed to be *inviscid* (that is, the fluid viscosity is negligible, or, equivalently, there are no energy losses due to friction), the loss head is also zero, and the energy equation reduces to the familiar *Bernoulli equation*, which describes ideal flow as

$$z_1 + \frac{V_1^2}{2g} + \frac{P_1}{\gamma} = z_2 + \frac{V_2^2}{2g} + \frac{P_2}{\gamma}$$

potential $\quad$ velocity $\quad$ pressure

where

z is called the *elevation* (or *potential*) *head*

$V^2/(2g)$ is called the *velocity head*

P/γ is called the *pressure* (or *static*) *head*

The sum of the elevation and pressure heads is called the *hydraulic grade line* (*HGL*) and can be measured with a *piezometer* tube. The sum of the HGL and the velocity head, the total energy of the system, is called the *energy grade line* (*EGL*) and can be measured with a *pitot* tube, often called a *stagnation* tube.

Energy loss in a system is generally due to viscous shear at the wall (called *friction loss* and given the symbol h_f), or due to fittings or other variations in the pipe network (called *minor losses* and given the symbol h_m). These can be shown to be a function of fluid properties (specifically, viscosity and density), flow properties (velocity), and system geometry (pipe diameter and length, and pipe material or surface roughness). Often the two losses are combined and expressed as the loss head as

[handwritten: friction + minor loss = head loss]

$$h_L = h_f + h_m$$

The most common equation used to calculate friction loss is the *Darcy-Weisbach equation*, which can be expressed as

$$h_f = f \frac{L}{D} \frac{V^2}{2g}$$

where

f = the *friction factor* (unitless)

L = the pipe length (m)

D = the pipe diameter (m)

V and g are as defined previously

The friction factor is dependent upon a dimensionless quantity called the *Reynolds number* (Re), a ratio of inertial forces to viscous forces in fluid flow, and the ratio of the pipe surface *roughness* (ε) to pipe diameter (D), called the *relative roughness*. The Reynolds number and the relative roughness can be expressed as

$$Re = \frac{\rho D V}{\mu} = \frac{D V}{v}$$

$$\text{relative roughness} = \frac{\varepsilon}{D}$$

where μ is the fluid dynamic (or absolute) viscosity (N $\cdot$ s/m^2), v is the kinematic viscosity (m^2/s), and all other variables are as defined previously. While equations exist to determine f for each of the flow regimes (laminar, transitional, and turbulent), it is far more common to use the Moody diagram to estimate the value of the friction factor, as seen in Figure 3.1. Values for pipe roughness are usually supplied by manufacturers, specified in problem statements, or may be found in tables accompanying the Moody diagram for a variety of materials, as seen in Table 3.3.

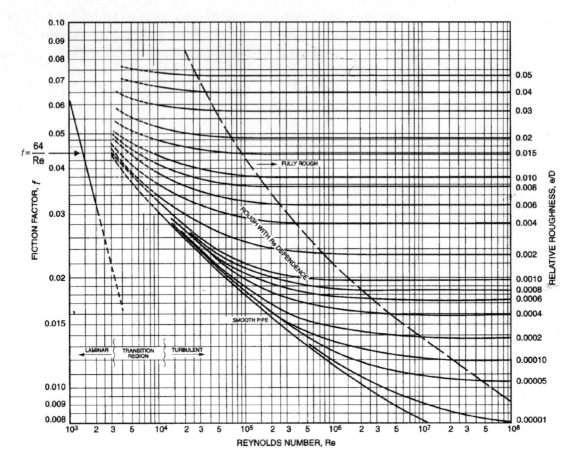

Figure 3.1   Moody diagram

Table 3.3   Surface roughness values for a variety of pipe materials

Material	ε [ft]	ε [mm]
Riveted steel	0.003 to 0.03	0.9 to 9.0
Concrete	0.001 to 0.01	0.3 to 3.0
Cast iron	0.00085	0.25
Galvanized iron	0.0005	0.15
Commercial steel	0.00015	0.046
Wrought iron	0.00015	0.046
Drawn tubing	0.000005	0.0015
Plastic and glass	0 (smooth)	0 (smooth)

The estimation of minor losses is determined experimentally and is generally expressed as

$$h_m = K_L \frac{V^2}{2g}$$

where K_L is called the loss coefficient (unitless) and is generally tabulated for a variety of pipe fittings, as presented in Table 3.4.

Table 3.4 Loss coefficients (K_L) for select pipe fittings

Component	K_L
Entrance, sharp	0.5
Entrance, rounded	0.2
Exit	1.0
Sudden contraction	0.0–0.5
Sudden expansion	0.0–1.0
90° elbow, flanged	0.3
90° elbow, threaded	1.5
Long radius 90°, threaded	0.7
45° bend	0.2–0.4
180° return bend, flanged	0.2
180° return bend, threaded	1.5
Threaded coupling	0.1
Through tee, flanged	0.2
Through tee, threaded	0.9
Globe valve, open	10
Gate valve, open	0.15
Gate valve, ¾ open	0.25
Gate valve, ½ open	2.1
Gate valve, ¼ open	17
Ball valve, open	0.05
Ball valve, ²/₃ open	5.5
Ball valve, ¹/₃ open	210

Finally, total losses for a system are determined by summing all of the friction and minor losses for the flow line under evaluation. For the case where pipe diameter is constant throughout the system, the system head loss can be expressed as

$$h_L = \left(f \frac{\Sigma L}{D} + \Sigma K_L \right) \frac{V^2}{2g}$$

Example 3.1

Energy equation

Determine the maximum flow rate of 60°F water (in gallons per minute) through 250 feet of a ¾-inch diameter garden hose if the elevation at location 2 is 16 feet higher than at location 1, and the pressure at the inlet is 85 psi. You may assume a friction factor of 0.02.

Solution

We start with the energy equation as follows:

$$h_P + Z_1 + \frac{V_1^2}{2g} + \frac{P_1}{\gamma} = Z_2 + \frac{V_2^2}{2g} + \frac{P_2}{\gamma} + h_f + h_T$$

This may be simplified by recognizing that there are no pumps or turbines, the flow exits to the atmosphere, the velocity through the system is constant, and the elevation change is given. This allows us to write

$$h_P = h_T = P_2 = (V_2 - V_1) = 0 \quad \text{and} \quad Z_2 - Z_1 = 16 \text{ ft} \quad \Rightarrow \quad \frac{P_1}{\gamma} = 16 - h_f$$

Using the data given in the problem statement and correcting for units yields

$$\frac{\left(85 \dfrac{\text{lb}}{\text{in}^2}\right)\left(144 \dfrac{\text{in}^2}{\text{ft}^2}\right)}{\left(62.4 \dfrac{\text{lb}}{\text{ft}^3}\right)} = 16 + h_f \quad \Rightarrow \quad h_f = 180.2 \text{ ft}$$

Now that the head loss is known, we may calculate velocity using the Darcy-Weisbach equation as follows:

$$h_f = f\frac{L}{D}\frac{V^2}{2g} = \frac{(0.02)(250 \text{ ft})V^2}{\left(\dfrac{0.75}{12}\text{ ft}\right)(2)\left(32.2 \dfrac{\text{ft}}{\text{s}^2}\right)} = 180.2 \text{ ft} \quad \Rightarrow \quad V = 12.0\frac{\text{ft}}{\text{s}}$$

Finally, volumetric flow rate can be calculated from velocity and hose diameter as follows:

$$Q = VA = \left(12.0\frac{\text{ft}}{\text{s}}\right)\left(\frac{\pi}{4}\right)\left(\frac{0.75}{12}\text{ ft}\right)^2 = 0.037\frac{\text{ft}^3}{\text{s}} \times \frac{7.48 \text{ gal}}{\text{ft}^3} \times \frac{60 \text{ s}}{\text{min}} = 16.5 \text{ gpm}$$

Let's check the assumption of $f = 0.02$. First we will need to calculate Re for water at 60°F as

$$\text{Re} = \frac{DV}{\nu} = \frac{(0.75 \text{ in})\left(\dfrac{1 \text{ ft}}{12 \text{ in}}\right)\left(12\dfrac{\text{ft}}{\text{s}}\right)}{1.21 \times 10^{-5}\dfrac{\text{ft}^2}{\text{s}}} = 62,000$$

Assuming the garden hose is considered smooth (plastic), the Moody diagram gives a friction factor at Re = 62,000 for smooth pipe of 0.02. So, the assumption in the problem statement was valid. If the problem did not specify a friction factor, an initial guess for f could have been made, and that guess validated, as was demonstrated here. If the assumption was not correct, the newly calculated value of f would be used to find a new velocity, and then validated as before. The process would be completed iteratively until the verified value closely matched the assumed value.

Another method to determine head loss in a system, commonly used in water supply design, is the *Hazen-Williams equation*, which can be expressed as

$$V = 0.849\, C R_H^{\ 0.63} S^{0.54} \quad \text{(SI units)}$$

$$V = 1.318\, C R_H^{\ 0.63} S^{0.54} \quad \text{(English units)}$$

where

V = the fluid velocity (m/s)

C = the Hazen-Williams roughness coefficient

R_H = the hydraulic radius (m)

S = the slope of the EGL (m/m)

The Hazen-Williams coefficient is a function of pipe material (like the value for ε discussed previously) and is often tabulated as seen in Table 3.5.

Table 3.5 Hazen-Williams coefficient values for a variety of pipe materials

Material	C_{new}	C_{old}
Brick	na	100
Cast iron	130–140	80–120
Concrete	130	120
Plastic	150	130
Riveted steel	110–130	100–110
Vitrified clay	110	110
Welded steel	120–140	110–120
Wood stave	120	110

The *hydraulic radius* (R_H) is defined as area divided by wetted perimeter, and expressed as

$$R_H = \frac{A}{P_w}$$

where A is the cross-sectional area of flow (m²) and P_w is the wetted perimeter (m), which is the portion of the conduit perimeter that is in contact with the fluid. Note that a pipe with a circular cross section that is flowing full has $R_H = D/4 = r/2$.

The slope of the EGL can be expressed as

full flowing circular pipe

$$S = \frac{h_L}{L}$$

where h_L is the energy loss term described previously (m), and L is the total length of conduit in the system being evaluated (m). Due to the widespread use in water distribution networks where water flows full through circular pipes, it is often desired to determine the volumetric flow rate for circular cross sections, in which case the Hazen-Williams equation can be expressed as

$$Q = 0.278 C D^{2.63} S^{0.54} \ \text{(SI units)}$$

$$Q = 0.432 C D^{2.63} S^{0.54} \ \text{(English units)}$$

where Q is the volumetric flow rate (m³/s), and D is the pipe diameter (m). It should be noted that the use of the Hazen-Williams equation is limited to water near room temperature under turbulent flow. Use of the Hazen-Williams equation is greatly enhanced through the use of the Hazen-Williams nomograph, as presented in Figure 3.2.

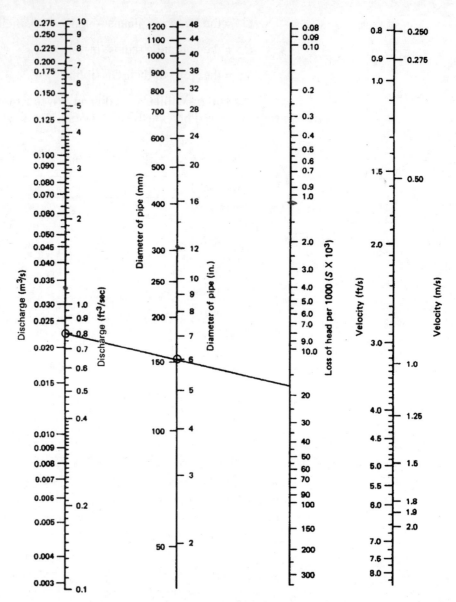

Figure 3.2   Hazen-Williams nomograph for $C = 100$

Example **3.2**

Hazen-Williams equation

Water at 1 cfs is conveyed through 8-inch cast iron ($C = 120$) pipe over a distance of two miles. Determine the pressure drop (in units of psi) over this distance.

Solution

Because a value for the Hazen-Williams coefficient (C) is given, start with the Hazen-Williams equation for volumetric flow:

$$Q = 0.432\,C D^{2.63}\,S^{0.54}$$

With Q, C, and D given, rearrange to solve for the slope of the EGL as

$$S^{0.54} = \frac{Q}{0.432\,CD^{2.63}} = \frac{1}{(0.432)(120)\left(\frac{8}{12}\right)^{2.63}} = 0.056$$

$$S = (0.056)^{1/0.54} = (0.056)^{1.852} = 0.0048\,\frac{\text{ft}}{\text{ft}}$$

Since slope is feet of head loss per foot of pipe length, head loss can be calculated as

$$S = \frac{h_L}{L} \quad \Rightarrow \quad h_L = SL = 0.0048\,\frac{\text{ft}}{\text{ft}} \times 2\,\text{mi} \times \frac{5280\,\text{ft}}{\text{mi}} = 50.7\,\text{ft}$$

The units of feet here refer to feet of water column, or ft H_2O. Since the problem statement asked for units of psi, convert ft H_2O to psi as follows:

$$H = 50.7\,\text{ft}\,H_2O \times \frac{14.7\,\text{psi}}{33.9\,\text{ft}\,H_2O} = 22.0\,\text{psi}$$

The alternative method is to use the Hazen-Williams nomograph. Start by drawing a straight line through the given flow and diameter on the nomograph and extend this line to the loss column. This yields a value for S of 6.7 ft per 1000 ft. However, this value is for $C = 100$. To convert this value to an equivalent value for this problem where $C = 120$, we need to evaluate the ratios of the slopes to Hazen-Williams coefficient with constant Q and D as

$$\frac{Q_1}{Q_2} = \frac{0.432\,C_1\,D_1^{2.63}\,S_1^{0.54}}{0.432\,C_2\,D_2^{2.63}\,S_2^{0.54}} \quad \text{or} \quad \left(\frac{S_2}{S_1}\right)^{0.54} = \frac{C_1}{C_2} \quad \Rightarrow \quad \frac{S_2}{S_1} = \left(\frac{C_1}{C_2}\right)^{1.85}$$

This may be rearranged to solve for S_2 using the information calculated previously as

$$S_2 = S_1\left(\frac{C_1}{C_2}\right)^{1.85} = 6.7\left(\frac{100}{120}\right)^{1.85} = 4.8\,\frac{\text{ft}}{1000\,\text{ft}}$$

This is exactly the same answer as calculated above, giving a pressure drop of 22 psi.

For the case where several pipes of different diameters or material properties are present in the system, connected in a *series configuration*, the equation of continuity dictates that the volumetric flow rates in each segment are equal, which can be expressed as

$$Q_1 = Q_2 = \cdots = Q_n$$

where n is the number of different pipes in the series, and the energy equation requires the total head loss for the system to be calculated as

$$\left(h_L\right)_{\text{TOTAL}} = \left(h_L\right)_1 + \left(h_L\right)_2 + \cdots + \left(h_L\right)_n$$

When several pipes are in a system connected in a *parallel configuration*, which can be defined as two or more pipes that diverge at some point in the system and converge at another point, the equation of continuity dictates that the volumetric flow rates can be expressed as

$$Q_{\text{TOTAL}} = Q_1 + Q_2 + \cdots + Q_n$$

and the energy equation requires the head loss for each parallel branch in the system to be equal, which can be expressed as

$$\left(h_L\right)_1 = \left(h_L\right)_2 = \cdots = \left(h_L\right)_n$$

Example **3.3**

A simple flow network

Water flows through the pipes in the simple network shown in Exhibit 1 below with $Q_{AB} = 1$ cfs. Determine Q_{BCD} given the following pipe data: $D_{BC} = D_{ED} = 10$ in; $D_{CD} = 6$ in; $D_{BE} = 8$ in; $L_{BC} = 1000$ ft; $L_{CD} = 400$ ft; $L_{BE} = 800$ ft; $L_{ED} = 600$ ft; $f = 0.018$.

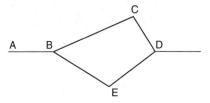

Exhibit 1

Solution

Based on the rules of parallel pipe networks, head loss in each branch must be equal, or, equivalently,

$$\left(h_L\right)_{BCD} = \left(h_L\right)_{BED}$$

Further, each parallel branch is comprised of two pipes in series; therefore, we can write

$$\left(h_L\right)_{BC} + \left(h_L\right)_{CD} = \left(h_L\right)_{BE} + \left(h_L\right)_{ED}$$

Substituting the expression for h_L above and adding the appropriate subscripts, we may write

$$h_L = \frac{fLV^2}{2gD} = \frac{fLQ^2}{2gDA^2} = \frac{8fLQ^2}{g\pi^2 D^5}$$

$$\frac{8fL_{BC}Q_{BC}^2}{g\pi^2 D_{BC}^5} + \frac{8fL_{CD}Q_{CD}^2}{g\pi^2 D_{CD}^5} = \frac{8fL_{BE}Q_{BE}^2}{g\pi^2 D_{BE}^5} + \frac{8fL_{ED}Q_{ED}^2}{g\pi^2 D_{ED}^5}$$

This may be simplified to

$$\frac{L_{BC}Q_{BC}^2}{D_{BC}^5} + \frac{L_{CD}Q_{CD}^2}{D_{CD}^5} = \frac{L_{BE}Q_{BE}^2}{D_{BE}^5} + \frac{L_{ED}Q_{ED}^2}{D_{ED}^5}$$

We also know that flow through pipes in series must be equal; therefore,

let $Q_{BC} = Q_{CD} = Q_1$ and $Q_{BE} = Q_{ED} = Q_2$

Substituting the values given we may write

$$\frac{(1000)Q_1^2}{\left(\dfrac{10}{12}\right)^5} + \frac{(400)Q_1^2}{\left(\dfrac{6}{12}\right)^5} = \frac{(800)Q_2^2}{\left(\dfrac{8}{12}\right)^5} + \frac{(600)Q_2^2}{\left(\dfrac{10}{12}\right)^5}$$

Simplifying, we can express Q_2 as a function of Q_1:

$$15,288\, Q_1^2 = 7,568\, Q_2^2 \quad \Rightarrow \quad Q_2 = 1.4213\, Q_1$$

Since total system flow is 1 cfs, we may write

$$Q_1 + Q_2 = 1\,\text{cfs} \quad \Rightarrow \quad Q_1 + 1.4213\, Q_1 = 2.4213\, Q_2 = 1\,\text{cfs} \quad \Rightarrow \quad Q_1 = 0.413\,\text{cfs}$$

Therefore the flow through the upper branch is approximately 0.4 cfs, or 40% of the total flow.

3.2 PUMP SELECTION

Pumps are machines that add energy to liquid systems (as compared to compressors, which add energy to gases), and may be classified as static-type or dynamic-type. Static-type pumps are often called *positive displacement* (or piston-style) pumps and produce flow through the static forces involved with changing the volume of the pump chamber. This type of pump is relatively uncommon in environmental applications, so it will not be discussed in detail in this review.

3.2.1 Types of Pumps

Dynamic-type pumps generally use a constant volume chamber, and flow is generated through the energy added by a set of blades (vanes, impellers) that are attached to a rotating shaft, which is turned by a motor. The most common dynamic device is the *centrifugal pump*, comprised of an *impeller* attached to a rotating shaft and a fixed *housing* (*casing*) enclosing the impeller.

Centrifugal pumps are further classified based upon the predominant direction of fluid flow within the housing, usually radial-flow, axial-flow, or mixed-flow. In *radial-flow pumps*, fluid inlet into the pump occurs at the center (*eye*) of the impeller, and the curved blades accelerate the liquid radially toward the end of the blade, where the fluid kinetic energy is converted to pressure head as the fluid impacts against the housing. This energy conversion is capable of developing large pressure increases, but flow rates are limited by the eye diameter. For a *single-stage pump*, fluid is then directed through an increasing volume channel toward the discharge opening. For systems where there is a demand for large pressure head, a *multistage pump* may be used where the discharge from one impeller is directed to the eye of a second impeller, and additional head is developed. *Axial-flow pumps* deliver fluid energy without substantial change in the fluid flow path, which is along the pump's primary axis. In this way, high flow rates can be accommodated; however, the pressure rise is limited. *Mixed-flow pumps* employ both radial and axial flow regimes to deliver reasonable flow rates at moderate pressure increases.

3.2.2 Pump Characteristics

Pump performance is usually based upon head delivered, pump efficiency, and brake horsepower, which are determined as a function of volumetric flow rate. The power gained by the fluid can be expressed as

$$\dot{P} = Q\gamma h_p$$

where $\dot{P}$ is the power gained by the fluid (N · m/s), γ is the fluid specific weight (N/m³), and Q and h_p are as defined previously. When calculated in English units (lb · ft/s) and divided by 550, the power has units of horsepower and is often referred to as the *water horsepower*. The overall pump efficiency (η) is a ratio of the power gained by the fluid to the power delivered to the pump by the rotating shaft and can be expressed as

$$\eta = \frac{\text{power gained by fluid}}{\text{shaft power delivered to pump}} = \frac{\dot{P}}{\dot{W}_S} = \frac{Q\gamma h_p / 550}{bhp}$$

where *bhp* is the pump *brake horsepower*, often supplied by pump manufacturers.

Example 3.4

Pump efficiency

Determine the pump efficiency for the flow through the garden hose in Example 3.1 if the brake horsepower of the pump is 1.25 hp.

Solution

Pump efficiency can be defined as follows:

$$\eta = \frac{\text{power gained by fluid}}{\text{shaft power delivered to pump}} = \frac{\dot{P}}{\dot{W}_S}$$

The shaft power delivered to the pump is also known as brake horsepower (bhp). The power gained by the fluid can be determined from the following expression:

$$P = Q\gamma h_p$$

where h_p is the pump head, which is equal to the total dynamic head (TDH). In this case, TDH may be calculated as either the sum of the friction head plus the elevation head, or it may be calculated from the pressure loss from the water inlet to the outlet. Since we know the friction head (180.2 ft) and the elevation head (16 ft), we may solve for the fluid power:

$$P = Q\gamma H = \left(0.037 \frac{\text{ft}^3}{\text{s}}\right)\left(62.4 \frac{\text{lb}}{\text{ft}^3}\right)(196.2 \text{ ft}) = 453 \frac{\text{ft}\cdot\text{lb}}{\text{s}}$$

This can be converted to units of horsepower as

$$P = 453 \frac{\text{ft}\cdot\text{lb}}{\text{s}} \times \frac{1 \dfrac{\text{ft}\cdot\text{lb}}{\text{s}}}{550 \text{ hp}} = 0.824 \text{ hp}$$

The pump efficiency can now be calculated as

$$\eta = \frac{0.824\ hp}{1.25\ hp} = 65.9\,\%$$

Performance characteristics for a given pump geometry and operating speed are usually presented graphically as *performance (characteristic) curves*, as presented in Figure 3.3.

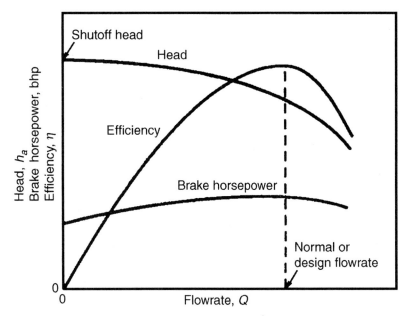

Figure 3.3 Typical centrifugal pump performance (characteristic) curve

Source: Munson, Young, and Okiishi, *Fundamentals of Fluid Mechanics*, 5th ed., © 2006 John Wiley & Sons, Inc. Reprinted by permission.

Variable speed drives may allow one pump to operate over a large range of flows without significant efficiency loss, offering an economical means to deliver varied demands. However, it is often necessary to use more than one pump in a system to provide the necessary head or flow rate, and multiple pumps offer redundancy for required or emergency maintenance. Analogous to the pipe flow analysis, it can be shown that multiple pumps in a *series* configuration must have equivalent flow rates, while the total head delivered is calculated as the sum of the individual heads delivered by each pump. Further, pumps in a *parallel* configuration must have equivalent head delivered, but the total volumetric flow rate is calculated as the sum of the individual flow rates delivered by each pump. The effect of pump configuration can be seen by examining the composite pump curves in Figure 3.4.

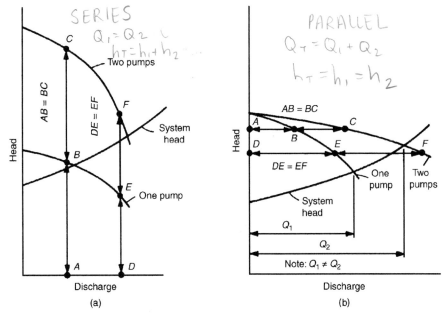

Figure 3.4 Composite pump curves for (a) series, and (b) parallel configurations

Source: Warren Viessman, Jr., and Mark J. Hammer, *Pollution Control*, 7th ed., © 2005. Reprinted by permission of Pearson Education, Inc., Upper Saddle River, N.J.

Example **3.5**

Pump selection

There is 800 gpm of water flowing through 1000 feet of 6-inch-diameter pipe with a friction factor of 0.02 and a loss coefficient for all minor losses equal to 10. Select the most appropriate pump from the choices offered on the pump curve in Exhibit 2.

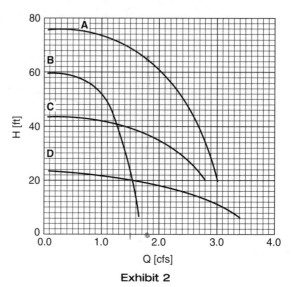

Exhibit 2

Solution

First, we need volumetric flow rate in cfs:

$$Q = \frac{800}{(7.48)(60)} = 1.78 \text{ cfs}$$

From the energy equation, if we assume there are no turbines, the elevation change is zero, and the velocity is constant in the pipe, then the head losses must be equal to the pressure drop, which is equal to the energy required by the pump to maintain flow. The head loss in the system, and therefore the head delivered by pump, may be calculated as

$$h_L = \frac{V^2}{2g}\left(f\frac{L}{D} + \sum K \right)$$

Problem statement gives f, L, D, and ΣK, but we need to calculate V:

$$V = \frac{Q}{A} = \frac{1.78}{\frac{\pi}{4}(0.5)^2} = 9.07\,\frac{\text{ft}}{\text{s}}$$

Now we can solve for head loss as

$$h_L = \frac{(9.07)^2}{(2)(32.2)}\left[(0.02)\left(\frac{1000}{0.5}\right)+10\right] = 63.9 \text{ ft}$$

Now we can go to the pump curves and find a pump that can supply a flow rate of 1.8 cfs at a head of 64 ft. The correct choice is pump A.

The inlet side of a pump is characterized by low pressure developed as liquid is drawn into the pump, which has the potential to cause *cavitation*. Cavitation is a phenomenon where the system pressure is less than or equal to the vapor pressure of the liquid, causing the liquid to form small bubbles of vapor, potentially damaging the structural components of the pump and substantially reducing pump efficiency. The minimum pressure head at the inlet to the pump is described by the required *net positive suction head (NPSH)* and can be expressed by the energy equation as

$$NPSH = \frac{P_i}{\gamma} + \frac{V_i^2}{2g} - \frac{P_v}{\gamma}$$

✱ use absolute pressures ✱

where P_i is the inlet pressure to the pump, V_i is the velocity at the inlet to the pump, and P_v is the vapor pressure of the fluid being pumped. Note that all pressures in this expression are required to be absolute instead of gage pressures. Applying the energy equation to the NPSH equation between the reservoir surface and the pump inlet yields

$$NPSH = \frac{P_{atm}}{\gamma} - \frac{P_v}{\gamma} - h_L - z_i$$

where P_{atm} is the prevailing atmospheric pressure, z_i is the elevation of the pump inlet, and all other variables are as defined previously.

Sometimes these two equations are denoted as *available NPSH*, or *NPSHA*. These equations are most often used to locate the pump inlet when all other parameters are known quantities. Values for NPSHA are to be compared to pump manufacturers' data that list the *required NPSH (NPSHR)* for a specific pump. For horizontal shaft pumps, NPSH is measured from the shaft for small pumps and from the top of the pump for large pumps. Care should be taken in locating any pump to ensure the NPSHR is exceeded at all times to ensure the absence of cavitation and therefore maximize the longevity of the pump.

Example **3.6**

NPSH

Determine the maximum elevation of the pump inlet for a pump that has an NPSHR specified by the manufacturer of 13.5 ft H_2O. You may assume head losses on the suction side of the pump are 8 ft H_2O, and the vapor pressure for water at the operating temperature of 50°F is 0.18 psi. Atmospheric pressure is 100 kPa.

Solution

With the data given in the problem statement, we only need to input the values, corrected for consistent units, into the NPSH equation, which incorporates the energy balance as follows:

$$NPSH = \frac{P_{atm}}{\gamma} - \frac{P_v}{\gamma} - h_L - z_i$$

$$\text{NPSH} = \frac{\left(100 \text{ kPa}\right)\left(\dfrac{14.696 \dfrac{\text{lb}}{\text{in}^2}}{101.325 \text{ kPa}}\right)\left(144 \dfrac{\text{in}^2}{\text{ft}^2}\right)}{\left(62.4 \dfrac{\text{lb}}{\text{ft}^3}\right)} - \frac{\left(0.18 \dfrac{\text{lb}}{\text{in}^2}\right)\left(144 \dfrac{\text{in}^2}{\text{ft}^2}\right)}{\left(62.4 \dfrac{\text{lb}}{\text{ft}^3}\right)} - 8 \text{ ft} - z_i$$

$$= 13.5 \text{ ft H}_2\text{O}$$

Solving for z_i yields a maximum elevation of 11.55 ft.

Primary pump characteristics depend upon a variety of geometric, fluid, and pump operating variables. Using dimensional analysis, a series of dimensionless groups can be developed to assist in the prediction of large pump performance based on laboratory observations of smaller, geometrically similar pumps. Although purely theoretical, these relationships are often referred to as pump *similitude* and can be represented by a series of *scaling laws,* which may be expressed as

$$\left(\frac{Q}{ND^3}\right)_1 = \left(\frac{Q}{ND^3}\right)_2 \quad \text{Flow coefficient}$$

$$\left(\frac{\dot{m}}{\rho ND^3}\right)_1 = \left(\frac{\dot{m}}{\rho ND^3}\right)_2 \quad \text{Mass flow coefficient}$$

$$\left(\frac{H}{N^2 D^2}\right)_1 = \left(\frac{H}{N^2 D^2}\right)_2 \quad \text{Head rise coefficient}$$

$$\left(\frac{P}{\gamma N^2 D^2}\right)_1 = \left(\frac{P}{\gamma N^2 D^2}\right)_2 \quad \text{Pressure rise coefficient}$$

$$\left(\frac{\dot{W}}{\rho N^3 D^5}\right)_1 = \left(\frac{\dot{W}}{\rho N^3 D^5}\right)_2 \quad \text{Power coefficient}$$

where

Q = volumetric flow rate

N = the pump rotational speed

D = the impeller diameter

$\dot{m}$ = mass flow rate

ρ = fluid density

H = the pump head (represented previously in this review as h_p)

P = the pressure rise through the pump

$\dot{W}$ = the shaft power

It is convenient to note that units must be consistent on both sides of the equality but need not be consistent within a single grouping. Also, as seen in the scaling laws, the flow coefficient is related to the mass flow coefficient by density, and the head rise coefficient is related to the pressure rise coefficient by the specific weight. However, it is often the case that the density or specific weight changes in the system and is usually dropped from the equality.

Example **3.7**

Scaling laws

An 8-inch-diameter centrifugal pump provides 2200 gpm of water at 42 ft of head, with an efficiency of 72% when rotating at 2400 rpm. Determine the shaft work, in hp, of a geometrically similar pump with a 10-inch impeller that provides the same volumetric flow rate.

Solution

In order to find the shaft work, we need to use the power coefficient equation, with the assumption that the fluid density is the same for both systems:

$$\left(\frac{\dot{W}}{\rho N^3 D^5}\right)_1 = \left(\frac{\dot{W}}{\rho N^3 D^5}\right)_2 \quad \Rightarrow \quad \dot{W}_2 = \dot{W}_1 \left(\frac{N_2}{N_1}\right)^3 \left(\frac{D_2}{D_1}\right)^5$$

First, we must determine the shaft work for pump 1 as follows:

$$Q = \frac{2200}{(7.48)(60)} = 4.9 \text{ cfs}$$

$$\dot{W}_1 = \frac{P_1}{\eta} = \frac{Q\gamma H}{\eta 550} = \frac{(4.9)(62.4)(42)}{(0.72)(550)} = 32.44 \text{ hp}$$

Now, using the flow coefficient scaling relationship:

$$\left(\frac{Q}{ND^3}\right)_1 = \left(\frac{Q}{ND^3}\right)_2$$

Setting the volumetric flow rate the same between the two pumps, we write

$$Q_1 = Q_2 \quad \text{or, equivalently,} \quad N_1 D_1^3 = N_2 D_2^3$$

Using the data given for D_1, D_2, and N_1, we can solve for N_2 as follows:

$$N_2 = N_1 \left(\frac{D_1}{D_2}\right)^3 = (2400)\left(\frac{8}{10}\right)^3 = 1230 \text{ rpm}$$

Finally, substituting values for $\dot{W}_1$, N_1, N_2, D_1, and D_2 into the power coefficient equation:

$$\dot{W}_2 = (32.44)\left(\frac{1230}{2400}\right)^3 \left(\frac{10}{8}\right)^5 = 13.33 \text{ hp}$$

One dimensionless group derived from the scaling laws commonly used in pump evaluations is the ratio of the flow coefficient to the head rise coefficient, which is called the *specific speed* (N_s) and is expressed as

$$\text{Specific Speed } (N_s) = \quad N_s = \frac{N Q^{1/2}}{(gH)^{3/4}}$$

PUMP

TYPE	SPEED (rpm)
RADIAL	<3000-4000
MIXED	4000 - 8000
AXIAL	>8000-10000

While the above expression is dimensionally homogeneous, it is more common in the United States to express specific speed in *US customary units* as follows:

$$N_S = \frac{NQ^{1/2}}{H^{3/4}}$$

where

N = rotational speed in revolutions per minute (rpm)

Q = volumetric flow rate in gallons per minute (gpm)

H = head in feet (ft)

Calculation of the specific speed in US customary units assists in the selection of the type of pump best suited for a particular system. Radial-flow pumps provide high head at low flow rates and are best suited to specific speeds less than 3000 or 4000, depending on the reference text you are reading. High capacity, low-head pumps are described by specific speeds greater than 8000 to 10,000 (again, depending on reference source), and generally require the use of axial-flow pumps. Mixed-flow pumps are usually selected for specific speeds in the range between the radial-flow and axial-flow regimes.

Example 3.8

Specific speed

A pump with an 8-inch impeller delivers 1 cfs of water with a delivered head of 55 ft when rotating at 2800 revolutions per minute. Is this pump most likely a radial-flow, axial-flow, or mixed-flow pump?

Solution

The equation for specific speed in US customary units as given above is

$$N_S = \frac{NQ^{1/2}}{H^{3/4}}$$

However, Q is given in units of cfs, not gpm as required. Converting to the correct units yields

$$N_S = \frac{NQ^{1/2}}{H^{3/4}} = \frac{(2800\ \text{rpm})\left[\left(1\,\frac{\text{ft}^3}{\text{s}}\right)\left(7.48\,\frac{\text{gal}}{\text{ft}^3}\right)\left(60\,\frac{\text{s}}{\text{min}}\right)\right]^{1/2}}{55^{3/4}} = 2937$$

Since this is less than 3000, it is safe to assume that this system would require a radial-flow pump. It might have been safe to assume a radial-flow pump without calculating the specific speed, given the relatively high head delivered (55 ft) at a relatively low flow rate (1 cfs is approximately 450 gpm). However, the calculation confirms the intuitive guess.

3.3 OPEN CHANNEL FLOW

Open channel flow is characterized by the presence of a *free surface*, or an interface between two fluids, where the fluid below is usually water and the fluid above is usually air. The fluid above is open to the atmosphere, and, as such, the driving force for flow is from gravity (the weight of the fluid) rather than pressure. The most common examples of open channel flow are rivers and streams; however, the environmental engineer has used gravity-induced flow for the benefit of mankind for centuries, as demonstrated by the Roman aqueduct and evidence of the use of irrigation channels in primitive societies.

Open channel flow is called *uniform* if the depth does not vary with distance. This is the case when the energy gained by the system due to the elevation change of the stream is directly offset by the friction losses due to friction at the channel bottom. *Varied flow* describes a flow condition where the depth of flow is not constant along a length of channel. Varied flow may be further categorized as *rapidly varying flow* when the change in depth is of the same order of magnitude as the change in distance along the channel, or *gradually varying flow* when there is a change in depth, but it is small or occurs over long distances.

3.3.1 Energy Considerations

If we apply the energy equation to open channel flow, we can assume that there are no pumps or turbines and set h_P and h_T equal to zero. Further, if we use the bottom of the channel as our basis for elevation, the slope of the channel (S_0) can be expressed as

$$S_0 = \frac{\text{rise}}{\text{run}} = \frac{\Delta z}{\Delta x} = \frac{\Delta z}{L}$$

where L is the length of the channel. Further, we know the pressure at the bottom of the channel is due to hydrostatic pressure and can be expressed as the depth of the water, usually given the symbol y to distinguish it from the elevation term z. We may now write the energy equation as

[handwritten: y = depth of flow]

$$y_1 - y_2 = \left(\frac{V_2^2 - V_1^2}{2g}\right) + \left(S_f - S_0\right)L$$

where S_f is often called the *friction slope* and is equal to h_L/L. If we define the *specific energy* (E) of a fluid as the sum of the hydrostatic pressure head and velocity head, we may express E as

$$E = y + \frac{V^2}{2g} = y + \frac{Q^2}{2gA^2}$$

[handwritten: Energy = pressure head + velocity head]

The energy equation may now be rewritten as

$$E_1 = E_2 + \left(S_f - S_0\right)L$$

If we further assume that most rivers can be approximated by a rectangular cross section of width b, we may express the flow rate per unit width (q) as

$$q = \frac{Q}{b} = \frac{Vyb}{b} = Vy$$

Note that the quantity q is constant for a given stream width, even if y varies. We may now express the specific energy for a stream of rectangular cross section and constant width as

$$E = y + \frac{q^2}{2gy^2}$$

Example 3.9

Specific energy

A stream that is 50 ft wide possesses the same specific energy when the depth of flow is 4 ft deep as it does when the depth of flow is 1 ft deep. Determine the volumetric flow rate (Q) of the stream in cfs.

Solution

First, we know that the specific energies are equal, so we can write

$$E_1 = E_2 \quad \Rightarrow \quad y_1 + \frac{q^2}{2gy_1^2} = y_2 + \frac{q^2}{2gy_2^2}$$

Substituting in the values given in the problem statement:

$$4 + \frac{q^2}{(2)(32.2)(4)^2} = 1 + \frac{q^2}{(2)(32.3)(1)^2}$$

Solving for the flow rate per unit width, we get

$$3 = 0.01456q^2 \quad \Rightarrow \quad q = 14.36 \, \frac{ft^2}{s}$$

Finally, solve for the volumetric flow rate as follows:

$$q = \frac{Q}{b} \quad \therefore \quad Q = qb = (14.36)(50) = 718 \text{ cfs}$$

Solution of this expression for E with respect to y is a cubic function in y, requiring three roots. For values of specific energy above a threshold (E_{min}), there is one negative root, which has no significant meaning and may be ignored in our analysis. To determine E_{min}, the derivative of the previous equation with respect to y is set equal to zero, and the critical value of y (y_c), defined as the *depth corresponding to the minimum energy state*, can be found as

$$y_c = \left(\frac{q^2}{g} \right)^{1/3}$$

Substitution of y_c into the specific energy expression yields

$$E_{min} = \frac{3}{2} y_c$$

Values of E that are above this minimum have two positive roots and correspond to the two possible flow depths that yield that specific energy value. One root is greater than y_c, indicating a stream that is deep and therefore moving at a relatively lower velocity in conformance with the EOC. This flow regime is termed *subcritical*, and the depth is given the symbol y_{sub}. The other root is smaller than y_c, indicating a stream that is shallow and therefore moving at a relatively greater velocity in conformance with the EOC. This flow regime is termed *supercritical*, and the depth is given the symbol y_{sup}. The *specific energy diagram* is a graphical representation of the specific energy as a function of depth, as shown in Figure 3.5, and is extremely useful in evaluating the potential flow conditions of a particular stream.

$y > y_c$ = subcritical

$y < y_c$ = supercritical

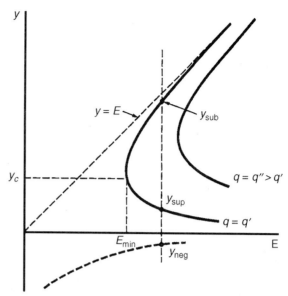

Figure 3.5 Specific energy diagram

Source: Munson, Young, and Okiishi, *Fundamentals of Fluid Mechanics*, 5th ed., © 2006, John Wiley & Sons, Inc. Reprinted by permission.

A useful dimensionless parameter to determine flow criticality is the *Froude number (Fr)*, which is a ratio of inertial forces to gravitational forces, and may be expressed as

$$Fr = \frac{V}{\left(gy\right)^{1/2}} = \frac{q}{g^{1/2} \, y^{3/2}}$$

where

V = the stream velocity at a flow per unit width q

g = the gravity constant

y = the depth of flow that corresponds to that velocity

When $Fr < 1$ the flow is subcritical, when $Fr > 1$ the flow is supercritical, and when $Fr = 1$ the flow is critical. Transition from supercritical to subcritical flow regimes is a common phenomenon and is known as a *hydraulic jump*, while transition from subcritical to supercritical flow is often characterized by flow under a sluice gate. In either case, the transition must occur along the specific energy curve and, therefore, must pass through the critical flow (as seen in the specific energy diagram). As such, both changes would be classified as rapidly varying flow.

Example **3.10**

Critical flow

Determine the critical flow rate of a stream 20 ft wide if the critical velocity is 9.6 fps.

Solution

Since we are given critical velocity, we know that $Fr = 1$ at this point, allowing us to calculate the critical depth as follows:

$$Fr = \frac{V}{(g\,y_c)^{1/2}} = 1 = \frac{9.6\,\frac{ft}{s}}{\left[\left(32.3\,\frac{ft}{s^2}\right)y_C\right]^{1/2}} \quad \Rightarrow \quad y_c = 2.86\,ft$$

Given a critical depth, we may calculate volumetric flow per unit stream width as follows:

$$y_c = \left(\frac{q^2}{g}\right)^{1/3} = 2.86\,ft = \left(\frac{q^2}{32.2\,\frac{ft}{s^2}}\right)^{1/3} \quad \Rightarrow \quad q = 27.45\,\frac{ft^2}{s}$$

Finally, this allows us to calculate the stream's critical flow rate as:

$$Q = q\,b = \left(27.45\,\frac{ft^2}{s}\right)(20\,ft) = 549\,\frac{ft^3}{s}$$

3.3.2 Manning's Equation for Uniform Flow

Most open channel flow encountered in environmental engineering is uniform flow in sewers and channels within the wastewater treatment plant. The most common relationship used to express stream velocity during uniform flow is *Manning's equation*, which can be expressed as

$$V = \frac{1.00}{n}\,R_H^{2/3}\,S^{1/2} \quad \text{(SI units)}$$

$$V = \frac{1.49}{n}\,R_H^{2/3}\,S^{1/2} \quad \text{(English units)}$$

where n is Manning's roughness coefficient, S is the slope of the EGL (m/m), which is often expressed as the slope of the channel bottom, and V and R_H are as defined previously. The volumetric flow rate is calculated by multiplying the above velocity expressions by the cross-sectional area of flow. Manning's coefficient is a function of pipe material and is often tabulated (like the value for C in the previous discussion) as presented in Table 3.6.

Table 3.6. Values of Manning's coefficient, n, for select materials

Material	n
Plastic (PVC)	0.009–0.011
Vitrified clay	0.010–0.017
Steel pipe	0.012–0.015
Concrete	0.012–0.016
Brick sewers	0.012–0.017

(continued)

Material	n
Earth, clean	0.018–0.022
Corrugated metal	0.022–0.028
Earth, with grass	0.025–0.035

To aid in the design of sewer systems that utilize circular pipes, Manning's equation may also be represented graphically as a nomograph, as shown in Figure 3.6. This version of the nomograph requires the use of a center pivot line to differentiate $Q/V/D$ values from n/S values, as shown on the figure. Therefore, it is necessary to have data that includes two of the three $Q/V/D$ quantities, plus either n or S to complete the evaluation. Usually, Q is calculated from flow data, and n is given based on material of construction. Values for V are then assumed from maximum and minimum limits, and a range of acceptable slopes and diameters is developed.

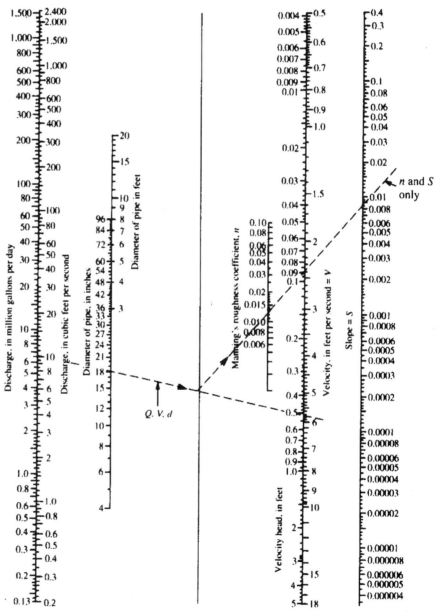

Figure 3.6   Manning's nomograph for all values of n for circular pipes flowing full

Often, the nomograph is simplified by representing a single value of n (usually 0.013), and allowing for slope and diameter to be calculated for given flow rates and velocity ranges using the ratio method demonstrated in Example 3.2 for the Hazen-Williams nomograph.

It should be noted that values from the nomograph represent the pipe flowing at capacity and are often designated by a subscript f to designate full flow rate (Q_f) or full velocity (V_f). Properly designed sewer systems will always flow at less than capacity, but, due to the circular cross section, calculating values for hydraulic elements may be difficult. This process is simplified through the use of the *partial flow nomograph*, or plot of hydraulic elements for circular cross sections. This nomograph allows the easy calculation of all hydraulic elements as a function of the full-flow value and a known depth of flow, as seen in Figure 3.7. Additional coverage on sewer hydraulics can be found in section 8.2.

Table of Hydraulic Parameters located in Lindeburg p.19-3

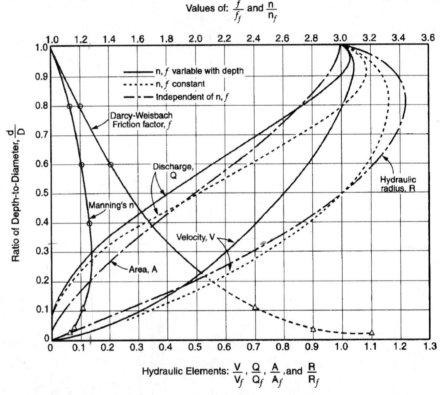

Figure 3.7 Partial flow nomograph of hydraulic elements for circular sewers

Source: *Design and Construction of Sanitary and Storm Sewers*; 1970, American Society of Civil Engineers. Reprinted by permission.

Example 3.11

Manning's equation

A flow of 1.5 cfs enters manhole 1, which has a ground elevation of 768 ft and an invert elevation of 760 ft above sea level. Manhole 2 is connected by 500 ft of 15-in sewer line ($n = 0.013$, varies with depth) and has a ground elevation of 762 ft. If the depth of flow is 32%, find the depth of cover (to the pipe crown) at manhole 2.

Solution

We start with Manning's equation in English units, solving for slope:

$$Q = \frac{1.486}{n} A R^{2/3} S^{1/2} \quad \Rightarrow \quad S^{1/2} = \frac{Qn}{(1.486)\,A\,R^{2/3}} \quad \Rightarrow \quad S = \left[\frac{Qn}{(1.486)\,A\,R^{2/3}} \right]^2$$

Now, from the partial flow nomograph at a depth of 32%, we can find each element:

$$\frac{A}{A_f} = 26\% \quad \Rightarrow \quad A = 26\%\, A_f = 0.26 \frac{\pi}{4} D^2 = (0.26)\frac{\pi}{4}\left(\frac{15}{12}\right)^2 = 0.32 \text{ ft}^2$$

$$\frac{R}{R_f} = 73\% \quad \Rightarrow \quad R = 73\%\, R_f = 0.73 \frac{D}{4} = (0.73)\left(\frac{15/12}{4}\right) = 0.23 \text{ ft}$$

$$\longrightarrow \quad \frac{n}{n_f} = 126\% \quad \Rightarrow \quad n = (1.26)(0.013) = 0.0164$$

Substituting these values, along with the value for Q given in the problem statement, into the expression for S above yields

$$S = \left[\frac{(1.5)(0.0164)}{(1.486)(0.32)(0.23)^{2/3}} \right]^2 = 0.019 \frac{\text{ft}}{\text{ft}}$$

Now, elevation change can be calculated from the slope and pipe length as

$$\Delta y = (0.019)(500) = 9.5 \text{ ft}$$

The elevation of the invert at manhole 2 can be calculated as follows:

invert at manhole 2 = invert at manhole 1 − elevation change
= 760 − 9.5 = 750.5 ft

Elevation to the crown can be calculated as

$$\text{elevation to invert} + \text{pipe diameter} = 750.5 + \frac{15}{12} = 751.75 \text{ ft}$$

Finally, the depth of cover is determined by the difference between ground elevation and crown elevation as

cover = 762 − 751.75 = 10.25 ft

3.3.3 Weir Equations

Flow measurement in open channels is often conducted through the use of weirs. *Sharp-crested weirs* are vertical, flat plates placed perpendicular to the flow path, often with notches cut into the plate to allow for a greater range of flow measurements with improved accuracy. *Broad-crested weirs* are obstructions, placed on the channel bottom, of specific relative dimensions in order to create critical flow over the weir.

Although more complex equations have been developed to estimate volumetric flow rate using weirs of many orientations, it is safe to combine weir parameters and simplify flow estimations as

$$Q = CbH^{3/2} \quad \text{Rectangular and trapezoidal}$$

$$Q = CH^{5/2} \quad \text{V-notch}$$

where

Q = the volumetric flow rate (m³/s)

H = the depth of water over the weir, measured at an upstream location (m)

b = the width of the base in the opening of the rectangular/trapezoidal weir (m) (that is, the width at the bottom of the rectangle or trapezoid)

C = the flow constant, which is dependent upon the type of weir and system of units

The flow constant is generally given the following values

$C = 1.84$ Rectangular, SI units

$C = 3.33$ Rectangular, English units

$C = 1.86$ Trapezoidal, SI units

$C = 3.367$ Trapezoidal, English units

$C = 1.40$ 90° V-notch, SI units

$C = 2.54$ 90° V-notch, English units

Example 3.12

Weir equation

You desire to measure flow rates ranging from 0.01 to 0.45 cfs using a 90° V-notch weir. Determine the maximum depth of flow above the bottom of the notch.

Solution

Start with the weir equation, and if we assume that maximum depth will occur at maximum flow rate, we may write

$$Q = CH^{5/2} = 0.45 \text{ cfs}$$

Find the weir coefficient for a 90° V-notch weir in English units, and rearrange to solve for H as

$$C_{90} = 2.54 \quad \Rightarrow \quad H^{5/2} = \frac{0.45}{2.54} \quad \Rightarrow \quad H = \left(\frac{0.45}{2.54}\right)^{2/5} = 0.50 \text{ ft} = 6.0 \text{ in}$$

Legal, Ethical, and Health Considerations

OUTLINE

The specifications for the PE exam in environmental engineering indicate that codes, standards, regulations, and guidelines are fair topics for questions on the exam. This chapter briefly discusses the environmental legislation and professional/ethical mindset useful in minimizing environmental impact, protecting occupational safety and health, and the evaluation of risks, especially those risks imposed through polluted environments. Regulatory and legal issues relevant to specific areas of environmental engineering are covered in other chapters as needed.

While enacting change in corporate management is always difficult, implementing change specifically to accommodate environmental stewardship may present additional challenges, unless decision makers are enlightened regarding the advantages of the proposed change. Design for the Environment (DfE) is a defined, systematic approach to selecting products and their manufacturing processes based upon environmental consequences. Two tools that are often used in the assessment process are the "cradle-to-grave" analysis provided by *life cycle assessments* (more recently applied "cradle-to-cradle" to reinforce the concept that all waste may be reprocessed into new raw materials), and *environmental impact assessments*. Life cycle assessments are discussed in detail in section 10.4 in the context of solid waste management, while environmental impact assessments are discussed in section 4.3 of this chapter.

Another ideology that is common in the corporate world is *Total Quality Management (TQM)*, which can be described as a process of delivering the highest quality product and services, while attempting to improve continuously in future deliverables. Recently, this has been extended to *Total Quality Management Systems and the Environment (TQEM)* by applying the TQM philosophy to environmental management and sustainable development. Possibly the most widely used environmental management system is the *International Standards Organization's (ISO) 14000* family of international standards, specifically ISO 14001. The ISO 14000 family consists of 26 standards in six classifications to assist corporations in the sound management of economic, social, and environmental issues.

4.1 ENVIRONMENTAL LEGISLATION

Environmental legislation and regulations are usually developed as a direct result of risk assessment data (covered in section 4.5) that identifies a potential negative health effect arising from a specific chemical that can be found in the environment. A complete history of environmental legislation is not possible within this review; you may want to consult any of several texts that provide substantial coverage of the history of environmental law. Table 4.1 summarizes the major pieces of environmental legislation in the United States, including several landmark acts of Congress.

Table 4.1 Summary of Landmark Environmental Legislation in the United States

Name	Date Enacted (Amended)	Major Provisions
Federal Insecticide, Fungicide, and Rodenticide Act (FIFRA)	1947 (1972)	Regulates the manufacture and use of all pesticides
Federal Water Pollution Control Act (FWPCA)	1956	Attempted to reduce water pollution by funding construction of municipal water treatment plants
Clean Air Act (CAA)	1963 (1970, 1990)	Established National Ambient Air Quality Standards (NAAQSs) for criteria pollutants
Solid Waste Disposal Act (SWDA)	1965	Promoted structured solid waste management through regulation of collection, transport, processing, recovery, and disposal systems
National Environmental Policy Act (NEPA)	1969	Required an environmental impact statement (EIS) for all projects receiving federal funding
Resources Recovery Act (RRA)	1970	Amended SWDA; established reuse and recycle as national priorities for managing solid waste; promoted energy recovery from solid waste
Occupational Safety and Health Act (OSHA)	1970	Established workplace safety standards for potential physical and chemical hazards
Clean Water Act (CWA)	1972 (1977)	Amended FWPCA; established the first EPA list of 65 priority pollutants (now at 127 pollutants); defined total maximum daily loads (TMDLs) and National Pollutant Discharge Elimination System (NPDES) permit program; required use of best available control technology (BACT)
Safe Drinking Water Act (SDWA)	1974 (1986, 1996)	First consumption law; established maximum contaminant levels (MCLs) for potable water

(continued)

Name	Date Enacted (Amended)	Major Provisions
Toxic Substances Control Act (TSCA)	1976	Regulates use of chemical substances; requires toxicity testing for newly introduced compounds
Resource Conservation and Recovery Act (RCRA)	1976	Regulates the generation, handling, storage, and disposal of solid and hazardous wastes
Comprehensive Environmental Response, Compensation, and Liability Act (CERCLA)	1980	Also known as "Superfund"; established the national priorities list (NPL) for remediation of hazardous waste sites using the hazard ranking system (HRS); addressed financial liability
Low-Level Waste Policy Act (LLWPA)	1980 (1985)	Regulates disposal of low-level radioactive wastes
Nuclear Waste Policy Act (NWPA)	1982 (1987)	Regulates storage of high-level radioactive wastes
Hazardous and Solid Waste Amendments (HSWA)	1984	Significantly increased the scope of RCRA; new technology standards for landfills; promotes waste minimization preference for hazardous materials
Superfund Amendments and Reauthorization Act (SARA)	1986	Amended CERCLA; provided cleanup fund and revised NPL and the hazard ranking system
Emergency Planning and Community Right-to-Know Act (EPCRA)	1986	Requires manufacturers to report on-site quantities of hazardous materials and develop emergency response plan in case of an uncontrolled release
Pollution Prevention Act (PPA)	1990	Established source reduction as a national priority

Regulations that are a direct result of the environmental legislation identified in Table 4.1 may vary from state to state, as some states pass more stringent laws, but that generally follow closely the provisions passed in the federal acts.

4.2 ENVIRONMENTAL ETHICS

The word *ethics* is derived from the Greek word *ethos*, which can be defined as the *character of a person as defined by their actions*. This character is a combination of individual personality traits we have from birth (nature), as well as the influence of our interactions with others throughout our lives (nurture). The relative role that each of these plays in developing an individual's ethic has been debated for centuries, and will not be decided here. However, it is important to look at ourselves both as individuals and as members of a society that is entirely dependent upon resources extracted from the environment in which we live.

The study of ethics is a personal journey, asking individuals to evaluate choices that they are willing to make in light of the impact of those choices on society. Individual *integrity* can be understood with the "no one's looking" litmus test, which asks the question, If you knew that nobody would ever find out what your actions were, what would you do in a particular situation? However, the society in which we live also develops standards that tend to model the ethic of the majority. For engineers, most professional societies have developed *codes of ethics* that describe expected behaviors for an array of situations. Generally, the primary standard set forth in all of the engineering codes strives to *protect the health and safety of the public*. However, recent codes of ethics have been developed worldwide for protecting environmental integrity. In addition, a 1996 update to the American Society of Civil Engineers (ASCE) Code of Ethics included the concept of "sustainable development" in the first of its seven Fundamental Cannons. You can read and obtain copies of the Codes of Ethics for ASCE, the National Society of Professional Engineers (NSPE), and the National Council of Examiners of Engineering and Surveying (NCEES) at the following Web sites:

- ASCE Code of Ethics: *www.asce.org/inside/codeofethics.cfm*

- NSPE Code of Ethics for Engineers: *www.nspe.org/ethics/eh1-code.asp*

- Model Rules of the NCEES, Section 240.15 *Rules of Conduct*: *www.ncees. org/introduction/about_ncees/ncees_model_rules.pdf*

The prevailing ethic in the area of environmental concerns has recently focused on the issue of *sustainability*, which can be defined as the *ability to meet society's current resource needs, without inhibiting future generations from doing the same.* Those of us living in the United States are, by far, the greatest per capita consumers of the Earth's resources in the world today. Even more disconcerting is the fact that several societies, specifically China and India, look toward the United States as a model for acquiring material possessions. The populations of those two countries alone far exceed that of the United States, and the Earth's remaining resources will not be able to satisfy the additional demands if they were to consume at our current rate. Making environmentally sustainable choices for development now may be the only hope for future generations.

4.3 ENVIRONMENTAL IMPACT

As discussed in section 2.3, ecology is the study of the interrelationships between living organisms and their surroundings. It has long been known that humans have an impact on their environment; however, the resiliency of nature has throughout time been remarkable in its ability to maintain balance. Population increases in the past few centuries, coupled with technological advances and the use of natural resources to meet the energy demands of those populations, have greatly increased the rate of impact, while nature has struggled to maintain the previous balance that was long enjoyed. The sustainability concepts introduced earlier in this chapter require current generations to mitigate the impact of their activities to ensure integral environments for those generations to come. Minimizing or eliminating the impact on our ecosystems of proposed human activities in the United States was greatly encouraged through federal legislation passed in 1969.

4.3.1 National Environmental Policy Act (NEPA)

Passed in 1969 and signed into law on January 1, 1970, the introduction to NEPA states:

> The purposes of this chapter are: To declare a national policy which will encourage productive and enjoyable harmony between man and his environment; to promote efforts which will prevent or eliminate damage to the environment and biosphere and stimulate the health and welfare of man; to enrich the understanding of the ecological systems and natural resources important to the Nation; and to establish a Council on Environmental Quality.

NEPA had two primary impacts on environmental policy in the United States. First, it set the stage for the establishment of the U.S. Environmental Protection Agency (U.S. EPA) by calling for the formation of a Council of Environmental Quality (CEQ). The CEQ was charged with the responsibility to oversee the review of required NEPA documentation on the impact that federal (or federally funded) projects had on their surroundings. The U.S. EPA was established in analogous fashion in the summer of 1970 and began operation by December 1970 as an agency to oversee all of the environmental programs that were gaining support on the national level.

The second consequence was the establishment of the "action forcing provisions," specific documentation that would be required prior to the undertaking of any federal project that might have an adverse impact on the ecosystem. This document is called an environmental impact statement (EIS) and will be discussed further in the next section. However, even federal projects not anticipating substantial negative impact are required to submit an environmental assessment (EA). In addition, both documents were to be made available for public and professional review, and the recommendations opened to public debate.

4.3.2 Environmental Impact Statements

As the name suggests, an environmental impact assessment is an evaluation tool that requires consideration of all potential impacts to an ecosystem for a current or proposed process. The document prepared for external review is called the *environmental impact statement* (*EIS*), as required by law in the United States under NEPA for any federally funded project. The process should be inclusive, not only in technical evaluation but also in the array of constituents represented in the data gathering and interpretation, impact prioritization, and project management. The four steps in performing an EIA are: (1) *screening*, (2) *scoping*, (3) *EIS preparation*, and (4) *review*.

The process of screening can be described as the *determination of which projects warrant an EIA*. As mentioned previously, some projects are required by law to complete an EIS, specifically when federal dollars are part of the funding source. Another method is the use of plus/minus lists, which often specify types of projects that always require an EIA or projects that usually require an EIA. Sometimes, the project site may be located in an environmentally sensitive area, and impact to that area due to additional human activity should be assessed. Often, a project threshold, such as land area used or total energy consumed, can be a good indicator if an assessment is warranted. Finally, an initial environmental evaluation (IEE), sometimes called a mini-EIA, can be used to identify anticipated impacts, without the quantification and evaluation required by a full EIA.

Scoping is used to identify the target (key or critical) impacts that will form the basis of the study. While it would appear important to quantify all potential impacts, the reality is that usually a few critical issues will determine if a project will proceed or be modified. The scoping process should identify all important issues; however, the assessment should focus only on those that have a potential to require mitigation or those that would instigate project modification. This is also a time where public involvement can assist in defining priorities and expedite the process by not having to face public challenges regarding the selection of critical issues after the submission of the EIS (that is, during the review process).

EIS preparation needs to consider not only the data to be presented but also *why* the data is important, *to whom* the EIS will be addressed, and *how* the information will be communicated. This requires knowledge of the prospective audience, a clear and defined purpose for the EIS, and a technical framework for the written document. Because the project advocate conducts the assessment, and the government agency that requested the EIA usually grants permission for the project to proceed, the EIS must be reviewed by a technically competent but impartial review team. This is done to assure the public that an independent decision was made that upholds the spirit of the EIA, namely, to protect the ecosystem from unnecessary harm. The government agency requesting the EIS is not required to abide by the opinion expressed by the technical review panel; however, it is most often the case that a well-performed review will carry substantial influence.

4.4 OCCUPATIONAL HEALTH AND SAFETY

Nearly all workplace environments can present the labor force with hazards on a regular basis. It is the job of those involved with the study of industrial and occupational health and safety to identify health and safety hazards, estimate the effects these threats pose on the labor force, educate employees and management on the risks associated with the presence of these hazards, and provide safeguards to mitigate the identified risks. In recent decades, environmental engineers have used these or similar processes to examine the potential risk of human activity on the health and safety of the general population, as well as on the world's ecosystems in general. The following sections will examine the Occupational Safety and Health Act of 1970, followed by a look at the specific risks and mitigation techniques involved with two common workplace hazards: noise pollution and radiological heath and safety.

4.4.1 Occupational Safety and Health Act (OSHA)

OSHA was signed into law on December 29, 1970. The following is OSHA's statement of purpose from the introduction to the act:

> To assure safe and healthful working conditions for working men and women; by authorizing enforcement of the standards developed under the Act; by assisting and encouraging the States in their efforts to assure safe and healthful working conditions; by providing for research, information, education, and training in the field of occupational safety and health; and for other purposes.

The OSH Administration (also known by the acronym OSHA) under the U.S. Department of the Labor states: "OSHA's mission is to assure the safety and health of America's workers by setting and enforcing standards; providing training, outreach, and education; establishing partnerships; and encouraging continual improvement in workplace safety and health." The administration attempts to ensure workplace safety with respect to physical, chemical, biological, and radiological hazards through regular site inspections, development of new regulations, and employer/employee education. Common workplace safety measures, such as the use of personal protective equipment (PPE) and abiding by permissible exposure levels (PELs), are among those enacted by OSHA.

The following sections will address two common environmental workplace hazards in greater detail. Noise pollution is discussed in section 4.4.2 and radiological health and safety will be covered in section 4.4.3.

4.4.2 Noise Pollution

Noise pollution has been defined as *unwanted sound* and can vary in intensity from a mild irritant (nuisance) to a hazard with the potential for permanent hearing loss.

In contrast to the material pollution that has been discussed previously and is most commonly addressed in environmental engineering, noise is an energy form and thus requires a different approach for evaluation and control. Relative levels of sound (measured in dB) are given for a variety of situations in Figure 4.1.

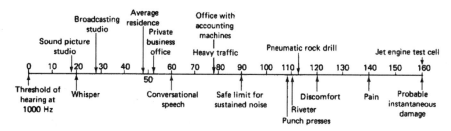

Figure 4.1   Relative scale for various sound pressure levels

Source: Mark J. Hammer, *Occupational Safety Management and Engineering*, 4th ed., ©1989. Reprinted by permission of Pearson Education, Inc., Upper Saddle River, N.J.

Sound is a result of the vibrations of solids or the dynamics of fluids as they interact with solid objects. These vibrations cause minute pressure variations that are able to be detected by the ear. It is generally assumed that the pressure fluctuations can be represented by a sinusoidal wave of *period* (*P*) traveling at a velocity commonly called the *speed of sound* (*c*), which may be calculated at atmospheric pressure as

$$c = 20.05\, T^{1/2}$$

where *c* has units of m/s, and *T* is the absolute temperature (K). The *frequency* (*f*) of the wave is the inverse of the period, and the *wavelength* (λ) of the sound wave can be calculated as

$$\lambda = \frac{c}{f} = c\,P$$

The amplitude is defined as the peak height (which is equivalent to the trough depth) measured from an arbitrary zero pressure line (that is, the prevailing pressure is assigned a value of zero). Since the average of the peak and trough would be zero, it is standard convention to describe the average pressure in terms of the *root mean squared pressure* (P_{rms}), which can be expressed as

$$P_{rms} = \left[\frac{1}{T} \int_0^T P^2(t)\, dt \right]^{1/2}$$

where *T* is the time over which the measurement was taken. Since the variation in pressures can vary greatly depending on the source, sound is measured on a log scale. The most common form of the expression of sound levels is the *sound pressure level* (*SPL* or L_p) in units of *decibels* (*dB*), which can be expressed as

$$SPL = 10 \log \frac{P_{rms}^{\,2}}{\left(P_{rms}\right)_0^{\,2}} = 20 \log \frac{P_{rms}}{\left(P_{rms}\right)_0}$$

where $(P_{rms})_0$ is a reference pressure set at 20 μPa (that is, 2×10^{-5} Pa). Most often, more than one source of noise must be considered simultaneously, in which case the SPL for each source should be determined and combined using the following expression:

$$SPL_{total} = \Sigma\, SPL = 10 \log\left[\Sigma\, 10^{(SPL_i/10)}\right]$$

where SPL_i denotes each of the individual sound pressure levels.

Example 4.1

Calculation of total SPL

A factory has four pieces of equipment in the same room with SPLs of 96, 108, 102, and 98 dB measured in the center of the room, which is a distance of 30 ft from each unit. Determine the total SPL from all four units.

Solution

The expression used to calculate the sum of individual noise levels can be found as

$$SPL_{total} = 10 \log\left[\Sigma\, 10^{SPL/10}\right]$$

Using the data given in the problem statement, we can solve for SPL_{total} as follows:

$$SPL_{total} = 10 \log\left[10^{\left(\frac{96}{10}\right)} + 10^{\left(\frac{108}{10}\right)} + 10^{\left(\frac{102}{10}\right)} + 10^{\left(\frac{98}{10}\right)}\right] = 109.5\ dB$$

It is often assumed that a wave emanates from a point. In as much, sound that is not reflected back toward the source will radiate in a spherical pattern, thus reducing the intensity inversely proportional to the square of the distance from the source. This change in SPL from a point source is often referred to as attenuation and can be expressed as

$$\left(\Delta SPL\right)_{point} = 10 \log\left(\frac{r_1}{r_2}\right)^2$$

where $(\Delta SPL)_{point}$ is the change in SPL for a point source (dB), and r_1 and r_2 are the distances between the source and two separate receptors.

Since the nonreflecting assumption is often not valid, another assumption that may be made is that the sound is emanating from a line source in a cylindrical pattern, and dispersion of waves occurs inversely proportional to distance from the source, expressed as

$$\left(\Delta SPL\right)_{line} = 10 \log\left(\frac{r_1}{r_2}\right)$$

where $(\Delta SPL)_{line}$ is the change in SPL for a line source (dB).

Example 4.2

Attenuation of SPLs

A piece of industrial equipment gives off sound at 105 dB measured at a distance 5 ft in front of the machine. At what distance is the sound level reduced to a safe level (90 dB), assuming the machine acts as a point source?

Solution

The change in SPL is a decrease of 15 dB (–15 dB) and is related to distance for a point source by the equation:

$$\Delta SPL = 10 \log \left(\frac{r_1}{r_2} \right)^2 = -15 \text{ dB}$$

Using the data given in the problem statement, the safe distance can be calculated as

$$-1.5 = \log \left(\frac{5}{r} \right)^2 \Rightarrow 0.0316 = \left(\frac{5}{r} \right)^2 \Rightarrow (0.0316)^{1/2} = \frac{5}{r} \Rightarrow r = 28.13 \text{ ft}$$

Noise control should always seek to reduce at the source whenever possible. This is usually accomplished by evaluating the source and applying corrective measures, such as machine balancing, friction reduction, or the use of barriers and application of dampening or sound absorbing materials. When potentially damaging sound levels persist after reduction and attenuation measures have been taken, *personal protective equipment* (*PPE*), such as earplugs or cup-style headgear, should be mandatory for operators and other employees.

4.4.3 Radiological Health and Safety

While there are many types of radiation present throughout the electromagnetic spectrum, radiological health generally focuses on *ionizing radiation*, a subset that possesses the potential for cellular damage. *Radioactivity* is the phenomenon whereby an unstable isotope undergoes nuclear disintegration, called *decay*, releasing a particle or electromagnetic radiation to carry off the excess energy. The three major classes of decay products are alpha particles, beta particles, and gamma radiation.

Alpha particles are equivalent to the nucleus of the helium-4 atom (represented as $_2^4 He$), comprised of 2 protons and 2 neutrons. Loss of an alpha particle must then reduce the mass of the isotope by 4 mass units and reduce the charge by 2. This is usually represented as

$$_Z^A X \rightarrow {_{Z-2}^{A-4}} X + {_2^4} He$$

where Z is the atomic number (number of protons) and A is the atomic mass number, which is equal to the number of neutrons plus the number of protons ($N + Z$). You will note that the original isotope (called a *parent*) yields a different element (called a *daughter*) upon alpha decay, due to the fact that the daughter nucleus contains two less protons than the parent. Alpha decay is common for elements of atomic number greater than 82.

Alpha particles have extremely high energy (exit velocities near 10^4 miles per second) but cannot penetrate dense materials (for example, they cannot penetrate farther than the epidermis) and can be shielded by ordinary clothing or a sheet of paper. Care must be taken for ingested material, where damage is local but intense. Emitted alpha particles, upon striking a surface, cause the substance to release electrons, which the alpha particle uses to form a stable helium atom. The atoms that have lost electrons remain ionized until they can replace their losses.

Another isotope instability results in the decay of a neutron into a proton, which is retained in the atom, and an electron, which is emitted. Because the loss of a neutron is balanced by the gain of a proton, the atomic mass number remains unchanged, but the daughter is a new element due to the change in atomic number. The emitted electron is called a *beta particle* (β), and the transformation is usually represented as

$$\prescript{A}{Z}{X} \rightarrow \prescript{A}{Z+1}{X} + \beta^-$$

Beta particles are fast-moving electrons, which have much greater penetrating power than alpha particles due to their small size but are less ionizing. They may be shielded through the use of thin metal plating; however, inappropriate choice of the plating material may result in the generation of X-rays. Appropriate shielding choices include lead, aluminum, and plastics.

Alpha and beta particle decay is often accompanied by *gamma ray* emission, which is a release of electromagnetic radiation energy as the newly formed element settles into a more stable state. Gamma rays are similar in energy and damage potential to X-rays, the latter of which is generated by high energy electrons striking a suitable target material. Both are highly penetrating energies that cause damage throughout a body, not solely on the surface, as is the case with alpha and beta particles. Due to their high penetration potential, gamma rays and X-rays require more substantial (thicker) shielding, with lead as the most common choice.

Radioactive decay is considered constant and is dependent on the original number of nuclei (N) present. The decay of radioactive isotopes can, therefore, be expressed as

$$N = N_0\, e^{-\lambda t}$$

where N_0 is the original number of nuclei, λ is the radioactive decay constant, and t is the time interval. Note that the use of λ in this expression should not be confused with the use of the same symbol previously for wavelength. Often, it is convenient to express the decay in terms of the species *half-life*, which can be defined as *the amount of time required for 50% of the nuclei to decay*. This can be determined from the above expression by solving for $t_{1/2}$ (sometimes represented as τ) when N is equal to one half of N_0, and is calculated as follows:

$$t_{1/2} = \tau = \frac{\ln 2}{\lambda} = \frac{0.693}{\lambda}$$

Solving for λ in the half-life expression and substituting into the previous one yields

$$N = N_0 \exp\left[\frac{-0.693\, t}{t_{1/2}}\right] = N_0 \exp\left[\frac{-0.693\, t}{\tau}\right]$$

Example **4.3**

Radiation Half-Life

How long must you store a 5 μCi/L solution of iodine-131 (half-life of 8.06 days) before safe disposal in a drain if the discharge limit is 10^{-6} μCi/mL?

Solution

The expression for radiation half-life is

$$N = N_0 \exp\left[\frac{-0.693\, t}{\tau}\right]$$

Using the data given in the problem statement, but correcting values for consistent volume units, we can determine the storage time as

$$10^{-6} \frac{\mu Ci}{mL} = 5 \times 10^{-3} \frac{\mu Ci}{mL} \exp\left[\frac{(-0.693)(t)}{8.06 \text{ day}}\right] \quad \Rightarrow \quad t = 99 \text{ days}$$

The dose for gamma and X-ray exposure is called the *roentgen* (R), while the dose unit that describes the amount of all radiation energies absorbed by the body is called the *rad*. The *relative biological effectiveness* (RBE) factor is a ratio of the absorbed dose of gamma radiation (in rads) to the absorbed dose of another type of radiation required to have an equal biological impact. Since all tissues and organs have unique sensitivity to different radiation types and sources, the *tissue weighting factor* (W_r) has been developed as a metric to assign high values of potential damage to sensitive tissues or organs, and low (or zero) values to other tissues or organs. Finally, the *roentgen equivalent man* (REM) is an indication of the extent of biological injury that is probable from a dose of a specific type of radiation to a specific tissue or organ and can be expressed as

dose in REMs = RBE × (dose in rads) × W

Since the radioactive source is usually considered a point source, gamma and X-ray radiation intensity decreases with the square of the inverse of the distance from the source. Often, separation distance is the most cost-effective form of protection from high dose exposure to a source. The inverse square law may be expressed as

$$I_2 = I_1 \left(\frac{r_1}{r_2}\right)^2$$

where I_1 is the radiation intensity at a distance r_1 from the source, and I_2 is the radiation intensity at a distance r_2 from the source. Other means of radiation attenuation generally utilize shielding, and tables and graphs have been created that provide the required thickness of various materials to achieve a desired percentage of radiation intensity reduction.

Example 4.4

Radiation attenuation

An accident has caused the release of radioactive materials with an intensity that is 25,000 times greater than a safe level when the measurement is taken 10 m from the source. At what distance must the safe perimeter be maintained during the clean-up process?

Solution

Radiation intensity varies with the square of distance as follows:

$$I_2 = I_1 \left(\frac{r_1}{r_2}\right)^2$$

Assuming the safe level at distance r_2 is I_2, and $I_1 = 25{,}000 \times I_2$ at a distance of 10 m, the safe distance may be calculated as

$$I_2 = (25,000) \, I_2 \left(\frac{10 \text{ m}}{r_2} \right)^2 \Rightarrow 0.00004 = \left(\frac{10 \text{ m}}{r_2} \right)^2 \Rightarrow r_2 = 1581 \text{ m}$$

So, the perimeter should be established at least 1.6 km (1.0 mile) from the accident site.

Finally, there are potential adverse physiological responses to other forms of energy in the electromagnetic spectrum, which are classified as *nonionizing radiation*. *Ultraviolet radiation* can have a mild-to-severe impact on the skin, depending on incident radiation intensity and exposure time, or can cause blindness if not managed properly from industrial sources (for example, electric arc welding). Although *visible light* is the most common, and generally the most safe, it can also have a negative impact on vision if not managed properly. In addition, *infrared radiation* may cause burns to exposed areas, as well as the potential for eye damage. Another source of thermal effects is *microwave radiation*, which causes burns and tissue damage by the creation of heat due to an increase in the kinetic energy of the affected tissue.

4.5 RISK ASSESSMENT

Risk is defined as the probability of a negative outcome from a specific event. Like any statistical probability, risk does not have units. No action is truly risk-free. Scientists and statisticians have quantified the risk associated with a number of activities, including smoking, riding on an airplane, receiving botox injections, and even eating peanut butter![1] Environmental engineers may need to estimate risks in order to set drinking water treatment standards or to determine when a contaminated site is "clean enough."

The risks allowed by the U.S. EPA are typically $1/10^4$, $1/10^5$, or $1/10^6$ (one in one million). This risk is the *incremental risk* allowed in addition to the *background risk*. The background risk is the risk due to all of the environmental factors to which the average person is exposed. For example, the background risk of contracting cancer is approximately 1 in 4, or $250,000/10^6$. *Toxicology* is the study of the nature, effects, and detection of contaminants in living organisms. A list of helpful definitions is provided in Table 4.2.

Table 4.2 Definitions associated with risk assessment

Acute toxicity	The adverse effect due to a single dose of a contaminant
Carcinogen	A substance capable of inducing cancer in an organism
Chronic toxicity	The adverse effects resulting from repeated doses of or long-term exposure to a substance
Mutagenic	The ability to induce mutations in genes or chromosomes of an organism
Teratogenic	The ability to develop defects in an embryo

---

1   Eating 40 tablespoons of peanut butter presents a 1 in 1 million risk, due to the possible presence of aflotoxins. (R. Wilson, "Analyzing the daily risks of life," *Technology Review*, vol. 82, 1979: 41-46.)

Before determining the risk that a certain compound presents to humans, the impact that the compound has on nonhuman animals must be determined. These tests are performed by subjecting a population of laboratory animals to varying doses of a contaminant and by observing the effects that the dose has on them. These results are often presented in a "dose-response" curve. A typical dose-response curve is shown in Figure 4.2.

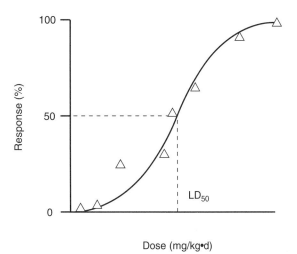

Figure 4.2 Dose-response curve

Note that the units of dose are mg/kg·d. In other words, the dose is the mass of chemical (expressed in mg) per mass of animal (in kg) per exposure time (in days). One quantity of interest that can be extracted from the dose-response curve is the LD_{50}, or the dose that is lethal to 50% of the population. The LD_{50} is shown in Figure 4.2. The laboratory results are extrapolated to humans using a series of safety factors, each one of which is one order of magnitude.

4.5.1 Exposure Assessment

In order to determine the risk facing humans, the extent of exposure must be determined. The first step is to determine the exposure pathway (inhalation, ingestion, or dermal contact). Second, the dose associated with the exposure pathway must be calculated. The dose may be estimated given values provided by the EPA, as shown in Table 4.3.

Table 4.3 EPA-recommended values for estimating intake

Parameter	Standard Value
Average body weight, adult	70 kg
Average body weight, child	
0–1.5 years	10 kg
1.5–5 years	14 kg
5–12 years	26 kg
Amount of water ingested daily, adult	2 L
Amount of water ingested daily, child	1 L
Amount of air breathed daily, adult	20 m³

(continued)

Parameter	Standard Value
Amount of air breathed daily, child	5 m^3
Amount of fish consumed daily, adult	6.5 g per day
Contact rate, swimming	50 mL per h
Skin surface available, adult male	1.94 m^2
Skin surface available, adult female	1.69 m^2
Skin surface available, child	
3–6 years (average for male and female)	0.720 m^2
6–9 years (average for male and female)	0.925 m^2
9–12 years (average or male and female)	1.16 m^2
12–15 years (average for male and female)	1.49 m^2
15–18 years (female)	1.60 m^2
15–18 years (male	1.75 m^2
Soil ingestion rate, children 1–6 years	200 mg per day
Soil ingestion rate, persons > 6 years	100 mg per day
Skin adherence factor, potting soil to hands	1.45 mg per cm^2
Skin adherence factor, kaolin clay to hands	2.77 g per cm^2
Exposure duration	
Lifetime	70 years
At one residence, 90th percentile	30 years
National median	5 years
Exposure frequency (EF)	
Swimming	7 days per year
Eating fish and shellfish	48 days per year
Exposure time (ET)	
Shower, 90th percentile	12 min
Shower, 50th percentile	7 min

Source: M. L. Davis and S. J. Masten, *Principles of Environmental Engineering and Science,* © 2004, McGraw-Hill Education. Reprinted by permission of the McGraw-Hill Companies.

The mass of contaminant ingested can be readily calculated using appropriate values from Table 4.3 and from knowledge of the contaminant concentration. By taking special care of units, these calculations are very straightforward, as shown in Example 4.5.

Example 4.5

Exposure assessment

A two-year-old child plays on a vacant ground that was once a small industrial site. The child plays on the site five days a week for three years. Soil tests show that the soil contains benzene (C_6H_6) at a concentration of 1.1 μg/kg. What mass of benzene does the child consume?

Solution

Given a soil ingestion rate of 200 mg/day (Table 4.3), the mass of C_6H_6 ingested is

$$\text{mass} = \left(1.1\times10^{-6}\,\frac{g}{kg}\right)\cdot\left(200\,\frac{mg}{day}\right)\cdot\left(\frac{1\,kg}{10^6\,mg}\right)\cdot\left(\frac{5\,day}{week}\right)\cdot\left(\frac{52\,week}{yr}\right)\cdot\left(3\,yr\right)$$

$$= 0.18\,\mu g$$

This may not seem like a lot of benzene, but consider how many molecules of benzene are ingested by the child. This calculation can be performed relatively easily by knowing the molecular weight of benzene (78 g/mole), recalling Avogadro's number, and taking care of units:

$$\left(0.18\,\mu g\right)\cdot\left(\frac{1\,g}{10^6\,\mu g}\right)\cdot\left(\frac{1\,\text{mole}}{78\,g}\right)\cdot\left(\frac{6\times10^{23}\,\text{molecules}}{\text{mole}}\right)=1.38\times10^{15}\,\text{molecules}$$

So the child has ingested 1380 trillion molecules of benzene. And given that benzene is "reasonably anticipated to be a human carcinogen," and that theoretically one molecule of a carcinogen can cause cancer, the original answer of 0.18 μg no longer seems insignificant. The *risk* to the child due to ingesting the benzene will be discussed in section 4.5.2.

The *dose* is the mass of contaminant ingested, inhaled, or absorbed through dermal contact, divided by the product of the person's body weight and an averaging time:

$$\text{dose} = \frac{\text{mass of contaminant ingested}}{(\text{body weight})\cdot(\text{averaging time})}$$

The averaging time varies depending on whether the contaminant is a carcinogen or noncarcinogen and is discussed further in sections 4.5.2 and 4.5.3, respectively. One type of dose, the CDI (chronic daily intake) uses an averaging time of 70 years.

| Example **4.6** |

Dose

What is the dose of benzene for the child eating the soil described in Example 4.5, setting the averaging time equal to the time of exposure?

Solution

The dose, averaged over the three years of exposure, is

$$\text{dose} = \frac{0.18\times10^{-3}\,mg\,C_6H_6}{\left(14\,kg\right)\cdot\left(3\,\text{years}\right)\cdot\left(\dfrac{365\,day}{yr}\right)} = 1.2\times10^{-8}\,\frac{mg}{kg\cdot day}$$

4.5.2 Characterizing the Risk for Carcinogens

CDI = Chronic Daily Intake

A contaminant's carcinogenic toxicity is related to its *potency factor* (or *slope factor*). The latter name arises from the dose-response curve, from whence the slope factor is extracted. Specifically, the slope factor is the slope of the dose-response curve *at very low doses*. Slope factors are available for either oral, dermal, or inhalation exposure routes.

Given that the potency factor is the slope of the dose-response curve at low doses, its units can be inferred to be $(mg/kg \cdot day)^{-1}$. By observing that these units are the reciprocal of the units of dose, it can readily be seen that risk is defined as

$$risk = potency\ factor \times dose$$

The dose used when calculating carcinogenic risk is the CDI, and thus the averaging time used is always 70 years.

Example 4.7

Cancer risk

What is the risk to the child that has ingested benzene given a potency factor of benzene of 0.029 kg·day/mg?

risk = 0.029 · dose · $\frac{3}{70}$ (84.6)

CDI

Solution

The mass of benzene calculated in Example 4.5 can be utilized in this example problem. However, the dose calculated in Example 4.6 cannot be used, as this dose used an averaging time of three years. Rather, the CDI is to be used. The CDI is calculated as

$$CDI = \frac{0.18 \times 10^{-3}\ mg\ C_6H_6}{\left(14\ kg\right) \cdot \left(70\ years\right) \cdot \left(\dfrac{365\ day}{yr}\right)} = 5.0 \times 10^{-10}\ \frac{mg}{kg \cdot day}$$

Thus, the incremental risk for the child to contract cancer due to the exposure of benzene is:

$$risk = (CDI)(slope\ factor) = \left(5.0 \times 10^{-10}\ \frac{mg}{kg \cdot day}\right)\left(0.029\ \frac{kg \cdot day}{mg}\right) = 1.45 \times 10^{-11}$$

The inverse of this risk is 6.9×10^{10}, so the risk can also be expressed as one in 69 billion.

For the case of exposure to multiple contaminants, the total risk is the sum of the individual risks:

$$total\ risk_{multiple\ contaminants} = \sum_i risk_i$$

4.5.3 Noncarcinogenic Impacts

Noncarcinogens differ from carcinogens in that noncarcinogens are not assigned a risk. As shown previously, a risk is associated with even minute doses of carcinogens. Rather than being assigned a risk, doses of noncarcinogens are characterized as being allowable or unallowable, of being a cause for concern or not. In essence, there is no middle ground. Whether a dose is allowable or not is characterized by comparing the dose to a benchmark dose, namely, the *Reference Dose* (RfD) for ingestion exposures or the *Reference Concentration* (RfC) for inhalation exposures. Just as the potency factor is an indicator of the degree of toxicity of a carcinogen, the reference dose is an indicator of the toxicity of a noncarcinogen.

The *hazard quotient*, or *hazard index* (HI), is used to show whether the exposure to a noncarcinogen is acceptable or not. It is defined as

$$HI = \frac{dose}{RfD}$$

When calculating the dose for noncarcinogens, the averaging time should be set equal to the duration of exposure.

If the hazard index is greater than unity, the dose can be characterized as unallowable or as a cause for concern. If a person is subjected to multiple contaminants, as is common in many circumstances, the individual hazard indices are summed up:

$$HI_{multiple\,contaminants} = \sum_i HI_i$$

This summed value is less than 1 for an acceptable exposure and greater than 1 for unacceptable exposures.

HI < 1 = Acceptable Exposure
HI > 1 = UNACCEPTABLE EXP.

Example 4.8

Hazard index

Assume that the playground from Example 4.5 also contains 4.2 µg/kg of PCBs. The reference doses for benzene and PCBs are 4.0×10^{-3} mg/kg·day and 7.0×10^{-6} mg/kg·day, respectively.

Solution

The dose of benzene, averaged over the exposure time, was calculated to be

$$1.2 \times 10^{-8} \frac{mg}{kg \cdot day}$$

In a similar manner, the dose of PCBs can be calculated to be

$$2.7 \times 10^{-7} \frac{mg}{kg \cdot day}$$

The hazard index for each contaminant is thus:

$$HI_{benzene} = \frac{dose}{RfD} = \frac{1.2 \times 10^{-8} \frac{mg}{kg \cdot day}}{4 \times 10^{-3} \frac{mg}{kg \cdot day}} = 3 \times 10^{-6}$$

$$HI_{PCB} = \frac{2.7 \times 10^{-7} \frac{mg}{kg \cdot day}}{7.0 \times 10^{-6} \frac{mg}{kg \cdot day}} = 0.039$$

The sum of these hazard indices is clearly controlled by the hazard index of the PCBs. However, since the sum (0.039) is less than unity, the exposure is said to be allowable in terms of noncarcinogenic impacts.

2

Applications

Water Resources

9/22/09

OUTLINE

The effect of engineering projects on the environment is often manifested in deleterious impacts to water resources. The Earth's water resources include surface water sources (for example, lakes, ponds, rivers, wetlands, and oceans) and groundwater sources (that is, aquifers). These various water resources are linked together naturally and through the effects of the built environment. Understanding how water moves in the environment (for example, as surface or subsurface flow) and realizing the special needs of our most valuable freshwater resources, such as wetlands and lakes, are essential to protecting our water resources.

5.1 SURFACE WATER HYDROLOGY

This section presents topics associated with surface water hydrology and focuses on the interrelationship of precipitation and surface water runoff. Surface water hydrology is separated in this review book from groundwater hydrology (section 5.4); the latter topic examines the fate of water after it enters the subsurface. After reading this section, you will be able to predict the surface runoff behavior (in terms of peak runoff flow, the total volume of runoff, or the time-dependent behavior of runoff) in response to a storm event.

5.1.1 Water Cycle and Hydrologic Budget

The *water cycle* is shown graphically in Figure 5.1. Water moves within the environment via such processes as precipitation and evaporation. Although water is found in solid, liquid, and vapor forms at different points in the water cycle, it is important to remember that the total mass of water is unchanged.

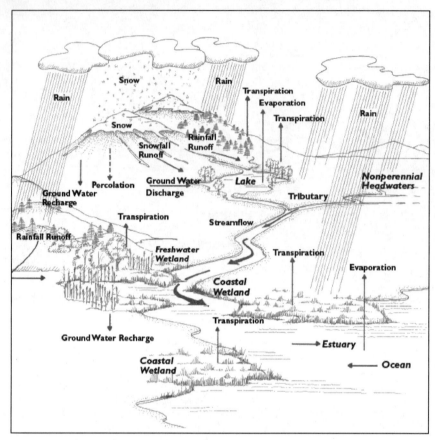

Figure 5. 1   Water cycle

Source: *Connecticut Storm Water Quality Management Manual,* Connecticut Department of Environmental Protection.

A *hydrologic budget* (or *water budget*) is a mass balance of water in a defined control volume (see section 1.4 for a discussion of mass balances). Thus, a hydrologic budget is a way of accounting for all the water moving into and out of the control volume and the change of volume of water stored in the control volume. In environmental engineering, the control volume used for a hydrologic budget analysis might be an aquifer, a watershed, a landfill, or perhaps a sewage lagoon, as demonstrated in Example 5. 1.

The general form for the hydrologic budget is:

change of storage of water in   =   volume entering the control volume   **(5.1)**
control volume   – volume exiting the control volume

A time span may be needed for this analysis, as often the water entering and exiting the control volume is provided as a flow rate ($L^3 \cdot T^{-1}$).

When solving hydrologic budget problems, the general form of the hydrologic budget equation can be modified to account for the specific forms of water movement. For example, consider Figure 5. 2, which shows a simplified water cycle for a lake.

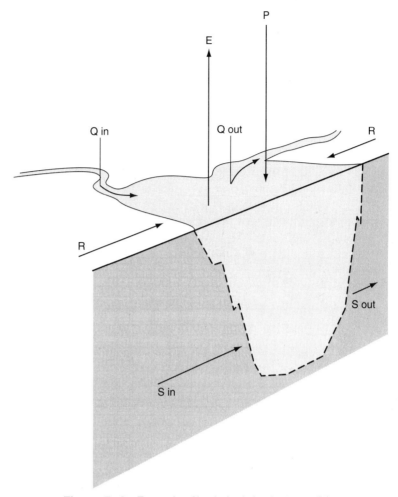

Figure 5. 2 Example of hydrologic budget on a lake

A hydrologic budget could be used to determine the change of storage in the lake for one year and would look like this:

$$\Delta \text{ storage} = (Q_{in} + P + R + S_{in}) - (Q_{out} + S_{out} + E)$$

where

Q = the volume of water flowing in the entering/exiting stream

P = the precipitation volume

R = the runoff volume

S = the seepage volume

E = evaporation volume

Example **5.1**

Hydrologic budget for a lagoon

A lined lagoon receives industrial wastewater and has no outlet structure; water is intended to leave the lagoon solely by evaporation. Assess whether the lined lagoon is leaking water into the groundwater using the following data:

Evaporation rate	60 in/year
Mean water depth, January 1, 2004	70.5 in
Mean water depth, January 1, 2005	88.5 in
Precipitation for 2005	40 in
Industrial effluent flow for 2005	1000 gal/day for five days each week
Lagoon surface area	15,000 ft²
Overland flow into pond	negligible

Solution

First, draw a schematic of the lagoon showing all flows into and out of the lagoon (Exhibit 1).

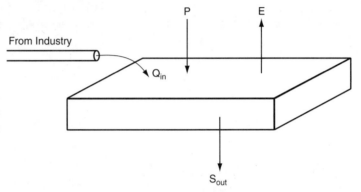

Exhibit 1 Lagoon schematic

Next, assuming no seepage into the pond, the seepage out of the pond can be solved for starting with the water budget equation (Equation 5.1):

$$\Delta \text{ storage } = \text{volume entering} - \text{volume exiting}$$
$$(Q_{in} + P) - (S_{out} + E)$$
$$S_{out} = (Q_{in} + P) - (E) - \Delta \text{ storage}$$

This equation also reflects the fact that the lagoon has no flow out.

Information has been provided to define each of the terms on the right-hand side of the equation. It is important to use consistent units throughout. For this problem, all terms will be expressed as volumes:

Q_{in} $= 1000 \text{ gal/d} \cdot 5 \text{ d/wk} \cdot 52 \text{ wk/yr} \cdot 1 \text{ yr} \cdot 1 \text{ ft}^3/7.48 \text{ gal}$
         $= 3.5 \cdot 10^4 \text{ ft}^3$

P $= 40 \text{ in} \cdot 15,000 \text{ ft}^2 \cdot (1 \text{ ft}/12 \text{ in})$
      $= 5.0 \cdot 10^4 \text{ ft}^3$

E $= 60 \text{ in/yr} \cdot 15,000 \text{ ft}^2 \cdot 1 \text{ yr} \cdot 1 \text{ ft}/12 \text{ in}$
      $= 7.5 \cdot 10^4 \text{ ft}^3$

$\Delta \text{ storage} = (88.5 \text{ in} - 70.5 \text{ in}) \cdot 15,000 \text{ ft}^2 \cdot 1 \text{ ft}/12 \text{ in}$
               $= 2.3 \cdot 10^4 \text{ ft}^3$

S_{out} $= (Q_{in} + P) - (E) - \Delta \text{ storage}$
      $= (3.5 \cdot 10^4 \text{ ft}^3 + 5.0 \cdot 10^4 \text{ ft}^3) - (7.5 \cdot 10^4 \text{ ft}^3) - 2.3 \cdot 10^4 \text{ ft}^3$
      $= -1.3 \cdot 10^4 \text{ ft}^3$

Since S_{out} has been defined as having a positive value when exiting the lagoon, a negative value infers that more flow has entered the lagoon in the year than has exited the lagoon. Perhaps the assumptions that there is no seepage into the lagoon and that runoff into the lagoon was negligible were inappropriate.

This example raises an important point concerning units in hydrology. All of the quantities were expressed in terms of a volume (that is, ft³), but the volume could also have been expressed in terms of a depth simply by dividing the volume by the pond area. Thus, in hydrology, quantities such as precipitation, runoff, and change in storage can all be expressed as depths or volumes. The conversion between the two sets of units simply requires an area, which may be the area of a watershed or pond/lagoon/lake.

2.2.2 Hydrologic Elements

The various transfer processes of the hydrologic cycle are sometimes referred to as hydrologic elements. The hydrologic elements described in this section include interception, rainfall, evapotranspiration, and infiltration.

Interception

Interception occurs when the surface area of vegetation captures rain before it hits the ground. In some native plantings, such as the prairies of the U.S. midwest, interception can abstract as much as 0.1 inch of precipitation.

Rainfall

A rainfall event is classified in terms of its *intensity,* duration, and *return period.* A rainfall intensity is expressed as a depth per time, such as in/hr or cm/hr. The return period (sometimes referred to as a *recurrence interval*) is the amount of time expected between rainfall events of the same intensity and duration. Thus, for a given locality, a storm that lasts one hour at an intensity of 2 in/hr might have a return period of five years. Such a storm would be expected every five years and is said to have a probability of 1/5. If this 5-year storm occurs in a given year, the probability of it occurring the following year is still 1/5. Table 5.1 shows the probability of occurrence for various storms of various return periods. Table 5.1 shows that there is a 4% chance of a 25-year storm occurring in any given year (1/25), and a 64% probability that the 25-year storm will occur in a 25-year period.

Table 5.1 Flood design frequency selection chart

Percent chance of equaling or exceeding such a flood at least once in this many years:					
100 years	**50 years**	**25 years**	**10 years**	**Any 1 year**	**Recurrence Interval (years)**
—	99	93	65	10	10
98	87	65	34	4	25
87	65	40	18	2	50
64	40	22	10	1	100

Source: State of Wisconsin Department of Transportation Facilities Development Manual, Procedure 13-10-1, Figure 2.

Intensities are most typically obtained from intensity-duration-frequency (IDF) curves, such as the curve shown in Figure 5. 3. Note that *frequency* is simply the inverse of return period. An IDF curve is typically used by first choosing a frequency, determining a duration, and then looking up the intensity from the graph. Determining the duration will be described in section 5.1.5.

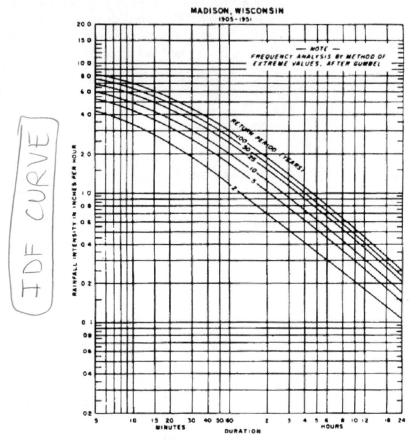

Figure 5.3 Example IDF curve

Source: *State of Wisconsin Department of Transportation Facilities Development Manual*, Procedure 13-10-5, Figure 4.

The average precipitation $\bar{p}$ over a region can be obtained from point data in one of three ways, all of which fit the formula

$$\bar{p} = \frac{1}{AT} \sum A_i p_i \qquad A_T = \sum A_i \tag{5.2}$$

in which p_i is a point precipitation value, A_i is a weighting factor, and A_T is the sum of the weighting factors:

(a) The simple arithmetic mean is appropriate when the individual values are all similar. In this case, set each $A_i = 1$, and then A_T is just the number of points.

(b) The widely used Thiessen average is a weighted average that, in effect, assumes the value p_i best represents the true precipitation at all locations that are closer to gage i than to any other gage. Each A_i is the area surrounding gage i, and A_T is the total gaged area. The boundary of each A_i is formed by lines that are the perpendicular bisectors of lines drawn between the gages themselves.

(c) The Isohyetal method is the only method that allows a knowledge of basin topography to enter the calculation. One begins the computation by drawing contour lines of equal precipitation (isohyets) throughout the region. Then, in Equation 5.2, A_i is the area between adjacent isohyets, p_i is the average precipitation between these adjacent isohyets, and A_T is the total gaged area.

In addition to varying with space, rainfall intensity also varies with time. A *hyetograph* shows how the storm intensity varies with time and is a graph of intensity versus time. Alternatively, the *y*-axis may be the depth of precipitation instead of intensity.

The TR-55 method (discussed in more detail in section 5.1.4) provides four generic rainfall distributions. The distributions vary according to geographic location in the United States and are shown in Figure 5.4.

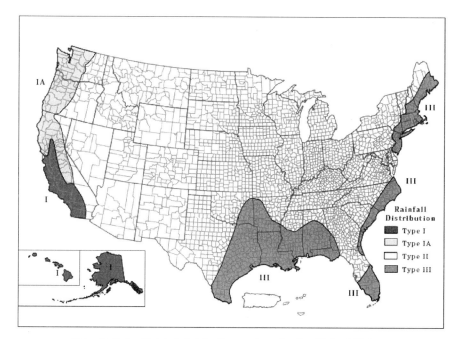

Figure 5.4 Rainfall distribution regions for the United States

Source: *Urban Hydrology for Small Watersheds,* U.S. Department of Agriculture, Figure B-3.

Figure 5.5 shows the four NRCS rainfall distributions. Three points are important to note about the distributions in Figure 5.5:

1. The distributions represent cumulative depth.

2. A 24-hour storm duration is assumed.

3. The ordinate of the graph represents a unit rainfall; thus, the actual distribution is obtained by multiplying the entire rainfall depth by the ordinate values corresponding to different times.

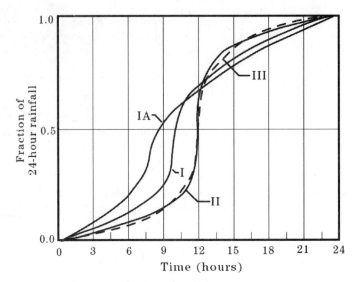

Figure 5.5 24-hour rainfall distributions

Source: *Urban Hydrology for Small Watersheds,* U.S. Department of Agriculture, Figure B-1.

Evapotranspiration

The quantification of evaporation or transpiration amounts (or of evapotranspiration ET, the sum of the two) can become important to engineers who conduct water supply studies. There are several computational approaches and one primary experimental method of estimating evaporation; each approach has its problems and leads to imprecision in the result:

(a) The water budget or mass conservation method attempts to account for all flows of water to and from the water body under study, including inflow, outflow, direct precipitation, and even seepage to the groundwater.

(b) The energy budget is like the water budget, except the energy flows rather than mass flows are the basic accounting medium.

(c) Direct empirical meteorological correlations are used to avoid the uncertainties of the first two methods, but attempts to avoid excessive complexity here usually lead to incomplete, and thus inaccurate, results.

(d) The National Weather Service Class A pan is four feet in diameter, 10 inches deep, made of unpainted galvanized iron, and used to measure evaporation by direct experiment. Multiplication of this result by a pan coefficient, typically about 0.7, then gives the evaporation from the adjacent larger water body. Difficult correlation studies are needed to ensure that the coefficient is appropriate to a particular application.

Infiltration

Infiltration of water into the soil is important in some studies. Horton's infiltration equation is widely used for this purpose, which is

$$f = f_c + (f_o - f_c)e^{-kt}$$

(5.3)

Here f is infiltration rate (in/hr); f_o and f_c are the initial and final infiltration rates, respectively; t is time (hr); and k (1/hr) is an empirically determined constant.

Another common way of characterizing infiltration is via the ϕ-index method. In this method, you plot the overall precipitation rate versus time; a horizontal line called the ϕ index is drawn on the plot, such that the volume of rainfall excess above this line is equal to the actual volume of observed runoff. Thus, the index indicates the average infiltration rate for the storm event.

5.1.3 Watershed Basics

A *watershed* is the portion of land surface on which all hydrologic calculations are based. It is defined as the area of land that drains to a single point. Watersheds are delineated based on topographic information. Boundaries are drawn such that a drop of water falling on any piece of land within the boundary will drain to a common point (that is, the watershed outlet).

The *time of concentration* (t_c) is the time required for a drop of water to travel from the hydrologically most distant point in the watershed to the watershed's outlet. Two common equations for calculating time of concentration are the Kirpich method and the SCS lag equation, as defined Table 5.2.

Table 5.2 Calculating time of concentration

Method	Equation for t_c (min)	Variables
Kirpich method	$t_c = 0.0078 \cdot L^{0.77} \cdot S^{-0.385}$	L = length of channel/ditch from headwater to outlet, ft S = average watershed slope, ft/ft
SCS lag equation	$t_c = \dfrac{0.00526 \cdot L^{0.8} \cdot [(1000/CN) - 9]^{0.7}}{S^{0.5}}$	L = hydraulic length of watershed, ft CN = curve number (see section 5.1.4) S = average watershed slope, %

Figure 5.6 provides a nomograph that offers a means of quickly estimating the time of concentration using the Kirpich method.

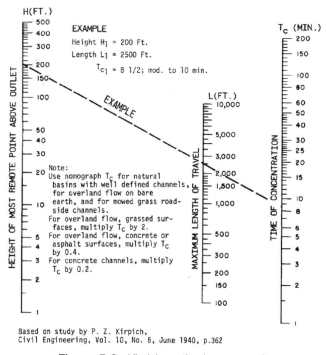

Figure 5.6 Kirpich method nomograph

Time of concentration can also be determined by summing up the flow time in various flow segments, that is, for n flow segments,

$$t_c = t_{\text{segment 1}} + t_{\text{segment 2}} + t_{\text{segment 3}} + \cdots t_{\text{segment } n}$$

Flow segments include *sheet flow, shallow concentrated flow,* channel flow, flow in storm sewer pipes, and so on. Sheet flow and shallow concentrated flow may be grouped together and termed *overland flow.* Each of these types of flow segments is discussed below:

■ *Sheet flow* is flow that occurs over planar surfaces. It is limited to a maximum of 300 feet at the upper reaches of a catchment. Manning's kinematic solution (Equation 5.4) can be used to estimate the sheet flow travel time. When using this equation, care must be taken to use the proper units.

$$T_t = \frac{0.007(nL)^{0.8}}{(P_2)^{0.5} S^{0.4}}$$

(5.4)

where

T_t = travel time (hr)

n = Manning's roughness coefficient (Table 5.3)

L = flow length (ft)

P_2 = 2-year, 24-hour rainfall (in)

S = land slope (ft/ft)

■ After a maximum of 300 feet, sheet flow typically transitions into *shallow concentrated flow.* Shallow concentrated flow velocities can be estimated from Figure 5.7.

Table 5.3 Manning's n for sheet flow

Surface Description	n^a
Smooth surfaces (concrete, asphalt, gravel, or bare soil)	0.011
Fallow (no residue)	0.05
Cultivated soils:	
Residue cover ≤ 20%	0.06
Residue cover > 20%	0.17
Grass	
Short grass prairie	0.15
Dense grasses[b]	0.24
Bermuda grass	0.41
Range (natural)	0.13
Woods[c]	
Light underbrush	0.40
Dense underbrush	0.80

[a]The n values are a composite of information compiled by Engman (1986).

[b]Includes species such as weeping lovegrass, bluegrass, buffalo grass, blue grama grass, and native grass mixtures.

[c]When selecting n, consider cover to a height of about 0.1 ft. This is the only part of the plant cover that will obstruct sheet flow.

Source: U.S. Department of Agriculture, *Urban Hydrology for Small Watersheds.*

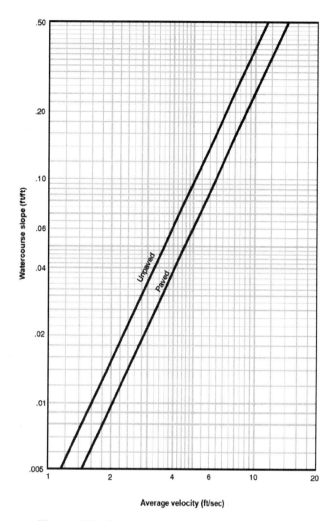

Figure 5.7 Shallow concentrated flow velocities

Source: U.S. Department of Agriculture, *Urban Hydrology for Small Watersheds.*

- *Open channel flow* follows shallow concentrated flow. Such flow may occur in roadside ditches, swales, culverts, storm sewer pipes that are not pressurized. Velocity in open channels can be estimated using Manning's equation, which is discussed in more detail in Chapter 3 and in this chapter.

- Another way to estimate flow velocity for a watershed (or a portion thereof) is to use Figure 5.8. Figure 5.8 is often used in conjunction with the rational method, which is discussed in section 5.1.5.

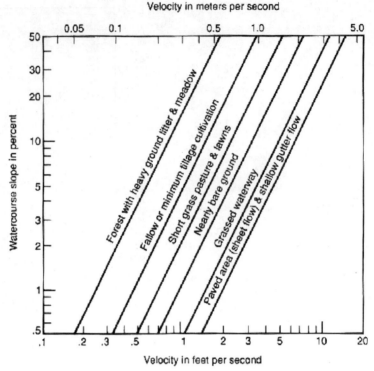

Velocity in meters per second

Figure 5.8   Flow velocities estimation

Source: U.S. Soil Conservation Service, *Urban Hydrology for Small Watersheds,* 1975

Example **5.2**

Impact of development on time of concentration

The drawing in Exhibit 2 shows a proposed layout for a subdivision. The engineers have proposed that the subdivision be served by 12-inch PVC storm sewer pipes. Before development, the site was wooded grassland. The 2-year, 24-hour rainfall is 3.5 inches. Determine the time of concentration to point A in Exhibit 2:

(a) Before development

(b) After development

Solution

The watershed is delineated as shown in Exhibit 3.

(a) The flow path for the pre-development runoff is also shown in Exhibit 3. The sheet flow segment is noted, and the remainder of the flow is assumed to be shallow concentrated flow. This assumes that the flow never transitions into open channel flow at the downstream portions of the watershed. The sheet flow segment is assumed to be 300 feet, and the remainder of the flow length is approximately 1900 feet.

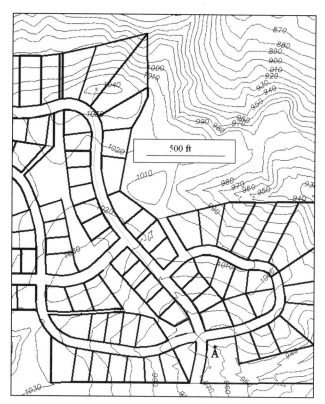

Exhibit 2   Proposed layout for subdivision

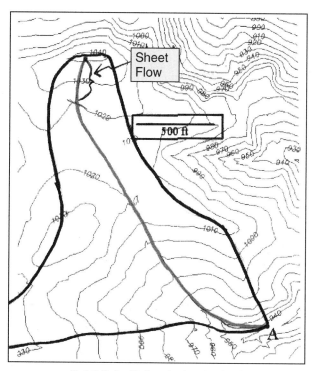

Exhibit 3   Delineated watershed

The travel time in the sheet flow segment is obtained from the kinematic Manning's equation (Equation 5.4), given the following values:

n = 0.4 (corresponding to woods with light underbrush, Table 5.3)

L = 300 ft

P_2 = 3.5 in

S = (1040 ft – 1020 ft)/300 ft

     = 0.067

Therefore, the travel time in the sheet flow segment is 0.51 hours or 30.6 minutes.

The travel time in the shallow concentrated flow portion is obtained from Figure 5.7 (given a land slope of [1020 ft – 940 ft]/1900 ft = 0.042) to be 3.4 ft/sec. Thus, the travel time is 1900 ft/3.4 fps = 560 sec = 9.3 min.

Therefore, the time of concentration before development is 30.6 minutes + 9.3 minutes = 39.9 minutes.

(b) The flow to point A in the developed parcel will consist of (in order): sheet flow; flow on the roadway until it is collected at the first stormwater inlet (assumed to be located 500 feet from the crest of the hill); flow in sewers; and flow in a grassed channel, assumed to be rectangular in cross section (10 ft wide × 1 ft deep). These flow paths are shown in Exhibit 4.

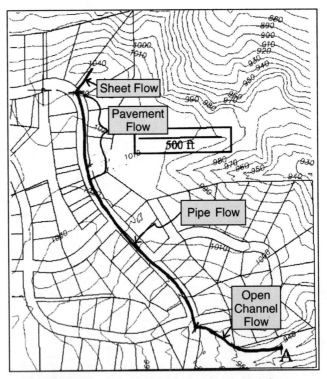

Exhibit 4   Flow segments for developed site

The sheet flow length is 150 feet. Assuming that all other values remain the same as for the predevelopment case, the time of travel in the sheet flow segment is 0.29 hours, or 17.5 minutes.

The velocity on the pavement can be obtained from Figure 5.7 to be approximately 5.1 fps. Given a flow length of 500 feet, the travel time is 98 seconds (1.6 min). (A more precise estimate of travel time could be obtained knowing the cross section of the gutter flow, as described in section 6.3.1.)

The pipe flow velocity can be obtained from Manning's equation (see section 3.3.2), assuming full flow conditions. Given a pipe length of 850 feet scaled from the drawing, a diameter of 12 inches, a Manning's n of 0.009, and a slope of 0.066 ft/ft ([1021 ft – 965 ft]/850 ft), the Manning's velocity is 16.8 fps. This corresponds to a travel time of 850 ft/16.6 fps = 51 sec = 0.85 min.

The velocity in the open channel (with a length of 600 ft as scaled from drawing) can also be obtained from Manning's equation, by using a Manning's n of 0.045, a hydraulic radius of 10/12 ft, and a channel slope of (965 ft – 940 ft)/600 ft = 0.042. A velocity of 6 fps is obtained (assuming full channel flow), which yields a travel time in the channel of 600 ft/6 fps = 100 sec (1.7 min).

Thus, the travel time of the developed site is equal to

$$t_c = t_{sheet} + t_{pavement} + t_{pipe} + t_{channel}$$

$$= 17.5 \text{ min} + 1.6 \text{ min} + 0.85 \text{ min} + 1.7 \text{ min}$$

$$= 21.7 \text{ min}$$

5.1.4 TR-55 (NRCS) Method

The TR-55 (NRCS)[1] runoff equation is used to determine the *depth* of runoff, Q, associated with a storm event:

$$Q = \frac{(P - 0.2S)^2}{(P + 0.8S)} \tag{5.5}$$

where

P = the rainfall depth, inches

S = the potential maximum retention after runoff begins, inches

$$S = \frac{1000}{CN} - 10$$

CN = the runoff curve number, and can be found from tables such as Tables 5.4a–c. The hydrologic soil group can be determined from soil borings or soil maps. The soil groups are summarized in Table 5.5.

A weighted curve number (CN_w) can also be estimated based on the fraction of imperviousness in the catchment using the following equation, which assumes a CN of the impervious area of 98:

$$CN_w = ([CN_{pervious} \cdot A_{pervious}] + [98 \cdot A_{impervious}])/A_{total}$$

---

1   The NRCS (Natural Resources Conservation Service) was formerly known as the SCS (Soil Conservation Service). However, this TR-55 (NRCS) method is still often referred to as the SCS method, the SCS TR-55 method, or the TR-55 method. In this review book, the method will be referred to as the NRCS method.

$$CN_W = [(CN_{perv} \cdot A_{perv}) + (98 \cdot A_{imper})] / A_{total}$$

where

$CN_{pervious}$ = curve number of the pervious area

$A_{pervious}$ = area of the pervious area

$A_{impervious}$ = area of impervious area

A_{total} = $A_{impervious} + A_{pervious}$

Given that the depth of runoff calculated by the NRCS method is a function of only two variables, the task of calculating the runoff depth via the NRCS method can be simplified through use of a chart, such as in Figure 5.9. In an exam situation, Figure 5.9 could be very effective for quickly estimating a value of the runoff depth, with adequate precision for an exam with multiple-choice format. Note that the *volume* of runoff is the product of the runoff depth and the watershed area.

Table 5.4a   Runoff curve numbers for urban areas[1]

Cover Type and Hydrologic Condition Impervious Area[2]	Average Percent	Curve Numbers for Hydrologic Soil Group			
		A	B	C	D
Fully developed urban areas (vegetation established)					
Open space (lawns, parks, golf courses, cemeteries, etc.)[3]:					
Poor condition (grass cover < 50%)		68	79	86	89
Fair condition (grass cover 50–75%)		49	69	79	84
Good condition (grass cover > 75%)		39	61	74	80
Impervious areas:					
Paved parking lots, roofs, driveways, etc. (excluding right-of-way)		98	98	98	98
Streets and roads:					
Paved; curbs and storm sewers (excluding right-of-way)		98	98	98	98
Paved; open ditches (including right-of-way)		83	89	92	93
Gravel (including right-of-way)		76	85	89	91
Dirt (including right-of-way)		72	82	87	89
Western desert urban areas:					
Natural desert landscaping (pervious areas only)		63	77	85	88
Artificial desert landscaping (impervious weed barrier, desert shrub with 1–2-inch sand or gravel mulch and basin borders)		96	96	96	96
Urban districts:					
Commercial and business	85	89	92	94	95
Industrial	72	81	88	91	93
Residential districts by average lot size:					
1/8 acre or less (town houses)	65	77	85	90	92
1/4 acre	38	61	75	83	87
1/3 acre	30	57	72	81	86
1/2 acre	25	54	70	80	85
1 acre	20	51	68	79	84
2 acres	12	46	65	77	82

(continued)

Cover Type and Hydrologic Condition Impervious Area[2]	Average Percent	Curve Numbers for Hydrologic Soil Group			
		A	B	C	D
Developing urban areas					
Newly graded areas (pervious areas only, no vegetation)	77	86	91	94	
Idle lands (CNs are determined using cover types similar to those in Table 5.4c)					

1. Average runoff condition, and $I_a = 0.2S$.
2. The average percent impervious area shown was used to develop the composite CNs. Other assumptions are as follows: impervious areas are directly connected to the drainage system, impervious areas have a CN of 98, and pervious areas are considered equivalent to open space in good hydrologic condition.
3. CNs shown are equivalent to those of pasture. Composite CNs may be computed for other combinations of open space cover type.

Source: U.S. Department of Agriculture, *Urban Hydrology for Small Watersheds.*

Table 5.4b Runoff curve numbers for cultivated agricultural lands[1]

Cover Description		Hydrologic Condition[3]	Curve Numbers for Hydrologic Soil Group			
Cover Type	Treatment[2]		A	B	C	D
Fallow	Bare soil	—	77	86	91	94
	Crop residue cover (CR)	Poor	76	85	90	93
		Good	74	83	88	90
Row crops	Straight row (SR)	Poor	72	81	88	91
		Good	67	78	85	89
	SR + CR	Poor	71	80	87	90
		Good	64	75	82	85
	Contoured (C)	Poor	70	79	84	88
		Good	65	75	82	86
	C + CR	Poor	69	78	83	87
		Good	64	74	81	85
	Contoured & terraced (C&T)	Poor	66	74	80	82
		Good	62	71	78	81
	C&T+ CR	Poor	65	73	79	81
		Good	61	70	77	80
Small grain	SR	Poor	65	76	84	88
		Good	63	75	83	87
	SR + CR	Poor	64	75	83	86
		Good	60	72	80	84
	C	Poor	63	74	82	85
		Good	61	73	81	84
	C + CR	Poor	62	73	81	84
		Good	60	72	80	83
	C&T	Poor	61	72	79	82
		Good	59	70	78	81
	C&T+ CR	Poor	60	71	78	81
		Good	58	69	77	80

(continued)

Cover Description		Hydrologic Condition[3]	Curve Numbers for Hydrologic Soil Group			
Cover Type	Treatment[2]		A	B	C	D
Close-seeded or broadcast legumes or rotation meadow	SR	Poor	66	77	85	89
		Good	58	72	81	85
	C	Poor	64	75	83	85
		Good	55	69	78	83
	C&T	Poor	63	73	80	83
		Good	51	67	76	80

1. Average runoff condition, and $I_a = 0.2S$.

2. Crop residue cover applies only if residue is on at least 5% of the surface throughout the year.

3. Hydrologic condition is based on combination factors that affect infiltration and runoff, including (a) density and canopy of vegetative areas, (b) amount of year-round cover, (c) amount of grass or close-seeded legumes, (d) percent of residue cover on the land surface (good ≥ 20%), and (e) degree of surface roughness. Poor: Factors impair infiltration and tend to increase runoff. Good: Factors encourage average and better than average infiltration and tend to decrease runoff.

Source: U.S. Department of Agriculture, *Urban Hydrology for Small Watersheds.*

Table 5.4c   Runoff curve numbers for other agricultural lands[1]

Cover Description	Hydrologic Condition	Curve Numbers for Hydrologic Soil Group			
Cover Type		A	B	C	D
Pasture, grassland, or range—continuous forage for grazing[2]	Poor	68	79	86	89
	Fair	49	69	79	84
	Good	39	61	74	80
Meadow—continuous grass, protected from grazing and generally mowed for hay	—	30	58	71	78
Brush—brush-weed-grass mixture with brush the major element[3]	Poor	48	67	77	83
	Fair	35	56	70	77
	Good	30[4]	48	65	73
Woods—grass combination (orchard or tree farm)[5]	Poor	57	73	82	86
	Fair	43	65	76	82
	Good	32	58	72	79
Woods[6]	Poor	45	66	77	83
	Fair	36	60	73	79
	Good	30[4]	55	70	77
Farmsteads—buildings, lanes, driveways, and surrounding lots	—	6559	74	82	86

1. Average runoff condition, and $I_a = 0.2S$.

2. *Poor:* < 50% ground cover or heavily grazed with no mulch.
   Fair: 50–75% ground cover and not heavily grazed.
   Good: > 75% ground cover and lightly or only occasionally grazed.

3. *Poor:* < 50% ground cover.
   Fair: 50–75% ground cover.
   Good: > 75% ground cover.

4. Actual curve number is less than 30; use CN = 30 for runoff computations.

5. CNs shown were computed for areas with 50% woods and 50% grass (pasture) cover. Other combinations of conditions may be computed from the CNs for woods and pasture.

6. *Poor:* Forest litter, small trees, and brush are destroyed by heavy grazing or regular burning. *Fair:* Woods are grazed but not burned, and some forest litter covers the soil. *Good:* Woods are protected from grazing, and litter and brush adequately cover the soil.

Source: U.S. Department of Agriculture, *Urban Hydrology for Small Watersheds.*

Table 5.5 NRCS soil classification

Soil Group	Description
A	Lowest runoff potential. Includes deep sands with very little silt and clay; also deep, rapidly permeable loess
B	Mostly sandy soils less deep than A, and loess less deep or less aggregated than A, but the type has above average infiltration after thorough wetting
C	Comprises shallow soils and soil containing considerable clay and colloid, though less than D
D	Highest runoff potential. Includes mostly clays of high swelling percent, but the group also includes some shallow soils with nearly impermeable subhorizons near the surface

Source: State of Wisconsin Department of Transportation Facilities Development Manual, Figure 13-10-5, Figure 6.

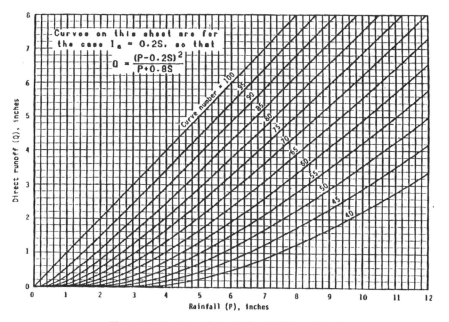

Figure 5.9 Runoff depth for NRCS method

Source: U.S. Department of Agriculture, *Urban Hydrology for Small Watersheds.*

5.1.5 Peak Runoff Estimation

In many instances, you may need to calculate the peak flow of stormwater runoff in addition to the total depth of runoff. This section describes two methods for estimating the peak runoff: the rational method and the NRCS method.

Rational Method

The rational method is the most widely used method for the estimation of peak discharge Q_p (ft^3/s) from runoff over small surface areas. In using it, you assume that a spatially and temporally uniform rainfall occurs for a time period that allows the entire catchment area to contribute simultaneously to the outflow. Clearly, the satisfaction of these limitations becomes more difficult as the basin size increases, so this equation is normally limited to basins that are below one square mile (640 acres) in size. The equation is

$$Q_p = CiA \qquad\qquad (5.6)$$

[handwritten: Rational Method A < 1 mi² < 640 Ac]

in which C is a nondimensional runoff coefficient that indicates the fraction of the incident rain that runs off the surface, i is the appropriate storm intensity (in/hr), and A is the watershed area (acres). Some add a dimensional conversion factor to this equation, but since 1 ft^3/s = 1.008 acre-in/hr, the conversion factor is usually ignored, as the other factors in the equation are not known with such accuracy. Table 5.6, adapted from reference 1 at the end of the chapter, gives reasonable ranges for C for various surfaces, as well as some guidance in selecting a value in the range.

The appropriate equation for the weighted runoff coefficient, C_w is

$$C_w = \frac{\sum_i C_i \cdot A_i}{\sum_i A_i} \qquad\qquad (5.7)$$

Table 5.6   Runoff coefficient, C

Description of Area	Runoff Coefficients		
Business			
Downtown	0.70	to	0.95
Neighborhood	0.50	to	0.70
Residential			
Single-family	0.30	to	0.50
Multiunits, detached	0.40	to	0.60
Multiunits, attached	0.60	to	0.75
Residential (suburban)	0.25	to	0.40
Apartment	0.50	to	0.70
Industrial			
Light	0.50	to	0.80
Heavy	0.60	to	0.90
Parks, cemeteries	0.10	to	0.25
Playgrounds	0.20	to	0.35
Railroad yard	0.20	to	0.35
Unimproved	0.10	to	0.30

(continued)

It often is desirable to develop a composite runoff coefficient based on the percentage of different types of surface in the drainage area. This procedure often is applied to typical "sample" blocks as a guide to selection of reasonable values of the coefficient for an entire area. Coefficients with respect to surface type currently in use are:

Character of Surface	Runoff Coefficients		
Pavement			
Asphaltic and Concrete	0.70	to	0.95
Brick	0.70	to	0.85
Roofs	0.75	to	0.95
Lawns, sandy soil			
Flat, 2%	0.05	to	0.10
Average, 2–7%	0.10	to	0.15
Steep, 7%	0.15	to	0.20
Lawns, heavy soil			
Flat, 2%	0.13	to	0.17
Average, 2–7%	0.18	to	0.22
Steep, 7%	0.25	to	0.35

The coefficients in these two tabulations are applicable for storms of 5- to 10-year frequencies. Less frequent, higher intensity storms require the use of higher coefficients, because infiltration and other losses have a proportionally smaller effect on runoff. The coefficients are based on the assumption that the design storm does not occur when the ground surface is frozen.

Source: ASCE, *Design and Construction of Sanitary and Storm Sewers* (Manual and Report No. 37); 1986, American Society of Civil Engineers. Reprinted by permission.

The intensity factor must also be chosen carefully. It is normally defined as the intensity of rainfall of a chosen frequency that lasts for a duration equal to the time of concentration t_c for the basin. Sometimes, the frequency will be dictated by policy (1-year, 5-year, or 10-year). Once the frequency has been chosen and the time of concentration has been picked, you usually consult an IDF plot to obtain i. Conceptually, this time is the time required for flow from the most remote point in the basin to reach the outlet, but in some cases it is simply estimated to be in the 5–15 minute range. Picking a shorter time usually leads to a higher intensity i and a larger Q_p; in one sense, this is conservative, but it may also be wasteful by causing you to design for an excessively large flow. The IDF plot, if developed properly, reports information that is the result of long-term statistical averages of many individual storms, not just the result of a compilation of relatively few data.

When the basin surface is not homogeneous, you should either subdivide the basin into smaller regions that are (nearly) uniform or compute a weighted average value for C, the weights being the areas.

Several other approaches to the estimation of peak discharge exist, the SCS methods being among the most prominent. If you want to apply these methods properly, however, then a lengthy description of the method and supporting data and charts are required. You should consult the references at the end of this chapter for an adequate description of the procedures.

Example **5.3**

Design flows for inlets

A storm drain is to be extended to serve two developing areas in a suburb. Exhibit 5 presents the intensity-duration-frequency plot for this region as well as a schematic diagram of the developments. Area A consists of 40 acres of mostly single-family residential units with some multiple family units; the time of concentration is 15 min. Area B drains to inlet 2 and contains several small businesses. The transit time for stormwater to move from Inlet 1 to inlet 2 is $T = 5$ min. Assuming a 5-year return period, estimate the peak discharges expected at the two inlets.

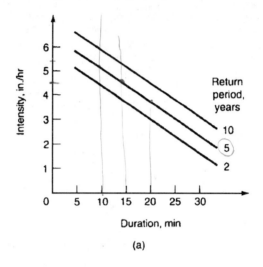

(a)

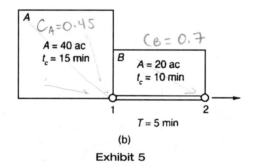

(b)

Exhibit 5

Solution

For Area A, assume a 15-min duration and find $i = 4.50$ in/hr for a 5-year return period from Exhibit 5. Referring to Table 5.6, it appears that $C = 0.45$ is reasonable for this residential area. For Point 1, the peak discharge should be about

$$Q_p = CiA = (0.45)(4.50)(40) = 81 \text{ ft}^3/\text{s}$$

This peak is expected to appear at the second inlet location at $15 + 5 = 20$ min after the storm begins.

If Area B is considered separately, then a 10-min duration leads, via Exhibit 5a, to $i = 5.17$ in/hr, the runoff coefficient may be nearly $C = 0.70$, and

$$Q_p = (0.70)(5.17)(20) = 72 \text{ ft}^3/\text{s}$$

at inlet 2 from Area B. However, the two computed peak discharges do not both arrive at Point 2 at the same instant. The peak flow from B arrives 10 min before the flow from A arrives.

To compensate for the fact that the two peak discharges do not coincide in time, the usual approximate procedure is to use an area-weighted coefficient C_w and a time of concentration that applies to the combination of the areas. Here, the time of concentration is 20 min. Thus

$$C_w = \Sigma C_i A_i \,/\, \Sigma A_i = [0.45(40) + 0.70(20)]/(40 + 20) = 0.53$$

For the 5-yr return period and a 20-min duration, Exhibit 5a gives $i = 3.83$ in/hr. and

$$Q_p = (0.53)(3.83)(60) = 122 \text{ ft}^3/\text{s}$$

which is lower by some 30 ft^3/s than the sum of the individual peak flows.

Peak Runoff with the NRCS method

The peak runoff can also be estimated from the NRCS method. (Recall that in section 5.1.4, the NRCS method was used to determine the depth of runoff.) The NRCS method can also be used to find the peak runoff rate using the concept of *unit peak discharges*. The following figures provide the unit peak discharge for the four rainfall distributions in the United States: Type I (Figure 5.10; Type IA (Figure 5.11); Type II (Figure 5.12); Type III (Figure 5.13). The value denoted I_a/P on these figures is the ratio of initial abstraction depth (I_a) to precipitation depth. The initial abstraction depth can be estimated based on the CN value as shown in Table 5.8. If the I_a/P value is less than the lowest value provided on the graphs in Figures 5.10–5.13, the lowest I_a/P value should be used.

Once a unit peak discharge has been obtained, the peak runoff q_p (cfs), can be estimated using this equation:

$$q_p = q_u \cdot A_m \cdot Q \cdot F_p \tag{5.8}$$

where

q_u = unit peak discharge (cfs/[mi$^2 \cdot$ in])

A_m = drainage area (mi^2)

Q = runoff depth (in)

F_p = the pond and swamp adjustment factor (see Table 5.7)

Table 5.7 Swamp and pond adjustment factor

% of Pond and Swamp Areas	F_p
0	1.00
0.2	0.97
1.0	0.87
3.0	0.75
5.0	0.72

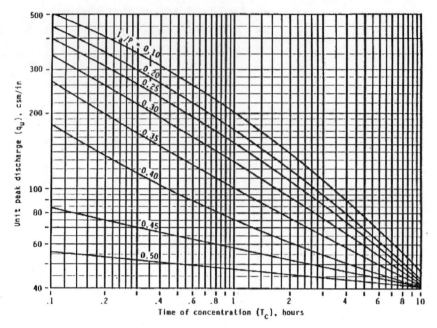

Figure 5.10 Unit peak discharge for Type I

Source: U.S. Department of Agriculture, *Urban Hydrology for Small Watersheds.*

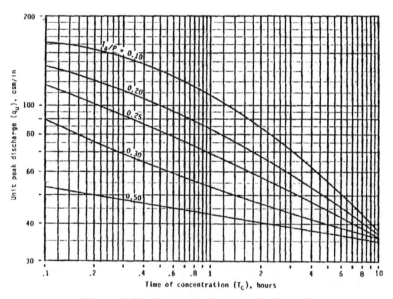

Figure 5.11 Unit peak discharge for Type IA

Source: U.S. Department of Agriculture, *Urban Hydrology for Small Watersheds.*

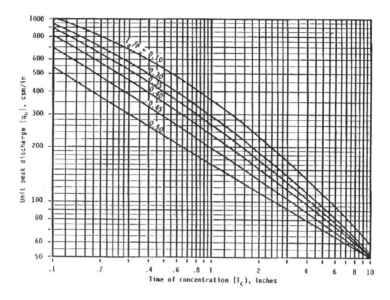

Figure 5.12 Unit peak discharge for Type II

Source: U.S. Department of Agriculture, *Urban Hydrology for Small Watersheds*, *www.wcc.nrcs.usda.gov/hydro/hydro-tools-models-tr55.html.*

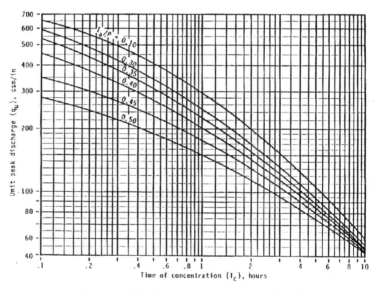

Figure 5.13 Unit peak discharge for Type III

Source: U.S. Department of Agriculture, *Urban Hydrology for Small Watersheds.*

Table 5.8 I_a values

Curve Number	I_a (in)	Curve Number	I_a (in)	Curve Number	I_a (in)
40	3.00	60	1.333	80	0.500
41	2.878	61	1.279	81	0.469
42	2.762	62	1.226	82	0.439
43	2.651	63	1.175	83	0.410
44	2.545	64	1.125	84	0.381
45	2.444	65	1.077	85	0.353
46	2.348	66	1.030	86	0.326
47	2.255	67	0.985	87	0.299
48	2.167	68	0.941	88	0.273
49	2.082	69	0.899	89	0.247
50	2.000	70	0.857	90	0.222
51	1.922	71	0.817	91	0.198
52	1.846	72	0.778	92	0.174
53	1.774	73	0.740	93	0.151
54	1.704	74	0.703	94	0.128
55	1.636	75	0.667	95	0.105
56	1.571	76	0.632	96	0.083
57	1.509	77	0.597	97	0.062
58	1.448	78	0.564	98	0.041
59	1.390	79	0.532		

Source: U.S. Department of Agriculture, *Urban Hydrology for Small Watersheds.*

Example 5.4

Impacts of development on stormwater runoff characteristics

A 50-acre field in Wisconsin planted in row crops has been sold to a developer, who will be building ¼-acre residential lots on 15 acres of the site and retail businesses on 30 acres of the site. The remainder of the site will be used for green space. The site can be modeled as a single catchment with a length of 1500 feet and a slope of 1%. Assume "good" hydrologic conditions and Hydrologic Soil Group B. Given a 6-inch rainfall depth, determine:

(a) The change in runoff depth from the site as a result of development

(b) The change in runoff volume from the site as a result of development

(c) The change in peak flow from the site as a result of development

Solution

(a) A curve number equal to 78 for the predeveloped site is selected from Table 5.4 (assuming straight row crops).

Table 5.4a Runoff curve numbers for urban areas[1] [Arc II]

Cover Type and Hydrologic Condition Impervious Area[2]	Average Percent	Curve Numbers for Hydrologic Soil Group			
		A	B	C	D
Fully developed urban areas (vegetation established)					
Open space (lawns, parks, golf courses, cemeteries, etc.)[3]:					
Poor condition (grass cover < 50%)		68	79	86	89
Fair condition (grass cover 50–75%)		49	69	79	84
Good condition (grass cover > 75%)		39	61	74	80
Impervious areas:					
Paved parking lots, roofs, driveways, etc. (excluding right-of-way)		98	98	98	98
Streets and roads:					
Paved; curbs and storm sewers (excluding right-of-way)		98	98	98	98
Paved; open ditches (including right-of-way)		83	89	92	93
Gravel (including right-of-way)		76	85	89	91
Dirt (including right-of-way)		72	82	87	89
Western desert urban areas:					
Natural desert landscaping (pervious areas only)		63	77	85	88
Artificial desert landscaping (impervious weed barrier, desert shrub with 1–2-inch sand or gravel mulch and basin borders)		96	96	96	96
Urban districts:					
Commercial and business	85	89	92	94	95
Industrial	72	81	88	91	93
Residential districts by average lot size:					
1/8 acre or less (town houses)	65	77	85	90	92
1/4 acre	38	61	75	83	87
1/3 acre	30	57	72	81	86
1/2 acre	25	54	70	80	85
1 acre	20	51	68	79	84
2 acres	12	46	65	77	82
Developing urban areas					
Newly graded areas (pervious areas only, no vegetation)	77	86	91	94	
Idle lands (CNs are determined using cover types similar to those in Table 5.4c)					

1. Average runoff condition, and $I_a = 0.2S$.

2. The average percent impervious area shown was used to develop the composite CNs. Other assumptions are as follows: impervious areas are directly connected to the drainage system, impervious areas have a CN of 98, and pervious areas are considered equivalent to open space in good hydrologic condition.

3. CNs shown are equivalent to those of pasture. Composite CNs may be computed for other combinations of open space cover type.

Source: U.S. Department of Agriculture, *Urban Hydrology for Small Watersheds*.

Table 5.4b   Runoff curve numbers for cultivated agricultural lands[1]   [Arc II]

Cover Description		Hydrologic Condition[3]	Curve Numbers for Hydrologic Soil Group			
Cover Type	Treatment[2]		A	B	C	D
Fallow	Bare soil	—	77	86	91	94
	Crop residue cover (CR)	Poor	76	85	90	93
		Good	74	83	88	90
Row crops	Straight row (SR)	Poor	72	81	88	91
		Good	67	78	85	89
	SR + CR	Poor	71	80	87	90
		Good	64	75	82	85
	Contoured (C)	Poor	70	79	84	88
		Good	65	75	82	86
	C + CR	Poor	69	78	83	87
		Good	64	74	81	85
	Contoured & terraced (C&T)	Poor	66	74	80	82
		Good	62	71	78	81
	C&T+ CR	Poor	65	73	79	81
		Good	61	70	77	80
Small grain	SR	Poor	65	76	84	88
		Good	63	75	83	87
	SR + CR	Poor	64	75	83	86
		Good	60	72	80	84
	C	Poor	63	74	82	85
		Good	61	73	81	84
	C + CR	Poor	62	73	81	84
		Good	60	72	80	83
	C&T	Poor	61	72	79	82
		Good	59	70	78	81
	C&T+ CR	Poor	60	71	78	81
		Good	58	69	77	80
Close-seeded or broadcast legumes or rotation meadow	SR	Poor	66	77	85	89
		Good	58	72	81	85
	C	Poor	64	75	83	85
		Good	55	69	78	83
	C&T	Poor	63	73	80	83
		Good	51	67	76	80

1. Average runoff condition, and $I_a = 0.2S$.

2. Crop residue cover applies only if residue is on at least 5% of the surface throughout the year.

3. Hydrologic condition is based on combination factors that affect infiltration and runoff, including (a) density and canopy of vegetative areas, (b) amount of year-round cover, (c) amount of grass or close-seeded legumes, (d) percent of residue cover on the land surface (good ≥ 20%), and (e) degree of surface roughness. Poor: Factors impair infiltration and tend to increase runoff. Good: Factors encourage average and better than average infiltration and tend to decrease runoff.

Source: U.S. Department of Agriculture, *Urban Hydrology for Small Watersheds.*

Table 5.4c Runoff curve numbers for other agricultural lands[1]

Cover Description Cover Type	Hydrologic Condition	Curve Numbers for Hydrologic Soil Group			
		A	B	C	D
Pasture, grassland, or range—continuous forage for grazing[2]	Poor	68	79	86	89
	Fair	49	69	79	84
	Good	39	61	74	80
Meadow—continuous grass, protected from grazing and generally mowed for hay	—	30	58	71	78
Brush—brush-weed-grass mixture with brush the major element[3]	Poor	48	67	77	83
	Fair	35	56	70	77
	Good	30[4]	48	65	73
Woods—grass combination (orchard or tree farm)[5]	Poor	57	73	82	86
	Fair	43	65	76	82
	Good	32	58	72	79
Woods[6]	Poor	45	66	77	83
	Fair	36	60	73	79
	Good	30[4]	55	70	77
Farmsteads—buildings, lanes, driveways, and surrounding lots	—	6559	74	82	86

1. Average runoff condition, and $I_a = 0.2S$.
2. *Poor:* < 50% ground cover or heavily grazed with no mulch.
   Fair: 50–75% ground cover and not heavily grazed.
   Good: > 75% ground cover and lightly or only occasionally grazed.
3. *Poor:* < 50% ground cover.
   Fair: 50–75% ground cover.
   Good: > 75% ground cover.
4. Actual curve number is less than 30; use CN = 30 for runoff computations.
5. CNs shown were computed for areas with 50% woods and 50% grass (pasture) cover. Other combinations of conditions may be computed from the CNs for woods and pasture.
6. *Poor:* Forest litter, small trees, and brush are destroyed by heavy grazing or regular burning. *Fair:* Woods are grazed but not burned, and some forest litter covers the soil. *Good:* Woods are protected from grazing, and litter and brush adequately cover the soil.

Source: U.S. Department of Agriculture, *Urban Hydrology for Small Watersheds*.

The curve number for the developed site can be estimated as a weighted average of three land uses:

¼-acre residential lots (CN = 75—Table 5.4a)

Retail business (CN = 92—Table 5.4a)

Green space (CN = 61—Table 5.4a)

$$CN_w = \frac{\sum_i CN_i \cdot A_i}{\sum_i A_i} = \frac{75*15+92*30+61*5}{50} = 84$$

From Figure 5.9, the runoff depths for the predeveloped and postdeveloped scenarios are 3.6 in and 4.2 in, respectively.

(b) Runoff volume is the product of runoff depth and area. Thus, the runoff volumes for the predeveloped and postdeveloped scenarios are 180 acre-in ($6.5 \cdot 10^5$ ft³) and 210 acre-inches ($7.6 \cdot 10^5$ ft³), respectively.

(c) To determine the peak flow using the NRCS method, the times of concentrations and I_a/P quantity must be calculated for both the pre- and postdevelopment scenarios. The times of concentration before development ($t_{c,pre}$) and after development ($t_{c,post}$) can be estimated from the SCS lag equation (presented in Table 5.2):

$$t_c = \frac{0.00526 \cdot L^{0.8} \cdot [(1000/CN) - 9]^{0.7}}{S^{0.5}}$$

$$t_{c,pre} = \frac{0.00526 \cdot 1500^{0.8} \cdot [(1000/78) - 9]^{0.7}}{0.01^{0.5}} = 132 \text{ min} = 2.2 \text{ hr}$$

$$t_{c,post} = \frac{0.00526 \cdot 1500^{0.8} \cdot [(1000/84) - 9]^{0.7}}{0.01^{0.5}} = 109 \text{ min} = 1.8 \text{ hr}$$

From Table 5.8, the I_a/P ratios for the predeveloped and postdeveloped scenarios are 0.564 and 0.381, respectively. According to Figure 5.4, a Type-II rainfall distribution applies to Wisconsin. Thus, Figure 5.12 can be used to determine the unit peak discharge values (q_u). The q_u values for the predeveloped and postdeveloped scenarios are 108 cfs/(mi² · min) and 155 cfs/(mi² · min), respectively. Given the area in mi² (50 acres · 1 mi²/640 acres = 0.078 mi²), the peak flows can be calculated using the rational method equation as:

$$q_{p,pre} = 108 \frac{\text{cfs}}{\text{mi}^2 \cdot \text{in}} \cdot 0.078 \text{ mi}^2 \cdot 3.6 \text{ in} = 30 \text{ cfs}$$

$$q_{p,post} = 155 \frac{\text{cfs}}{\text{mi}^2 \cdot \text{min}} \cdot 0.078 \text{ mi}^2 \cdot 4.2 \text{ in} = 51 \text{ cfs}$$

5.1.6 Hydrographs

When a precipitation event (a storm) occurs over the watershed, it causes several processes within the basin. First is the initial moistening of the land surface and the vegetation, followed by the local filling of small surface indentations (depression storage) and the buildup of some depth of water on the land surface (initial detention storage) before the flow of water over the land begins. At the same time, infiltration begins. For the larger storms, some, possibly even most, of the precipitations enters a stream and flows out of the basin. The discharge past this outflow point is a time-variant process. A plot of the outflow versus time is called a hydrograph; Figure 5.14 is a definition sketch of a hydrograph.

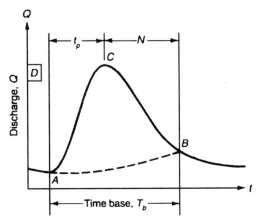

Figure 5.14 Hydrograph sketch

There are two components to any perennial stream flow: a relatively short-term component, which is the storm-induced surfaced water outflow, and a longer-term slowly varying component called base flow, which is the contribution from the groundwater to the flow. In Figure 5.14, the storm hydrograph is caused by an effective storm precipitation of D hours, causing first the increasing discharge on the rising limb of the hydrograph from A to the crest C, and then the recessional limb from C to B, when the storm-related discharge ceases. The base flow, below line AB in the figure, can be separated from the storm flow in any of several ways:

(a) The constant discharge method assumes that the base flow remains unchanged, and is demonstrated in Figure 5.15. This is the simplest method, but in reality the base flow increases as the storm event progresses due to the increase in groundwater recharging the stream.

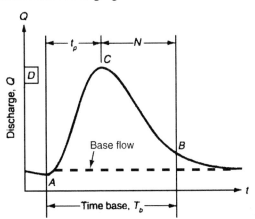

Figure 5.15 Base flow separation—constant discharge method

(b) The constant slope method assumes that the base flow increases at a constant rate. There are several methods to determine where exactly the base flow curve intersects the direct runoff hydrograph. The constant slope method assumes that the intersection occurs at the inflection of the receding limb of the hydrograph, denoted as point B in Figure 5.16.

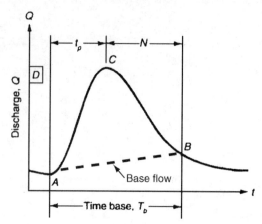

Figure 5.16 Base flow separation—constant slope method

(c) A third method is the concave method, which assumes that the base flow continues to decrease following the beginning of a storm until the direct runoff hydrograph peaks, at which point the base flow hydrograph increases to the inflection point on the receding limb. This is demonstrated in Figure 5.17.

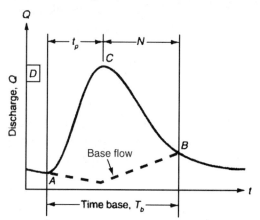

Figure 5.17 Base flow separation—concave method

Example 5.5

Comparison of base flow separation techniques

The following data is for a stream hydrograph after a storm event. Use this hydrograph data to compute the volume of direct runoff using

(a) the constant discharge method and

(b) the constant slope method.

Time (hr)	q (cfs)
0	5
1	20
2	35
3	25
4	14
5	9
6	7
7	6
8	5.5

Solution

The volume of runoff is obtained from the area under the direct runoff hydrograph and does not include base flow. The procedure to solve this problem is to separate out the base flow and then find the area under the resulting direct runoff hydrograph.

(a) For the constant discharge method, assume a constant base flow (q_b) of 5 cfs. The direct runoff hydrograph is found by subtracting this base flow from the provided hydrograph.

The volume of runoff is found using the trapezoidal method for finding the area under the curve as shown in Exhibit 6.

		Constant Discharge		
Time (hr)	**q (cfs)**	**q_b (cfs)**	**q_{direct} (cfs)**	**Volume (ft³)**
0	5	5	0	
1	20	5	15	27,000
2	35	5	30	81,000
3	25	5	20	90,000
4	14	5	9	52,200
5	9	5	4	23,400
6	7	5	2	10,800
7	6	5	1	5,400
8	5.5	5	0.5	2,700
			Sum:	292,500

Exhibit 6 Tabular results for constant discharge method

Thus, the total volume of runoff is calculated to be 292,500 ft³.

(b) For the constant slope method, the inflection point is assumed to occur at hour 4. Thus, the constant slope base flow starts at 0 hours with a flow of 5 cfs and increases to 14 cfs at hour 4 as shown in Exhibit 7. The slope of this line is 9 cfs/4 hr, which is used to plot the constant slope base flow separation line.

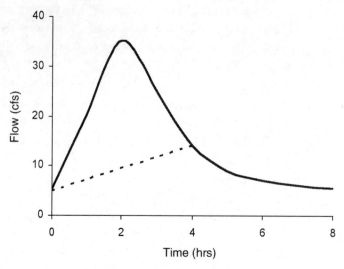

Exhibit 7 Constant slope method

The data for the base flow line as well as the direct runoff hydrograph ordinates and volume of runoff is shown in Exhibit 8.

		Constant Slope		
Time (hr)	q (cfs)	q_b (cfs)	q_{direct} (cfs)	Volume (ft^3)
0	5	5	0	
1	20	7.25	12.75	22,950
2	35	9.5	25.5	68,850
3	25	11.75	13.25	69,750
4	14	14	0	23,850
5	9	9	0	0
6	7	7	0	0
7	6	6	0	0
8	5.5	5.5	0	0
			Sum:	185,400

Exhibit 8 Tabular results for constant slope method

Thus, the total volume of runoff is calculated to be 185,400 ft^3.

5.1.7 Unit Hydrographs

A unit hydrograph (unitgraph) has a volume of 1 inch (1 cm) of direct runoff over the drainage basin as a result of a storm of D hours effective duration. Effective duration is the time interval when excess rainfall exists and direct runoff occurs. Any direct storm runoff has a volume

$$V = \int Q dt$$

(5.9)

which may also be written as $V = xA$, in which x is in inches (cm) and A is the basin area. Determination of the volume V, which is the area ABC in Figure 5.14, can be computed efficiently and accurately by use of the trapezoidal rule. Assume the time base T_b is divided into m intervals $\Delta t = T_b/m$ and the direct runoff ordinates Q_i, $i = 1$ to $m + 1$, are known with $Q_1 = Q_{m+1} = 0$. By the trapezoidal rule

$$V = \int Q dt = (Q_1 + Q_2)\frac{\Delta t}{2} + (Q_2 + Q_3)\frac{\Delta t}{2} + \cdots + (Q_m + Q_m + 1)\frac{\Delta t}{2}$$

or

$$V = \Delta t \sum_{i=2}^{m} Q_i \qquad\qquad (5.10)$$

Normally, x will not be 1 in (cm). The unit hydrograph is simply obtained by dividing each of the ordinates Q_i of the direct storm runoff plot by x. The unitgraph has a variety of applications.

The suitability of the unitgraph for these uses, however, depends on the appropriateness of several assumptions, including the following:

■ Rainfall excesses of one duration D will always produce hydrographs with the same time base, independent of the intensity of the excess.

■ The time distribution of the runoff does not change from storm to storm, so long as D is unchanged; thus, an increase in runoff volume by $P\%$ increases each hydrograph ordinate Q_i by $P\%$. Moreover, the distribution is not affected by prior precipitation.

The development of a unit hydrograph that produces reliable results in applications will be enhanced if you follow some experience rules:

■ Basin sizes should be between 1000 acres and 1000 square miles.

■ The direct storm runoff should preferably the within a factor of 2 of 1 inch, and the storm structure should be relatively simple.

■ The unitgraph should be derived from several storms of the same duration; in other words, compute several unit hydrographs and then average them.

If one does not have sufficient storm data to derive a unitgraph, then theoretical or empirical methods may be used to develop a "synthetic" unitgraph based on information such as peak flow values and basin characteristics. Numerous such methods have been proposed. Two of the more commonly used synthetic methods are Synder's method, originally developed for Appalachian watersheds, and the SCS method, developed by the Soil Conservation Service. They must be applied with care for best results; space does not permit an explanation here of these methods in the detail that is needed, so the reader may consult the chapter references for the complete methods.

Change of Unitgraph Duration

Each unit hydrograph is associated with an effective storm duration D. If one wants a unitgraph for some other effective storm duration without developing it directly from storm data, this can be done. (If the new storm duration differs from the existing one by no more than 25%, then normal practice is to use the existing one without alteration.) Two methods are used:

1. *Lagging.* This method can be used to construct a new unitgraph for a storm of effective duration nD, given the unitgraph for the storm having effective duration D, where n is an integer only. Simply add together n of the original

unit hydrographs, starting each successive unitgraph D hours after the beginning of the preceding one. This step produces a hydrograph associated with an effective duration of nD hours and with a runoff volume of n inches over the basin. Now divide all the hydrograph ordinates Q_i by n to obtain the new unitgraph. The method is easily set up in a table.

2. *S-curve*. This method is much more general and can be used to construct a unitgraph for either a shorter or a longer effective storm duration than the original. Say the desired new effective storm duration is D_{new}. First, one constructs the *S*-curve (which is a summation curve, or a sum of unitgraphs, and which also takes the general shape of an *S*) by successively lagging by D hours and summing (adding together) the ordinates of a total of T_b/D original unitgraphs. Next, draw a second *S*-curve, lagged D_{new} hours after the first *S*-curve. The differences in ordinates of these two S-curves, each multiplied by the ratio D/D_{new}, will be the ordinates of the new unitgraph for the storm of effective duration D_{new}.

Figure 5.18a–e demonstrates the use of the S-hydrograph method by creating a four-hour unit hydrograph from a two-hour unit hydrograph.

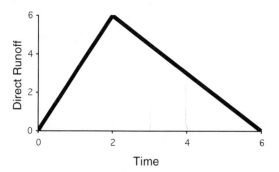

Figure 5.18a   The original unit hydrograph for a storm of duration equal to two time units

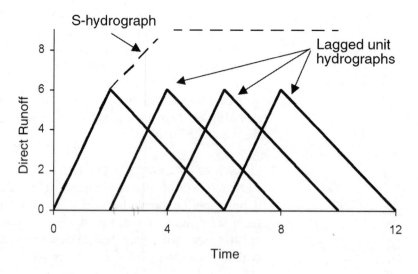

Figure 5.18b   An S-hydrograph is the sum of an infinite number of unit hydrographs

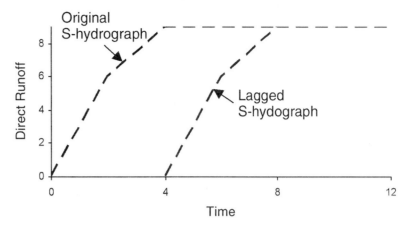

Figure 5.18c The original S-hydrograph is lagged by four hours (the desired duration of the final unit hydrograph)

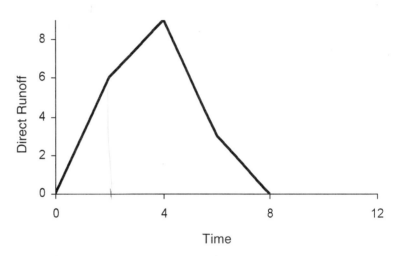

Figure 5.18d The ordinates of the lagged S-hydrograph are subtracted from the ordinates of the original S-hydrograph

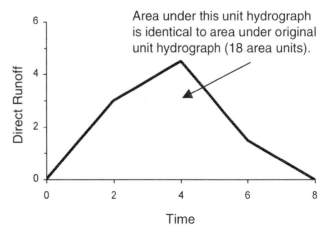

Figure 5.18e The final unit hydrograph is obtained by multiplying each of the ordinates by D/Dn_{ew} (2 in this case)

Example **5.6**

Unit hydrograph manipulation

Stream runoff from a 1500-acre watershed is plotted in Exhibit 9 for a storm having an effective duration D of two hours.

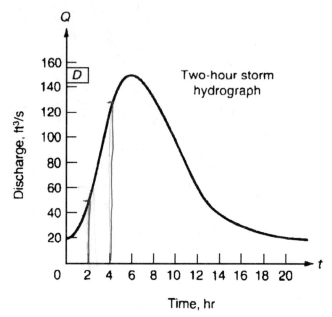

Exhibit 9   Two hour storm hydrograph

(a) Compute the ordinates of and plot the two-hour unit hydrograph.

(b) Use the information for the two-hour unit hydrograph to construct a three-hour unit hydrograph.

(c) Construct the composite storm hydrograph caused by 1.5 inches of excess precipitation falling in the first two hours, followed immediately by 0.7 inches of excess precipitation in the next two hours.

Solution

(a) The computations are presented in Exhibit 10. First the amount of the base flow must be identified and separated from the overall runoff. Since little information is available in this problem and also because the runoff duration is relatively short, it is assumed that the base flow is a constant 20 ft^3/s.

Time, hr	Stream Flow, ft^3/s	Storm Flow Q_i, ft^3/s	Unitgraph Ord. U_i, ft^3/s
(1)	(2)	(3)	(4)
0	20	0	0
2	60	40	58
4	113	93	135
6	150	130	188
8	127	107	155
10	96	76	110

Time, hr	Stream Flow, ft³/s	Storm Flow Q_i, ft³/s	Unitgraph Ord. U_i, ft³/s
12	65	45	65
14	43	23	33
16	27	7	10
18	20	0	0

Exhibit 10

The data in column 2 comes directly from the hydrograph in Exhibit 9. The storm flow Q_i, column 3, is the stream flow minus the base flow. Selecting a time interval $\Delta t = 2$ hr for use in Equation 5.10, the storm runoff volume is

$$V = \Delta t \sum_{i=2}^{m} Q_i = (2 \text{ hr})(521 \text{ ft}^3/\text{s}) = 1042 \frac{\text{ft}^3}{\text{s}} - \text{hr}$$

$$V = \left[1042 \frac{\text{ft}^3}{\text{s}} - \text{hr} \right]\left[60^2 \frac{\text{s}}{\text{hr}} \right] = 3.75 \times 106 \text{ ft}^3$$

This storm runoff volume is equivalent to a depth x of water over the basin of

$$x = \frac{V}{A} = \frac{(3.75 \times 106 \text{ ft}^3)(12 \text{ in/ft})}{(1500 \text{ acres})(43,560 \text{ ft}^2/\text{acre})} = 0.69 \text{ in}$$

The unitgraph ordinates $U_i = Q_i/x$ are tabulated in column 4, and the unit hydrograph is plotted in Exhibit 11.

(b) One constructs the S-curve by repeatedly lagging the 2-hr unitgraph, whose ordinates are listed in column 4 in Exhibit 10, and adding together all the values that are associated with each time instant. The individual ordinates S_i of the S-curve are

$$S_i = \sum_{n=1}^{i} Q_n$$

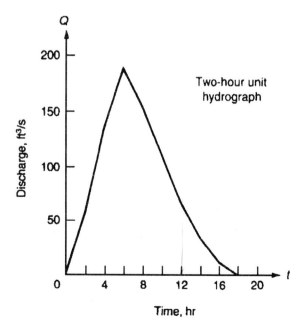

Exhibit 11

The 2-hr *S*-curve is plotted in Exhibit 12. Also shown is this same *S*-curve lagged three hours; the differences in ordinates of these two *S*-curves are then multiplied by the ratio $D/D_{new} = 2/3$ to scale the volume of the new hydrograph properly to end with the 3-hr unitgraph plotted in Exhibit 13.

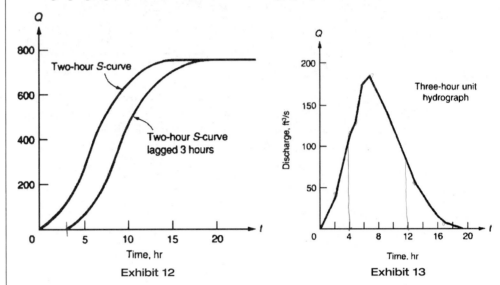

Exhibit 12   Exhibit 13

Scrutiny of this computational sequence would show that the peak discharge in the new unitgraph is slightly smaller than is the peak of the 2-hr unitgraph, as you would expect.

(c) Computations will be tabulated in Exhibit 14. Time will be measured from the start of the storm. The 2-hr unitgraph will be multiplied by 1.5 for the first portion of the runoff, followed by a second unitgraph scaled by 0.7. Finally, the base flow is added.

Time, hr	Unitgraph Ord., U_i, ft³/s	$1.5 \times U_i$, ft³/s	$0.7 \times U_i$, lag 2 hr, ft³/s	Sum, with *BF*, ft³/s
0	0	0	—	10
2	58	87	0	97
4	135	203	41	254
6	188	282	95	387
8	155	233	132	375
10	110	165	109	284
12	65	98	77	185
14	33	50	46	106
16	10	15	23	48
18	0	0	7	17
20			0	10

Exhibit 14

The composite storm hydrograph is plotted in Exhibit 15.

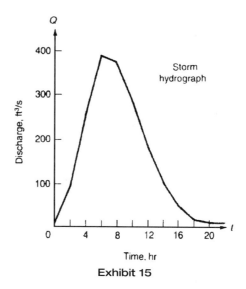

Exhibit 15

Unit hydrographs can also be generated using the NRCS procedure. Figure 5.19 shows the generic NRCS unit hydrograph. This figure shows four curves: the unit hydrograph, a triangular approximation of the unit hydrograph, "mass curves" of the unit hydrograph, and the triangular hydrograph. A mass curve is a cumulative distribution of the runoff *depth* of the hydrograph.

The abscissa on Figure 5.19 is dimensionless time, expressed as t/t_p. The ordinate is dimensionless flow rate (q/q_p) and dimensionless runoff depth (Q_o/Q).

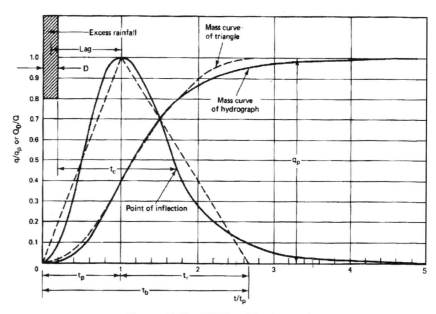

Figure 5.19 NRCS unit hydrograph

Source: Richard H. McCuen, *Hydrologic Analysis and Design*, 3d ed., © 2004. Reprinted by permission of Pearson Education, Inc., Upper Saddle River, N.J.

It is important to understand the nomenclature in Figure 5.19.

t_p = time to peak

   = $2/3 \cdot t_c$

t_r = duration of rainfall excess

t_b = time base of triangular unit hydrograph

t_c = time of concentration

t_l = lag time

D = rainfall duration

Q_o = cumulative runoff volume

q_p = peak flow

The value for the peak flow of the unit hydrograph can be estimated from Equation 5.11 using values obtained from Figure 5.19:

$$q_p = \frac{484 \cdot A \cdot Q}{t_p}$$ (5.11)

where

A = the area of the catchment (mi^2)

Q = depth of runoff (inch)

  = 1 in for a unit hydrograph

t_p = time to peak (hr)

The curves in Figure 5.19 can be expressed in tabular form, as shown in Table 5.9.

Table 5.9 Values for dimensionless unit hydrograph and mass curve

Time Ratios, t/T_p	Discharge Ratios, q/q_p	Mass Curve Ratios, Q_d/Q
0	0.000	0.000
0.1	0.030	0.001
0.2	0.100	0.006
0.3	0.190	0.012
0.4	0.310	0.035
0.5	0.470	0.065
0.6	0.660	0.107
0.7	0.820	0.163
0.8	0.930	0.228
0.9	0.990	0.300
1.0	1.000	0.375
1.1	0.990	0.450
1.2	0.930	0.522
1.3	0.860	0.589
1.4	0.780	0.650
1.5	0.680	0.700
1.6	0.560	0.751

(continued)

Time Ratios, t/T_p	Discharge Ratios, q/q_p	Mass Curve Ratios, Q_d/Q
1.7	0.460	0.790
1.8	0.390	0.822
1.9	0.330	0.849
2.0	0.280	0.871
2.2	0.207	0.908
2.4	0.147	0.934
2.6	0.107	0.953
2.8	0.077	0.967
3.0	0.055	0.977
3.2	0.040	0.984
3.4	0.029	0.989
3.6	0.021	0.993
3.8	0.015	0.995
4.0	0.011	0.997
4.5	0.005	0.999
5.0	0.000	1.000

Source: U.S. Department of Agriculture Soil Conservation Service (1972).

Example 5.7

Unit hydrograph development

This example will continue with Example 5.4 by developing a unit hydrograph (using 0.5-hour increments for the abscissa) for the postdevelopment catchment.

Solution

The time of concentration for the post-development catchment was determined to be 1.8 hours in the Example 5.4 solution, and thus the time to peak, t_p, is $2/3 \cdot 1.8$ hr, which equals 1.2 hours.

The value for the peak flow rate for this unit hydrograph is

$$q_p = \frac{484 \cdot A \cdot Q}{t_p} = \frac{484 \cdot 0.078\,\text{mi}^2 \cdot 1\,\text{in}}{1.2\,\text{hr}} = 31 \text{ cfs}$$

Given the values for t_p and q_p, it is relatively simple to create the unit hydrograph, as demonstrated in Exhibit 16. Note that columns 1 and 3 are obtained directly from Table 5.9, with the use of linear interpolation for the flow ratio at times equal to 2.5 hours and 3.5 hours. Columns 2 and 4 are obtained by multiplying the values in columns 1 and 3 by $t_p = 1.2$ hr and $q_p = 31$ cfs, respectively.

(1) Time Ratios (t/t_p)	(2) Time (hr)	(3) Flow Ratios (q/q_p)	(4) Flow (cfs)
0	0	0.000	0.0
0.5	0.6	0.470	14.6
1	1.2	1.000	31.0
1.5	1.8	0.680	21.1
2	2.4	0.280	8.7
2.5	3	0.127	3.9
3	3.6	0.055	1.7
3.5	4.2	0.025	0.0
4	4.8	0.011	0.8
4.5	5.4	0.005	0.2
5	6	0.000	0.0

Exhibit 16   Unit hydrograph data

The resulting unit hydrograph is shown in Exhibit 17.

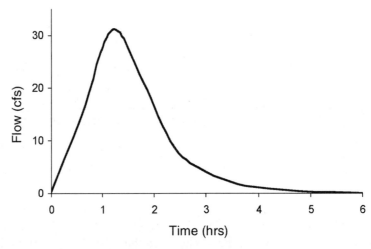

Exhibit 17   Final hydrograph

Example 5.8

Unit hydrogaph creation

Create a unit hydrograph based on the hydrograph in Exhibit 18 for direct runoff from a 50 hectare catchment.

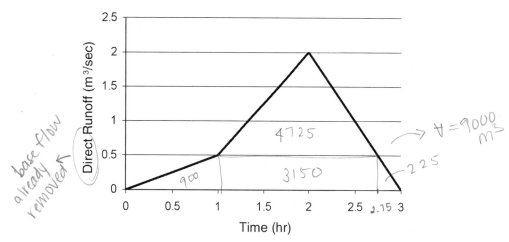

Exhibit 18   Direct runoff hydrograph

Solution

Since the ordinate is direct runoff, base flow has already been separated from the original runoff hydrograph. The steps involved in solving this problem include determining the volume of runoff; determining the depth of runoff; and dividing each ordinate of the original hydrograph by this depth of runoff.

The volume of runoff is the area under the curve, which can readily be estimated by dividing the hydrograph into triangular and rectangular sections as shown in Exhibit 19.

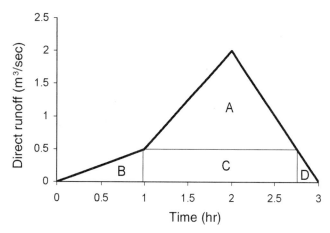

Exhibit 19   Runoff volume determination

The runoff volumes corresponding to each of these areas is calculated as follows:

$$V_A = 0.5 \cdot (1.75 \text{ hr}) \cdot \left(1.5 \frac{\text{m}^3}{\text{sec}} \right) \cdot \left(3600 \frac{\text{sec}}{\text{hr}} \right) = 4725 \text{ m}^3$$

$$V_B = 0.5 \cdot (1 \text{ hr}) \cdot \left(0.5 \frac{\text{m}^3}{\text{sec}} \right) \cdot \left(3600 \frac{\text{sec}}{\text{hr}} \right) = 900 \text{ m}^3$$

$$V_C = (1.75 \text{ hr}) \cdot \left(0.5 \frac{\text{m}^3}{\text{sec}}\right) \cdot \left(3600 \frac{\text{sec}}{\text{hr}}\right) = 3150 \text{ m}^3$$

$$V_D = 0.5 \cdot (0.25 \text{ hr}) \cdot \left(0.5 \frac{\text{m}^3}{\text{sec}}\right) \cdot \left(3600 \frac{\text{sec}}{\text{hr}}\right) = 225 \text{ m}^3$$

Thus, the total runoff volume is 9000 m³.

The depth of runoff, x, is the volume of runoff divided by the watershed area.

$$x = \frac{9000 \text{ m}^3}{50 \text{ ha}} \cdot \frac{1 \text{ ha}}{10{,}000 \text{ m}^2} = 0.018 \text{ m} = 1.8 \text{ cm}$$

Since a unit hydrograph in the SI system of units has a depth of runoff equal to 1 cm, each of the ordinates of the original hydrograph will be divided by 1.8. The results may be found in Exhibit 20.

Time (hr)	q_{original} (cfs)	q_{UH} (cfs)
0	0	0
1	0.5	0.28
2	2	1.1
3	0	0

Exhibit 20   Example 5.6 results

Hydrologic Routing

Routing methods track water masses as a function of the time that they course through streams, rivers, and reservoirs. Hydrologic routing is based on conservation of mass, supplemented by a relation between storage and discharge. It is an incomplete, approximate computation because it ignores momentum considerations, but it is often used because it can produce sufficiently accurate results with far less computational effort than is required in hydraulic routing (which does include the momentum equation). In this section, the hydrologic routing of flows through both reservoirs and rivers will be reviewed.

When the inflow hydrograph to either a reservoir or river reach is compared with the subsequent outflow hydrograph at the other end, two characteristic features are normally present: (1) the peak discharge of the inflow is attenuated, or reduced, in the outflow; and (2) the peak outflow occurs later, or lags, the peak inflow. The difference between inflow I and outflow Q at any instant is equal to the rate of change of the storage S of water in the region between the inflow and outflow stations, or

$$I - Q = \frac{dS}{dt} \tag{5.12}$$

Usually, this equation is integrated between two time instants t_n and t_{n+1}, and the trapezoidal rule is applied over the interval $\Delta t = t_{n+1} - t_n$ to obtain

$$(I_{n+1} + I_n)\frac{\Delta t}{2} - (Q_{n+1} + Q_n)\frac{\Delta t}{2} = S_{n+1} - S_n \tag{5.13}$$

The typical routing problem begins with an inflow hydrograph given (a set of values I_n, $i = 1$, N). The value of the initial outflow must also be known. The remaining two unknowns in the equation are Q_{n+1} and S_{n+1}. Once the relation between storage and outflow is specified, the new outflow can be computed, and the computation can progress to the next time increment. This storage relation differs, however, depending on the application.

Reservoir Routing

Reservoir outflow either is controlled by gages and/or valves or it is not controlled, owing to their absence. In uncontrolled reservoirs, the storage relation is either of the form $S = f(Q)$ when the reservoir water surface has no slope, as in short or deep reservoirs, or it is $S = f(Q, I)$ when the surface does slope, as in shallow reservoirs. For controlled reservoirs, the storage representation may again be of either type, with the added problem that a separate storage relation must be determined for each combination of gate/valve setting. When $S = f(Q, I)$, the routing method is similar to river routing.

The storage indication, or Puls, method of hydrologic routing is commonly applied to reservoirs. When storage is assumed to be a function only of outflow, the method uses the following steps:

■ Equation 5.13 is rearranged to give

$$I_{n+1} + I_n + \left(\frac{2}{\Delta t} S_n - Q_n \right) = \left(\frac{2}{\Delta t} S_{n+1} - Q_{n+1} \right)$$

(5.14)

■ From whatever data are given, a table or graph of $(2S/\Delta t + Q)$ versus Q is prepared; it is called a storage indication curve.

■ The storage indication curve and inflow data are used in applying Equation 5.14 sequentially over time increments until the outflow has been computed as a function of time.

The Puls method is applied in Example 5.9.

Example **5.9**

Reservoir routing

Some elevation-discharge and elevation-area data for a small reservoir with an ungated spillway are given below. An inflow sequence to the reservoir for part of a flood is given in a second table.

Elev., ft	0	1	2	3	4	5	6
Area, acres	1000	1020	1040	1050	1060	1080	1100
Outflow, ft³/s	0	525	1490	2730	4200	5880	7660

Data	Hour	Inflow, ft³/s
4/23	12 PM	1500
4/24	12 AM	1600
	12 PM	3100
4/25	12 AM	9600

Determine by routing the outflow discharge and reservoir water surface elevation at 12 AM on 25 April. Arrange the computations in a tabular form. Use a 12-hour routing period and assume that the reservoir water level just reaches the spillway crest (elevation 0.0) at 12 PM on 23 April.

Solution

Since the outflow Q is given directly as a function of elevation, the first task is to determine the reservoir storage S as a function of elevation also. The given areas are the surface areas of the reservoir water surface; integrating these areas over the incremental elevation changes produces the incremental changes in storage. This computation will be tabulated along with the compilation of data points for the storage indication curve in Exhibit 21. Elevation values will also be used as the index n in the equations. The equations used in computing the table entries follow Exhibit 21.

Elev., n, ft	Area A, acres	Avg. Area, $\overline{A}$ acres	S_{AF}, acre-ft	$\dfrac{2}{\Delta t}S+Q$, ft³/s
0	1000		0	0
		1010		
1	1020		1010	2560
		1030		
2	1040		2040	5600
		1045		
3	1050		3085	8950
		1055		
4	1060		4140	12,550
		1070		
5	1080		5210	16,400
		1090		
6	1100		6300	20,400

Exhibit 21

$$\overline{A}=\frac{1}{2}(A_n+A_{n+1}) \quad \Delta S=\overline{A}\Delta h \quad S_{AF}=\sum \Delta S$$

$$\frac{2}{\Delta t}S+Q=\frac{2S_{AF}(43,560)}{12(60^2)}+Q=2.02S_{AF}+Q,\ \text{ft}^3/\text{s}$$

The resulting storage indication curve is plotted in Exhibit 22.

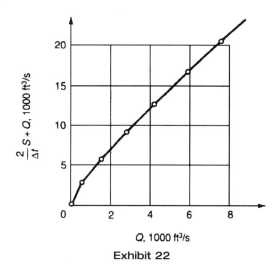

Exhibit 22

Now Equation 5.14 can be applied sequentially in Exhibit 23, with all flows in ft³/s.

n	Time	I	$\frac{2}{\Delta t}S+Q$	$\frac{2}{\Delta t}s+Q$	Q
(a)	**(b)**	**(c)**	**(d)**	**(e)**	**(f)**
1	4/23 12 PM	1500	0	—	0
2	4/24 12 AM	1600	1700	3100	700
3	4/24 12 PM	3100	2800	6400	1800
4	4/25 12 AM	9600	—	15,500	5500

Exhibit 23

All inflows were given data. Also, $Q_1 = 0$ was given. Thus, the value $(2S/\Delta t - Q)_1$ can then be computed to be zero. Now all terms on the left side of Equation 5.14 are known for $n = 1$, and this equation gives $2S/\Delta t + Q = 3100$ for $n + 1 = 2$ in column (e). Entering the storage indication curve, Exhibit 22, with this value gives $Q = 700$ ft³/s ($n = 2$, column (f)). Since $(2S/\Delta t + Q) - 2Q = 2S/\Delta t - Q$, column (d) with $n = 2$ is $3100 - 2(700) = 1700$. Applying Equation 5.14 with $n = 2$ then yields $1600 + 3100 + 1700 = 6400$ in column (e) for $n = 3$. And these operations are cyclically repeated until the solution is completed. Thus, the outflow from the reservoir at 12 AM, 25 April, is $Q = 550$ ft³/s. Using this discharge and the outflow-discharge data, the water surface elevation E at that time is, using interpolation,

$$E = 4.00\left(\frac{5,500-4,200}{5,880-4,200}\right) \times 1.00 = 4.77 \text{ ft}$$

above the spillway crest.

River Routing

All forms of hydrologic river routing begin with the assumption of some relation between storage in the river section and the inflow and outflow at the ends of the section. The most common of these methods is the Muskingum method, which assumes that this relation is a weighted linear relation between storage, inflow, and outflow taking the form

$$S = K[xI + (1 - x)Q] \tag{5.15}$$

in which K is a proportionality factor with units of time, and x is the weighting factor giving the relative importance of the inflow and outflow contributions to storage. For example, for a simple reservoir you expect $S = f(Q)$ only so $x = 0$ could be chosen; if inflow and outflow are of equal importance, then $x = 0.5$ should be selected. For most streams, x is between 0.2 and 0.3. The parameters K and x can be determined for a specific routing application if suitable data are available, so that $[xI + (1 - x)Q]$ can be plotted versus storage S for several values of x between 0 and 0.5. The value of x that most nearly collapses the plotted data onto a single fitted straight line is used in the routing application, and K is the slope of that fitted line. The final form of the Muskingum routing equation is

$$Q_{n+1} = C_o I_{n+1} + C_1 I_n + C_2 Q_n \tag{5.16}$$

in which

$$C_0 = (\Delta t/2 - Kx)/D \tag{5.17}$$

$$C_1 = (\Delta t/2 + Kx)/D \tag{5.18}$$

$$C_2 = (K - Kx - \Delta t/2)/D \tag{5.19}$$

and

$$D = K - Kx - \Delta t/2 \tag{5.20}$$

*[handwritten: Maybe $D = K - Kx + \frac{\Delta t}{2}$ * example is unclear!]*

Observe that you must always have $C_0 + C_1 + C_2 = 1$. These equations can be derived by using Equation 5.15 to express S_n and S_{n+1}, inserting the results in Equation 5.14 and rearranging the terms.

Example 5.10

Thirty-six hours of data for stream flow are given in the following table:

Time	6 AM	12 AM	6 PM	12 PM	6 AM	12 AM	6 PM
I, ft³/s	10	30	70	50	40	32	25

The Muskingum parameters have been determined to be $K = 10$ hr, $\Delta t = 6$ hr, and $x = 0.23$. The flow is steady in the reach at 6 AM on the first day. Determine the outflow hydrograph from this stream reach.

Solution

Direct computation using first Equation 5.20 and then Equations 5.17–5.19 will lead to

$$D = 10.70, \quad C_0 = 0.065, \quad C_1 = 0.495, \text{ and } C_2 = 0.440$$

Use of a table is an aid in organizing the computations:

Time	I_n, ft³/s	$C_0 I_{n+1}$	$C_1 I_n$	$C_2 Q_n$	Q_{n+1}, ft³/s
6 AM	10				10
12 AM	30	2.0	5.0	4.4	11.4
6 PM	70	4.6	14.9	5.0	24.5
12 PM	50	3.3	34.7	10.8	48.8
6 AM	40	2.6	24.8	21.5	48.9
12 AM	32	2.1	19.8	21.5	43.4
6 PM	25	1.6	15.8	19.1	36.5

The inflow data are reproduced in the first two columns. The next three columns contain the terms that appear on the right side of Equation 5.16; the last column is the sum of the three previous column entries, as Equation 5.16 indicates, and is the outflow hydrograph. According to these computations, the peak outflow is 48.9 ft³/s and occurs at 6 AM on the second day.

5.2 LIMNOLOGY

Limnology is the study of freshwater bodies such as lakes and ponds. Water quality in lakes is characterized by turbidity, secchi disk depth,[2] nutrient concentration (for example, phosphorus and nitrogen), temperature, dissolved oxygen concentration, and so on.

These water quality characteristics vary with the age of a lake. As a lake ages, it transitions from being an *oligotrophic* (literally "few foods") lake, to a *mesotrophic* lake to a *eutrophic* lake ("well fed"). Oligotrophic lakes are generally clear, are low in nutrients, and do not support large fish populations. Eutrophic lakes arc high in nutrients and support a large amount of biomass. Eutrophic lakes can support large fish populations but are also susceptible to oxygen depletion. Mesotrophic lakes have characteristics (such as nutrient concentration, fish population, and water clarity) that lie somewhere between those of oligotrophic and eutrophic lakes.

As a lake transitions from an oligotrophic lake to a eutrophic lake, it is said to become more *productive,* in the sense that the eutrophic lake produces more biomass. Eutrophic lakes are considered undesirable from many viewpoints, as the resulting algal growth can severely limit recreational use, decrease aesthetics, cause taste and odor problems if the lake is a source of drinking water, and decrease the dissolved oxygen concentration in the lakes. Algae are plants (and therefore photosynthetic), and thus the decrease in dissolved oxygen typically occurs in the night. During the night no oxygen is produced, but a large oxygen demand is exerted due to the decay of dead algae.

The production of a lake is directly proportional to the concentration of the *limiting nutrient.* A limiting nutrient is the nutrient with the lowest concentration

---

2   A secchi disk is a black-and-white patterned disk (usually 12 inches in diameter) that is lowered into the water until the observer can no longer see the disk. The depth at which the disk cannot be seen is directly related to the water clarity.

relative to the concentration needed for plant growth. Thus, addition of the limiting nutrient will stimulate additional growth. In nearly all freshwater lakes, phosphorus is the limiting nutrient for algal growth. In many salt water systems, nitrogen is the limiting nutrient.

Naturally, the aging of a lake can take thousands of years, but human impacts (for example, runoff containing excess phosphorus) can rapidly speed up this process. This more rapid process is termed *cultural eutrophication.*

The water quality of lakes as a function of their trophic state is shown in Table 5.10.

Table 5.10   Lake classification based on productivity

Lake Classification		Chlorophyll *a* Concentration ($\mu g \cdot L^{-1}$)	Secchi Depth (m)	Total Phosphorus Concentration ($\mu g \cdot L^{-1}$)
Oligotrophic	Average	1.7	9.9	8
	Range			
		0.3–4.5	5.4–28.3	3.0–17.7
Mesotrophic	Average	4.7	4.2	26.7
	Range			
		3–11	1.5–8.1	10.9–95.6
Eutrophic	Average	14.3	2.5	84.4
	Range			
		3–78	0.8–7.0	15–386
Hypereutrophic		> 50	< 0.5	Often > 100

Note: Classification for oligotrophic, mesotrophic, and eutrophic lakes from R. G. Wetzel, *Limnology* (W. B. Saunders, 1983), 767. Classification for hypereutrophic lakes from N. R. Kevern, D. L. King, R. Ring, "Lake classification systems—Part 1," *The Michigan Riparion*, February 1996, last updated December 1999. *www.mlswa.org/lkclassill.htm.*

Source: M. L. Davis and S. J. Masten, *Principles of Environmental Engineering and Science*, © 2004, McGraw-Hill Education. Reprinted by permission of the McGraw-Hill Companies.

| Example **5.11** |

Algae can be represented as $C_{106}H_{263}O_{110}N_{16}P$. Given this chemical formula, if the concentration of total nitrogen in a lake is 10 ppb and the concentration of total phosphorus is 0.7 ppb, which of the two is the limiting nutrient?

Solution

The stoichiometric ratio of N/P in bacteria is 16:1. Given the atomic weights of N and P (14 and 31, respectively), the *mass ratio* of N:P in the algae is:

$$\left(\frac{N}{P}\right)_{algae} = \frac{16 \text{ moles N} \cdot \dfrac{14g}{1 \text{ mole N}}}{1 \text{ mole P} \cdot \dfrac{31g}{1 \text{ mole P}}} = 7.2$$

The mass ratio of N:P in the water is:

$$\left(\frac{N}{P}\right)_{water} = \frac{\dfrac{10\mu g}{1\,L}}{\dfrac{0.07\mu g}{1\,L}} = 14.3$$

Thus, this body of water is phosphorus limited. Adding more nitrogen to the water would not result in greater growth of algae, since there is already more nitrogen in the water than they need to grow. A rule of thumb is that when the mass concentration ratio of nitrogen to phosphorus in the water is greater than 10, the water will be phosphorus limited.

Stratification

Lakes in temperate regions undergo a phenomenon known as *thermal stratification.* Water is most dense at a temperature of 4°C (39°F), and this temperature-density relationship drives the transition of lakes from being stratified to being nonstratified. As a result of thermal stratification, the lake is split into layers:

- An upper layer of warm, lighter water called the *epilimnion*

- A middle, transition zone that prevents mixing, called the *metalimnion*

- A bottom layer of cool, heavier water called the *hypolimnion*

The transition between stratification and complete mixing occurs during a process known as *turnover.* In temperate regions, turnover occurs in the spring and fall. In spring, turnover occurs as the surface water warms to 4°C (and density increases). This water then sinks, bringing colder, deeper water to the surface. As summer progresses, the surface water warms, the deeper waters remain cold, and a *thermocline* sets up between the epilimnion and hypolimnion. This thermocline resists any mixing between the upper and lower layers. Fall turnover occurs as the surface water cools, becomes increasingly dense, and sinks to the bottom of the lake, thus promoting mixing of the lake.

Stratification traps nutrients released from bottom sediments in the hypolimnion. Also, under some circumstances, the hypolimnion can become anoxic as the oxygen is depleted and is not replenished due to the presence of the thermocline. The steep temperature gradient of the metalimnion prevents any surface water with dissolved atmospheric oxygen from reaching the bottom waters.

Water quality in lakes can be modeled using the mass balance approach described in sections 1.4 and 1.5. Many lakes can be modeled as CSTRs, especially shallow lakes in which the wind can effectively mix the water. In the case of a stratified lake, the epilimnion may be treated as a CSTR.

Example **5.12**

Mass balance application to a lake

A regulatory agency needs to set a limit for the phosphorus concentration in a wastewater treatment plant's effluent. The goal is to ensure that the lake's phosphorus concentration remains below 10 µg/L. The lake has an area of 5 ha and a mean depth of 8 m. The wastewater treatment plant discharges wastewater into the lake at a rate of 0.5 m³/sec. The lake is also fed by a stream having a phosphorus concentration of 3 µg/L and a flow of 0.04 m³/sec. Homes adjacent to the lake add an additional 1 kg/day of phosphorus. Phosphorus settles out in the lake according to a rate constant of 0.35 day⁻¹. Determine an effluent phosphorus limit for the wastewater treatment plant assuming steady state conditions apply.

Solution

Exhibit 24 provides a diagram of the lake.

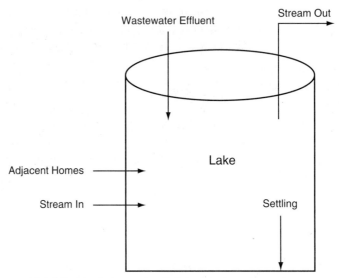

Exhibit 24   Schematic of lake with phosphorus inputs

The solution begins with the generic mass balance equation for steady state conditions:

$$0 = \dot{m}_{in} - \dot{m}_{out} \pm \dot{m}_{rxn}$$

This can be rewritten for the specific case at hand as:

$$0 = (Q_{stream} \cdot C_{stream} + L + Q_{wwtp} \cdot C_{wwtp}) - (Q_{out} \cdot C_{out}) - (V \cdot k \cdot C_{out})$$

Solving for C_{wwtp} yields:

$$C_{wwtp} = \frac{-(Q_{stream} \cdot C_{stream} + L - Q_{out} \cdot C_{out} - V \cdot k_1 \cdot C_{out})}{Q_{wwtp}}$$

where

Q_{stream} = the flow rate of the stream entering the pond (0.04 m³/sec)

C_{stream} = the concentration of phosphorus in the stream entering the pond (3 µg/L)

L = the mass loading of phosphorus to the pond from the adjacent homes (1 kg/day)

Q_{wwtp} = the flow rate of the wastewater treatment plant effluent (0.5 m³/sec)

C_{wwtp} = the concentration of phosphorus in the wastewater treatment plant effluent (unknown)

Q_{out} = the flow rate of water exiting the pond

= $Q_{stream} + Q_{wwtp}$ (0.54 m³/sec)

C_{out} = concentration of phosphorus at any point in the lake, including the stream draining the lake (10 µg/L)

V = the volume of the lake

= (area of lake) · (depth of lake)

= (5 ha) · (10,000 m²/ha) · (8 m) = $4 \cdot 10^5$ m³

k_1 = first-order rate constant for settling (0.35 day⁻¹)

Substituting these values into the governing equation (and being careful of units) yields an allowable discharge from the wastewater treatment plant of 19.8 µg/L.

5.3 WETLANDS

The U.S. Army Corps of Engineers defines a wetland as "those areas that are inundated or saturated by surface or ground water at a frequency and duration sufficient to support, and that under normal circumstances do support, a prevalence of vegetation typically adapted for life in saturated soil conditions. Wetlands generally include swamps, marshes, bogs, and similar areas."

Alternatively, the U.S. EPA Web site states that a wetland is "an area that is regularly saturated by surface water or groundwater and is characterized by a prevalence of vegetation that is adapted for life in saturated soil conditions." These definitions highlight the three traits by which a wetland is characterized: hydrologic characteristics; soil characteristics; and types of vegetation.

1. **Hydrology**. The wetland soil must be saturated at some time during the growing season, but the depth of water must be less than two meters. The seasonal pattern of water levels in a wetland is known as the *hydroperiod*. Examples of hydroperiods for different types of wetlands are shown in Figure 5.20.

2. **Soils**. Wetland soils are classified as *hydric*. A hydric soil is a "soil that formed under conditions of saturation, flooding, or ponding long enough during the growing season to develop anaerobic conditions in the upper part. The concept of hydric soils includes soils developed under sufficiently wet conditions to support the growth and regeneration of hydrophytic vegetation. Soils that are sufficiently wet because of artificial measures are included in the concept of hydric soils. Also, soils in which the hydrology has been artificially modified are hydric if the soil, in an unaltered state, was hydric. Some soil series, designated as hydric, have phases that are not hydric, depending on water table, flooding, and ponding characteristics." [3] Hydric soils are identified based on their color, permeability, texture, and smell.

3. **Vegetation**. The types of vegetation are characteristics of the vegetation adapted to the soils found in wetlands.

---

3   *http://soils.usda.gov/use/hydric/intro.html*, accessed on 6/30/06.

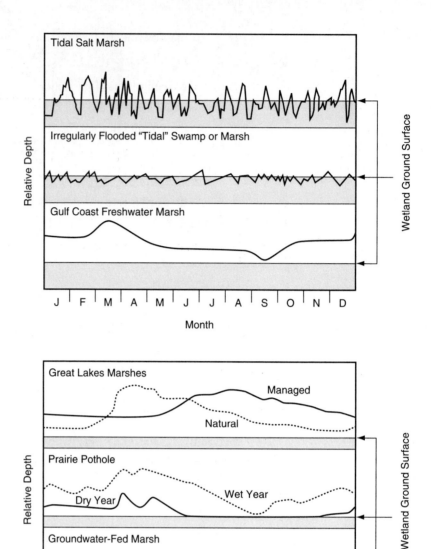

Figure 5.20 Hydroperiods for several different types of wetlands

Source: William J. Mitsch and James G. Gosselink, *Wetlands*, © 2000, John Wiley & Sons, Inc. Reprinted by permission.

Section 404 of the Clean Water Act regulates the disposal of fill material into waterways, including wetlands. The U.S. Army Corps of Engineers administers the program and is in charge of enforcement. The U.S. EPA develops and interprets policies and criteria used in reviewing permit applications. In general, agricultural and forestry practices are exempt from wetland regulations.

Wetlands can be categorized in several ways, but one common classification scheme is to divide wetlands into four categories: marshes, swamps, bogs, and fens.

1. *Marshes* are frequently or continually inundated with water. They are characterized by soft-stemmed vegetation including reeds, sedges, and cattails.

2. *Swamps* are any wetland dominated by woody plants.

3. *Bogs* are characterized by spongy peat deposits, acidic waters, and a floor covered by a thick carpet of sphagnum moss.

4. *Fens* are peat-forming wetlands that receive nutrients from sources other than precipitation or runoff. Sources of recharge include seeps, springs, and groundwater. Fens are less acidic than bogs.

Wetlands serve many purposes for society. Benefits of wetlands include:

■ Preventing or reducing the risk of floods

■ Improving water quality through nutrient and suspended solids removal

■ Recharging groundwater

■ Providing habitat for diverse plant and animal communities

■ Providing recreation opportunities

Wetlands have also been constructed for the use of treating wastewater. A schematic of such a constructed wetland is provided in Figure 5.21.

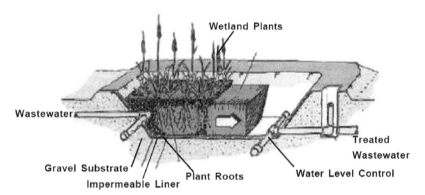

Figure 5.21 Constructed wetland for wastewater treatment

Source: *Constructed Treatment Wetlands*, EPA 843-F-03-013, accessed from *www.epa.gov/owow/wetlands/pdf/ConstructedW.pdf.*

Chapter 404 of the Clean Water Act also addresses the use of *compensatory mitigation.* Compensatory mitigation is defined by the U.S. Army Corps of Engineers as "the restoration, creation, enhancement, or in exceptional circumstances, preservation of wetlands and/or other aquatic resources for the purpose of compensating for unavoidable adverse impacts which remain after all appropriate and practicable avoidance and minimization has been achieved." In practice, this means that if a developer adversely impacts an existing wetland in the process of developing a portion of land, the developer must replace the affected wetland with another wetland. Of course, this infers that the regulatory agency allows the adverse effect to the wetland in the first place.

The Army Corps of Engineers and the U.S. EPA have established a three-part process, known as *mitigation sequencing.* The process prioritizes the response to the request for mitigation from a developer.

1. Avoid adverse impacts if a reasonable alternative exists

2. Minimize adverse impacts

3. Provide compensatory mitigation if steps 1 and 2 cannot be realized

The Army Corps of Engineers (or approved state authority) is responsible for determining the appropriate amount of mitigation required.

The most common mitigation techniques include the following:

- *Establishment* is the development of a wetland where a wetland did not previously exist.

- *Restoration* is the reestablishment or rehabilitation of an existing wetland with the goal of increasing wetland function or the number of wetland acres.

- *Enhancement* is the improvement of an existing wetland's function (for example, improving water quality, stormwater retention, or habitat quality)

Compensatory mitigation banking allows a third party to establish, restore, or enhance a wetland. A certain number of credits are assigned to this bank, and the third party can then sell these credits to developers needing to provide compensatory mitigation due to adverse effects of their development on an existing wetland.

5.4 GROUNDWATER

Groundwater refers to water that is stored beneath the Earth's surface, as contrasted to *surface water,* which is stored above the Earth's surface in ponds, lakes, rivers, and oceans. Groundwater is stored in geologic formations called *aquifers.* An aquifer is defined as a geologic formation that is able to hold large amounts of water *and* is able to conduct that water.

Aquifers may be either confined or unconfined. A *confined aquifer* is one that is bounded on top by a confining layer, sometimes called an *aquitard.* Thus, a confined aquifer is under pressure; the pressure at the top of the confined aquifer is due to the head of water above the top of the aquifer. An unconfined aquifer is one in which the upper level of the aquifer is at atmospheric pressure. The surface of an unconfined aquifer is called the *water table.* Figure 5.22 demonstrates these two types of aquifers.

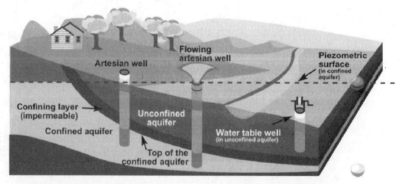

Figure 5.22   Confined and unconfined aquifers

Source: Environment Canada's Freshwater Web site (*www.ec.gc.ca/water*). Reproduced with the permission of the Minister of Public Works and Government Services, 2007.

Figure 5.23 also shows the *piezometric surface.* The piezometric surface is the imaginary level to which the water in an aquifer would rise. For an unconfined aquifer, the piezometric surface is the water surface. For a confined aquifer, the piezometric surface is some distance above the bottom of the confining layer, and this distance is equal to the pressure head in the confined aquifer.

The pressure head in a confined aquifer can be large enough such that the piezometric surface is above the ground surface. If a pathway exists from the confined aquifer to the ground surface, the groundwater will flow from the aquifer to above the ground surface. Such a phenomenon is known as an *artesian well* (Figure 5.23).

Of increasing importance to environmental engineers is the *recharge area*, also shown in Figure 5.23. This is the area in which water seeps into the ground to supply the confined aquifer. The recharge area is significant in that it provides an opportunity for potential contamination of the aquifer. Also, increasing the imperviousness of the recharge area via land development can negatively impact the ability of the aquifer to be recharged.

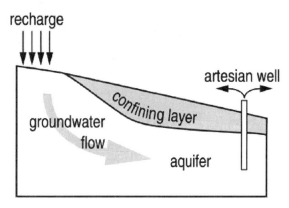

Figure 5.23 Artesian well

Source: "Water Quantity Issues at Chickasaw National Recreation Area," *Park Science* 19, no. 2, December 1999. Reprinted by permission of the National Park Service.

As previously defined, aquifers are characterized by their ability to transport water and to store water. The ability to store and transport water is characterized by the following properties:

- *Porosity, n*, is defined as $n = \dfrac{V_v}{V_t}$

  where V_v = the volume of voids

  V_t = the total volume

- *Void ratio, e* is defined as $e = \dfrac{V_v}{V_s} = \dfrac{n}{1-n}$

  where V_s = the volume of the solids (that is, soil particles)

- *Specific yield* is the percentage of an aquifer's water that will drain due to gravity.

- The *storage coefficient* (or *storativity*) is the volume of water that an aquifer gains or loses in response to a unit change in head. The values of storativity range from 10^{-5} to 10^{-3} for confined aquifers, and between 10^{-2} and 0.35 for unconfined aquifers. Storativity is illustrated for confined and unconfined aquifers in Figure 5.24.

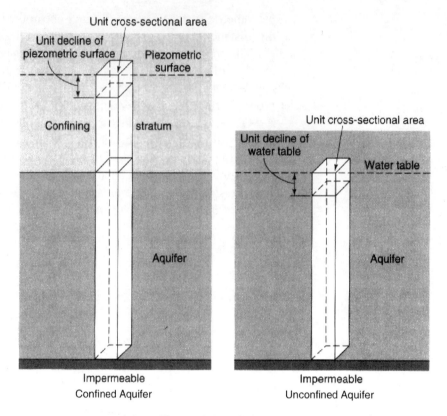

Figure 5.24   Storativity

Source: Todd, *Groundwater Hydrology*, 2nd ed., © 1980, John Wiley & Sons, Inc.
Reprinted by permission.

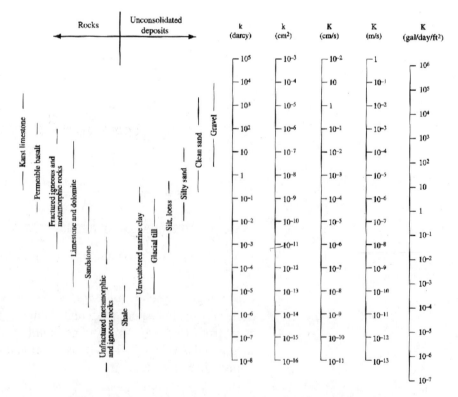

Figure 5.25   Values of hydraulic conductivity and permeability

Source: Freeze and Cherry, *Groundwater*, © 1979. Reprinted by permission of Pearson
Education, Inc., Upper Saddle River, N.J.

■ The *hydraulic gradient* is the slope of the piezometric surface. The rate at which groundwater travels is directly related to this gradient.

■ *Hydraulic conductivity* is the ability of a geologic formation to transport water. It is the rate of water flow through a cross section of an aquifer in response to a unit hydraulic gradient. Values for hydraulic conductivity K are provided in Figure 5.25.

■ *Permeability* is a function of the geologic formation and not a function of the fluid properties. Permeability is related to the square of the grain size diameter. Values for permeability k are provided in Figure 5.25.

Groundwater Flow

The simplest equation for groundwater flow is the one-dimensional Darcy's law. Darcy's law is:

$$Q = -K \cdot A \frac{dh}{dx} \qquad (5.21)$$

where

K = hydraulic conductivity $[L/T]$

A = cross-sectional area of aquifer $[L^2]$

dh/dx = hydraulic gradient $[L/L]$

The negative sign in Darcy's law arises from the fact that the slope of the piezometric surface is negative in the direction of flow.

The variables found in Darcy's law are illustrated in Figure 5.26. In this diagram, the hydraulic gradient could be written as $\frac{dh}{dx} = \frac{h_1 - h_2}{L}$. The cross-sectional area is the product of b, the thickness of the aquifer, and the distance of the aquifer into the paper. Figure 5.26 shows Darcy's law for a confined aquifer, but the analysis is the same for an unconfined aquifer. For the unconfined aquifer, the hydraulic gradient is the slope of the actual water surface.

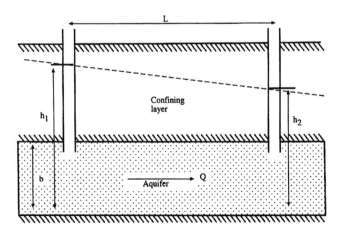

Figure 5.26 Darcy's law

Source: ENV 302, Environmental Hydrogeology, Northern Arizona University.

The *Darcy velocity* is defined as:

$$v_{darcy} = \frac{Q}{A} = -K \cdot \frac{dh}{dx} \qquad (5.22)$$

where Q is the flow rate calculated by the Darcy equation. The Darcy velocity is much lower than the true velocity, as the cross-sectional area through which the groundwater flows is much smaller than A; that is, the groundwater is flowing through the pores between the soil particles. Of more use to environmental engineers is the *pore velocity,* as this better characterizes the flow of water within the pores.

$$v_{pore} = \frac{v_{darcy}}{n} \qquad (5.23)$$

Example 5.13

Two observation wells are drilled 2000 feet apart as shown in Exhibit 25.

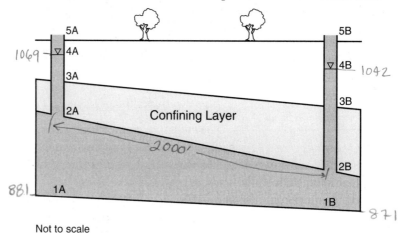

Not to scale

Exhibit 25

Soil boring data were obtained when drilling the wells, and some elevations are provided in Exhibit 26. The soil has a porosity of 22% and a hydraulic conductivity of $2 \cdot 10^{-3}$ m/s. Find the pore velocity. $n = 0.22$

$K = 2 \times 10^{-3} \; \frac{ft}{s}$

Location	Elevation
1A	881
1B	871
2A	970
2B	846
3A	1031
3B	1010
4A	1069
4B	1042
5A	1123
5B	1135

Exhibit 26 Soil boring data

Solution

A large amount of extraneous information is provided from the soil boring results. The only data relevant to this example problem are the elevations of the piezometric surface at the upstream and downstream wells (1069 ft and 1042 ft, respectively). Given this information, the Darcy velocity can be calculated as:

$$v_{darcy} = -K \cdot \frac{dh}{dx} = -2 \cdot 10^{-3} \, \frac{\text{ft}}{\text{sec}} \cdot \frac{-(1069 \, \text{ft} - 1042 \, \text{ft})}{2000 \, \text{ft}} = 2.7 \cdot 10^{-5} \, \text{ft/s}$$

The pore velocity is

$$v_{pore} = \frac{v_{darcy}}{n} = \frac{2.7 \cdot 10^{-5} \, \text{ft/s}}{0.22} = 1.23 \cdot 10^{-4} \, \text{ft/s}$$

Thus, the groundwater in this instance travels approximately 10.6 feet each day.

Well Hydraulics

The basic equation describing local, steady groundwater movement is Darcy's law, which can be written

$$V = -K_i = -Ki = -K \, \frac{dH}{dL} \qquad \textbf{(5.24)}$$

In this equation, V is the average velocity of a discharge Q that occurs through a soil cross-sectional area A. Darcy's law indicates that V is the product of the local hydraulic conductivity K, which depends on the local soil or rock properties, and the local gradient i of the piezometric head $H = p/\gamma + z$, or $i = dH/dL$. This may also be interpreted as a difference in fluid energy between points, because the kinetic energy associated with groundwater flow is negligible. To obtain the actual fluid velocity, called the seepage velocity, in the subsurface saturated zone, divide the average velocity by the local porosity. A variety of units are used in describing groundwater parameters, so you should take care to use consistent units in all computations.

Steady Flow

Equations for steady flow from a well in either an unconfined or a confined aquifer can be derived from Darcy's law. These simple equations only have meaning and are accurate when several simplifying assumptions are valid, including the following: (1) the aquifer, which is a geologic formation that contains enough saturated permeable material to yield significant quantities of water, must be large in extent and have uniform hydraulic properties (for example, K is constant); (2) the pumping must occur at a constant rate for an extremely long time so that startup transients no longer exist; (3) the well fully penetrates the aquifer; (4) the well depth is much larger than the drawdown near the well; and (5) the estimate of the gradient i is a good one.

An aquifer is called unconfined if the upper edge of the saturated zone (ignoring capillary effects) is at atmospheric pressure; this edge is called the water table. Figure 5.27 is a schematic cross section of a well in an unconfined, horizontal aquifer. A cylindrical coordinate system (x, y) is placed at the base of the well; the drawdown of the undisturbed water table is s; the gradient of the piezometric head is $i = dy/dx$ at the water table and is assumed to apply to the entire water column below it. The radius of the well is $x = r_w$. Applying Darcy's law gives

$$V = \frac{Q}{Q} = \frac{Q}{2\pi xy} = -K\frac{dy}{dx} \tag{5.25}$$

Rearranging this expression and integrating it between points (r_1, h_1) and (r_2, h_2) along the water table yields

Unconfined
Aquifer

$$Q = \frac{\pi K\left[h_2^{\,2} - h_1^{\,2}\right]}{\ln(r_2 / r_1)} \tag{5.26}$$

as the expression for the steady pumping rate, or discharge, for this case.

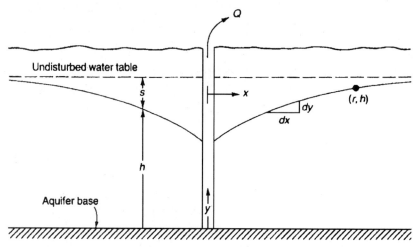

Figure 5.27   Cross section of a well in a ~~confined~~ aquifer
                                        unconfined

The case for steady pumping from a confined, horizontal aquifer of thickness m is similar to the first case, as shown in Figure 5.28. However, the gradient i is determined from the local slope of the piezometric head curve (shown dashed), which is no longer the same as the edge of the saturated zone. Equation 5.25 still applies to this case if the area through which flow occurs is corrected to $A = 2\pi xb$. Now, the integration between points (r_1, h_1) and (r_2, h_2) on the piezometric surface results in

Confined
Aquifer

$$Q = \frac{2\pi Kb(h_2 - h_1)}{\ln(r_2 / r_1)} \tag{5.27}$$

for the discharge. Sometimes, the transmissivity $T = Kb$ is introduced into this equation.

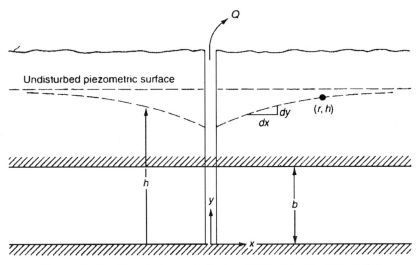

Figure 5.28 Steady pumping from a confined, horizontal aquifer

Unsteady Flow

The first significant solution for unsteady flow to a well was originally developed by Theis for a confined aquifer. It expresses the drawdown s as

$$s = \frac{Q}{4\pi T} W(u) \qquad T = Kb \qquad (5.28)$$

in which

$$u = \frac{r^2 S}{4Tt} \qquad S = storage \\ constant/ht \qquad (5.29) \\ specific\ yield$$

and

$$W(u) = \int_u^\infty \frac{e^{-u} du}{u} = -0.5772 - \ln(u) + u - \frac{u^2}{2 \times 2!} + \cdots \qquad (5.30)$$

is called the well function of u. Table 5.11 presents tabulated values for this function. The discharge Q is constant over the pumping period, and r is the radius at which s is computed (to find the drawdown at the well, use $r = r_w$) at a time t after pumping began. The solution depends on a knowledge of two aquifer properties: the transmissivity T and the storage constant S. The storage constant is the amount of water removed from a unit volume of the aquifer when the piezometric head is lowered one unit. Two methods for the determination of these aquifer properties will be described next.

Table 5.11 Values of the function $W(u)$ for various values of u

u	$W(u)$	u	$W(u)$	u	$W(u)$	u	$W(u)$
1×10^{-10}	22.45	7×10^{-8}	15.90	4×10^{-5}	9.55	1×10^{-2}	4.04
2	21.76	8	15.76	5	9.33	2	3.35
3	21.35	9	15.65	6	9.14	3	2.96
4	21.06	1×10^{-7}	15.54	7	8.99	4	2.68
5	20.84	2	14.85	8	8.86	5	2.47
6	20.66	3	14.44	9	8.74	6	2.30

(continued)

u	$W(u)$	u	$W(u)$	u	$W(u)$	u	$W(u)$
7	20.50	4	14.15	1×10^{-4}	8.63	7	2.15
8	20.37	5	13.93	2	7.94	8	2.03
9	20.25	6	13.75	3	7.53	9	1.92
1×10^{-9}	20.15	7	13.60	4	7.25	1×10^{-1}	1.823
2	19.45	8	13.46	5	7.02	2	1.223
3	19.05	9	13.34	6	6.84	3	0.906
4	18.76	1×10^{-6}	13.24	7	6.69	4	0.702
5	18.54	2	12.55	8	6.55	5	0.560
6	18.35	3	12.14	9	6.44	6	0.454
7	18.20	4	11.85	1×10^{-3}	6.33	7	0.374
8	18.07	5	11.63	2	5.64	8	0.311
9	17.95	6	11.45	3	5.23	9	0.260
1×10^{-8}	17.84	7	11.29	4	4.95	1×10^{0}	0.219
2	17.15	8	11.16	5	4.73	2	0.049
3	16.74	9	11.04	6	4.54	3	0.013
4	16.46	1×10^{-5}	10.94	7	4.39	4	0.004
5	16.23	2	10.24	8	4.26	5	0.001
6	16.05	3	9.84	9	4.14		

Source: Bedient and Huber, *Hydrology and Floodplain Analysis*, 3d ed., © 2002. Reprinted by permission of Pearson Education, Inc., Upper Saddle River, N.J.

The first method of determining T and S is by using the original Theis equations. An examination of these equations shows that a plot of $W(u)$ vs. u, called a type curve, will have the same shape as a plot of s vs. r^2/t on log-log graph paper. The two curves are plotted, and one graph is laid over the other so the curves lie on one another. Then, a so-called match point, which is a set of data for u, $W(u)$, s, and r^2/t, is taken from the plots, inserted in Equations 5.20 and 5.21, and the resulting relations are solved for S and T. Since the match point is used to establish a connection between the data plot and the other plots, the match point need not be on the curve itself, although most practitioners do choose the match point atop the superimposed curves.

The second method, called the Cooper-Jacob method, is appropriate when u is small (for example, $u \le 0.01$ is a common rule). In this method, s is plotted against pumping time on semi-logarithmic paper; the curve eventually becomes linear. A fitted straight line is then extended to the point $s = 0$, where the value $t = t_0$ is noted. You then solve for the aquifer properties from

$$u \le 0.01 \qquad T = \frac{2.3Q}{4\pi(s_2 - s_1)} \log 10 \left(\frac{t_2}{t_1} \right) \qquad (5.31)$$

and

$$S = \frac{2.25 T t_0}{r^2} \qquad (5.32)$$

Use of these equations is simplified if points 1 and 2 are chosen so that $t_2/t_1 = 10$; of course, $s_2 - s_1$ is the difference in drawdown over this same time interval. The result in Equation 5.32 must be nondimensional.

Several approaches are possible for unsteady, unconfined well flow, but the simplest is to use the Theis method, Equation 5.28, with modified definitions of T and S. Now $T = Kb$ is based on the saturated thickness when pumping commences, and S is the specific yield, the volume of water released when the water table drops one unit. This approach is accurate when the drawdown is small in comparison with the saturated thickness of the aquifer.

Example 5.14

A well has been pumped at a steady rate for a very long time. The well has a 12-inch diameter and fully penetrates an unconfined aquifer that is 150 feet thick. Two small observation wells are 70 and 150 feet from the well, and the corresponding observed drawdowns are 24 and 20 feet. If the estimated hydraulic conductivity is 10 ft/day (sandstone), what is the discharge?

Solution

The saturated aquifer thicknesses at the observation wells are $h_1 = 150 - 24 = 126$ ft, $h_2 = 150 - 20 = 130$ ft, and the use of Equation 5.26 leads directly to

$$Q = \frac{\pi K \left[h_2^2 - h_1^2 \right]}{\ln(r_2 / r_1)} = \frac{\pi(10)\left[(130)^2 - (126)^2 \right]}{\ln(150 / 70)} = 42,200 \text{ ft}^3/\text{day}$$

This is equivalent to 0.49 ft³/s or 220 gal/min.

Example 5.15

Calculating transmissivity and specific yield I

Data on time t since pumping began versus drawdown s were collected from an observation well located 400 ft from a well that fully penetrated a confined aquifer that is 80 ft thick and is pumped at 200 gal/min. The data are presented in Exhibit 27.

t, min	s, ft	t, min	s, ft
35	2.82	103	4.43
41	3.12	131	4.60
48	3.25	148	5.00
60	3.60	205	5.35
80	3.98	267	5.80

Exhibit 27

Determine the aquifer properties T and S.

Solution

The data in Exhibit 27 have been used with $r = 400$ ft to compute s vs. r^2/t, which have been plotted in Exhibit 28 on a sheet of log-log paper and the plot placed on top of a log-log plot of $W(u)$ vs. u (the type curve). The two plots have been moved around until the closest fit between the curves was found, taking care that the coordinate axes are parallel. If the match point is chosen as shown on the figure, then $s = 5$ ft,

$r^2/t = 10^2$ ft^3/min, $u = 0.0175$, and $W(u) = 350$. Rearranging Equation 5.28,

$$T = Q \frac{W(u)}{4\pi s} = \frac{200 \text{ gal/min} \cdot (3.50)}{7.48 \text{ gal/ft}^3 \cdot 4\pi(5 \text{ ft})} = 1.49 \text{ ft}^2/\text{min}$$

Then from Equation 5.29

$$S = \frac{4Tu}{r^2/t} = \frac{4(1.49)(0.0175)}{10^3} = 1.04 \times 10^{-4}$$

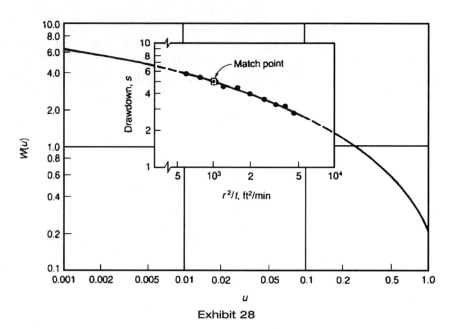

Exhibit 28

Example **5.16**

Calculating transmissivity and specific yield II

Last April pumping began at an 8-inch-diameter well at a steady rate of 300 gal/min, while observations were made at a well 100 ft away. Values of elapsed time and drawdown were taken for 16 hours and plotted (see Exhibit 29). Use the plotted data to determine values for the transmissivity and storage constant of this aquifer. In addition, estimate the drawdown in the observation well after four months of steady pumping at 300 gal/min.

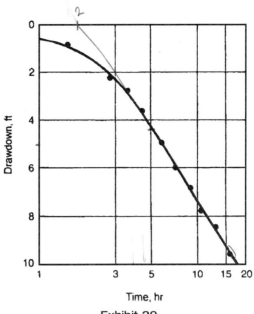

Exhibit 29

Solution

Exhibit 29 is a semi-logarithmic plot of data in the form needed to apply the Cooper-Jacob (or modified Theis, as it is also called) method. If you extend the straight-line portion of the plot to $s = 0$, as shown in Exhibit 30, you can read from the plot the value $t_0 = 2$ hrs. You also need a pair of data points (s_1, t_1) and (s_2, t_2) for use in Equation 5.31. For example, at $t_1 = 4.0$ hr, $s_1 = 3.3$ ft, and at $t_2 = 10.0$ hr, $s_2 = 7.4$ ft. Then

$$T = \frac{2.3Q}{4\pi(s_2 - s_1)} \log_{10}\left(\frac{t_2}{t_1}\right) = \frac{2.3(300)}{4\pi(7.4 - 3.3)} \log_{10}\left(\frac{10.0}{4.0}\right) = 5.33 \text{ gal/min/ft}$$

$$T = \frac{5.33 \frac{\text{gal/min}}{\text{ft}}}{7.48 \frac{\text{gal}}{\text{ft}^3}}\left(60 \frac{\text{min}}{\text{hr}}\right)\left(24 \frac{\text{hr}}{\text{day}}\right) = 1026 \text{ ft}^2/\text{day}$$

The storage constant can then be found as

$$S = \frac{2.25Tt_0}{r^2} = \frac{2.25(1026 \text{ ft}^2/\text{day})(\frac{2}{24} \text{ day})}{(100 \text{ ft})^2} = 0.0192$$

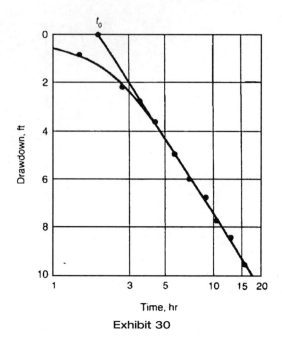

Exhibit 30

Equation 5.31 can also be used to find the drawdown after four months. If s_2 is the drawdown after four months, then t_2 is four months, or approximately 122 days. If you pick the other point to be $s_1 = 0$ and $t_1 = t_0 = 2.0$ hr, then

$$s_2 = \frac{2.3Q}{4\pi T} \log_{10}\left(\frac{t_2}{t_1}\right) = \frac{2.3(300)}{4\pi(5.33)} \log_{10}\left(\frac{122(24)}{2}\right) = 32.6 \text{ ft}$$

is the predicted drawdown.

ADDITIONAL RESOURCES

1. ASCE. *Design and Construction of Sanitary and Storm Sewers*. Manual No. 37. American Society of Civil Engineers, 1986.

2. Fetter, C.W. *Applied Hydrogeology*. Merrill Publishing Co., 1988.

3. Natural Resource Conservation Service. *Urban Hydrology for Small Watersheds*. Accessed from *www.wcc.nrcs.usda.gov/hydro/hydro-tools-models-tr55.html*.

4. Maidment, D.R. *Handbook of Hydrology*. McGraw-Hill, 1993.

5. Mays, L.W. *Water Resources Engineering*. John Wiley and Sons, 2004.

6. Mitsch, W.J., and J.G. Gosselink. *Wetlands*. 3d ed. John Wiley and Sons, 2000.

7. Wetzel, R.G. *Limnology: Lake and River Ecosystems*. 3d ed. Academic Press, 2001.

Stormwater Management

Stormwater is surface runoff in response to a precipitation event or snow melt. As such, stormwater generation is governed by the principles of the hydrologic cycle and the hydrologic budget discussed in Chapter 5.

6.1 IMPACT OF URBANIZATION ON THE HYDROLOGIC CYCLE

The process of transforming land from its natural state to a more urbanized state dramatically alters the hydrologic budget for a watershed and for drinking water aquifers. Specifically, land development often

▪ decreases the amount of infiltration by increasing imperviousness and replacing native, deep-rooted plants with shallow-rooted grasses;

▪ decreases the amount of interception by decreasing the amount of vegetation leaf area;

▪ increases the amount of runoff due to increasing the curve number (or runoff coefficient);

■ decreases the amount of water that can be stored in depressions as a result of site grading;

■ increases stream flow in response to storm events, which in turn increases stream velocity and the amount of stream bank erosion;

■ decreases the amount of aquifer recharge due to increased imperviousness;

■ decreases the amount of evapotranspiration due to the net decrease in vegetation; and

■ decreases the time of concentration.

In essence, the only aspect of the hydrologic cycle not impacted by land development is the rate of precipitation.

One purpose of *low impact development* is to help ensure sustainability by developing land in such a way as to mitigate the effects noted above.[1] For example, the amount of imperviousness can be decreased through careful road and lot layout, native deep-rooted plantings can be specified for landscaped areas, *porous pavements* can be specified for paved areas, etc.

The impacts of land development on the hydrologic budget are demonstrated in Figure 6.1. The development of the land also greatly affects stream flow hydrographs, as shown in Figure 6.2.

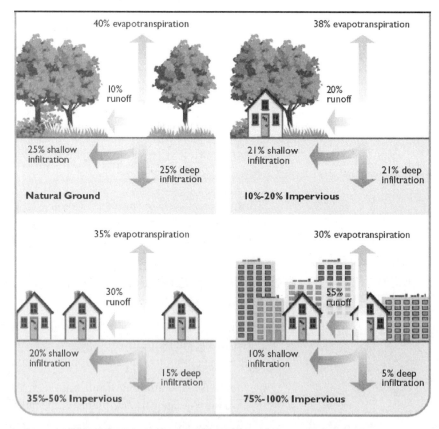

Figure 6.1 Impacts of urbanization on the hydrologic cycle

Source: *Connecticut Storm Water Quality Management Manual*, Connecticut Department of Environmental Protection.

---

1   The American Society of Civil Engineers (ASCE) Code of Ethics defines sustainable development as "the challenge of meeting human needs for natural resources, industrial products, energy, food, transportation, shelter, and effective waste management while conserving and protecting environmental quality and the natural resource base essential for future development."

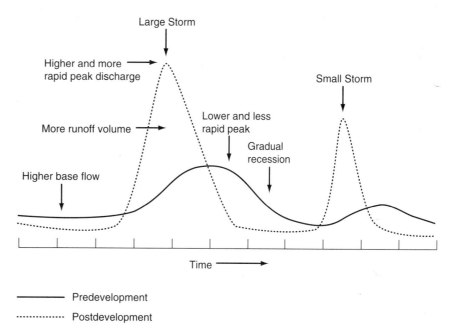

Figure 6.2   Impacts of development on stream hydrographs

Source: *Connecticut Storm Water Quality Management Manual*, Connecticut Department of Environmental Protection.

It is important to note from Figure 6. 2 that the following impacts of urbanization can be *quantified* from a comparison of pre- and postdevelopment stream flow hydrographs:

- Increased peak flow

- Increased volume of runoff

- Shorter t_p (time to peak—see discussion in section 5.1.6 on hydrographs)

- Decreased baseflow

Example **6.1**

Development impacts on infiltration

Pre- and postdevelopment runoff hydrographs for a watershed are shown in Exhibit 1. Assume that both hydrographs are the result of a 0.5-inch, 2-hour storm. Estimate the change in the volume of infiltration as a result of development.

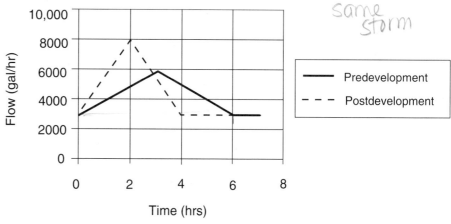

Exhibit 1   Impact of development on hydrographs

Solution

This problem begins with a simplified diagram (Exhibit 2) that shows a control volume on which a mass balance can be applied. In Exhibit 2, *P* represents precipitation, *I* represents infiltration, and *Q* represents direct runoff.

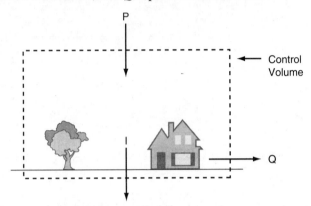

Exhibit 2 Control volume approach

By applying the hydrologic budget equation (Equation 5.1), it is seen that

$$Q + I = P, \text{ or } I = P - Q$$

Thus, the infiltration depths for predevelopment and postdevelopment conditions are

$$I_{pre} = P_{pre} - Q_{pre} \text{ and } I_{post} = P_{post} - Q_{post}, \text{ respectively.}$$

The difference between the predevelopment and postdevelopment infiltration depth is found by subtracting the postdevelopment equation from the predevelopment equation. Given that the identical precipitation depth fell in each case, the resulting equation is

$$I_{pre} - I_{post} = Q_{post} - Q_{pre}$$

From the hydrograph, we see that the baseflow is 3000 gal/hr. By separating out this baseflow, the contribution of surface runoff to the hydrograph can be calculated.

$$
\begin{aligned}
Q_{pre} \quad &= \quad \text{area under the hydrograph but above the baseflow} \\
&= \quad 0.5 \cdot (6 \text{ hr}) \cdot (6000 \text{ gal/hr} - 3000 \text{ gal/hr}) \\
&= \quad 9000 \text{ gal} \\
Q_{post} \quad &= \quad 0.5 \cdot (4 \text{ hr}) \cdot (8000 \text{ gal/hr} - 3000 \text{ gal/hr}) \\
&= \quad 10{,}000 \text{ gal}
\end{aligned}
$$

Thus, the change in infiltration volume is 1000 gallons.

6.2 STORMWATER QUALITY

Table 6.1 shows the various types of pollutants found in stormwater, the deleterious impacts the contaminants can have on the receiving stream, and strategies to assist in removal of the contaminants.

The impacts of stormwater on receiving streams include:

- Channel scour, widening, and downcutting

- Stream bank erosion and increased sediment loads

- Shifting bars of coarse sediment

- Burying of stream substrate

- Loss of pool/riffle structure

- Man-made stream enclosure or channelization

- Floodplain expansion

Table 6.1 Contaminants in stormwater

Stormwater Pollutant	Potential Sources	Receiving Water Impacts	Removal Promoted by[1]
Excess Nutrients Nitrogen, Phosphorus (soluble)	Animal waste, fertilizers, failing septic systems, landfills, atmospheric deposition, erosion and sedimentation, illicit sanitary connections	Algal growth, nuisance plants, ammonia toxicity, reduced clarity, oxygen deficit (hypoxia), pollutant recycling from sediments, decrease in submerged aquatic vegetation (SAV)	Phosphorus: High soil exchangeable aluminum and/or iron content, vegetation and aquatic plants Nitrogen: Alternating aerobic and anaerobic conditions, low levels of toxicants, near neutral pH (7)
Sediments Suspended, Dissolved, Deposited, Sorbed Pollutants	Construction sites, streambank erosion, washoff from impervious surfaces	Increased turbidity, lower dissolved oxygen, deposition of sediments, aquatic habitat alteration, sediment and benthic toxicity	Low turbulence, increased residence time
Pathogens Bacteria, Viruses	Animal waste, failing septic systems, illicit sanitary connections	Human health risk via drinking water supplies, contaminated swimming beaches, and contaminated shellfish consumption	High light (ultraviolet radiation), increased residence time, media/soil filtration, disinfection
Organic Materials Biochemical Oxygen Demand, Chemical Oxygen Demand	Leaves, grass clippings, brush, failing septic systems	Lower dissolved oxygen, odors, fish kills, algal growth, reduced clarity	Aerobic conditions, high light, high soil organic content, low levels of toxicants, near neutral pH (7)
Hydrocarbons Oil and Grease	Industrial processes; commercial processes; automobile wear, emissions, and fluid leaks; improper oil disposal	Toxicity of water column and sediments, bioaccumulation in food chain organisms	Low turbulence, increased residence time, physical separation or capture techniques
Metals Copper, Lead, Zinc, Mercury, Chromium, Aluminum (soluble)	Industrial processes, normal wear of automobile brake linings and tires, automobile emissions and fluid leaks, metal roofs	Toxicity of water column and sediments, bioaccumulation in food chain organisms	High soil organic content, high soil cation exchange capacity, near neutral pH (7)
Synthetic Organic Chemicals Pesticides, VOCs, SVOCs, PCBs, PAHs (soluble)	Residential, commercial, and industrial application of herbicides, insecticides, fungicides, rodenticides; industrial processes; commercial processes	Toxicity of water column and sediments, bioaccumulation in food chain organisms	Aerobic conditions, high light, high soil organic content, low levels of toxicants, near neutral pH (7), high temperature and air movement for volatilization of VOCs
Deicing Constituents Sodium, Calcium, Potassium Chloride Ethylene Glycol Other Pollutants (soluble)	Road salting and uncovered salt storage. Snowmelt runoff from snow piles in parking lots and roads during the spring snowmelt season or during winter rain on snow events	Toxicity of water column and sediments, contamination of drinking water, harmful to salt intolerant plants. Concentrated loadings of other pollutants as a result of snowmelt.	Aerobic conditions, high light, high soil organic content, low levels of toxicants, near neutral pH (7)

(continued)

Stormwater Pollutant	Potential Sources	Receiving Water Impacts	Removal Promoted by[1]
Trash and Debris	Litter washed through storm drain network	Degradation of aesthetics, threat to wildlife, potential clogging of storm drainage system	Low turbulence, physical straining/capture
Freshwater Impacts	Stormwater discharges to tidal wetlands and estuarine environments	Dilution of the high marsh salinity and encouragement of the invasion of brackish or upland wetland species such as Phragmites	Stormwater retention and volume reduction
Thermal Impacts	Runoff with elevated temperatures from contact with impervious surfaces (asphalt)	Adverse impacts to aquatic organisms that require cold and cool water conditions	Use of wetland plants and trees for shading, increased pool depths

[1] Factors that promote removal of most stormwater pollutants include:
• Increasing hydraulic residence time
• Low turbulence
• Fine, dense, herbaceous plants
• Medium-fine textured soil

Source: *Connecticut Storm Water Quality Management Manual*, Connecticut Department of Environmental Protection.

6.3 STORMWATER CONVEYANCE AND COLLECTION

Three aspects of stormwater conveyance and collection will be presented in this section:

1. The flow of stormwater in the street gutter

2. The collection of the gutter flow by an inlet

3. The conveyance of the stormwater in underground pipes

6.3.1 Flow of Stormwater in Roadways

As stormwater flows in gutters, it spreads out into the street until it comes under the influence of a stormwater inlet. As shown in Figure 6.3 (a plan view of the roadway, which slopes downward from right to left), the extent of this *spread* is typically not allowed to encroach on more than half of the driving lane. By leaving half of a travel lane open in each direction, an opening of dimension A (shown in Figure 6.3) is provided, equal in width to one travel lane. This allows for the passage of at least one vehicle at a time during the design storm. Consequently, inlets are spaced such that the maximum spread is not violated. Flow that is not captured by an inlet is termed *bypass* flow.

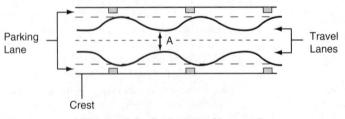

Figure 6.3 Illustration of "spread"

Stormwater flowing in a gutter can be readily modeled by Manning's equation. Recall from section 3.3.2 that Manning's equation is

$$V = 1.49 \cdot n^{-1} \cdot R^{2/3} \cdot S^{0.5}$$

where the 1.49 factor applies to U.S. customary units. Given that most gutters are triangular in cross section (or are a composite of more than one triangular shape), Manning's equation has been rewritten in a form to take into account the parameters typically found in triangular shapes; the flow rate for this version of the Manning's equation can be obtained from the nomograph in Figure 6. 4. Such a nomograph can greatly speed up the calculation process and limit the potential for calculation errors.

Example **6.2**

Stormwater flow in a typical road cross section

A road cross section is as shown in Exhibit 3.

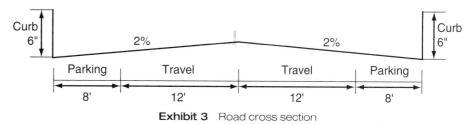

Exhibit 3 Road cross section

The road has an average slope along its length of 0.3%. What is the maximum amount of flow that the gutter can carry before the maximum spread is violated? Assume a Manning's *n* of 0.013.

Solution

Since the road is symmetrical, only one side needs to be considered.

The maximum allowable spread is 14 feet (the parking lane plus half of the driving lane). The values needed for the modified Manning's equation are as follows:

$$z = 1/0.02 = 50$$
$$n = 0.013$$
$$z/n = 3800$$
$$S = 0.003 \text{ ft/ft}$$
$$y = 2\% * 14 \text{ ft} = 0.28 \text{ ft}$$

From the equation or nomograph, the flow is found to be approximately 4 cfs.

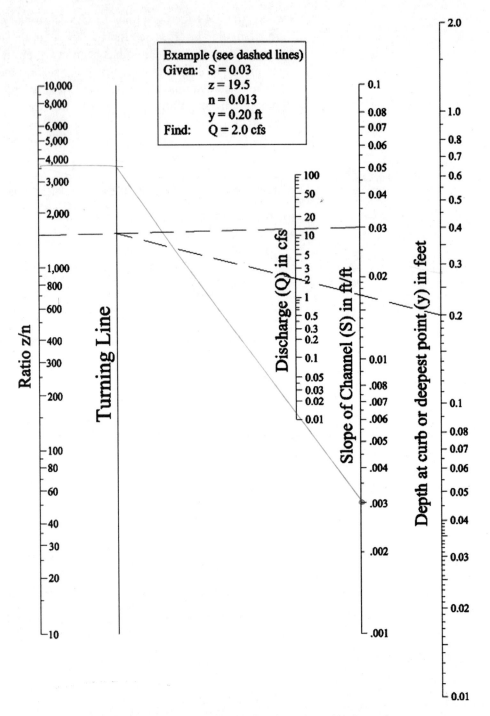

Example (see dashed lines)
Given: S = 0.03
       z = 19.5
       n = 0.013
       y = 0.20 ft
Find: Q = 2.0 cfs

Figure 6.4 Nomograph for flow in a triangular channel

Composite channels can also be analyzed with the modified Manning's equation. A common example of a composite section is that found in a "depressed gutter," that is, a culvert in which the cross slope of the gutter section varies from the cross slope of the road. To readily find the flows in these sections, the section is divided into triangles, some of which most likely will overlap. The flow can be determined for each triangular section and combined as appropriate. This concept is illustrated in Example 6.3.

Example **6.3**

Flow in a composite channel

For the flow cross section shown in Exhibit 4, estimate the amount of flow conveyed if the street has a longitudinal slope of 1%.

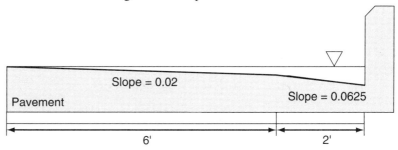

Pavement

Slope = 0.02

Slope = 0.0625

6'   2'

Exhibit 4 Flow cross section

Solution

The cross section can be divided into three polygons, a, b, and c, as shown in Exhibit 5. Moreover, the polygons can be combined into three triangles (a + c, b + c, and c). The various dimensions can be obtained relatively easily through geometric relationships. Thus, the total flow in the cross section, Q_{total}, is found to be

$$Q_{total} = Q_{a+c} + Q_{b+c} - Q_c$$

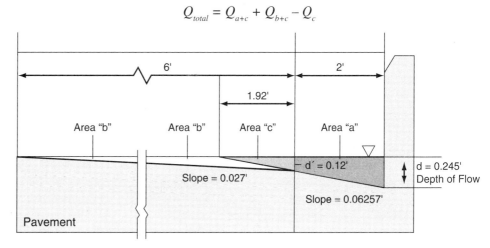

6'   2'

1.92'

Area "b"   Area "b"   Area "c"   Area "a"

Slope = 0.027'   d' = 0.12'   d = 0.245'
Depth of Flow

Slope = 0.06257'

Pavement

Type "A" Curb and Gutter

Exhibit 5 Computing flows in sub-cross sections

For finding Q_{a+c}: y = 0.245 ft
   n = 0.013
   z = 1/0.625 = 16
   z/n = 1230
   S = 0.01
   Q_{a+c} = 1.5 cfs (from Figure 6.4)

For finding Q_{b+c}: y = 0.12 ft
   n = 0.013
   z = 1/0.02 = 50
   z/n = 4000
   S = 0.01
   Q_{a+c} = 0.8 cfs (from Figure 6.4)

For finding Q_c:

$$y = 0.12 \text{ ft}$$
$$n = 0.013$$
$$z = 16$$
$$z/n = 1230$$
$$S = 0.01$$
$$Q_c = 0.25 \text{ cfs (from Figure 6.4)}$$

Thus, the total flow = 2.05 cfs.

3.3.2 Inlet Design

A stormwater inlet is a structure that intercepts stormwater flowing in the roadway gutter and transfers it to the storm sewer pipe network. Inlets may either be located *on grade,* such as the inlets in Figure 6.3, or in *sag* locations (see Figure 6.5). Inlets on grade typically do not capture all of the flow and allow the excess flow to be bypassed. Inlets located in sag locations (sometimes called the *sump* location) are at the low point of a road profile, and thus water may pond above them. There is no bypass for a sag inlet, although excess flow may overtop the curb.

The primary criterion used in spacing inlets is to ensure that the spread of water does not exceed the maximum spread allowable. In addition, inlets should be placed upstream of all intersections, bridges, pedestrian ramps, commercial drive aprons, etc.

There are four main types of inlets, as shown in Figure 6.6:

1. *Grate inlets* have a slotted or screened pattern, the top of which is flush with the roadway surface. They are effective at capturing gutter flow as they are placed directly at the bottom of the gutter but are prone to clogging.

2. *Curb-opening inlets* have an opening in the *face* of the curb. They are not as prone to clogging as grate inlets but are not as effective in on-grade locations. However, they are much more effective in sag locations.

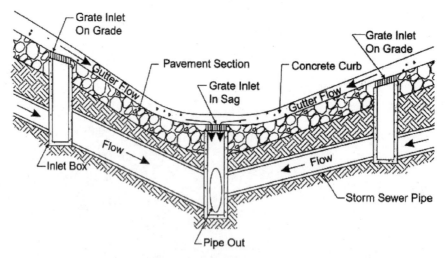

Figure 6.5 Inlets on grade and in sag

Source: *Stormwater Conveyance Modeling and Design.* Reprinted with permission of the Bentley Institute Press.

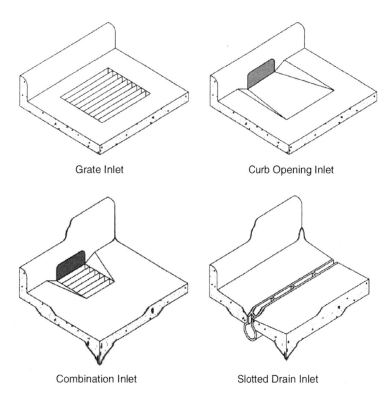

Grate Inlet   Curb Opening Inlet

Combination Inlet   Slotted Drain Inlet

Figure 6.6 Common inlet types

Source: *Stormwater Conveyance Modeling and Design.* Reprinted with permission of the Bentley Institute Press.

3. *Combination inlets* are a one-piece unit consisting of a grate inlet and a curb-opening inlet. They thus combine the best qualities of the grate inlets (high performance on grade) and the curb-opening inlet (resistance to clogging, high performance in sags).

4. *Slotted drain inlets* have a grate opening oriented along the length of a pipe. They are used to intercept sheet flow before it becomes problematic and are often oriented perpendicular to the direction of flow.

Grates are characterized by their *efficiency, E,* which is defined as

$$E = Q_{intercepted}/Q_{total}$$

where

$Q_{intercepted}$ = the flow intercepted by the inlet
Q_{total} = the total flow approaching an inlet

Bypass flow (sometimes called *carryover*) is the flow that is not intercepted by an inlet.

The design of all four types of inlets in their various locations (that is, sag vs. on grade) is beyond the scope of this review book. The references at the end of the chapter provide more in-depth details on this subject. For this review book, the design of the most common inlets (grate inlets on grade and curb opening inlets in sag locations) will be briefly discussed.

The flow approaching a grate inlet may be characterized as *frontal flow* or *side flow.* These two types of flows are depicted in Figure 6.7.

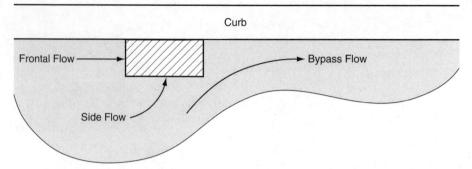

Figure 6.7 Frontal flow and side flow

The flow intercepted by a grate inlet on grade, Q_i, is obtained from the following equation:

$$Q_i = Q[R_f E_o + R_s(1-E_o)] \tag{6.1}$$

where

Q = flow in the gutter approaching the inlet

R_f = the ratio of frontal flow intercepted to total frontal flow
(obtained from Figure 6.8)

E_o = ratio of frontal flow to total gutter flow

$\quad = 1 - (1 - W/T)^{2.67}$

where W = width of depressed gutter or grate (ft)

$\quad T$ = total spread of water in the gutter (ft)

$\quad R_s$ = the ratio of side flow intercepted to total side flow can be obtained
from Figure 6.9.

Example **6.4**

Inlet efficiency

This example continues Example 6.2. A four-foot curved vane grate inlet is installed and is located at the point where the allowable spread is reached (corresponding to a flow of 4 cfs). How much flow will bypass the inlet?

Solution

The first step is to determine E_o, the ratio of frontal flow to total gutter flow.

$\quad E_o \ = 1 - (1 - W/T)^{2.67}$
$\qquad = 1 - (1 - 4 \text{ ft}/14 \text{ ft})^{2.67}$
$\qquad = 0.6$

From Figure 6.8, a value of R_f can be obtained given

$L \ = 4 \text{ ft}$
$V \ = \text{velocity in cross section}$
$\quad = Q/A \text{ (where } A \text{ is the cross-sectional area of flow)}$
$\quad = (3 \text{ ft}^3/\text{sec})/(0.5 \cdot 0.28 \text{ ft} \cdot 14 \text{ ft})$
$\quad = 2.0 \text{ ft/sec}$

From Figure 6.8, R_f is determined to be equal to 1.0. Therefore, all of the frontal flow is captured by this grate.

From Figure 6.9, a value for R_s can be obtained given

$$S_x = 0.02$$
$$L = 4 \text{ ft}$$
$$V = 2.0 \text{ ft/s}$$

From Figure 6.9, R_s is determined to be 0.48.
Next, the flow intercepted by this grate is calculated.

$$
\begin{aligned}
Q_i &= Q[R_f E_o + R_s(1-E_o)] \\
&= 4 \text{ cfs} \cdot [1.0 \cdot 0.6 + 0.48 \cdot (1-0.6)] \\
&= 3.2 \text{ cfs}
\end{aligned}
$$

Thus, the flow bypassed around this inlet = 4 cfs – 3.2 cfs = 0.8 cfs.

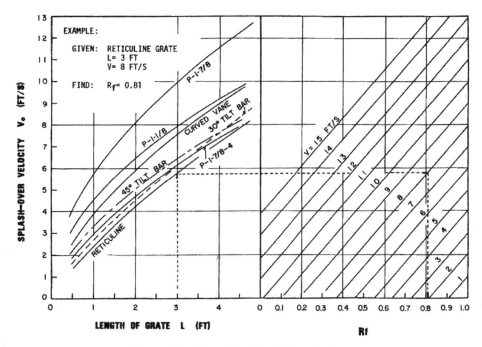

Figure 6.8 Grate inlet frontal flow efficiency

Source: Federal Highway Administration.

Curb opening inlets in a sump location can be modeled either as weirs or as orifices, depending upon the depth of water at the weir opening. If the water depth is equal to the curb opening height, the inlet acts as a weir. If the water depth is greater than 1.4 times the opening height, the inlet acts as an orifice. For water depths greater than the curb opening height but less than 1.4 times the opening height, the flow is said to be in transition and is not readily modeled. The flow capacity of a curb-opening inlet in a sump location can be obtained from Figure 6.10. To use Figure 6.10 to determine the flow rate, draw a horizontal line across from the depth of water d to where the line intersects an appropriate curve and read the corresponding flow rate from the abscissa of the graph.

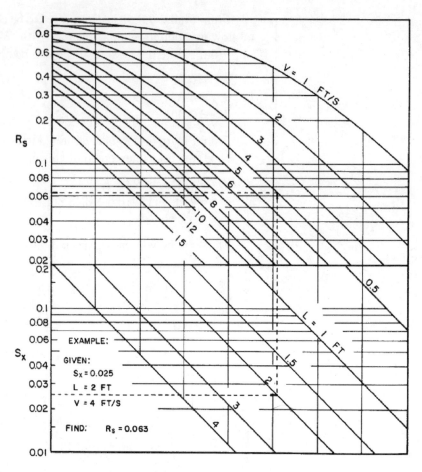

Figure 6.9 Grate inlet side flow interception efficiency

Source: Federal Highway Administration.

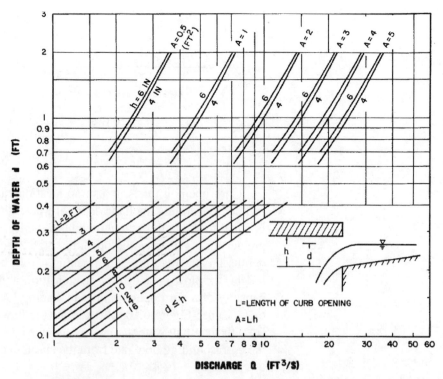

Figure 6.10 Curb-opening inlet capacity in sump locations

Source: Federal Highway Administration.

Example **6.5**	

Flow capacity of a curb-opening inlet

A 6-foot-long curb-opening inlet with an opening height of 4 inches is placed in a sump location. The cross section of the roadway is identical to that shown in Example 6. 3. Determine the flow capacity of the inlet:

(a) Given that the maximum allowable spread is obtained

(b) Given a ponding depth of 9 inches

Solution

(a) Given a depth of water d of 0.245 feet and a curb opening length L of 6 feet, a discharge of approximately 2.3 cfs is obtained from Figure 6.10. Note that the inlet is operating as a weir.

(b) The area of the inlet is the area of a rectangle that is 4 feet long and 4 inches tall, or 2 ft². Given d of 0.75 feet and an opening height h of 4 inches, the discharge is found to be 8.5 cfs from Figure 6.10. Note that the inlet will act as an orifice for this ponding depth.

The design of combination inlets can be completed based on the design of a grate inlet on grade or a curb opening inlet in sag locations. Combination inlet design can be accomplished by assuming that a combination inlet on grade acts solely as a grate inlet (because the curb opening captures a negligible amount of flow) and that a combination inlet in a sag location acts as a curb opening (because the grate inlet is apt to clog).

6.3.3 Culverts

A culvert is a relatively short conduit that conveys water underground, most commonly under a roadway or railway embankment. Culverts are typically aligned perpendicular to the roadway they are crossing. Culverts may be made from a variety of materials, including PVC (polyvinyl chloride), steel, concrete, and HDPE (high density polyethylene). Cross-sectional shapes include circular, elliptical, arched, or rectangular (see Figure 6.11). Various culvert end sections can be used (see Figure 6.12) and should be selected by considering structural stability of the embankment, aesthetics, erosion control, etc. These end controls will also affect the hydraulic efficiency of a culvert (for example, less energy is lost when a stream transitions to a culvert through a beveled end section than through a squared-off end section).

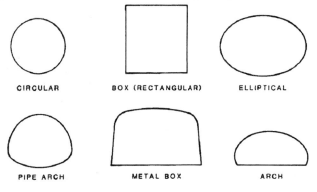

Figure 6.11 Typical culvert cross-sectional shapes

Source: U.S. Department of Transportation/Federal Highway Administration, *Hydraulic Design of Highway Culverts*, 2005.

The hydraulic design of a culvert should consider the following:

- Allowable headwater

- Type of flow control (inlet control versus outlet control)

- Permissible barrel and outlet velocities

- Location and orientation of the barrel(s)

- Inlet and outlet protection for scour control

- Debris control

- Any necessary fish passage criteria

The type of flow control is very iportant when designing culverts and has a significant impact on the predicted performance. Culverts are either *inlet controlled* or *outlet controlled.* In other words, either the inlet conditions will control how much water a pipe can convey or some aspect of the outlet will control the discharge. Under inlet control, the culvert is capable of conveying more flow than can enter the culvert. Flow for this type of control is supercritical. A culvert under inlet control is depicted in Figure 6.13.

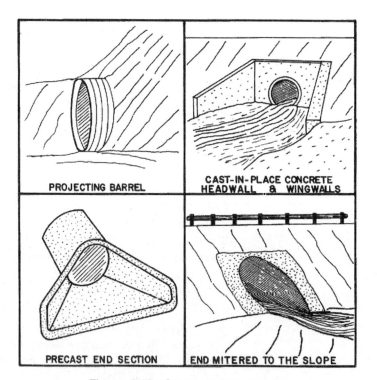

Figure 6.12 Culvert end treatments

Source: U.S. Department of Transportation/Federal Highway Administration, *Hydraulic Design of Highway Culverts*, 2005.

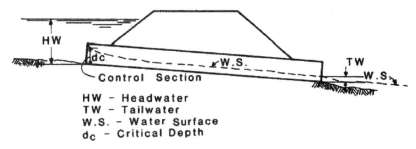

Figure 6.13 Inlet control example

Source: U.S. Department of Transportation/Federal Highway Administration, *Hydraulic Design of Highway Culverts*, 2005.

Outlet control occurs when the culvert barrel is incapable of conveying as much flow as the inlet will accept. Examples of conditions causing outlet control are shown in Figure 6.14. The *control section* referred to in Figure 6.14 is a section where the depth of flow and discharge are known.

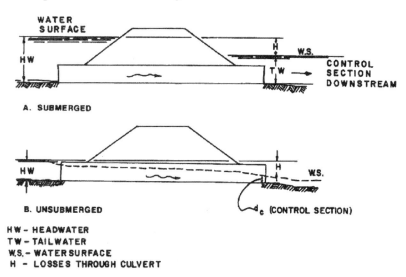

Figure 6.14 Outlet control examples

Source: U.S. Department of Transportation/Federal Highway Administration, *Hydraulic Design of Highway Culverts*, 2005.

Culvert design may require complex computer algorithms to complete. For the purposes of the PE exam, a series of nomographs for the most common cases is presented in Figures 6.15–6.20.

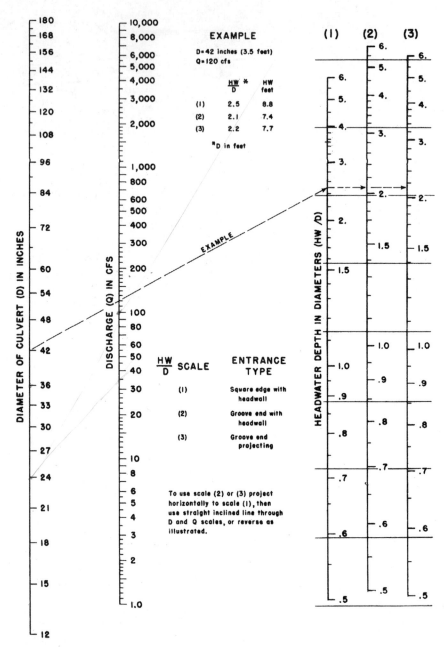

Figure 6.15 Headwater depth for concrete pipe culverts with inlet control

Source: U.S. Department of Transportation/Federal Highway Administration, *Hydraulic Design of Highway Culverts*, 2005.

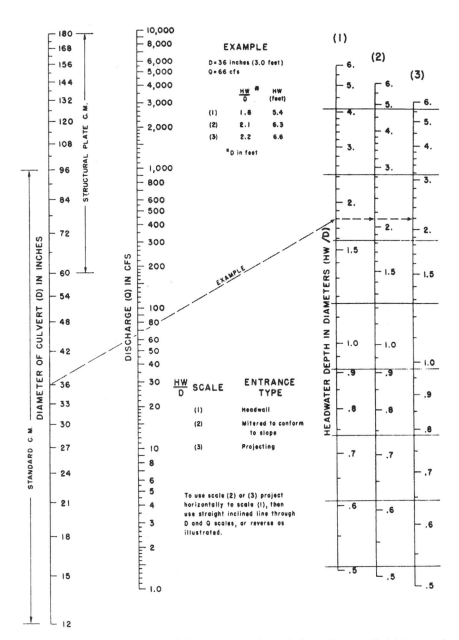

Figure 6.16 Headwater depth for corrugated metal pipe culverts with inlet control

Source: U.S. Department of Transportation/Federal Highway Administration, *Hydraulic Design of Highway Culverts*, 2005.

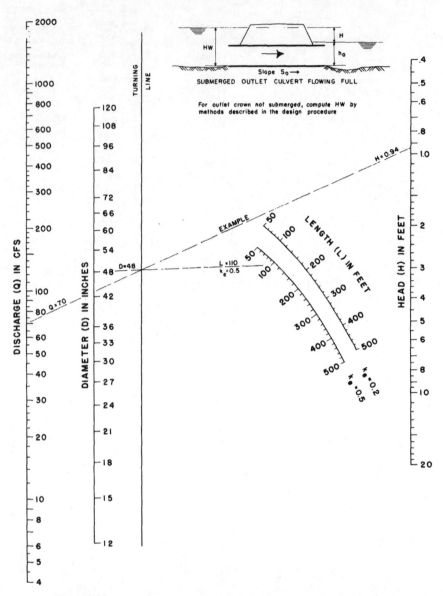

Figure 6.17   Head for concrete pipe culverts flowing full, $n = 0.012$

Source: U.S. Department of Transportation/Federal Highway Administration, *Hydraulic Design of Highway Culverts*, 2005.

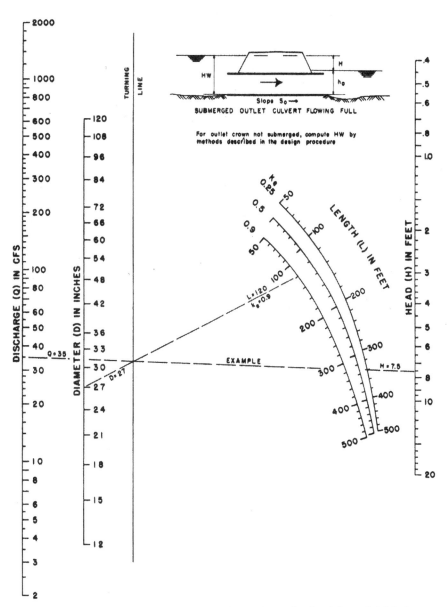

Figure 6.18 Head for concrete pipe culverts flowing full, $n = 0.024$

Source: U.S. Department of Transportation/Federal Highway Administration, *Hydraulic Design of Highway Culverts*, 2005.

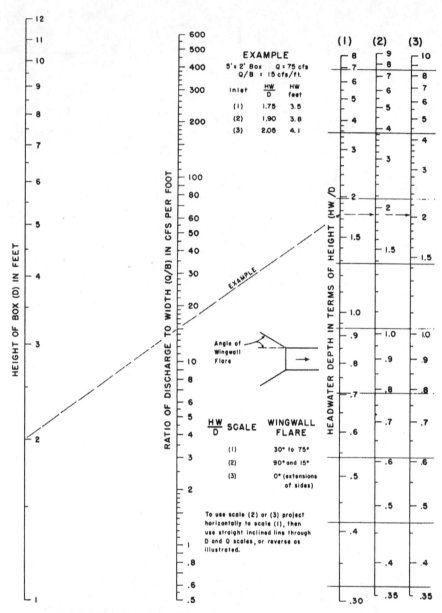

Figure 6.19   Headwater depth for box culverts with inlet control

Source: U.S. Department of Transportation/Federal Highway Administration, *Hydraulic Design of Highway Culverts*, 2005.

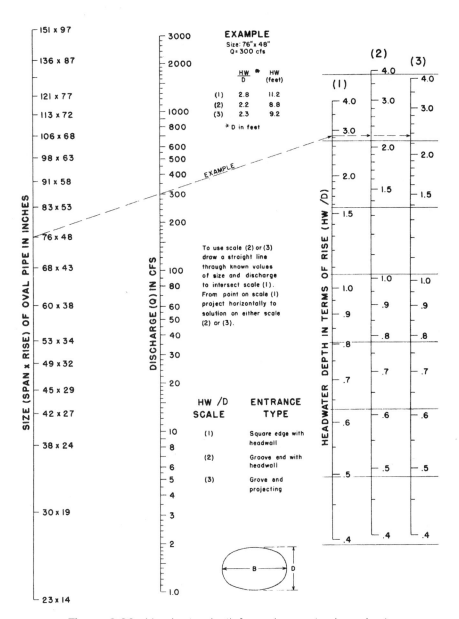

Figure 6.20 Headwater depth for oval concrete pipe culverts

Source: U.S. Department of Transportation/Federal Highway Administration, *Hydraulic Design of Highway Culverts*, 2005.

| Example **6.6** | Culvert design |

A culvert must be designed for a roadway crossing. The invert of the upstream end of the new culvert must be set at an elevation of 100 feet. The elevation of the roadway shoulder is 105 feet. The presence of other utilities under the roadway require that the pipe diameter be less than or equal to 24 inches. The pipe is specified to be circular concrete pipe. Use a *freeboard* of 1.5 feet and a design flow of 100 cfs.[2] Assume inlet control.

---

2   *Freeboard* is the vertical distance from the desired high water mark to the flood elevation. Freeboard is used as a safety factor.

Solution

From the problem statement, an allowable headwater depth of 3.5 feet can be obtained. According to Figure 6.15, a 36-inch pipe or larger is required to convey this flow, given the restriction of a headwater of 3.5 feet. This assumes the most favorable of the end wall treatments provided on the nomograph (square edge with headwall). However, space restrictions preclude the use of a 36-inch pipe. Thus, more than one 24-inch pipe will be necessary.

Figure 6.15 shows that a 24-inch pipe with 3.5 feet of headwater depth can convey approximately 35 cfs. Thus, specifying three such pipes will convey the required 100 cfs of flow while not violating the requirements for headwater height.

6.3.4 Storm Sewer Design

The design of a storm sewer (which in this review manual refers to that portion of the stormwater management system that is underground) consists of selecting pipe diameters, materials, lengths, and cross-sectional shapes; specifying upstream and downstream elevations of pipes; placing manhole structures; and designing outlet structures. Occasionally, lift stations (pumping stations) need to be sized and placed, although their use is not as widespread as in wastewater collection. The following criteria should be heeded when designing a storm sewer system:

■ Pipe velocities for full flow should be kept between 3 fps and 15 fps to prevent sediment accumulation and pipe scour, respectively.

■ Typically three feet of cover should be provided to prevent crushing of the pipe.

■ Pipe crowns should be matched at manholes and other junction structures.

■ Often, a 12-inch minimum diameter is specified to reduce the likelihood of clogging.

■ Pipe diameters should never decrease in a downstream direction.

■ Horizontal separation from potable water distribution systems is required by regulatory agencies.

Surcharging occurs when the hydraulic grade line of the system is higher than the pipe crown. A pipe that is surcharged conveys more flow than a pipe that is not surcharged, all other factors remaining the same. However, surcharging may occur because of a poor design, clogging in the pipe, deteriorating networks, an increase in runoff due to urbanization of the contributing land area, and increased water elevation of receiving water, which then backs up into the storm sewer.

When the hydraulic grade line increases such that its elevation is higher than the ground level, an overflow occurs. Of increasing concern to environmental engineers lately is the occurrence of *combined sewer overflows,* or *CSOs.*[3] However,

---

3   A *combined sewer* is one in which the sanitary sewage and stormwater are combined. Combined sewers are relatively common in many older cities. Two reasons for their popularity were the cost savings (one pipe system instead of two) and hydraulic advantages (flushing of the pipes in response to storm events). Combined sewers are no longer allowed in new construction.

when a CSO occurs, the water that overflows the system is a mixture of stormwater and untreated domestic wastewater.

Full flow in storm sewer pipes that are not surcharged can be estimated using Manning's equation. A nomograph has been provided in Figure 6.21. Note that Manning's equation applies to pipes flowing full due to the effects of gravity, and that Figure 6.21 applies only to pipes with a circular cross.

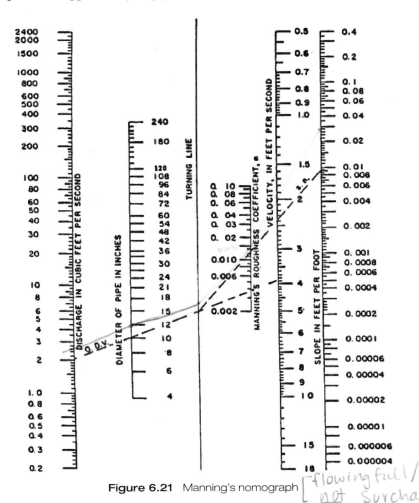

Figure 6.21 Manning's nomograph [flowing full/ not surcharged]

Model Drainage Manual, 1991, by the American Association of State Highway and Transportation Officials, Washington, D.C. Reprinted with permission.

The "hydraulic elements chart," sometimes called the "partial flow diagram" as shown in Figure 6.22 is very useful when evaluating pipes that are not flowing full. Such partial flow conditions are very common in stormwater and wastewater conveyance systems. When pipes flow partially full, the hydraulic radius varies; although this variation with depth is only a matter of geometry, calculating the wetted perimeter and cross-sectional area for a portion of a circular cross section can become labor intensive; thus, Figure 6.22 becomes very useful.

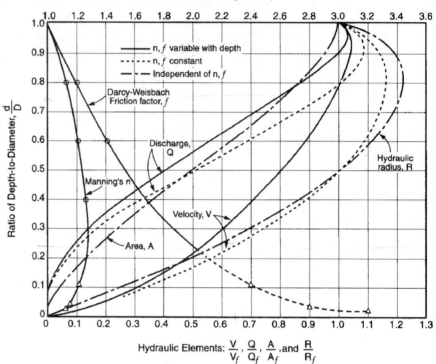

Figure 6.22 Hydraulic elements chart

Design and Construction of Sanitary and Storm Sewers; 1970, American Society of Civil Engineers. Reprinted by permission.

The variables in Figure 6.22 include:

V_f = velocity of full pipe discharge

Q_f = full pipe discharge

A_f = cross-sectional area of full pipe

R_f = hydraulic radius of full pipe flow

D = diameter of pipe

d = depth of water in pipe

There are three families of curves in Figure 6.22. The dashed line applies if n and f (the Manning's roughness coefficient and the Darcy-Weisbach friction factor, respectively) are assumed to be constant, regardless of the depth of water in the pipe. The solid line is used otherwise. The line type "Independent of n, f" is used for those pipe properties (cross-sectional area and hydraulic radius) whose values do not vary as n and f vary. Note that a curve exists for each quantity on the abscissa (that is, V/V_f, Q/Q_f, A/A_f, and R/R_f), and the correct curve must be used when determining the corresponding value on the abscissa.

Figure 6.22 might be used in the following instances:

■ Given a pipe diameter and a value for the full pipe flow (or velocity), determine the depth of water in the pipe.

■ Given a depth of water in a pipe of known diameter, estimate the flow.

Figure 6.22 verifies that, given constant n and f, the discharge and cross-sectional area of water in a half-full circular pipe are equal to half of the full-flow discharge and half of the full-flow area, respectively. Also, for the half-full pipe, the hydraulic radius and velocity are the same as their full-flow values ($R = D/4$ for a circle or semicircle, and velocity equals flow divided by area, which each have changed by a factor of ½). Also, Figure 6.22 shows that for a range of water depths, a pipe can carry more flow when it is partly full than when it is completely full.

Two examples will clarify the use of Figure 6.22.

Example **6.7**

Hydraulic elements I

Given a 12-inch pipe that is carrying 3 inches of water, calculate:

(a) V/V_f = 0.65 $\dfrac{d}{D} = \dfrac{3}{12} = \dfrac{1}{4} = 0.25$

(b) Q/Q_f = 0.12

(c) A/A_f 0.2

(d) R/R_f 0.6

Assume that n and f do not vary with depth. ‐ ‐ ‐ ‐ ·

Solution

In all cases, d/D is 3/12 in, or 0.25. A horizontal line can be drawn corresponding to $d/D = 0.25$ on the hydraulic elements chart (Figure 6.22), and the line remains constant for this entire problem. Thus, the requested values can be found by determining where the $d/D = 0.25$ line intersects each of the other lines on the chart.

(a) $V/V_f = 0.67$

(b) $Q/Q_f = 0.12$

(c) $A/A_f = 0.18$

(d) $R/R_f = 0.60$

Example **6.8**

Hydraulic elements II

= 0.008

A 400-foot length of 12-inch concrete pipe is set at a slope of 0.8%. The depth of water in the pipe is 9 inches. $\dfrac{d}{D} = \dfrac{9}{12} = 0.75$

(a) What is the discharge in this pipe?

(b) What is the velocity of the water in this pipe? $\dfrac{V}{V_f} = 1.15 (26) =$

Solution

The following values will help in solving the problem:

$d/D = 9/12 = 0.75$

$Q_f = 3.1$ cfs (from Figure 6.21) = $\dfrac{Q}{Q_f} = 0.95$ $Q = (.95)(31)$

$\llcorner n = 0.01$

$$A_f = \pi*(1\ \text{ft})^2/4 = 0.79\ \text{ft}^2$$

$$V_f = 4.1\ \text{ft/s (from Figure 6.21)}$$

(a) From Figure 6.22, $Q/Q_f = 0.94$; thus $Q = 0.94 \cdot 3.1\ \text{cfs} = 2.9\ \text{cfs}$

(b) From Figure 6.22, $V/V_f = 1.15$; thus $V = 1.15 \cdot 4.1\ \text{ft/s} = 4.7\ \text{ft/s}$

Alternatively, given the partly filled pipe flow from (a), the velocity in the partly filled pipe is this flow rate divided by the area corresponding to 9 inches of water. This area can be determined, since $A_f = 0.79\ \text{ft}^2$ and $A/A_f = 0.82$ (from Figure 6.22); thus $A = 0.82 \cdot 0.79\ \text{ft}^2 = 0.65\ \text{ft}^2$ and $V = Q/A = 2.9\ \text{cfs}/0.65\ \text{ft}^2 = 4.5$ ft/sec.

This answer for velocity differs slightly from the 4.7 ft/s obtained using the first method, but this variance is acceptable given the uncertainty inherent in using a graphical solutions approach.

Design of storm sewer systems is most commonly completed using computer models; however, a firm understanding of the principles underlying the design is required to interpret model results (and to answer questions on the PE exam). Design of storm sewer systems lends itself to tabular methods.

Example **6.9**

Storm sewer flow

A stormwater collection system near Madison, Wisconsin, drains the three sub-catchments as shown in Exhibit 6. Inlet X captures all of the flow from subcatchment 1, inlet Y captures all of the flow from subcatchment 2, and inlet Z captures all of the flow from subcatchment 3. Pipe Y-out discharges to a water elevation at 486 feet. An IDF (intensity-duration-frequency) curve for Madison is provided in Exhibit 7. The subcatchments and pipes ($n = 0.013$) are described in Exhibits 8 and 9, respectively.

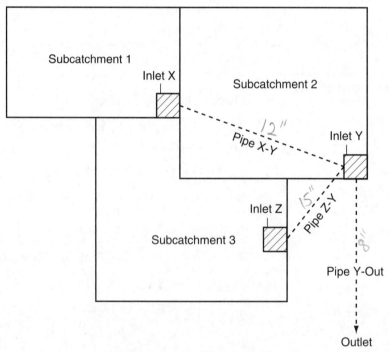

Exhibit 6 Problem schematic

When installing the system, the contractor inadvertently installed an 8-inch pipe for pipe Y-out. Determine the impact that the 100-year storm will have on this undersized pipe's velocity.

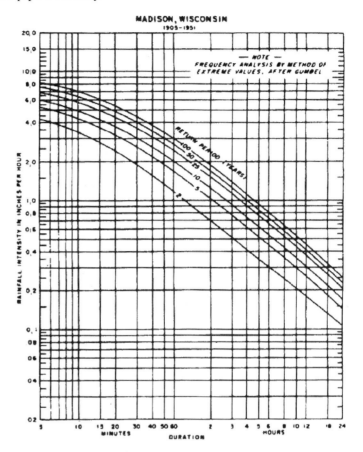

Exhibit 7 IDF curve

Source: State of Wisconsin Department of Transportation, *Facilities Development Manual*, Procedure 13-10-5, Figure 4.

Subcatchment	C	A (acres)	t_c (min)	I (in/hr)
1	0.57	1.5	10 → 7.0	
2	0.7	3	15 → 6.0	
3	0.7	0.7	10 → 7.0	

Exhibit 8 Subcatchment characteristics

Pipe	Length (ft)	Upstream Invert Elevation (ft)	Downstream Invert Elevation (ft)	Diameter (in)	slope
X-Y	500	505	490	12	0.03
Z-Y	1000	494	490	15	0.004
Y-Out	750	490	485	8	0.0067

Exhibit 9 Pipe characteristics

Solution

Some preliminary quantities can be found. First, the peak flow for subcatchments 1 and 2 can readily be obtained from the rational method. The intensities are found from the IDF curve, corresponding to a duration equal to the catchment's time of concentration and a 100-year return period.

$$Q_{p,1} = C_1 \cdot I_1 \cdot A_1$$
$$= 0.57 \cdot 7 \text{ in/hr} \cdot 1.5 \text{ acre}$$
$$= 6.0 \text{ cfs}$$

$$Q_{p,3} = C_3 \cdot I_3 \cdot A_3$$
$$= 0.7 \cdot 7 \text{ in/hr} \cdot 0.7 \text{ acre}$$
$$= 3.4 \text{ cfs}$$

Also, the discharge in each pipe under full-flow conditions can be obtained from Manning's nomographs (Figure 6.21).

$$\text{Pipe X-Y: Slope} = \frac{505 \text{ ft} - 490 \text{ ft}}{500 \text{ ft}} = 0.03 \text{ ft / ft}, D = 12 \text{ in}$$

$$Q_f = 6 \text{ cfs}$$

$$\text{Pipe Z-Y: Slope} = \frac{494 \text{ ft} - 490 \text{ ft}}{1000 \text{ ft}} = 0.004 \text{ ft / ft}, D = 15 \text{ in}$$

$$Q_f = 4.1 \text{ cfs}$$

$$\text{Pipe Y-Out: Slope} = \frac{490 \text{ ft} - 485 \text{ ft}}{750 \text{ ft}} = 0.007 \text{ ft/ft}, D = 8 \text{ in}$$

$$Q_f = 0.96 \text{ cfs}$$

To determine the velocity in pipe Y-out, the flow in the pipe needs to be determined. This flow is determined by applying the rational method to the subcatchments. The flow to inlet Y is obtained using the entire area of the three subcatchments (8.5 acres), based on a weighted runoff coefficient. The intensity is based on the time of concentration, which in this case is the *maximum* of one of the following options:

Option 1 : $t_{c,2}$

Option 2 : $t_{c,3}$ + travel time in pipe Z-Y

Option 3 : $t_{c,1}$ + travel time in pipe X-Y

The travel time in pipe is equal to the pipe's length divided by the water's velocity in the pipe. Moreover, the peak flow in pipes X-Y and Z-Y is simply the peak flow of subcatchments 1 and 3, respectively. The velocity in pipe X-Y can be found from Figure 6.21 since the pipe is flowing full. (This can be verified by comparing the actual flow in the pipe $Q_{p,1}$ to the full pipe flow.) Thus, $V_{X-Y} = 7.8$ fps. The velocity in pipe Z-Y can be found from the hydraulic elements chart (Figure 6.22) given that $Q/Q_f = 3.4$ cfs / 4.1 cfs = 0.83. Assuming that n and f do not vary, $d/D = 0.7$ and $V/V_f = 1.14$. Given $V_f = 3.3$ fps (from Figure 6.21 or by calculating $V_f = Q_f/A_f$), $V_{Z-Y} = 3.8$ fps. The travel times in pipes Z-Y and X-Y are then

$$V_{Z-Y} = \frac{1000 \text{ ft}}{3.8 \text{ fps}} = 263 \text{ sec} = 4.4 \text{ min}$$

$$V_{X-Y} = \frac{500 \text{ ft}}{7.8 \text{ fps}} = 64 \text{ sec} = 1.1 \text{ min}$$

The time of concentration to use for inlet Y is the maximum of the three options:

Option 1 : $t_{c,2} = 15$ min

Option 2 : $t_{c,3}$ + travel time in pipe Z-Y = 10 min + 4.4 min = 14.4 min

Option 3 : $t_{c,1}$ + travel time in pipe X-Y = 10 min + 1.1 min = 11.1 min

Thus, a time of concentration of 15 minutes will be used. This time corresponds to an intensity of 6 in/hr (Exhibit 7).

The weighted runoff coefficient is found by this equation:

$$C_w = \frac{\sum C_i A_i}{\sum A_i} = \frac{0.57 \cdot 1.5 \text{ acre} + 0.7 \cdot 3.0 \text{ acre} + 0.7 \cdot 0.7 \text{ acre}}{1.5 \text{ acre} + 3.0 \text{ acre} + 0.7 \text{ acre}} = 0.66$$

The peak flow for inlet Y (and thus in pipe Y-out) is

$$Q_{p,Y} = C_{weighted} \cdot I \cdot A_{total}$$
$$= 0.66 \cdot 6 \text{ in/hr} \cdot 5.2 \text{ acre}$$
$$= 20.6 \text{ cfs}$$

Given that this pipe can only convey 0.96 cfs at full flow, the pipe is clearly flowing pressurized.

Assuming that the inlet and outlet conditions allow a flow of 20.6 cfs, the velocity in the pipe is calculated by

$$V = \frac{Q}{A} = \frac{20.6 \text{ cfs}}{\pi \cdot \dfrac{(0.667 \text{ ft})^2}{4}} = 59 \text{ fps}$$

This is much larger than the maximum velocity allowed.

Example **6.10**

Surcharging

Continue Example 6.9 and estimate the extent of surcharging.

Solution

The elevation of the HGL at inlet Y will reveal the extent of surcharging and can be determined by writing the energy equation between inlet Y and the outlet:

$$\frac{P_Y}{\gamma} + \frac{V_Y^2}{2g} + Z_Y = \frac{P_{Out}}{\gamma} + \frac{V_{Out}^2}{2g} + Z_{Out} + h_L$$

The energy equation can be simplified by assuming that the velocity on both ends of the pipe are identical and that P_{out} = atmospheric pressure (0 psig). The equation then becomes

$$\frac{P_Y}{\gamma} = (Z_{Out} - Z_Y) + h_L$$

The quantity $\dfrac{P_Y}{\gamma}$ is the elevation of the HGL above the pipe centerline. The following information can be inserted into this simplified equation:

Z_{out} = elevation of the tailwater
   = 486 ft

Z_Y = elevation of pipe centerline
   = invert elevation + pipe radius
   = 490 ft + 0.33 ft
   = 490.33 ft

$h_L = S_f \cdot L$ (See section 3.3.2 for Manning's equation for friction head loss)

$$= \frac{4.66 \cdot \left(n \cdot Q\right)^2}{D^{5.33}} \cdot L$$

$$= \frac{4.66 \cdot \left(0.013 \cdot 20.6 \text{ cfs}\right)^2}{0.667^{5.33}} \cdot 750 \text{ ft}$$

   = 2170 ft

In other words, the HGL is more than 2000 feet above the pipe invert. So, the system would need 2000 feet of head to be able to push 20.6 cfs of flow through an 8-inch diameter pipe! Clearly, surcharging has occurred.

The solutions to Examples 6.9 and 6.10 are based on many unstated assumptions. These assumptions include:

- Uniform flow conditions where the flows from pipes Z-Y and X-Y join

- Uniform and steady flow for all pipes

- Absence of clogging

- 100% capture efficiency of the inlets

- Constant tail water elevation

- Neglecting of tail water effects on flow in pipe Y-out

6.4 STORMWATER TREATMENT

After stormwater has been captured by inlets and conveyed via the storm sewer system, it may need to be treated, depending on the governing regulations. Stormwater treatment regulations focus on three primary areas:

1. Reduction of postdevelopment peak flows to predevelopment peak flows for a given return period

2. Infiltration of stormwater from the developed site that is equivalent to some fraction (for example, 90%) of the infiltration of the predevelopment site

3. Removal of contaminants, most commonly suspended solids, but also oils and greases, nutrients, pathogens, synthetic organics, and so on.

A variety of treatment technologies are available to treat the stormwater. A summary of the mechanisms by which contaminants are removed is provided in Table 6.2.

The treatment technologies most commonly used to treat stormwater are provided in Table 6.3. The most common treatment methods include ponds, infiltration basins, and infiltration trenches. These can be designed to allow standing water for a given amount of time and can be enhanced with vegetation that will optimize the removal of various contaminants via filtration, adsorption, or other removal processes. Often, of utmost importance is the treatment of the *first flush,* which is the stormwater that is produced in the early stages of a storm and has a high concentration of contaminants that have accumulated on the land since the last storm event.

Table 6.2 Stormwater treatment mechanisms

Mechanism	Pollutants Affected
Gravity settling of particulate pollutants	Solids, BOD, pathogens, particulate COD, phosphorus, nitrogen, synthetic organics, particulate metals
Filtration and physical straining of pollutants through a filter media or vegetation	Solids, BOD, pathogens, particulate COD, phosphorus, nitrogen, synthetic organics, particulate metals
Infiltration of particulate and dissolved pollutants	Solids, BOD, pathogens, particulate COD, phosphorus, nitrogen, synthetic organics, particulate metals
Adsorption on particulates and sediments	Dissolved phosphorus, metals, synthetic organics
Photodegradation	COD, petroleum hydrocarbons, synthetic organics, pathogens
Gas exchange and volatilization	Volatile organics, synthetic organics
Biological uptake and biodegradation	BOD, COD, petroleum hydrocarbons, synthetic organics, phosphorus, nitrogen, metals
Chemical precipitation	Dissolved phosphorus, metals
Ion exchange	Dissolved metals
Oxidation	COD, petroleum hydrocarbons, synthetic organics
Nitrification and denitrification	Ammonia, nitrate, nitrite
Density separation and removal of floatables	Petroleum hydrocarbons

Source: *Connecticut Storm Water Quality Management Manual*, Connecticut Department of Environmental Protection.

Table 6.3 Common stormwater treatment technologies

Conventional Practices	Reasons for Limited Use	Suitable Applications
Dry Detention Ponds	■ Not intended for water quality treatment. Designed to empty out between storms; lack the permanent pool or extended detention required for adequate stormwater treatment ■ Settled particulates can be resuspended between storms	Flood control and channel protection
Catch Basins	■ Limited pollutant removal ■ No volume control ■ Resuspension of settled particulates	■ Pretreatment or in combination with other stormwater treatment practices ■ Stormwater retrofits

(continued)

Conventional Practices	Reasons for Limited Use	Suitable Applications
Conventional Oil/Particle Separators	■ Limited pollutant removal ■ No volume control ■ Resuspension of settled particulates	■ Pretreatment or in combination with other stormwater treatment practices ■ Highly impervious areas with substantial vehicle traffic
Underground Detention Facilities	■ Not intended for water quality treatment ■ Particulates can be resuspended between storms	■ Flood control and channel protection ■ Space-limited or ultra-urban sites
Permeable Pavement	■ Reduced performance in cold climates due to clogging by road sand and salt ■ Porous asphalt or concrete recommended for limited use in some areas	Modular concrete paving blocks, modular concrete or plastic lattice, or cast-in-place concrete grids are suitable for use in spillover parking, parking aisles, residential driveways, and roadside rights-of-way
Dry Wells	■ Not intended as stand-alone stormwater runoff quality or quantity control ■ Potential for clogging/failure ■ Applicable to small drainage areas ■ Potential groundwater quality impacts	■ Infiltration of clean rooftop runoff ■ Stormwater retrofits ■ Space-limited ultra-urban ■ Pretreatment or in combination with other stormwater treatment practices
Vegetated Filter Strips	Typically, cannot alone achieve the 80% TSS removal goal	■ Pretreatment or in combination with other treatment practices ■ Limited groundwater recharge ■ Outer zone of a stream buffer ■ Residential applications and parking lots
Grass Drainage Channels	Typically, cannot alone achieve the 80% TSS removal goal	■ Part of runoff conveyance system to provide pretreatment ■ Replace curb and gutter drainage ■ Limited groundwater recharge
Level Spreaders	Typically, cannot alone achieve the 80% TSS removal goal	■ Pretreatment or in combination with other treatment practices ■ Use with filter strips and at outlets of other treatment practices to distribute flow ■ Groundwater recharge

Source: *Connecticut Storm Water Quality Management Manual*, Connecticut Department of Environmental Protection.

The quantity of flow exiting a stormwater *BMP (best management practice)* is selected based on the level of protection required and the intended purpose of the BMP. Five possible purposes of stormwater treatment are described as follows:

1. *Recharge.* Capture rainfall events in order to replenish the groundwater (especially important for low-intensity events)

2. *Water quality.* Capture and treat a significant volume of water such that an appreciable positive impact is made on water quality

3. *Channel protection.* Detain or otherwise capture runoff such that flows are attenuated to an extent that the integrity of the channel is not compromised

4. *Overbank flood protection.* Detain stormwater such that as a result of development, the overtopping of stream banks is no more frequent than it was prior to development

5. *Extreme flood protection.* Contain the largest and most infrequent storm events

As BMPs are designed to offer additional levels of protection, the required size (and cost) also increases as shown in Figure 6.23.

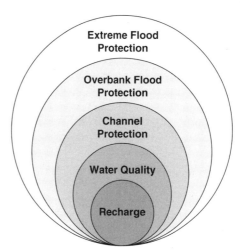

Figure 6.23 BMP sizing

Source: *Minnesota Stormwater Manual.* Reprinted courtesy of the Minnesota Pollution Control Agency.

6.4.1 Detention Ponds

Detention ponds allow for the storage of stormwater. As such, they attenuate the peak flows, provide some removal of suspended solids by settling, and provide removal of other contaminants by processes such as adsorption to sediments. A pond can either be classified as a *wet pond* or a *dry pond.* A wet pond retains water between storm events by virtue of a relatively impermeable liner or due to a high groundwater table. Dry ponds do not retain water for a significant length of time, but they do help to attenuate peak flows through the geometry of the outlet structure and can provide some removal of contaminants. Wet detention ponds may serve as an amenity to a developed site if properly designed and integrated.

Figure 6.24 provides a plan view of a detention pond. Note the irregular shape that is desired to make the engineered pond look "natural." A *forebay* is illustrated in Figure 6.24, which is a small pretreatment area designed to remove a significant portion of suspended solids and intended to be much more easily cleaned as compared to the larger pond. An *aquatic bench* is provided to allow pond vegetation to flourish. A *safety bench* is provided on ponds with relatively steep sides (4:1 and steeper) to provide a safe spot for someone who has entered the pond. Perimeter fences may also be required. Native landscaping is often required around ponds. The *riser structure* is an outlet device that controls the rate at which water leaves the pond.

Figure 6.25 provides a section view of a typical detention pond. Note the riser with the single square orifice, which empties out from the pond to a stable outfall. Various water elevations and their intended functions are also shown in Figure 6.25. An emergency spillway should be provided in the event of a storm event much larger than the return period used in designing the pond. In such a case, the emergency spillway provides a path for the water to travel that will not adversely affect the integrity of the embankment that retains the water.

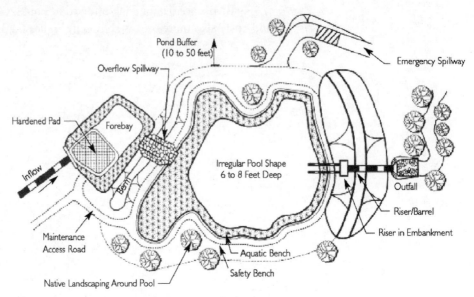

Figure 6.24 Plan view of a detention pond

Source: *Connecticut Storm Water Quality Management Manual*, Connecticut Department of Environmental Protection.

Ponds are perhaps the most common forms of stormwater management, but they do have limitations. They tend to use a large amount of land area and may pose a safety hazard. Unlined ponds may adversely affect groundwater. Also, it is important to note that ponds do not decrease the amount of stormwater *volume,* other than a small portion that evaporates and infiltrates.

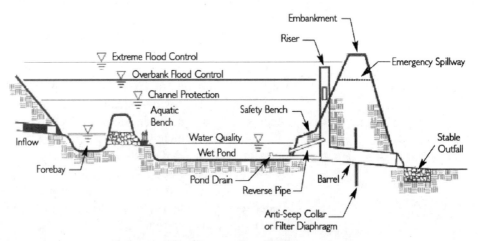

Figure 6.25 Section view of a detention pond

Source: *Connecticut Storm Water Quality Management Manual*, Connecticut Department of Environmental Protection.

A pond can be drained by any number of outlet devices. For example, a riser is depicted in Figure 6.26. Other options include broad-crested weirs, sharp crested weirs, stone weepers, level spreaders, etc. The discharge equations for various outlet devices are provided in Table 6.4.

Two curves that are of great use when analyzing pond performance are the *stage-storage* curve (storage in pond vs. stage elevation) and the *stage-discharge* curve (discharge from pond vs. stage elevation). The stage-storage curve only requires information on the geometry of the pond. The storage for various "slices" of the pond can be determined using the *end area* method as shown in Example 6.11. Creating a stage-discharge curve requires knowing information about the discharge device geometry.

Table 6.4 Discharge equations for outflow devices

Stone weeper	$Q = \dfrac{h^{3/2} W}{\left[L/D + 2.5 + L^2 \right]^{0.5}}$	Q = total flow through dam (cfs) h = ponding depth (ft) W = total length of dam (ft) L = horizontal flow path length (ft) D = average rock diameter (ft)
Orifice	$Q = C_d \cdot A \cdot (2\,gh)^{0.5}$	Q = discharge (L³/T) C_d = coefficient of discharge (typically 0.6) h = height of water above orifice (L)
Broad-crested weir	$Q = 0.385\ C \cdot L \cdot (2\,g)^{0.5} \cdot H^{1.5}$	L = length of weir (L) C = discharge coefficient (see Figure 6.27) H = head of water (L)
Sharp-crested weir	$Q = K \cdot (2\,g)^{0.5} \cdot H^{1.5}$	$K = 0.4 + 0.05(H/P)$ H = head of water (L) P = height of weir

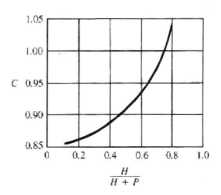

Figure 6.26 23 C-value for broad-crested weirs

Source: J. A. Roberson, J. J. Cassidy, and M. H. Chaudhry, *Hydraulic Engineering*, 2nd Edition © 2001, John Wiley & Sons, Inc. Reprinted by permission.

Example **6.11**

Stage-storage relationship

Compute the stage-storage relationship for the pond geometry in Exhibit 10.

Elevation (ft)	Area (ft²)
101	20,000
103	40,100
105	65,700
107	108,300

Exhibit 10 Stage-area information

Solution

The volume of storage in a "slice" between two adjacent elevation contours can be determined by knowing the areas corresponding to each contour. The volume of a slice bounded by two areas A_1 and A_2 is given as

$$V_{slice} = \frac{A_1 + A_2}{2} \times (\Delta E)$$

where ΔE is the difference in elevation of the contours corresponding to A_1 and A_2.

The volumes for each slice are computed and entered into Exhibit 11. The volumes for each slice are summed up in the final column to represent the total storage in the pond as a function of water elevation.

Elevation (ft)	Area (ft²)	Slice Volume (acre/ft)	Pond Storage Volume (acre/ft)
101	20,000		
103	40,100	0.92	0.92
105	65,700	1.18	2.10
107	108,300	1.96	4.06

Exhibit 11   Calculating storage volume

Example 6.12

Discharge-elevation curve

For the discharge device shown in Exhibit 12, create a discharge-elevation curve. The riser pipe includes a 4-inch-diameter orifice with an invert elevation of 533 feet and a rectangular opening (4 ft wide by 2 ft tall, invert at 536 ft).

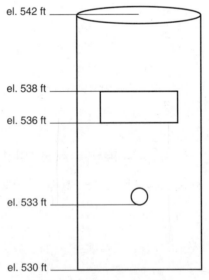

el. 542 ft

el. 538 ft

el. 536 ft

el. 533 ft

el. 530 ft

Exhibit 12   Discharge device

Solution

Before attempting to solve this problem numerically, it is important to understand conceptually how the flow from this device will vary with elevation. This variation is shown in Exhibit 13.

Elevation Range	Type of Flow
530 ft – 533 ft	No flow
533 ft – 536 ft	Orifice flow through 4-inch orifice
536 ft – 538 ft	Orifice flow through 4-inch orifice and weir flow through rectangular opening
538 ft – 542 ft	Orifice flow through 4-inch orifice and orifice flow through rectangular opening

Exhibit 13   Description of flow as depth varies

The flow through the orifice is governed by the orifice equation provided in Table 6.4:

$$Q = C_d \cdot A \cdot (2 \; g \cdot h)^{0.5}$$

where

$$C_d = 0.6$$

$$A = \pi \cdot (4/12 \; \text{ft})^2/4 \cdot (2 \cdot 32.2 \; \text{ft/s}^2 \cdot h)^{0.5}$$

Thus, the flow through the 4-inch orifice can be expressed as

$$Q_{4\text{-inch orifice}} = 0.7 \; h^{0.5}$$

where h is measured in feet above the invert of the orifice.

The flow through the rectangular opening operating under weir conditions is provided in Table 6.4, assuming the opening acts as a broad-crested weir:

$$Q = K \cdot (2 \; g)^{0.5} \cdot h^{1.5} \quad \leftarrow \text{sharp crested}$$

where

$$K = 0.4 + 0.05(h/P)$$

$$P = 6 \; \text{ft}$$

Thus, the flow through the rectangular opening when acting as a weir, $Q_{top, weir}$, can be expressed as

$$Q_{top, weir} = (0.4 + 0.0083 \; h) \cdot (8) \cdot h^{1.5}$$

or

$$Q_{top, weir} = 3.2 \; h^{1.5} + 0.066 \; h^{2.5}$$

where h is in feet above the invert of the rectangular opening.

The flow through the rectangular opening operating under orifice flow, $Q_{top, orifice}$, can be obtained in a similar manner to that used for the 4-inch diameter orifice. In the case of the weir, $A = 8 \; \text{ft}^2$. Thus, $Q_{top, orifice} = 0.6 \cdot 8 \; \text{ft}^2 \cdot (2 \; gh)^{0.5} = 38.5 \; h^{0.5}$, where h is in ft above the invert of the rectangular orifice.

The three discharge equations can now be solved for by inserting in appropriate values of h. Values are shown in Exhibit 14.

Elevation	4-inch Orifice		Rectangular Opening Operating as a Weir		Rectangular Opening Operating as an Orifice		Total Discharge (cfs)
	h (ft)	Q (cfs)	h (ft)	Q (cfs)	h (ft)	Q (cfs)	
530							0.00
531							0.00
532							0.00
533	0	0.0					0.00
534	1	0.7					0.70
535	2	1.0					0.99
536	3	1.2	0	0.0			1.21
537	4	1.4	1	3.3			4.67
538	5	1.6	2	9.4			10.99
539	6	1.7			0	0.0	1.71
540	7	1.9			1	38.5	40.35
541	8	2.0			2	54.4	56.43
542	9	2.1			3	66.7	68.78

Exhibit 14   Stage-discharge data

Sizing Detention Ponds

Several methods exist for sizing detention ponds, many of them performed readily by computers. However, useful estimates for pond volumes can be obtained from noncomputerized methods.

The concept of volume requirements is relatively straightforward and is presented graphically in Figure 6.27. This figure shows the predevelopment hydrograph and the postdevelopment hydrograph without any detention. The excess runoff volume before time t^* of the postdevelopment hydrograph as compared to the predevelopment hydrograph, denoted by the shaded area, must be stored if the peak flow after development is to be equal to the peak flow before development. After time t^* has passed, the excess volume can be released in a controlled manner based on the design of the outlet structure.

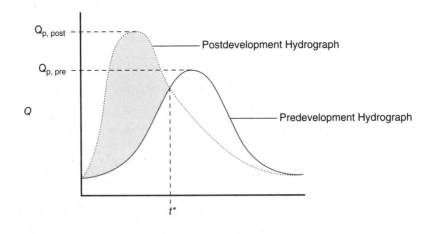

Figure 6.27   Illustration of detention storage volume

Maidment (1993) presents an equation to estimate the volume required for stormwater detention:

$$V_{est} = 1.1 \cdot V_{in} \cdot \left(\frac{Q_{p,in}}{Q_{p,out}} \right) \tag{6.2}$$

where

V_{est} = an estimate of the storage volume (L^3)

V_{in} = the volume corresponding to the inflow hydrograph (equivalent to the postdevelopment hydrograph) (L^3)

$Q_{p,in}$ = the peak flow of the inflow hydrograph (L^3/T)

$Q_{p,out}$ = the desired peak flow exiting the pond (L^3/T)

This equation provides only an estimate for sizing the pond. Actual pond design must take into account the site topography and the design of the outlet structure. The inflow hydrograph then needs to be routed through the pond to determine the actual time variant flow of stormwater leaving the pond.

A more common method of estimating pond volume requirements is the NRCS method (section 9.1.4). The basis of the NRCS method is shown in Figure 6.28. The method may be summarized as follows:

■ Decide on a value for q_o, the desired peak outflow discharge.

■ Estimate the peak inflow discharge, q_i. This is the peak flow, for a given return period, for the *developed* watershed. The NRCS method for determining peak flow (section 5.1.5) can be used.

■ Compute q_o/q_i. Given this quantity, look up V_s/V_r in Figure 6.29.

■ Determine the runoff depth Q for postdeveloped conditions and convert this depth to a volume by multiplying by the watershed area. This product is V_r, the runoff volume in Figure 6.29.

■ Then determine V_s, the volume of storage: $V_s = V_r \cdot (V_s/V_r)$.

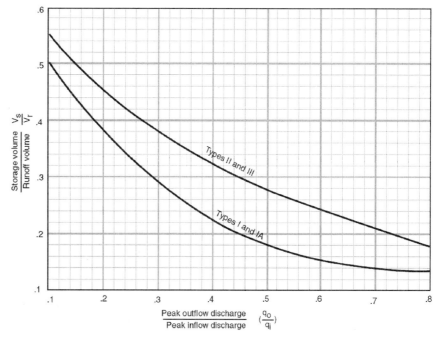

Figure 6.28 Approximate pond sizing using the NRCS method

Source: U. S. Department of Agriculture, *Urban Hydrology for Small Watersheds*.

Example **6.13**

Detention pond sizing

A 60-acre catchment drains to the detention pond examined in Example 6.11. Assess whether the size of the pond is appropriate if the peak flows for the 2-year and 10-year storms are to be equal before and after development of the watershed. Assume a Type II rainfall distribution. Appropriate information is summarized in Exhibit 15.

	Predevelopment	Postdevelopment
q_p (cfs), $T = 2$ yr	43	122
q_p (cfs), $T = 10$ yr	71	162
Q (in), $T = 2$ yr	0.8	2.0
Q (in), $T = 10$ yr	1.4	3.0

Exhibit 15   Predevelopment and postdevelopment flows

Solution

(a) For the 2-year storm, q_o/q_i = 43 cfs/122 cfs = 0.35. That is, the peak flow entering the pond will equal 122 cfs (the postdevelopment peak runoff), and the peak flow exiting the pond is forced to be the same as the flow that exited the watershed prior to development.

Given a Type II rainfall distribution, a value of V_s/V_r of 0.35 is obtained from Figure 6.29.

The next step is to determine the volume of storage required. First, the volume of runoff for the developed site is determined: V_r = (2 in) · (60 acre) · (1 ft/12 in) = 10 acre-ft. Given that V_s/V_r = 0.35, the volume of storage required is 3.5 acre-ft.

By inspecting the stage-storage relationship from Example 6.11, it is noted that the pond will be able to store 3.5 acre-ft. By interpolation, the maximum elevation of water in the pond will be 106.4 ft.

(b) For the 10-year storm, q_o/q_i = 71 cfs/162 cfs = 0.44.

A value of V_s/V_r of 0.30 is obtained from Figure 6.29.

V_r = (3 in) · (60 acre) · (1 ft/12 in) = 15 acre-ft. Given that V_s/V_r = 0.3, the volume of storage required is 4.5 acre-ft.

By inspecting the stage-storage relationship from Example 6.11, it is noted that the pond will not be able to store 4.5 acre-ft and that overtopping will most likely occur. In other words, no matter what type of outlet device is designed, the 10-year pre- and postdevelopment peak flows can never match for this given pond geometry.

6.4.2 Infiltration Techniques

Infiltration increasingly is specified as a means to lower runoff volumes, but, more important, it offers a means of directly recharging the groundwater. In operation, stormwater is diverted to the infiltration system, where it is infiltrated within less than 24–48 hours. Permeable soils are required, and the bottom of the infiltration system should be a safe distance above the seasonally high groundwater table, such as 3 feet. Infiltration may be accomplished using trenches, basins, beds, dry wells, and underground storage/infiltration units. Pretreatment is required for infiltration systems and may take the form of filter strips, sedimentation basins, etc. Plan and profile views of an infiltration basin are shown in Figure 6.29 and Figure 6.30, respectively.

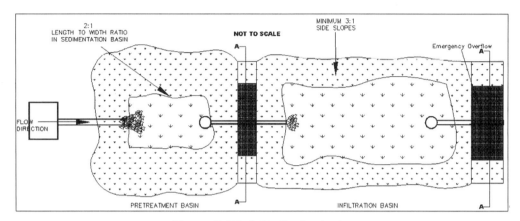

Figure 6.29   Infiltration basin plan

Source: *Minnesota Stormwater Manual.* Reprinted courtesy of the Minnesota Pollution Control Agency.

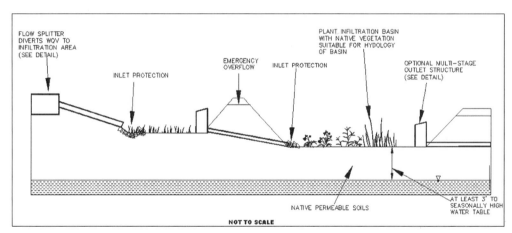

Figure 6.30   Infiltration basin profile

Source: *Minnesota Stormwater Manual.* Reprinted courtesy of the Minnesota Pollution Control Agency.

To design an infiltration system, the first step is to determine the volume of water to be infiltrated. The appropriate regulatory agency may express this in terms of a volume (for example, watershed-inches), as a percentage of the quantity of infiltration prior to development, etc.

Next, the maximum depth of the system is specified as follows:

$$D = i \times t$$

where

i = infiltration rate (in/hr)

t = maximum drawdown time (hr)

The infiltration rate is a function of the soil type. The maximum drawdown time is typically set by code, and 24 or 48 hours are typical.

The area of the infiltration system is determined based on the volume required for infiltration and the depth of the system. For a basin, the area is

$$A_{basin} = (\text{design volume})/D$$

For a trench that is filled with porous material, such as gravel, the required area is

$$A_{trench} = (\text{design volume})/(n \cdot D)$$

where n is the porosity of the filter media.

As compared to detention ponds, infiltration trenches are less apt to increase stormwater temperatures, directly recharge the groundwater, and increase water quality. However, they are prone to clogging, which greatly reduces their effectiveness. They also offer a quick conduit of surface water into the aquifer and thus pose the potential for groundwater contamination.

6.5 REGULATORY FRAMEWORK

In 1987, Congress required the U.S. Environmental Protection Agency to amend the Clean Water Act to require NPDES (National Pollutant Discharge Elimination System) permits for stormwater. Such requirements represented a significant addition to the Clean Water Act, as it had previously focused on industrial wastewater effluents and municipal sewage treatment. Phase I of these requirements were addressed in 40 CFR 122.26 for *MS4s (municipal separate storm sewer systems)* serving populations of 100,000 or more. Phase II requirements were addressed in 40 CFR 122.30–122.37 and targeted MS4s serving between 10,000 and 100,000 people.

Permit applications were to propose a stormwater management program that would reduce stormwater pollutants to the "maximum extent practicable" (MEP). These programs were to accomplish the following:

- Identify major outfalls and pollutant loadings

- Detect and eliminate nonstormwater discharges to the system

- Reduce pollutants in runoff from industrial, commercial, and residential areas

- Control stormwater discharges from new development and redevelopment area

Of great importance to practicing civil and environmental engineers in the land development area has been the phase II requirements for stormwater management from construction site. A management plant must be submitted for sites disturbing more than one acre of land. Most states are authorized by the U.S. EPA to implement the phase II requirements for new construction sites. State regulations vary with respect to submission of stormwater management plan, extent of treatment required, hydrologic return periods that apply to these quantities, etc.

ADDITIONAL RESOURCES

1. Atlanta Regional Commission. *Georgia Stormwater Management Manual.* Accessed from *www.georgiastormwater.com.*

2. Connecticut Department of Environmental Protection. *Connecticut Stormwater Quality Manual.* Accessed from *http://dep.state.ct.us/wtr/ stormwater/strmwtrman.htm#chapter.*

3. Haestad Methods and S. R. Durrans. *Stormwater Conveyance Modeling and Design.* The Bentley Institute Press, 2003.

4. Maidment, David R., ed. *Handbook of Hydrology.* McGraw-Hill, 1993.

5. U.S. Department of Transportation, *Hydraulic Design of Highway Culverts.* Publication No. FHWA-NHI-01-020, 2001.

Drinking Water Treatment and Conveyance

The purpose of drinking water treatment is to create *potable water*. Potable water is water that is safe for human consumption. Although water-borne diseases are relatively rare in the United States, they are responsible for millions of deaths worldwide. Some of the most common water-borne diseases are shown in Table 7.1.

Table 7.1 Common water-borne diseases

Disease	Effects	Scope
Arsenic poisoning	Cancers of skin, bladder, kidneys, and lungs	Currently no reliable worldwide estimate; up to 77 million in Bangladesh affected
Cholera	Severe diarrhea and vomiting leading to extreme dehydration and death	5000 deaths in year 2000
Campylobacteriosis	Diarrhea (often including the presence of mucus and blood), abdominal pain, malaise, fever, nausea, and vomiting	Approximately 5%–14% of all diarrhea worldwide is thought to be caused by Campylobacter
Malaria	Fever, chills, headache, muscle aches, tiredness, nausea and vomiting, diarrhea, anemia, and jaundice (yellow coloring of the skin and eyes). Convulsions, coma, severe anemia, and kidney failure can also occur.	300–500 million cases of malaria, with over 1 million deaths each year
Methaemoglobinemia	Reduced ability of blood to transport oxygen; highest risk for bottle-fed infants	Relatively rare
Typhoid	Symptoms can be mild or severe and include sustained fever as high as 39°C–40°C, malaise, anorexia, headache, constipation or diarrhea, rose-colored spots on the chest area, and enlarged spleen and liver.	17 million cases worldwide

Source: World Health Organization,*www.who.int/water_sanitation_health/diseases/diseasefact/en/index.html.*

The diseases described in Table 7.1 are quite rare in developed countries thanks to effective wastewater treatment *and* effective drinking water treatment. However, water-borne disease outbreaks do occur, such as the cryptosporidium outbreak in Milwaukee, Wisconsin, in 1993, that sickened an estimated 400,000 people and killed more than 100 people.

7.1 CONTAMINANTS IN DRINKING WATER TREATMENT

Contaminants in drinking water are categorized as *primary contaminants* or *secondary contaminants*. Primary contaminants are those that are believed to have deleterious effects on human health, and the corresponding *primary standards* are legally enforceable standards that apply to public water systems. Secondary contaminants are those contaminants that may cause cosmetic effects (such as skin or tooth discoloration) or aesthetic effects (such as taste, odor, or color) in drinking water. The U.S. Environmental Protection Agency (U.S. EPA) recommends secondary standards to water systems but does not require systems to comply. However, states may choose to adopt secondary standards as enforceable standards.

Vocabulary associated with drinking water treatment standards from the U.S. EPA are defined as follows:

■ Maximum Contaminant Level Goal (MCLG). The level of a contaminant in drinking water below which there is no known or expected risk to health. MCLGs allow for a margin of safety and are nonenforceable public health goals.

■ Maximum Contaminant Level (MCL). The highest level of a contaminant that is allowed in drinking water. MCLs are set as close to MCLGs as feasible using the best available treatment technology and taking cost into consideration. MCLs are enforceable standards.

▪ Maximum Residual Disinfectant Level Goal (MRDLG). The level of a drinking water disinfectant below which there is no known or expected risk to health. MRDLGs do not reflect the benefits of the use of disinfectants to control microbial contaminants.

▪ Maximum Residual Disinfectant Level (MRDL). The highest level of a disinfectant allowed in drinking water.

There are six main type of primary contaminants: microorganisms, disinfection by-products, disinfectants, inorganic chemicals, organic chemicals, and radionuclides.

Microorganisms include bacteria, viruses, and parasitic protozoa. Bacteria are single-celled organisms and are the most primitive living things. Viruses are infectious agents that cannot reproduce themselves but "reprogram" host cells to reproduce viral material. Viruses are the smallest of the infectious organisms (submicron in size). Hepatitis A is an example of a virus. Protozoa are also single-celled organisms, more complex than bacteria, with giardia and cryptosporidium being two of the more common protozoa. Protozoa range in size from 1 μm to 2000 μm. They form cysts with thick cell walls that increase the difficulty of their removal. The U.S. EPA's primary standards for microorganisms are provided in Table 7.2.

Table 7.2   Primary standards for microorganisms

Contaminant	MCLG (mg/L)	MCL (mg/L)	Potential Health Effects from Ingestion of Water	Sources of Contaminant in Drinking Water
Cryptosporidium	zero	99% removal (as of 1/1/02 for systems serving > 10,000 and 1/14/05 for systems serving < 10,000)	Gastrointestinal illness (e.g., diarrhea, vomiting, cramps)	Human and fecal animal waste
Giardia lamblia	zero	99.9% removal/inactivation	Gastrointestinal illness (e.g., diarrhea, vomiting, cramps)	Human and animal fecal waste
Heterotrophic plate count (HPC)	n/a	No more than 500 bacterial colonies per milliliter.	HPC has no health effects; it is an analytic method used to measure the variety of bacteria that are common in water. The lower the concentration of bacteria in drinking water, the better maintained the water system is.	HPC measures a range of bacteria that are naturally present in the environment
Legionella	zero	No limit, but EPA believes that if *Giardia* and viruses are removed/inactivated, *Legionella* will also be controlled.	Legionnaire's disease, a type of pneumonia	Found naturally in water; multiplies in heating systems
Total coliforms (including fecal coliform and *E. coli*)	zero	5.0%	Not a health threat in itself; it is used to indicate whether other potentially harmful bacteria may be present	Coliforms are naturally present in the environment, as well as feces; fecal coliforms and *E. coli* only come from human and animal fecal waste.

(continued)

Contaminant	MCLG (mg/L)	MCL (mg/L)	Potential Health Effects from Ingestion of Water	Sources of Contaminant in Drinking Water
Turbidity	n/a	At no time can turbidity (cloudiness of water) go above 5 nephelolometric turbidity units (NTU); systems that filter must ensure that the turbidity go no higher than 1 NTU (0.5 NTU for conventional or direct filtration) in at least 95% of the daily samples in any month. As of January 1, 2002, turbidity may never exceed 1 NTU, and must not exceed 0.3 NTU in 95% of daily samples in any month.	Turbidity is a measure of the cloudiness of water. It is used to indicate water quality and filtration effectiveness (e.g., whether disease-causing organisms are present). Higher turbidity levels are often associated with higher levels of disease-causing microorganisms, such as viruses, parasites, and some bacteria. These organisms can cause symptoms such as nausea, cramps, diarrhea, and associated headaches.	Soil runoff
Viruses (enteric)	zero	99.99% removal/inactivation	Gastrointestinal illness (e.g., diarrhea, vomiting, cramps)	Human and animal fecal waste

Source: *www.epa.gov/safewater/mcl.html#mcls.*

Disinfection by-products (DBPs) are pollutants that are created as a result of the reaction between a disinfectant (for example, chlorine or ozone) and naturally occurring organic materials in water. DBPs provide a challenge to water utilities, as the utility must balance the risk of too little disinfection (that is, increased risk of microbial-related disease) with too much disinfection (creation of disinfection by-products). Chlorine, which is one of the most common and effective disinfectants, can react with naturally occurring organic material to form trihalomethanes (THMs) and haloacetic acids (HAAs). Trihalomethanes include four chemicals: chloroform, bromoform, bromodichloromethane, and dibromochloromethane. Table 7.3 provides further information about disinfection by-products, including their primary standards.

Table 7.3 Primary standards for disinfection by-products

Contaminant	MCLG (mg/L)	MCL (mg/L)	Potential Health Effects from Ingestion of Water	Sources of Contaminant in Drinking Water
Bromate	zero	0.01	Increased risk of cancer	By-product of drinking water disinfection
Chlorite	0.8	1.0	Anemia; infants and young children: nervous system effects	By-product of drinking water disinfection
Haloacetic acids (HAAs)	n/a[a]	0.06	Increased risk of cancer	By-product of drinking water disinfection
Total trihalomethanes (TTHMs)	none[b] ---------- n/a[a]	0.10 ---------- 0.08	Liver, kidney ,or central nervous system problems; increased risk of cancer	By-product of drinking water disinfection

[a] Although there is no collective MCLG for this contaminant group, there are individual MCLGs for some of the individual contaminants:

- Trihalomethanes: bromodichloromethane (zero); bromoform (zero); dibromochloromethane (0.06 mg/L). Chloroform is regulated with this group but has no MCLG.
- Haloacetic acids: dichloroacetic acid (zero); trichloroacetic acid (0.3 mg/L). Monochloroacetic acid, bromoacetic acid, and dibromoacetic acid are regulated with this group but have no MCLGs.

[b] MCLGs were not established before the 1986 Amendments to the Safe Drinking Water Act. Therefore, there is no MCLG for this contaminant.

Source: *www.epa.gov/safewater/mcl.html#mcls.*

The EPA also regulates disinfectants found in drinking water. A relatively high concentration of disinfectant can cause adverse health impacts in addition to the formation of disinfection by-products. Primary standards are provided in Table 7.4.

Table 7.4 Primary standards for disinfectants

Contaminant	MRDLG (mg/L)	MRDL (mg/L)	Potential Health Effects from Ingestion of Water	Sources of Contaminant in Drinking Water
Chloramines (as Cl_2)	4	4.0	Eye/nose irritation; stomach discomfort, anemia	Water additive used to control microbes
Chlorine (as Cl_2)	4	4.0	Eye/nose irritation; stomach discomfort	Water additive used to control microbes
Chlorine dioxide (as ClO_2)	0.8	0.8	Anemia; infants and young children: nervous system effects	Water additive used to control microbes

So.urce: *www.epa.gov/safewater/mcl.html#mcls.*

The primary standards also address a variety of inorganic chemicals. Many of these substances are required in trace amounts in a healthy diet, but a wide range of adverse health impacts occur due to high doses (see Table 7.5). Many of these contaminants are naturally occurring but are released directly into the environment via human activity (for example, mining and industrial processes) or are released indirectly (for example, overpumping of groundwater aquifers that exposes arsenic in geologic formations to air, which increases arsenic solubility in water).

Table 7.5 Primary standards for selected inorganic chemicals

Contaminant	MCLG (mg/L)	MCL (mg/L)	Potential Health Effects from Ingestion of Water	Sources of Contaminant in Drinking Water
Antimony	0.006	0.006	Increase in blood cholesterol; decrease in blood sugar	Discharge from petroleum refineries; fire retardants; ceramics; electronics; solder
Arsenic	0[a]	0.010 as of 01/23/06	Skin damage or problems with circulatory systems, and may have increased risk of getting cancer	Erosion of natural deposits; runoff from orchards, runoff from glass and electronics production wastes
Asbestos (fiber > 10 micrometers)	7 million fibers per liter	7 MFL	Increased risk of developing benign intestinal polyps	Decay of asbestos cement in water mains; erosion of natural deposits
Barium	2	2	Increase in blood pressure	Discharge of drilling wastes; discharge from metal refineries; erosion of natural deposits
Beryllium	0.004	0.004	Intestinal lesions	Discharge from metal refineries and coal-burning factories; discharge from electrical, aerospace, and defense industries

(continued)

Contaminant	MCLG (mg/L)	MCL (mg/L)	Potential Health Effects from Ingestion of Water	Sources of Contaminant in Drinking Water
Cadmium	0.005	0.005	Kidney damage	Corrosion of galvanized pipes; erosion of natural deposits; discharge from metal refineries; runoff from waste batteries and paints
Chromium (total)	0.1	0.1	Allergic dermatitis	Discharge from steel and pulp mills; erosion of natural deposits
Copper	1.3	Treatment technique[b]; action level = 1.3	Short-term exposure: gastrointestinal distress Long-term exposure: liver or kidney damage People with Wilson's disease should consult their personal doctor if the amount of copper in their water exceeds the action level	Corrosion of household plumbing systems; erosion of natural deposits
Cyanide (as free cynanide)	0.2	0.2	Nerve damage or thyroid problems	Discharge from steel/metal factories; discharge from plastic and fertilizer factories
Fluoride	4.0	4.0	Bone disease (pain and tenderness of the bones); children may get mottled teeth	Water additive which promotes strong teeth; erosion of natural deposits; discharge from fertilizer and aluminum factories
Lead	zero	Treatment technique[b]; action level = 0.015	Infants and children: delays in physical or mental development; children could show slight deficits in attention span and learning abilities Adults: kidney problems; high blood pressure	Corrosion of household plumbing systems; erosion of natural deposits

[a] MCLGs were not established before the 1986 Amendments to the Safe Drinking Water Act. Therefore, there is no MCLG for this contaminant.

[b] Lead and copper are regulated by a Treatment Technique that requires systems to control the corrosiveness of their water. If more than 10% of tap water samples exceed the action level, water systems must take additional steps. For copper, the action level is 1.3 mg/L, and for lead it is 0.015 mg/L.

Source: *www.epa.gov/safewater/mcl.html#mcls.*

A wide range of organic chemicals are also regulated by the U.S. EPA, and a few of the more common contaminants are provided in Table 7.6. Many of these chemicals are synthetic chemicals, purposely created to benefit society (for example, herbicides to improve crop yields). Many of these standards are set extremely low (on the order of µg/L) due to their high toxicity (see section 4.5 for a discussion on risk assessment).

Table 7.6 Primary standards for selected organic chemicals

Contaminant	MCLG (mg/L)	MCL (mg/L)	Potential Health Effects from Ingestion of Water	Sources of Contaminant in Drinking Water
Acrylamide	zero	Treatment technique[a]	Nervous system or blood problems; increased risk of cancer	Added to water during sewage/wastewater treatment
Alachlor	zero	0.002	Eye, liver, kidney, or spleen problems; anemia; increased risk of cancer	Runoff from herbicide used on row crops
Atrazine	0.003	0.003	Cardiovascular system or reproductive problems	Runoff from herbicide used on row crops
Benzene	zero	0.005	Anemia; decrease in blood platelets; increased risk of cancer	Discharge from factories; leaching from gas storage tanks and landfills
Benzo(a)pyrene (PAHs)	zero	0.0002	Reproductive difficulties; increased risk of cancer	Leaching from linings of water storage tanks and distribution lines
Carbon tetrachloride	zero	0.005	Liver problems; increased risk of cancer	Discharge from chemical plants and other industrial activities
Chlordane	zero	0.002	Liver or nervous system problems; increased risk of cancer	Residue of banned termiticide
Chlorobenzene	0.1	0.1	Liver or kidney problems	Discharge from chemical and agricultural chemical factories
2,4-D	0.07	0.07	Kidney, liver, or adrenal gland problems	Runoff from herbicide used on row crops
Epichlorohydrin	zero	Treatment technique[a]	Increased cancer risk, and over a long period of time, stomach problems	Discharge from industrial chemical factories; an impurity of some water treatment chemicals
Lindane	0.0002	0.0002	Liver or kidney problems	Runoff/leaching from insecticide used on cattle, lumber, gardens
Polychlorinated biphenyls (PCBs)	zero	0.0005	Skin changes; thymus gland problems; immune deficiencies; reproductive or nervous system difficulties; increased risk of cancer	Runoff from landfills; discharge of waste chemicals
Tetrachlorethyleneo	zero	0.005	Liver problems; increased risk of cancer	Discharge from factories and dry cleaners
Toluene	1	1	Nervous system, kidney, or liver problems	Discharge from petroleum factories
Trichloroethylene	zero	0.005	Liver problems; increased risk of cancer	Discharge from metal degreasing sites and other factories

(continued)

Contaminant	MCLG (mg/L)	MCL (mg/L)	Potential Health Effects from Ingestion of Water	Sources of Contaminant in Drinking Water
Vinyl chloride	zero	0.002	Increased risk of cancer	Leaching from PVC pipes; discharge from plastic factories
Xylenes (total)	10	10	Nervous system damage	Discharge from petroleum factories; discharge from chemical factories

ª Each water system must certify, in writing, to the state (using third-party or manufacturer's certification) that when acrylamide and epichlorohydrin are used in drinking water systems, the combination (or product) of dose and monomer level does not exceed the levels specified, as follows:

- Acrylamide = 0.05% dosed at 1 mg/L (or equivalent)
- Epichlorohydrin = 0.01% dosed at 20 mg/L (or equivalent)

Source: *www.epa.gov/safewater/mcl.html#mcls.*

Radionuclides are compounds that emit radiation and, as shown in Table 7.7, can increase the risk of contracting cancer. They are often found in bedrock and dissolve readily in groundwater.

Table 7.7   Primary standards for radionuclides

Contaminant	MCLG (mg/L)	MCL (mg/L)	Potential Health Effects from Ingestion of Water	Sources of Contaminant in Drinking Water
Alpha particles	noneª ---------- zero	15 picocuries per Liter (pCi/L)	Increased risk of cancer	Erosion of natural deposits of certain minerals that are radioactive and may emit a form of radiation known as alpha radiation
Beta particles and photon emitters	noneª ---------- zero	4 millirems per year	Increased risk of cancer	Decay of natural and man-made deposits of certain minerals that are radioactive and may emit forms of radiation known as photons and beta radiation
Radium 226 and radium 228 (combined)	noneª ---------- zero	5 pCi/L	Increased risk of cancer	Erosion of natural deposits
Uranium	zero	30 µg/L as of 12/08/03	Increased risk of cancer, kidney toxicity	Erosion of natural deposits

ª MCLGs were not established before the 1986 Amendments to the Safe Drinking Water Act. Therefore, there is no MCLG for this contaminant.

Source: *www.epa.gov/safewater/mcl.html#mcls.*

The types of contaminants found in drinking water depend on the source of water. (Drinking water sources include groundwater or surface water.) For example, microorganisms are less likely to be found in groundwater as the natural filtration process of water moving through the subsurface can remove many of the microorganisms. On the other hand, radionuclides are predominantly found in groundwater sources, as the radionuclides found in subsurface geologic formations dissolve into the groundwater.

As mentioned previously, secondary contaminants are not enforced by the U.S. EPA, but some of the contaminants are enforced at the state level. Secondary standards are shown in Table 7.8.

Table 7.8 Standards for secondary contaminants

Contaminant	Secondary Standard
Aluminum	0.05 to 0.2 mg/L
Chloride	250 mg/L
Color	15 (color units)
Copper	1.0 mg/L
Corrosivity	noncorrosive
Fluoride	2.0 mg/L
Foaming agents	0.5 mg/L
Iron	0.3 mg/L
Manganese	0.05 mg/L
Odor	3 threshold odor number
pH	6.5–8.5
Silver	0.10 mg/L
Sulfate	250 mg/L
Total dissolved solids	500 mg/L
Zinc	5 mg/L

Source: *www.epa.gov/safewater/mcl.html#mcls.*

7.2 TREATMENT TECHNOLOGIES

A wide variety of unit processes are available to transform surface water and groundwater into potable water. This section will discuss the following treatment processes: lime softening, ion exchange, coagulation, and filtration. Disinfection processes, including UV disinfection, chlorination, and ozonation, will also be described. Additional processes that may be used for hazardous waste treatment as well as drinking water treatment are discussed in Chapters 11 and 12.

7.2.1 Lime Softening

The purpose of lime softening is to remove *hardness* from the water. Hardness does not cause any known adverse health impacts and is beneficial for corrosion control. However, water that is too hard is disliked by consumers because it leaves scale on plumbing fixtures and within hot water heaters and requires more detergents and soaps because it does not lather well.

Hardness is defined as the sum of all divalent cations in water. The predominant cations that cause hardness are calcium and magnesium. Hardness consists of *carbonate hardness* and *noncarbonate hardness,* depending on whether the hardness is associated with the carbonate ion (CO_3^{-2}). The sum of carbonate and noncarbonate hardness is known as *total hardness.* Carbonate hardness is equal to the alkalinity or total hardness, whichever is less. These terms are illustrated in Figure 7.1.

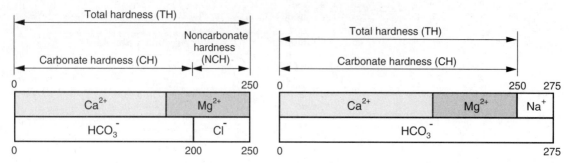

Figure 7.1   Hardness relationships

Source: M. L. Davis and S. J. Masten, *Principles of Environmental Engineering and Science*, © 2004, McGraw-Hill Education. Reprinted by permission of the McGraw-Hill Companies.

Softening is rarely required by regulatory agencies but may be implemented by a water utility for water with hardness greater than 150 mg/L.

The lime softening step is a precipitation reaction. The pH is increased by adding lime, typically to about 10.3 to remove hardness due to calcium and to about 11 to precipitate out magnesium hardness. A typical configuration for lime softening in practice is shown in Figure 7.2.

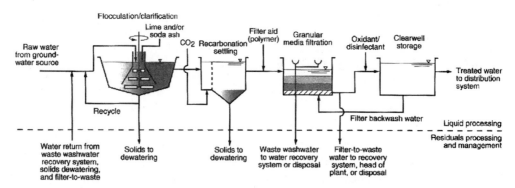

Figure 7.2   Typical lime softening process

Source: MWH, *Water Treatment: Principles and Design*, 2d ed., © 2005, John Wiley & Sons, Inc. Reprinted by permission.

Davis and Masten describe the lime softening process as consisting of five steps:

Step 1:   Add enough lime ($Ca(OH)_2$ in the following reaction) to neutralize free acids. This step will produce solid calcium carbonate precipitate, or lime sludge.

$$H_2CO_3 + Ca(OH)_2 \leftrightarrow CaCO_{3(s)} + 2H_2O$$

Step 2:   Precipitate carbonate hardness due to calcium by adding more lime to increase the pH.

$$Ca^{2+} + 2HCO_3^- + Ca(OH)_2 \leftrightarrow 2CaCO_{3(s)} + 2H_2O$$

Step 3: Precipitate carbonate hardness due to magnesium, if necessary.

$$Mg^{2+} + 2HCO_3^- + Ca(OH)_2 \leftrightarrow MgCO_{3(s)} + CaCO_{3(s)} + 2H_2O$$

$$Mg^{2+} + CO_3^{2+} + Ca(OH)_2 \leftrightarrow Mg(OH)_{2(s)} + CaCO_{3(s)}$$

Step 4: Remove noncarbonate hardness due to calcium by adding soda ash (Na_2CO_3).

$$Ca^{2+} + Na_2CO_3 \leftrightarrow CaCO_{3(s)} + 2Na^+$$

Step 5: Remove noncarbonate hardness due to magnesium by adding more lime.

$$Mg^{2+} + Ca(OH)_2 \leftrightarrow Mg(OH)_{2(s)} + Ca^{2+}$$

The quantity of lime required can be determined from stoichiometric relationships of the above equations. Alternatively, Kawamura (2000) provides the following relationship to determine the lime dosage required for removing carbonate hardness:

$$\text{Lime (CaO in mg/L)} = (A + B + C) \times D/E \tag{7.1}$$

$A = CO_2$ as CO_2 (mg/L) $\times 56/44$

$B = HCO_3$ alkalinity as $CaCO_3$ (mg/L) $\times 56/100$

$C = Mg$ as Mg (mg/L) $\times 56/24.3$

D = excess lime required to raise pH to 11 to remove magnesium; it is generally 1.1 to 1.2 (that is, an additional 10% to 20% of the sum of A, B, and C)

E = purity of the quick lime (0.88–0.95)

For slaked lime ($Ca(OH)_2$), the relationship is:

$$Ca(OH)_2 = (A' + B' + C') \times D'/E' \tag{7.2}$$

$A' = CO_2$ as CO_2 (mg/L) $\times 74/44$

$B' = HCO_3$ alkalinity as $CaCO_3 \times 74/100$

$C' = Mg$ as Mg (mg/L) $\times 74/24.3$

D' = excess lime for magnesium removal (1.1 to 1.2)

E' = purity of slaked lime (0.93–0.95)

Calculations involving softening equations often require the molecular weights of a few compounds. These weights are provided in Table 7.9.

Table 7.9 Molecular weights (g/mol) commonly used in hardness and alkalinity calculations

CaO	Ca(OH)₂	Na₂CO₃	CaCO₃	NaOH	CO₂	Ca	Mg	HCO₃⁻	CO₃⁻²
56	74	106	100	40	44	40	24.3	61	60

Example 7.1

Lime softening

A water sample has the characteristics shown in the Exhibit 1.

Ca^{2+}	5 eq/m³
Mg^{2+}	18 mg/L
HCO_3^-	180 mg/L
pH	7.9
CO_3^{-2}	0.02 eq/m³
CO_2	10^{-5} mol/L

Exhibit 1   Water sample characteristics

(a) Determine the alkalinity of the sample

(b) Estimate the annual mass of slaked lime required to treat this water, if the water is treated at a rate of 0.74 MGD

Solution

(a) Alkalinity is defined as

$$Alk = (HCO_3^-) + (CO_3^{-2}) + (OH^-) - (H^+)$$

For this problem

$$(HCO_3^-) = 180 \text{ mg/L} \cdot (1 \text{ mol}/60{,}000 \text{ mg}) \cdot (1 \text{ equivalent/mol})$$

$$= 0.003 \text{ eq/L}$$

$$(CO_3^{-2}) = 0.00002 \text{ eq/L}$$

$$(OH^-) = 10^{-(14-7.9)} \text{mol/L} \cdot (1 \text{ equivalent/mol})$$

$$= 7.9 \cdot 10^{-7} \text{eq/L}$$

$$(H^+) = 10^{-(7.9)} \cdot (1 \text{ equivalent/mol})$$

$$= 1.2 \cdot 10^{-8} \text{ mol/L} \cdot (1 \text{ equivalent/mol})$$

$$= 1.2 \cdot 10^{-8} \text{ eq/L}$$

Clearly, the alkalinity is due primarily to the HCO_3^- and equals 0.003 eq/L. Expressing this in terms of mg/L as $CaCO_3$:

$$0.003 \text{ eq/L} \cdot (1 \text{ mol}/2 \text{ equivalents}) \cdot (100 \text{ g/mol}) \cdot (1000 \text{ mg/g}) = 150 \text{ mg/L}$$

(b) The annual mass of lime required is the product of the flow rate and the concentration. The concentration can be obtained from Equation 7.2:

$$Ca(OH)_2 = (A' + B' + C') \times D'/E'$$

$$A' = 10^{-5} \text{ mol/L} \cdot 44 \text{ g/mol} \cdot (1000 \text{ mg/g}) \cdot 74/44$$

$$= 5.8 \text{ mg/L}$$

$$B' = 150 \text{ mg/L}$$

$$C' = 18 \text{ mg/L}$$

$$D' = 1.15 \text{ (assumed)}$$

$$E' = 0.94 \text{ (assumed)}$$

Thus, the concentration of $Ca(OH)_2$ = (5.8 mg/L + 150 mg/L + 18 mg/L) · (1.15/0.94), or 213 mg/L.

The daily mass is obtained using the relationship derived in section 1.2.1, or more precisely, Example 1.3.

$$\text{Daily mass of lime} = 8.34 \cdot 213 \text{ mg/L} \cdot 0.74 \text{ MGD}$$

$$= 1315 \text{ lb/day}$$

$$\text{Annual mass} = 240 \text{ tons}$$

7.2.2 Ion Exchange

Ion exchange is used widely in home water softeners and in commercial applications for removal of hardness but is less common in municipal drinking water treatment. The concept behind ion exchange is that opposite charges attract. An *ion exchange resin* has a charged surface, and undesirable ions in the water of the opposite charge (such as the divalent cations that cause water hardness) are attracted to the resin. Eventually, the resin surface becomes filled up (that is, its exchange capacity is *exhausted*), at which point the resin is washed with a regenerating solution. Once regenerated, the exchange resin can be reused for ion exchange.

Resins are either cationic (with a negative surface charge that attracts positive ions) or anionic (with a positive surface charge that attracts negative ions). Exchange capacity is expressed in terms of meq/mL or perhaps as kgr $CaCO_3$/ft^3 (kgr = kilograins, where 1 grain = 75 mg; 21.8 meq/mL in 1 kgr/ft^3).

Simplified ion exchange equations are provided below for the removal of carbonate hardness, noncarbonate hardness, and regeneration (U.S. EPA, 1996). In all cases, the ion exchange resin is denoted as X, and sodium is assumed to be the ion that exchanges from the resin.

Carbonate Hardness
$$Ca(HCO_3)_2 + Na_2X \rightarrow CaX + 2NaHCO_3$$
$$Mg(HCO_3)_2 + Na_2X \rightarrow MgX + 2NaHCO_3$$

Noncarbonate Hardness
$$CaSO_4 + Na_2X \rightarrow CaX + Na_2SO_4$$
$$CaCl_2 + Na_2X \rightarrow CaX + 2NaCl$$
$$MgSO_4 + Na_2X \rightarrow MgX + Na_2SO_4$$
$$MgCl_2 + Na_2X \rightarrow MgX + 2NaCl$$

Regeneration
$$CaX + 2NaCl \rightarrow CaCl_2 + Na_2X$$
$$MgX + 2NaCl \rightarrow MgCl_2 + Na_2X$$

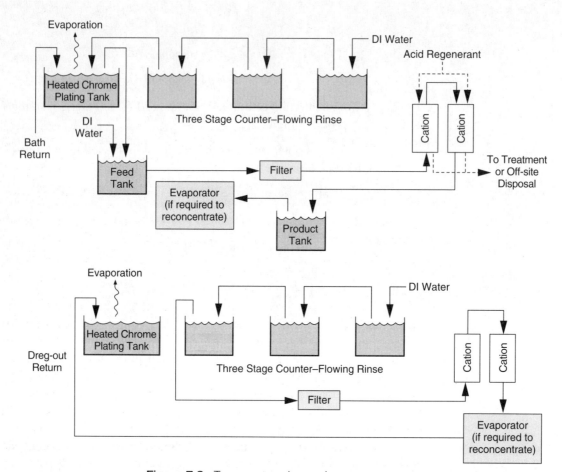

Figure 7.3 Two common ion exchange processes

Source: U.S. Environmental Protection Agency Office of Solid Waste, *Waste Minimization for Metal Finishing Facilities*, 1995.

A schematic of an ion exchange process in practice is shown in Figure 7.3.

Engineering considerations for an ion exchange system include:

- Characterizing the water source

- Planning for discharge of the waste brine

- Arranging for bench-scale and/or pilot-scale studies

- Specifying type and quantity of resin

- Specifying column details, such as number of columns, height of columns, and pressure drop

---

Example **7.2**

Ion exchange resin requirements

An ion exchange unit containing a resin with an exchange capacity of 1.1 meq/mL is to be used to remove nitrate from a groundwater source. The flow rate of the groundwater is 0.1 MGD, and the nitrate concentration is to be decreased from 30 mg/L to 0 mg/L. The goal is to regenerate the unit every other day. Estimate a volume of resin required.

Solution

The removal of nitrate will first be expressed in terms of eq/L:

$$(25 \text{ mg } NO_3^-/L) \cdot (1 \text{ mol}/62{,}000 \text{ mg}) \cdot (1 \text{ eq}/1 \text{ mol}) = 4 \cdot 10^{-4} \text{ eq/L}$$

The total equivalents of nitrate removed in two days is:

$$4 \cdot 10^{-4} \text{ eq/L} \cdot (0.1 \cdot 10^6 \text{ gal/day}) \cdot 2 \text{ days} \cdot (3.785 \text{ L}/1 \text{ gal}) = 303 \text{ eq}$$

The volume of resin required is:

$$303 \text{ eq} \cdot \frac{\text{mL}}{1.1 \text{ meq}} \cdot \frac{1000 \text{ meq}}{\text{eq}} \cdot \frac{1 \text{ m}^3}{10^6 \text{ mL}} = 0.28 \text{ m}^3$$

This value is a useful estimate to begin the design; a full-scale design covers many more details, some of which are mentioned above. The reader is referred to Chapter 16 of *Water Treatment Principles and Design,* listed in the additional resources at the end of this chapter.

7.2.3 Coagulation and Flocculation

Many of the turbidity-causing solids in water sources are colloidal in size. A *colloid* is a particle with a size between 1 nm and 1 μm ($10^{-9} - 10^{-6}$ m). Due to their small size, they remain in suspension and do not settle out on their own accord. One simple measure of the colloidal concentration in a water sample is its *turbidity,* or cloudiness. Turbidity units are NTU; refer to Table 7.2 for the U.S. EPA standard for turbidity

Colloids are said to be *stable* when they remain as individual colloids in suspension; that is, they do not attach to one another and form larger aggregates. The primary reason for colloid stability is the surface charges on the colloids. For example, many colloidal-sized particles of clay origin have a negatively charged surface, and these negatively charged surfaces repel one another.

The purpose of *coagulation* is to *destabilize* the colloidal material through the addition of chemical *coagulants*. The purpose of *flocculation* is to form relatively large *flocs* (on the order of 10^{-3} m) such that the flocs can be readily removed in subsequent settling and filtration steps. Coagulation processes occur in a matter of seconds, while flocculation requires times on the order of minutes.

In a typical coagulation/flocculation treatment train, water enters the treatment train and is rapidly mixed with the coagulant. The destabilized particles are slowly agglomerated in the flocculation basins by paddles or other devices. The flocs are removed by clarification, which is typically followed by filtration and disinfection.

Three mechanisms have been identified for the removal of colloidal material via coagulation: adsorption and charge neutralization, adsorption and interparticle bridging, and precipitation and enmeshment (MWH, 2005).

1. In the *adsorption and charge neutralization* mechanism, coagulant ions with a charge opposite of that of the colloidal material attach to the colloids. When the colloidal surface charge is neutralized, the colloids are destabilized and can form aggregates.

2. For the *adsorption and interparticle bridging* mechanism, charged sites on polymeric coagulants attach to oppositely charged sites on the colloids. These relatively large polymers attach to many colloids, creating a bridge between them, and grow large enough such that sedimentation is possible.

3. *Precipitation and enmeshment* occurs when aluminum and ferric coagulants at high doses form precipitates. As these large precipitates move about via flocculation and eventually settle, they entrap colloidal materials. This process is also known as *sweep floc.*

The jar test is a very useful and simple test to determine the most appropriate type of coagulant, the proper coagulant dose (that is, concentration), and optimal pH. The jar test simulates the rapid mix, coagulation, flocculation, and settling steps in a water treatment facility. The apparatus consists of six beakers that are each stirred with individual, variable-speed paddles. A typical jar test is run in four steps:

Step 1: *Rapid mix*—the addition of chemical followed immediately by 30–60 seconds of stirring at 100 rpm

Step 2: *Flocculation*—stirring at 30 rpm for 15 minutes

Step 3: *Settling*—no stirring for 60 minutes

Step 4: *Analysis*—collection and analysis of samples from each beaker

The times provided in the steps outlined above are known as *detention times.* Detention time is often represented with the variables θ or τ. Detention time is defined as

$$\theta = \frac{V}{Q}$$

where

V = the volume of the reactor

Q = the flow rate through the reactor

Typical jar test results are shown in Figure 7.4. Note that for this sample data, a distinct increase in turbidity is observed for coagulant concentrations greater than 50 mg/L. This restabilization could be due to *charge reversal,* in which so much coagulant has been added that the charge on the colloid surface has changed from negative to positive, and the colloids again repel each other.

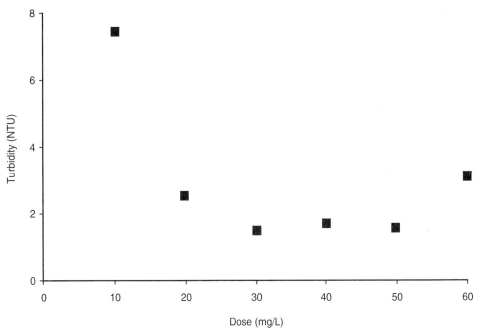

Figure 7.4 Typical jar test results for alum

Two types of coagulants are commonly used in water treatment: metal salts and polymers. The most common metal salts are salts of iron and aluminum.

Alum, or aluminum sulfate, forms aluminum hydroxide precipitate according to the following equation:

$$Al_2(SO_4)_3 \cdot 14H_2O + 6HCO_3^- \leftrightarrow 2\,Al(OH)_{3(s)} + 6CO_2 + 14H_2O + 3SO_4^{-2}$$

This equation assumes that the water has sufficient alkalinity. If the water source does not have sufficient alkalinity, sulfuric acid will be produced and the pH will decrease significantly, greatly decreasing the efficiency of the coagulation process. Also, the actual precipitate formed is much more complex than $Al(OH)_{3(s)}$, and some have suggested it may take the form of $[Al_8(OH)_{20} \cdot 28H_2O]^{4+}$ (Davis and Cornwell, 2006).

Ferric chloride is another common coagulant, and the reaction, with alkalinity present, is provided as follows:

$$FeCl_3 \cdot 7H_2O + 3HCO_3^- \leftrightarrow Fe(OH)_{3(s)} + 3CO_2 + 3Cl^- + 7H_2O$$

It is important to note that the mechanism by which each of these coagulants removes turbidity depends on a number of variables, including the colloidal concentration, pH, temperature, coagulant dose, and so on. Figure 7.5 illustrates how the coagulation mechanism varies with pH and dose, and Figure 7.6 illustrates the same for iron coagulants.

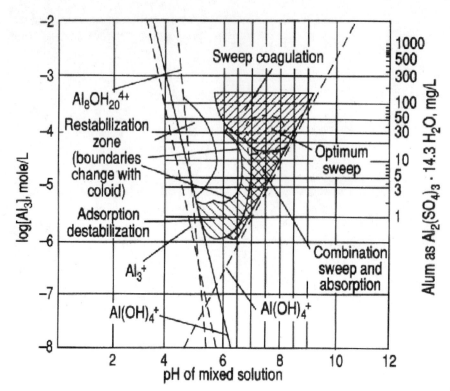

Figure 7.5   Operating range of alum

Source: MWH, *Water Treatment: Principles and Design*, 2nd ed., © 2005, John Wiley & Sons, Inc. Reprinted by permission.

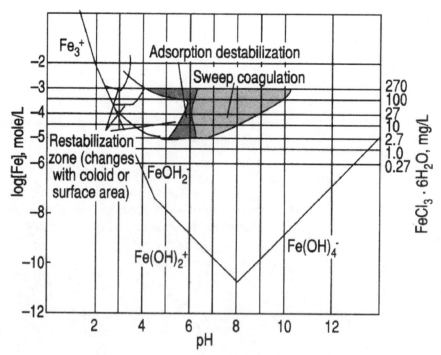

Figure 7.6   Operating range of iron coagulants

Source: MWH, *Water Treatment: Principles and Design*, 2nd ed., © 2005, John Wiley & Sons, Inc. Reprinted by permission.

Example **7.3**	

Estimating alum requirements

Estimate the quantity of alum a 1 MGD treatment plant needs to purchase annually if it wishes to operate in the sweep floc range, assuming enough alkalinity is present to keep the pH at 8. Express your answer in units of

(a) Tons

(b) Gallons of 5% alum solution

Solution

From Figure 7.5, an alum concentration of 35 mg/L should create sweep floc conditions at a pH of 8. The ability of this concentration to produce a sweep floc for the actual water supply at hand should be verified with a jar test.

(a) the daily mass of alum required is

35 mg/L · 1 MGD · 8.34 = 292 lb/day

Thus, the annual requirement is 53 tons.

(b) Assuming that the 5% alum solution has the same density as water, its concentration may be expressed as 0.42 lb/gal, as follows:

$$\frac{5\text{ g}}{100\text{ g}} \quad \frac{2.205 \cdot 10^{-3}\text{ lb}}{\text{gal}} \quad \frac{1000\text{ g}}{\text{L}} \quad \frac{3.785\text{ L}}{\text{gal}} = 0.42\text{ lb/gal}$$

The daily requirement is

$$292\text{ lb/day} \cdot (1\text{ gal}/0.42\text{ lb}) = 695\text{ gal/day}$$

The annual requirement is then 254,000 gallons.

Organic polymers may also be used as a coagulant. Polymers may be either synthetic or natural, but the former are more common in drinking water treatment. Polymers may promote coagulation via particle bridging or charge neutralization. They may be used as the primary coagulant but are more commonly used as a secondary coagulant, with the purpose of creating larger and stronger flocs.

After being stabilized by the coagulation process, flocs are formed in the flocculation process. Flocculation is promoted by gentle mixing, and the mixing may be promoted by paddles (horizontal or vertical shaft) or hydraulically. The methods are compared in Table 7.10.

Table 7.10   Comparison of flocculation methods

Process Issue	Horizontal Shaft with Paddles	Vertical-Shaft Turbines	Hydraulic Flocculation
Type of floc produced	Large and fluffy	Small to medium, dense	Very large and fluffy
Head loss	None	None	0.05–0.15 m
Operational flexibility	Good, limited to low G	Excellent	Moderate to poor
Capital cost	Moderate to high	Moderate	Low to moderate
Construction difficulty	Moderate	Easy to moderate	Easy to difficult

(continued)

Process Issue	Horizontal Shaft with Paddles	Vertical-Shaft Turbines	Hydraulic Flocculation
Maintenance effort	Moderate	Low to moderate	Low to moderate
Compartmentalization	Moderate	Excellent	Excellent, some designs nearly plug flow
Advantages	■ Generally produces large floc ■ Reliable ■ No head loss ■ One shaft for several mixers	■ Flocculators can be maintained or replaced without basin shutdown ■ No head loss ■ Very flexible, reliable ■ Highest energy input potential	■ Simple and effective ■ Easy, low-cost maintenance ■ No moving parts ■ Can produce very large flocs
Disadvantages	■ Compartmentalization more difficult ■ Replacement and some maintenance requires shutdown of basin	Difficult to specify proper impellers and reliable gear drives in competitive bidding process	Little flexibility

Source: MWH, *Water Treatment: Principles and Design*, 2nd ed., © 2005, John Wiley & Sons, Inc. Reprinted by permission.

The design of a flocculation basin includes the size of the basins, type of flocculation device, the speed of the flocculators, the power requirements, and so on. One parameter very important in the design of flocculation systems is the *velocity gradient, G*. The velocity gradient allows particles to move at different speeds, increasing the likelihood for collisions among destabilized particles and thus the creation of flocs. If G is too large, too much shear will be induced such that particles will not form large flocs. G (sec^{-1}) is defined generically as

$$G = \sqrt{\frac{P}{\mu V}} \qquad (7.3)$$

where

P = power input to water ($M \cdot L^2 \cdot T^{-3}$)

μ = fluid viscosity ($M \cdot L^{-1} \cdot T^{-1}$)

$\quad$ = 10^{-3} N·s/m^2 for water at 20°C

V = volume of the reactor (L^3)

For specific types of flocculation systems, the equations for velocity gradient are (Kawamura, 2000) provided in Equations 7.4 and 7.5.

Baffled channel: $\qquad\qquad G = \sqrt{\frac{gh}{vt}} \qquad (7.4)$

where

g = acceleration due to gravity (L/T^2)

h = headloss across tank (L)

v = kinematic viscosity (L^2/T)

$\quad$ = 10^{-6} m^2/s for water at 20°C

t = mean detention time (T)

Mechanical mixers with paddles: $G = \sqrt{\dfrac{C_d \cdot A \cdot v^3}{2vV}}$ (7.5)

where

C_d = drag coefficient based on paddle shape and flow conditions

A = cross-sectional area of the paddles (L^2)

v = relative velocity of the paddles with respect to the fluid (L/T)

v = kinematic viscosity (L^2/T)

V = volume of tank (L^3)

The design of flocculation systems includes the criteria shown in Table 7.11.

Table 7.11   Design parameters for flocculators

Design Parameter	Unit	Horizontal Shaft with Paddles	Vertical-Shaft Turbines
Velocity gradient, G	s^{-1}	20–50	10–80
Tip speed, maximum	m/s	1	2–3
Rotational speed	rev/min	1–5	10–30
Compartment dimensions (plan)			
Width	m	3–6	6–30
Length	m	3–6	3–5
Number of compartments	No.	2–6	4–6
Variable-speed drives	—	Usually	Usually

Source: MWH, *Water Treatment: Principles and Design*, 2nd ed., © 2005, John Wiley & Sons, Inc. Reprinted by permission.

Example 7.4

Preliminary design of a flocculation system

A flocculation system is required for a 5000 m³/day water treatment utility. The flocculation system is to consist of two treatment trains in parallel, with each train consisting of two compartments in series. Each compartment is desired to have a detention time of 10 minutes. A preliminary selection of vertical shaft flocculators has been made, and information describing the flocculators is as follows:

Area of paddles = 5 m²

Relative speed = 0.25 m/s

C_d = 1.9

(a) Determine the size of each compartment

(b) Estimate the power input to the water

Solution

(a) Determine the size of each compartment using the definition of detention time and remembering that the flow is split between the two flocculation trains.

$$\text{Volume} = \theta \cdot Q$$

$$= 10 \text{ min} \cdot 2500 \text{ m}^3/\text{day} \cdot (1 \text{ day}/1440 \text{ min})$$

$$= 17.4 \text{ m}^3$$

(b) First, compute the velocity gradient, G (Equation 7.5).

$$G = \sqrt{\frac{C_d \cdot A \cdot v^3}{2vV}}$$

where

$C_d = 1.9$

$A = 5 \text{ m}^2$

$v = 0.25 \text{ m/s}$

$v = 10^{-6} \text{ m}^2/\text{s}$

$V = 17.4 \text{ m}^3$

This produces a velocity gradient of 65 sec^{-1}, which is consistent with the values in Table 7.11.

Obtain the power by rewriting the generic velocity gradient equation (Equation 7.3) as

$$P = G^2 \cdot V \cdot v$$

$$= 74 \text{ W}$$

7.2.4 Sedimentation

Gravity-induced settling is called *sedimentation* or *clarification* and is the process in which solid particles are allowed to fall out of suspension in relatively quiescent water. Tanks are designed to greatly reduce flow velocities, such that particles are not being agitated by movement of fluid and are allowed to collect at the bottom of the tank. Sedimentation in water treatment generally occurs for high turbidity surface waters as a first treatment step prior to chemical addition, or following coagulation and flocculation.

The rate at which particles move in the vertical direction is described by their *terminal settling velocity* (V_t), which can be expressed as

$$V_t = \left[\frac{4 \, g \left(\rho_p - \rho_f \right) d}{3 \, C_D \, \rho_f} \right]^{1/2}$$

where g is the gravity constant; ρ_p and ρ_f are the densities of the particle and the fluid, respectively; d is the diameter of the particle; and C_D is the drag coefficient. C_D is dependent on Reynolds number (Re) and can be determined as

$$C_D = \frac{24}{\text{Re}} \qquad\qquad \text{Re} \leq 1$$

$$C_D = \frac{24}{\text{Re}} + \frac{3}{\sqrt{\text{Re}}} + 0.34 \qquad 1 \leq \text{Re} \leq 10^4$$

$$C_D = 0.4 \qquad\qquad Re > 10^4$$

where the Reynolds number is defined as

$$Re = \frac{\rho\, d\, V_t}{\mu}$$

Since V_t and Re are dependent on each other, an iterative solution is required where a velocity is assumed to determine Re, and that value is used to determine C_D, which is used to calculate V_t. The Reynolds number is corrected and the process repeated until the error in V_t is small. To reduce the complexity of the solution method, it is often assumed that *Stoke's law* is applicable, which is used to describe external flow around spherical bodies at low values for Re. Low Re values are possible for very small particles (on the order of micrometers in diameter) and low settling velocities (on the order of centimeters per second) and simplifies the expression for V_t by setting $C_D = 24/Re$. In these cases, V_t can be expressed as

$$V_t = \frac{g\left(\rho_p - \rho_f\right)d^2}{18\,\mu}$$

where g, ρ_p, ρ_f, d, and μ are as defined in section 8.3.

The flow of water in the clarifier can be characterized by the fluid velocity in the vertical direction, called the *surface settling rate*, and is expressed as

$$V_0 = \frac{Q}{A}$$

where V_0 has units ft/s or m/s, Q is the volumetric flow rate into the clarifier (cfs or m³/s), and A is the surface area of the clarifier (ft² or m²). Any particles possessing $V_t > V_0$ are collected in the clarifier, while particles with $V_t < V_0$ are collected only if they enter the clarifier at a height less than the water depth in the basin.

For design purposes, the surface settling rate is more commonly expressed in the common English units for Q, gallons per day (gpd), and is called the *overflow rate*, with units of gpd/ft². Hydraulic residence time (t_R), as described previously, is also used for clarifier design. Another design parameter is the *side-water depth*, which usually describes the depth of water not including the sludge blanket that forms on the bottom of the tank. Because the three parameters above are not independent, care must be taken when using them in clarifier design to keep from overspecifying the design. An additional design parameter is the *weir loading* (*wl*), sometimes called the *weir overflow rate* (*WOR*), which is a measure of the rate of effluent flow over the discharge weir, and can be expressed as

$$wl = \frac{Q}{L_w}$$

where *wl* has common units of gpd/ft, L_w is the length of the outlet weir (ft), and Q is as above.

Clarifier shape is usually rectangular or circular, with the size of the tanks determined through the use of empirical design criteria, which are based on operating conditions that have been found to be most effective in practice. For circular tanks, influent is in the center of the tank, and water flows radially towards the weir, located at the tank perimeter, with weir length equal to the tank perimeter. Rectangular tanks generally have a L:W ratio of 3–5:1, and weirs are set inside the tank at one end in a box or finger arrangement.

Design criteria are often listed in tables and are provided for several water (and wastewater) treatment applications in the environmental engineering section of the *FE-Supplied Reference Handbook* (6th ed., p. 127). In general, V_0 for water treatment can range from 300 to 2500 gpd/ft^2 (approximately equal to 12–100 m^3 per m^2 per day, or equivalently 12–100 m/d) and *horizontal* (or *flow-through*) *velocities* (V_h), defined as the volumetric flow rate divided by the cross-sectional area of the tank perpendicular to the velocity vector, should be kept below 0.5 ft/min in all cases. Design criteria for detention times vary from one to eight hours, side-water depths should be maintained greater than 10 feet, and weir loadings are generally less than 20,000 gpd/ft.

Design criteria for sedimentation basins are provided in Table 7.12.

Table 7.12   Design criteria for sedimentation basins

Type of Basin	Overflow Rate (gpd/ft²)	Hydraulic Residence Time (hr)
Water Treatment		
Presedimentation	300–500	3–4
Clarification following coagulation and flocculation		
1. Alum coagulation	350–550	4–8
2. Ferric coagulation	550–700	4–8
3. Upflow clarifiers	1500–2200	1
a. Groundwater	1000–1500	4
b. Surface water		
Clarification following lime-soda softening		
1. Conventional	550–1000	2–4
2. Upflow clarifiers	1000–2500	1
a. Groundwater	1000–1800	4
b. Surface water		
Wastewater Treatment		
Primary clarifiers	600–1200	2
Fixed film reactors		
Intermediate and final clarifiers	400–800	2
Activated sludge	800–1200	2
Chemical precipitation	800–1200	2

Source: Metcalf and Eddy, *Wastewater Engineering: Treatment and Reuse*, 3rd ed., © 1991, McGraw-Hill Education. Reprinted by permission of the McGraw-Hill Companies.

High rate clarification refers to processes that allow for increased settling performance for a given tank volume. The two most common methods are parallel plate settlers and tube settlers. These methods can easily be used to retrofit existing sedimentation basins.

The principle behind parallel plate and tube settlers is that they decrease the distance a floc has to settle. Rather than make the entire sedimentation basin shallower (which would decrease sludge storage, increase wind effects, etc.), the solids settle a short distance between parallel plates or within tubes. The tubes or plates are inclined to allow the accumulated solids to slide down from the tube or plate surface. The net effect is to increase the loading rate to a sedimentation basin without increasing the land area required. Surface loading rates for alum flocs in tube settlers are between 1 and 2.5 gpm/ft² (2.5–6.25 m/h). An example of a tube settler installation is provided in Figure 7.7.

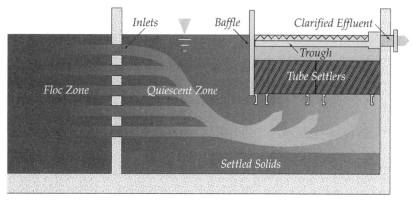

Figure 7.7 Tube settler schematic

Source: Illustration courtesy of Kerry Dissinger, Tube Settler Product Manager, Brentwood Industries, Inc., Reading, PA.

7.2.5 Filtration

Removal of nonsettleable solids in water treatment is generally accomplished through *gravity filtration*, usually using a *granular media*. Filtration mechanisms vary; however, characteristics of appropriate filter media are consistent: (1) fine enough to strain particles, (2) coarse enough to resist clogging, (3) deep enough to allow for large quantities of floc collection and long run times between cleaning cycles, and (4) shallow enough to minimize pressure drop and to allow for efficient backwash cleaning. Media characteristics are described with respect to bed depth and the diameter of the media particles, which may vary over a wide range of values. The standard notation to describe particle size distribution is d_x, which is defined as the *diameter that has X% of the material mass below the stated size*. The *effective size* is defined as d_{10}, while the *uniformity coefficient* is defined as d_{60}/d_{10}.

Often, a *sand bed* is the media of choice and can be effective as long as the uniformity coefficient is low. High uniformity coefficients can lead to size segregation during the backwash process, where coarse particles settle faster and leave a layer of fine sand on the top of the bed. This often leads to rapid clogging of the filter and the need for frequent cleaning. In order to avoid rapid surface plugging, it is common to use a *dual media* system where the sand bed is overlaid with an *anthracite cap* layer. The specific gravity of the anthracite (~1.5) is lower than the specific gravity of the sand (2.65), which allows a sand media with a smaller d_{10} settle faster than the anthracite with a slightly larger d_{10}. This system allows for straining of larger floc in the large-diameter anthracite cap without media fouling, while smaller particles can be removed in the lower, small-diameter sand layers.

Design of a typical dual media rapid sand filter is based on hydraulic loading and varies on the characteristics of the particulate that is being removed. In water treatment, loading rates range from 2 to 10 gpm/ft² of bed area with a typical value of 5 gpm/ft², and maximum filter areas of around 1000 ft² (30 ft × 30 ft) for each unit. Water from the coagulation and flocculation units is distributed over several rapid sand filters to a depth of 3–5 feet. The anthracite cap is usually 18 inches deep with a d_{10} of approximately 1 mm and a uniformity coefficient < 1.7. The sand layer has the same uniformity coefficient but has a d_{10} closer to 0.5 mm and a depth of 12 inches. The sand is supported above the underdrain system by 1.5–2.5 feet of sand and gravel, which is stratified in several layers of increasing grain size with increasing filter depth.

Pressure drop through a packed bed is due to the friction losses associated with the media and affects the rate of flow that is possible through a filter bed for a given, fixed head (that is, depth of water over the bed). Generally, clean media has a head loss on the order of 2–3 feet, while head loss of 8–10 feet may signal the need for backwashing. Head loss (h_f) through a clean filter containing a single diameter of filter media can be described by the Rose equation as follows:

$$h_f = \frac{1.067 \, V_S^2 \, L \, C_D}{g \, \eta^4 \, d}$$

where V_S is the approach velocity (volumetric flow rate divided by bed area, Q/A_S, where A_S is the area of the empty bed at plan view), L is the filter depth, C_D is the drag coefficient as defined in section 3.2.3 of this review, g is the gravity constant, η is the bed porosity (void volume/total bed volume), and d is the mean diameter of the filter media grain. The mean diameter may be determined from either the arithmetic average ($[S_1 + S_2]/2$) or geometric average ($[S_1 \times S_2]^{1/2}$), where S_1 is the size of the sieve opening on which the particle was retained and S_2 is the next larger sieve opening.

Another expression for head loss through a clean filter containing a single diameter of filter media can be described by the Carmen-Kozeny equation as follows:

$$h_f = \frac{f \, L \left(1 - \eta\right) V_S^2}{\eta^3 \, g \, d}$$

where L, η, V_S, g, and d are as defined previously, and f is the friction factor, which can be expressed as:

$$f = \frac{150 \left(1 - \eta\right)}{\text{Re}} + 1.75$$

For the more realistic circumstance where filter media are not of a single type or size, the head loss equations above can be modified to account for the change in the friction loss due to different media types and grain sizes as follows:

$$h_f = \frac{1.067 \, V_S^2 \, L}{g \, \eta^4 \, d} \sum \frac{C_{Dij} \, x_{ij}}{d_{ij}} \quad \text{Rose Equation}$$

$$h_f = \frac{L \left(1 - \eta\right) V_S^2}{\eta^3 \, g \, d} \sum \frac{f_{ij} \, x_{ij}}{d_{ij}} \quad \text{Carmen-Kozeny Equation}$$

where subscript i refers to the media type, subscript j refers to the size fraction, and x is the mass fraction of media of the specified type and size (m_{ij}/m_{tot}). Note that the $\sum x_{ij}$ must be equal to 1.

$\sum x_{ij} = 1$

Particulate matter that becomes trapped within the filter bed decreases the pore area available for flow, which increases frictional resistance (that is, head loss). This particulate matter must be removed and the filter bed reestablished in order for the filtration unit to be effective on a continuous basis. This is accomplished through a process called *backwashing*, where the flow of water to the filter is stopped and water and/or air is pumped through the bed in the upwards direction (that is, opposite the flow direction during filter operations) on a regular, periodic basis. This cleansing is made more effective by the *fluidization* of the filter media, where the flow velocity in the upwards direction is sufficient to lift and separate the media grains, and the mixing action of the particles scour trapped particulate matter.

The minimum velocity required to obtain fluidization has been calculated using various approaches in the literature. If fluid velocities during backwashing are too large, it is possible that the filter material may be flushed with the removed floc. One means to determine effective fluidization is to examine the bed *expansion*, which is a measure of the bed depth during backwashing relative to the bed depth during operation. An estimate of the depth of the fluidized bed (L_{fb}) during backwashing may be obtained from the following expression:

$$L_{fb} = \frac{L_0\left(1-\eta_0\right)}{1-\eta_{fb}} = \frac{L_0\left(1-\eta_0\right)}{1-\left(\dfrac{V_B}{V_t}\right)^{0.22}} \quad \text{Single type, diameter material}$$

$$L_{fb} = L_0\left(1-\eta_0\right)\sum \frac{x_{ij}}{1-\left(\dfrac{V_B}{V_{t,ij}}\right)^{0.22}} \quad \text{Multiple types, diameters of material}$$

where L_0 is the original bed depth of the fluidized fraction, η_0 is the initial bed porosity, η_{fb} is the porosity of the fluidized bed, V_B is the backwash velocity (defined as Q/A_S, where Q is the backwash flow rate and A_S is as defined above), V_t is the terminal settling velocity (calculated as described in section 3.2.3. of this review), and x_{ij} is as defined previously.

Expansion of the bed during backwashing is approximately 50% (that is, $L_{fb} = 1.5\ L_0$). Volumetric flow rates around 15 gpm/ft^2 of filter area maintained for 5–10 minutes have been shown to provide effective cleaning. Maximum backwashing flow rates are set to maintain fluidizing velocities that will not wash the filter media from the bed. After the backwashing flow is stopped, the filter media will settle by specific gravity and size, with the larger sand particles settling first, followed by the smaller sand particles, and the lighter anthracite particles on the top. Water that runs through the filter in the first 3–5 minutes after backwashing is often discarded because it may still contain particulates suspended during the wash.

The time-variant behavior of effluent turbidity and headloss for a filter run are shown in Figure 7.8. The origin of the abscissa corresponds to the time immediately after completion of backwashing. The *ripening* or *maturation* process occurs immediately after backwashing. The turbidity peak is caused by residual backwash water and the fact that the filter media become more efficient at trapping particles in the first 15 minutes to two hours after completion of backwashing.

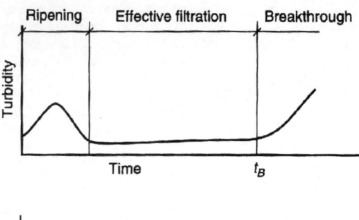

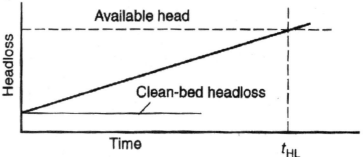

Figure 7.8   Performance characteristics of a rapid sand filter

Source: MWH, *Water Treatment: Principles and Design*, 2nd ed., © 2005, John Wiley & Sons, Inc. Reprinted by permission.

7.2.6 Membranes

Membranes sieve out particles in a similar fashion to a rapid sand filter but use a semipermeable synthetic material rather than relying on trapping within void spaces. The semipermeable quality of the membrane is designed to prevent undesirable components to pass through the membrane, while other components are allowed to pass through. A comparison between membrane filtration and rapid granular (sand) filtration is provided in Table 7.13.

Membrane processes are categorized as *microfiltration*, with 0.1 μm pores and *ultrafiltration*, with 0.01 μm pores. As such, microfiltration is effective in removing particles and bacteria, while ultrafiltration additionally removes viruses and smaller colloids.

In practice, *hollow fiber membranes* are the most common method of applying membrane filtration. Each hollow fiber is on the order of 1 mm in diameter (OD), and bundles of hollow fibers are included in a hollow fiber module. Figure 7.9 shows a portion of one such module. The relatively pure water that travels across the membrane is termed the *permeate,* while the water that is retained on the feed side of the membrane and continues to travel through the hollow membrane is termed the *retentate.* Figure 7.9 is an example of an *inside-out* process; other manufacturers have designed outside-in reactors.

Design of membranes also varies with respect to the *flow regime.* Possible flow configurations are shown in Table 7.14.

Table 7.13 Comparison of membrane processes and rapid granular filtration

Criteria	Units	Membrane Filtration	Rapid Granular Filtration
Filtration rate (permeate flux)	m/h[a]	0.03–0.17	5–15
	gpm/ft^2	0.01–0.07	2–6
Operating pressure	bar	0.2–2	0.18–0.3
	ft	7–70	6–10
Filtration cycle duration	min	30–90	
	d		1–4
Backwash cycle duration	min	1–5	10–15
Ripening period	min	None	15–120
Recovery	%	> 95	> 95
Filtration mechanism	—	Straining	Depth filtration

[a] Conventional units for membrane permeate flux are L/m^2 · h and gal/ft^2 · d. The conversions to the units shown in this table are 1 L/m^2 · h = 103 m/h and 1 gal/ft^2 · d = 1440 gpm/ft^2.

Source: MWH, *Water Treatment: Principles and Design*, 2nd ed., © 2005, John Wiley & Sons, Inc. Reprinted by permission.

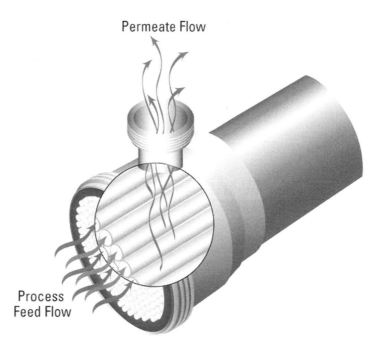

Figure 7.9 Hollow fiber membrane module

Source: Illustration courtesy of Koch Membrance Systems, Inc.

Transmembrane pressures are generally between 3 and 15 psi. Like rapid sand filters, the pressure drop across membranes and the effluent concentration increase with use, and backwashing is required. Such backwashing occurs using air and/or water. Eventually, the membrane performance decreases after repeated use cycles due to *fouling*. Fouling is caused by adsorption of compounds and clogging of membrane pores that cannot be reversed by backwashing Fouling is remedied by periodic chemical cleaning. The effects of backwashing and chemical cleaning on transmembrane pressure is demonstrated in Figure 7.10.

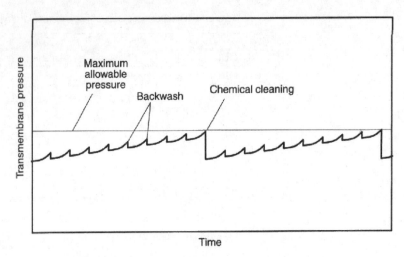

Figure 7.10   Effects of backwashing and fouling

Source: MWH, *Water Treatment: Principles and Design*, 2nd ed., © 2005, John Wiley & Sons, Inc. Reprinted by permission.

Table 7.14   Comparison of hollow-fiber membrane configurations

Configuration	Advantages	Disadvantages
Outside in	■ Can treat more water at same flux because outside of fiber has more surface area ■ Less sensitive to presence of large solids in the feed water	Cannot be operated in cross-flow mode
Inside out (dead-end mode)	Less expensive to operate than inside out in cross-flow mode	■ Large solids in feed water can clog lumen ■ Can treat less water at same flux because inside of fiber has less surface area
Inside out (cross-flow mode)	Can be operated at higher flux with high-turbidity feed water because cross-flow velocity flushes away solids and reduces impact of particles forming cake at membrane surface	■ Large solids in feed water can clog lumen ■ Can treat less water at same flux because inside of fiber has less surface area ■ Pumping costs associated with recirculating feed water through lumen can be expensive

Source: MWH, *Water Treatment: Principles and Design*, 2nd ed., © 2005, John Wiley & Sons, Inc. Reprinted by permission.

7.2.7 Reverse Osmosis

Reverse osmosis (RO) processes include nanofiltration (NF), with pore sizes on the order of 0.001 μm, and RO. Reverse osmosis membranes are nonporous, and treatment efficiency relies on concentration gradients, pressure, and water flux rate. RO is used widely for desalination of seawater in the Middle East and is becoming increasingly popular in the United States. The primary difference between reverse osmosis and membrane filtration is that RO removes dissolved substances, while the focus of membrane processes is the removal of particulate substances.

Unlike microfiltration or nanofiltration, which exclude contaminants based on size, an RO membrane works by excluding contaminants based on their molecular weights. Recall the concept of *osmotic pressure* from basic chemistry. Consider the system in Figure 7.11, in which a solution with a low concentration of solute is separated from a solution of high concentration by a semipermeable membrane. A semipermeable membrane only allows certain compounds to move through it. For example, the membrane in Figure 7.13 only allows water to be transported across it. Due to osmotic pressure, water will move across the membrane until the concentration on each side of the membrane is the same.

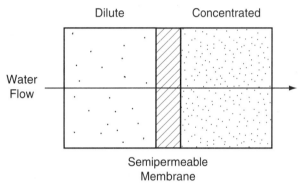

Figure 7.11 Osmosis concept

Source: "Reverse Osmosis Treatment of Drinking Water," Cornell Cooperative Extension, Colllege of Human Ecology. Reprinted by permission of the artist, Karen English, © 1995.

By contrast, in reverse osmosis the process shown in Figure 7.11 is reversed. Figure 7.12 illustrates that as a pressure is exerted on the more highly concentrated side, water is forced through the membrane. Contaminants remaining on the concentrated side of the membrane are periodically washed off.

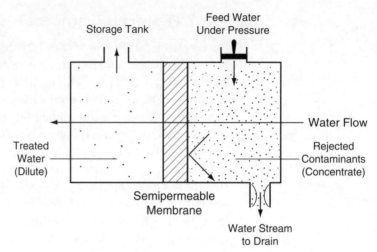

Figure 7.12   Reverse osmosis concept

Source: "Reverse Osmosis Treatment of Drinking Water," Cornell Cooperative Extension, Colllege of Human Ecology. Reprinted by permission of the artist, Karen English, © 1995.

Figure 7.13 illustrates the basic components of an RO system in practice. The pressure booster pump is necessary to produce the very high pressures needed in RO systems (150–800 psi). The cartridge filters remove particulates that could foul the RO membranes.

The solvent recovery rate, r is defined as

$$r = \frac{Q_p}{Q_f} \tag{7.6}$$

where

Q_p = permeate stream flow ($L^3 \cdot T^{-1}$)

Q_f = feed stream flow ($L^3 \cdot T^{-1}$)

The rate of rejection, Rej, is defined as

$$\text{Rej} = \frac{C_f - C_p}{C_f} = 1 - \frac{C_p}{C_f} \tag{7.7}$$

where

C_p = concentration in permeate ($M \cdot L^{-3}$)

Q_f = concentration in feed ($M \cdot L^{-3}$)

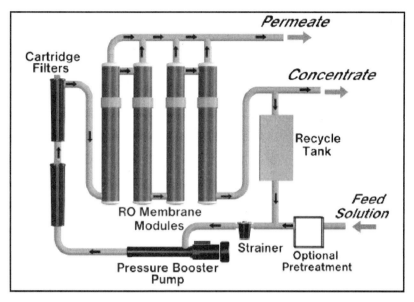

Figure 7.13 Schematic of an RO system

Source: "Reverse Osmosis Applications for Metal Finishing Operations, *www.epa.gov/ region9/waste/p2/projects/metal-reverseosmosis.pdf.*

7.2.8 Oxidation

Redox (reduction–oxidation) reactions are used in water treatment for many purposes. As discussed in section 1.3.3, redox reactions involve the transfer of electrons between chemical constituents. An *oxidant* (or *oxidizing agent*) is a material that gains electrons by causing another compound (the *reducing agent*) to lose electrons.

Common oxidants include chlorine, chlorine dioxide, ozone, potassium permanganate, and hydrogen peroxide. Common applications include removal of iron and manganese, removal of taste and odor causing compounds hydrogen sulfide removal, and disinfection.

To illustrate one use of oxidation, the chemical equations for the removal of iron by various oxidants are provided below. The choice of oxidant would need to be made based on bench-scale or pilot-scale testing, as many factors in "real" drinking water sources may hinder the performance of any one of the oxidants. For example, the existence of natural organic matter (NOM) hinders the removal of iron by oxygen, permanganate, chlorine dioxide, and free chlorine due to the fact that the NOM and iron form strong complexes. Rate effects also need to be considered.

Oxidation of ferrous iron by aeration

$$4Fe(HCO_3)_2 + O_2 + 2H_2O \rightarrow 4Fe(OH)_{3(s)} + 8CO_2$$

Oxidation of iron by chlorination

$$2Fe(HCO_3)_2 + Ca(HCO_3)_2 + Cl_2 \rightarrow 2Fe(OH)_{3(s)} + CaCl_2 + 6CO_2$$

Oxidation of iron by chlorine dioxide

$$Fe(HCO_3)_2 + NaHCO_3 + ClO_2 \rightarrow Fe(OH)_{3(s)} + NaClO_2 + 3CO_2$$

Oxidation by potassium permanganate

$$3Fe(HCO_3)_2 + KMnO_4 + 7H_2O \rightarrow 3Fe(OH)_{3(s)} + MnO_2 + KHCO_3 + 5H_2CO_3$$

Oxidation by ozone

$$2Fe(HCO_3)_2 + O_3 + 2H_2O \rightarrow 2Fe(OH)_{3(s)} + O_2 + 4CO_2 + H_2O$$

Example **7.5***

Oxidation in a batch reactor

Determine how much Mn(II) and permanganate will remain after 10 seconds of oxidation with permanganate in a completely mixed batch reactor. The initial concentration of Mn(II) is 0.5 mg/L and the initial concentration of permanganate is 1.5 times as much as is required stoichiometrically. Assume that the second order rate constant is 10^5 L/mole·s. The reaction is provided.

$$3Mn^{2+} + 2MnO_4^- + 2H_2O \rightarrow 5MnO_2 + 4H^+$$

Solution

The initial molar concentration of manganese $C_{m,o}$ needs to be determined.

$$C_{m,o} = (0.5 \cdot 10^{-3} \text{ g/L}) / (55 \text{ g/mol})$$

$$= 9.1 \cdot 10^{-6} \text{ mol/L}$$

By stoichiometry, and given the fact that an excess permanganate concentration of 50% is used, the initial concentration of permanganate $C_{p,o}$ is

$$C_{p,o} = 1.5 \cdot (2/3) \cdot 9.1 \cdot 10^{-6} \text{ mol/L}$$

$$= 9.1 \cdot 10^{-6} \text{ mol/L}$$

To determine the concentration of permanganate and manganese after 1 minute, a mass balance equation needs to be written for a batch reactor. From Chapter 1, the governing equation is

$$\left(\frac{dm}{dt}\right)_{cv} = \pm \dot{m}_{rxn}$$

For the case of a second order reaction, the equation simplifies to

$$dC_m/dt = -k \cdot C_m \cdot C_p$$

The integration of this equation results in the following:

$$\frac{C_m}{C_{p,o} - \frac{2}{3}(C_{m,o} - C_m)} = \frac{C_{m,o}}{C_{p,o}} \exp\left[-(C_{p,o} - \frac{2}{3}C_{m,o})kt\right]$$

The only unknown is C_m. Inserting all the known values into the previous equation results in the following:

$$\frac{C_m}{9.1 \cdot 10^{-6} - \frac{2}{3}(9.1 \cdot 10^{-6} - C_m)} = \frac{9.1 \cdot 10^{-6}}{9.1 \cdot 10^{-6}} \exp\left[-(9.1 \cdot 10^{-6} - \frac{2}{3}9.1 \cdot 10^{-6}) \cdot 10^5 \cdot 10\right]$$

Solving we find that C_m, the concentration of manganese after 10 seconds, is $1.63 \cdot 10^{-7}$ M.

The reaction has resulted in a decrease of manganese, Δ_m of

$$\Delta_m = 9.1 \cdot 10^{-6} \text{ mol/L} - 1.63 \cdot 10^{-7} \text{ mol/L}$$

$$= 8.9 \cdot 10^{-6} \text{ mol/L}$$

By stoichiometry, every mole of manganese that has been oxidized required 2/3 mole of permanganate, thus the change in concentration of permanganate is $6 \cdot 10^{-6}$ mol/L (2/3 $\cdot$ 8.9 $\cdot$ 10^{-6} mol/L). The final concentration of permanganate is

$$C_p = 9.1 \cdot 10^{-6} \text{ mol/L} - 6 \cdot 10^{-6} \text{ mol/L}$$

$$= 3.1 \cdot 10^{-6} \text{ mol/L}$$

*Adapted from MWH, *Water Treatment Principles and Design*, 2d ed. (Wiley, 2005), Example 8-8, p. 557.

7.2.9 Corrosion Control

Corrosion is caused by a combination of phenomena, including electrochemical, chemical, biological, and metallurgical. The ability of a water sample to induce corrosion can be estimated by the value of the water's Langelier Index (LI) or its Stability Index (SI), which are defined as follows:

$$\text{Langelier Index} = pH - pH_s \qquad \textbf{(7.8)}$$

$$\text{Stability Index} = 2pH_s - pH \qquad \textbf{(7.9)}$$

The saturation pH, pH_s, can be obtained from Figure 7.14. This figure is used by entering the graph given hardness as mg/L of $CaCO_3$ and by traveling up, across, and down as shown by the arrowed line (hardness = 240 mg/L as $CaCO_3$, alkalinity = 200 mg/L as $CaCO_3$, temperature = 70°F for the example shown).

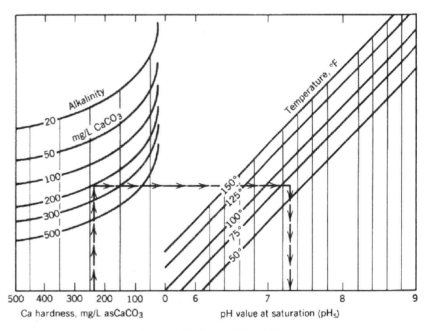

Figure 7.14 Saturation pH

Source: S. Kawamura, *Integrated Design of Water Treatment Facilities*, © 1993, John Wiley & Sons, Inc. Reprinted by permission.

The ability to corrode is related to the LI or SI using criteria provided in Table 7.15.

Table 7.15 Corrosion characteristics

Corrosion Characteristic	LI	SI
Medium to heavy scaling	+0.5 to +1.0	4–5
Slight scale formation	+0.2 to +0.3	5–6
Equilibrium	0	6–7
Slightly corrosive	–0.2 to –0.3	7–7.5
Medium to heavy corrosion	–0.5 to –1.0	7.5–8.5

Source: S. Kawamura, *Integrated Design of Water Treatment Facilities*, © 1993, John Wiley & Sons, Inc. Reprinted by permission.

For the example values shows in Figure 7.15 (that is, $pH_s = 7.3$), the Langelier Index equals –0.5 and the Stability Index equals 7.8. Thus, this water sample would be characterized as "medium to heavy corrosion."

In practice, the extent of corrosion can be quantified through the use of samples called *corrosion coupons*. Coupons come in a variety of sizes, shapes, and materials.

A coupon test is performed using the following steps:

Step 1: Weigh new coupons

Step 2: Install coupons into water system for 30–100 days

Step 3: Remove, clean, and analyze coupons. Analysis may consist of weighing and measuring the depth of or otherwise quantifying the pitting.

Step 4: Report results, typically in units of mpy (mils per year; 1 mil = 0.001 in)

Corrosion can be minimized via the following:

- Selection of materials. Materials to use are stainless steel, plastic piping, or lining, but avoid *galvanic corrosion,* which occurs when dissimilar metals are joined.

- Use of corrosion inhibitors. The most common chemicals used for corrosion inhibition are polyphosphate, zinc orthophosphate, and sodium silicate.

- Cathodic protection. Buried pipes and tanks can be protected by inducing a small electrical voltage difference between the structure and the ground.

- pH control. Increasing the pH is a very effective and simple means of minimizing corrosion.

7.2.10 Disinfection

This section will discuss the use of free chlorine, combined chlorine, chlorine dioxide, and ozone as disinfectants. The choice of which disinfectant to use will depend on the following considerations:

- Magnitude of contact time. This is governed by the $C \cdot t$ value (product of concentration and contact time) as discussed below.

- Formation of disinfection by-products. All disinfectants form by-products, but certain disinfectants produce higher concentrations of more toxic by-products.

■ Quality of water. Disinfectants will react with organics in water, regardless of whether the organics are inert or living. Thus, water with a high turbidity requires a relatively high concentration of chlorine, while also increasing the formation of disinfection by-products.

■ Safety. By their nature, all disinfectants are hazardous to human health. For example, many form potentially hazardous gases, and some have fire hazards associated with them.

The classical, and most simple, model for disinfection is Chick's law, which describes a first-order inactivation of bacteria. Chick's law is

$$\frac{dN}{dt} = -KN \tag{7.10}$$

where

dN/dt = rate of change in number of organisms with time (# of organisms $\cdot\, L^{-3} \cdot T^{-1}$)

N = concentration of organisms (# of organisms $\cdot\, L^{-3}$)

K = Chick's law rate constant (T^{-1})

The first-order equation can be solved by integration to yield:

$$\ln\left(\frac{N}{N_0}\right) = -Kt$$

where N_0 = the number of organisms at time = 0.

The desired amount of treatment is often expressed as the *log removal* or *log inactivation*, and varies for each microbial species or class of species. One log removal would leave 10% of the original organisms, or equivalently, inactivate 90% of the present microbes. Two-log removal would leave 1% of the original population, inactivating 99%, while 3-log removal and 4-log removal would inactivate 99.9% and 99.99% of the original organisms present, respectively. Although many species are of concern, chlorine doses are usually based on the species most resistant to disinfection.

In the case of surface waters, that species would be *Giardia lamblia* cysts, which requires 3-log removal to meet regulatory constraints. However, conventional water treatment (coagulation-flocculation-sedimentation-filtration) has been shown to offer 2–2.5-log removal of *Giardia lamblia*, leaving only a 0.5–1-log removal requirement from the disinfection process. Biological species present in groundwater sources depend on the depth to the aquifer and its connectivity with surface waters, generally focusing on virus removal. Removal of viruses is generally regulated at 4-log removal, although conventional treatment has been shown to offer 2-log removal, leaving a 2-log removal requirement for the disinfection process.

The ability to achieve a specified removal of a microbial species is a function of both the *chlorine concentration* as well as the amount of *contact time* the organism must experience at that concentration. The combination of these two factors is often referred to as the $C \cdot t$ *product*, which carries units of (mg/L) $\cdot$ min and is often tabulated for specific microorganisms at specified temperatures and pH values. $C \cdot t$ values for removal of *Giardia lamblia* are given in Table 7.16. Note that the values for viruses are typically lower than those for *Giardia lamblia*, and, therefore, the *Giardia lamblia* disinfection concentration generally controls chlorine dosing.

Design of a chlorine contact basin is based on the contact time necessary to achieve the $C \cdot t$ value at a specified residual chlorine concentration, usually assumed to be 0.5–1.0 mg/L. Often a safety factor of 2–3 is applied to the calculated contact time to ensure all water has sufficient contact. This is also enhanced through the use of a 40 or 50:1 L:W ratio in the contact basin (plug flow tank), with square channel cross sections and placement of multiple tank baffles to encourage mixing and reduce short circuiting.

Table 7.16 CT values for conventional disinfectants

Disinfectant	Units	Inactivation		
		2-log	3-log	4-log
For inactivation of viruses				
Chlorine[a]	mg · min/L	3	4	6
Chloramine[b]	mg · min/L	643	1067	1491
Chlorine dioxide[c]	mg · min/L	4.2	12.8	25.1
Ozone	mg · min/L	0.5	0.8	1.0
UV	mW · s/cm²	21	36	not available

	Inactivation					
	0.5-log	1-log	1.5-log	2-log	2.5-log	3-log
For inactivation of *Giardia* Cysts						
Chlorine[d]	17	35	52	69	87	104
Chloramine[c]	310	615	930	1230	1540	1850
Chlorine dioxide[e]	4	7.7	12	15	19	23
Ozone[e]	0.23	0.48	0.72	0.95	1.2	1.43

[a] Values are based on a temperature of 10°C, pH range of 6 to 9, and a free chlorine residual of 0.2 to 0.5 mg/L.
[b] Values are based on a temperature of 10°C and a pH of 8.
[c] Values are based on a temperature of 10°C and a pH range of 6 to 9.
[d] Values are based on a free chlorine residual less than or equal to 0.4 mg/L, temperature of 10°C, and a pH of 7.
[e] Values are based on a temperature of 10°C and a pH of 6 to 9.

Source: U.S. EPA, *Alternative Disinfectants and Oxidants Guidance Manual*, EPA 815-R-99-014, 1999.

Disinfectants have other uses in water treatment in addition to the inactivation of pathogenic organisms. Such uses include (U.S. EPA, 1999):

- Minimization of DBP formation

- Control of nuisance Asiatic clams and zebra mussels

- Oxidation of iron and manganese

- Prevention of regrowth in the distribution system

- Removal of taste and odors via chemical oxidation

- Improvement of coagulation and filtration efficiency

■   Prevention of algal growth in sedimentation basins and filters

■   Removal of color

Disinfection by Chlorination

Chlorine is the most common disinfectant in the United States, despite growing concerns about the formation of disinfection by-products (DBPs). Also, the safety issues associated with chlorine gas are not insignificant. However, chlorine is popular due to the fact that it is a mature technology and is reliable.

Chlorine is typically applied as an aqueous solution containing hypochlorous acid (HOCl) and the hypochlorite ion (OCl⁻). The chemical reaction is

$$HOCl \leftrightarrow H^+ + OCl^-$$

and thus the equilibrium constant is defined as

$$K_a = \frac{\left[H^+\right] \cdot \left[OCl^-\right]}{\left[HOCl\right]}; pK_a = 7.54 .$$

The combination of hypochlorous acid and the hypochlorite ion are termed *free chlorine*. The equilibrium relationship is shown graphically in Figure 7.15 and is useful as it shows the relative proportions of HOCl and OCl⁻ at various pHs and temperatures. Hypochlorous acid exhibits faster disinfection kinetics than the hypochlorite ion, and at a pH of 7, nearly all the free chlorine exists as hypochlorous acid.

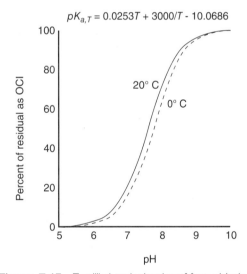

Figure 7.15 Equilibrium behavior of free chlorine

Source: MWH, *Water Treatment: Principles and Design*, 2nd ed., © 2005, John Wiley & Sons, Inc. Reprinted by permission.

One benefit of disinfection by free chlorine is that it provides a *residual* concentration of disinfectant in the system; that is, if added in sufficient quantities, free chlorine will be found at the extents of the water distribution system, at which point it is able to oxidize organics. This is a valuable attribute of free chlorine in terms of public safety.

Example **7.6**

Residual chlorine in a distribution system

Given a first-order decay constant of free chlorine of 0.4 day $^{-1}$, estimate the concentration at the most distant point of the water distribution system. After leaving the 0.5 MGD plant, the water is stored in a 200,000-gallon clear well, is piped 4000 feet to a 150,000-gallon elevated storage tank (the only such elevated storage tank in the system), and then travels 700 feet to the most distant point in the system. The concentration of free chlorine entering the clear well is 1 mg/L. Assume a velocity in all pipes of 2 ft/s.

Solution

This system can be simplified as shown in Exhibit 2.

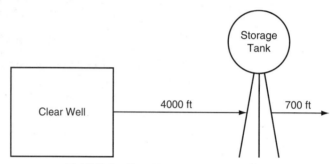

Exhibit 2   Simplified system schematic

The system can be expressed as a series of CSTRs and PFRs as shown in Exhibit 3.

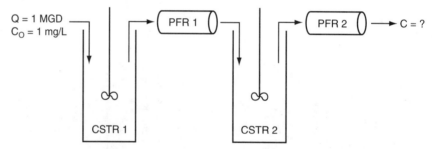

Exhibit 3   System as series of PFRs and CSTRs

The governing equation for a PFR, from section 1.5.3, is

$$C_{effluent} = C_o e^{-k \cdot \theta}$$

The governing equation for a CSTR with one input, one outflow, and first-order decay, was derived in Example 1.4:

$$C_{out} = C_{in} \frac{1}{1 + \theta \cdot k}$$

Given these governing equations, the concentrations exiting each reactor can be determined. The effluent concentration exiting the clear well, C_{CSTR1}, is computed given

$C_{in} = 1$ mg/L

$k = 0.4$ day^{-1}

$\theta = V/Q$

= 200,000 gal/0.5 MGD

= 0.4 day

Thus, C_{CSTR1} = 0.86 mg/L.

The effluent concentration leaving the first length of pipe, C_{PFR1}, is computed given

C_o = 0.86 mg/L

k = 0.4 day^{-1}

θ = travel time in pipe

= Length/Velocity

= 4000 ft/(2 ft/s)

= 2000 s (or 0.023 day)

Thus, C_{PFR1} = 0.85 mg/L.

In a similar manner, the concentrations exiting CSTR2 and PFR2 can be calculated:

$$C_{CSTR2} = 0.85 \text{ mg/L} \frac{1}{1 + 0.3 \text{ day} \cdot 0.4 \text{ day}^{-1}} = 0.76 \text{ mg/L}$$

$$C_{PFR2} = 0.76 \frac{\text{mg}}{\text{L}} e^{-0.4 \text{day}^{-1} \cdot 0.004 \text{day}} = 0.76 \text{ mg/L}$$

The advantages and disadvantages of chlorine disinfection are listed as follows (U.S. EPA, 1999):

Advantages

■   Chlorine is the easiest and least expensive disinfection method, regardless of system size.

■   Chlorine is the most widely used disinfection method.

■   Chlorine is available as calcium and sodium hypochlorite. Use of these solutions is more advantageous for smaller systems than chlorine gas because they are easier to use, are safer, and need less equipment compared to chlorine gas.

■   Chlorine provides a residual.

Disadvantages

■   Chlorine may cause a deterioration in the ability to coagulate and filter dissolved organic substances.

■   Chlorine forms halogen-substituted by-products.

■   Finished water could have taste and odor problems, depending on the water quality and chlorine dosage.

■   Chlorine gas is a hazardous corrosive gas.

■   Chlorine is less effective at high pH.

Chlorine can be used in practice in other forms, including *combined chlorine* and chlorine dioxide.

Combined chlorine refers to the complexes that chlorine forms when it reacts with ammonia in the water. The three chloramine species are

$$\text{Monochloramine: } NH_3 + HOCl \rightarrow NH_2Cl + H_2O$$

$$\text{Dichloramine:} \quad NH_2Cl + HOCl \rightarrow NHCl_2 + H_2O$$

$$\text{Trichloramine:} \quad NHCl_2 + HOCl \rightarrow NCl_3 + H_2O$$

The term *total chlorine* is used to refer to the summation of combined chlorine and free chlorine. Although combined chlorines are not as effective as hypochlorous acid, the use of combined chlorine forms insignificant amounts of THMs as compared to the use of free chlorine.

Chlorine dioxide is nearly as effective as hypochlorous acid. It requires the pH to be greater than 8, which minimizes corrosion. However, chlorine dioxide leads to the formation of chlorite (ClO_2) and chlorate (ClO_3) ions. Chlorine dioxide is never shipped due to its explosive nature and is produced on-site. Its production is described in the following reactions:

$$2NaClO_2 + Cl_{2(g)} = 2ClO_{2(g)} + 2NaCl$$

$$2NaClO_2 + HOCl = 2ClO_{2(g)} + NaCl + NaOH$$

$$5NaClO_2 + 4HCl = 4ClO_{2(g)} + 5NaCl + 2H_2O$$

The effectiveness of chlorine dioxide is not nearly as dependent on pH as is free chlorine. Also, its effectiveness increases with increasing temperature and decreases as turbidity increases.

Disinfection by Ozone

Ozone is the most powerful oxidant and disinfectant commonly used in municipal drinking water treatment. The primary drawback of ozone is that it does not provide a residual of disinfectant in the distribution system; thus, it is often combined with another disinfectant such as free chlorine or chloramines. Contact times of ozone are the smallest of all conventional disinfectants (see Table 7.16).

Ozone reacts very quickly in water and can create the hydroxyl free radical, which is among the most reactive oxidizing agents in water, with reaction rates on the order of 10^{10}–10^{13} M^{-1} s^{-1}.

Ozone water treatment systems have four basic components: a gas feed system, an ozone generator, an ozone contactor, and an off-gas destruction system. Contact between ozone and the dissolved organics occurs in the ozone contactor. Production of ozone on-site is required due to its very short half-life. Off-gas destruction converts the ozone back into oxygen due to the toxicity of ozone. A simplified schematic is provided in Figure 7.16.

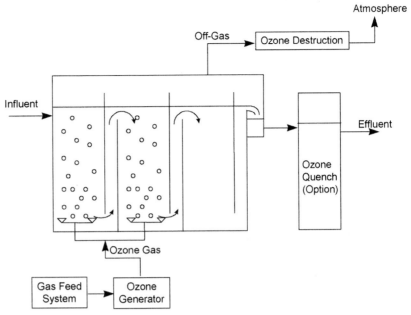

Figure 7.16 Ozone System Schematic

Source: U.S. EPA, *Alternative Disinfectants and Oxidants Guidance Manual*, EPA 815-R-99-014, 1999.

The EPA report *Alternative Disinfectants and Oxidants Guidance Manual* (U.S. EPA, 1999) lists the following advantages and disadvantages of ozone:

Advantages

- Ozone is more effective than chlorine, chloramines, and chlorine dioxide for inactivation of viruses, cryptosporidium, and Giardia.

- Ozone oxidizes iron, manganese, and sulfides.

- Ozone controls color, taste, and odors.

- One of the most efficient chemical disinfectants, ozone requires a very short contact time.

- Upon decomposition, the only residual is dissolved oxygen.

- Biocidal effectiveness is not influenced by pH.

Disadvantages

- DBPs are formed, particularly by bromate and bromine-substituted DBPs, in the presence of bromide, aldehydes, ketones, etc.

- The initial cost of ozonation equipment is high.

- The generation of ozone requires high energy inputs.

- Ozone is highly corrosive and toxic.

- Ozone provides no residual.

- Ozone requires a higher level of maintenance and operator skill.

Disinfection with Ultraviolet (UV) Radiation

As UV rays penetrate a cell wall of a microorganism, it reacts with nucleic acids and other vital cell functions, thereby inactivating the microorganism. According to the EPA, UV is effective on inactivating bacteria and viruses at low dosages (5–25 mW · s/cm² for 2-log removal and 90–140 mW · s/cm² for 4-log removal). However, UV radiation is not as effective on protozoa such as giardia or cryptosporidium, and much higher dosages are required (100–8000 mW · s/cm² for 2-log removal). A UV dose, D, is defined as

$$D = I \cdot t$$

where

D = UV dose, mW × s/cm²

I = UV intensity, mW/cm²

t = exposure time, s

The formation of DBPs by UV radiation is not a significant concern, although some ozone is formed by the process; thus, the DBPs formed by ozonation may be associated with UV radiation. Also, UV radiation does not create a disinfectant residual.

7.2.11 THM Removal

Trihalomethanes (THMs) are a type of DBP and are formed by a reaction between free chlorine and dissolved organic material. The EPA requirement for THMs in drinking water is provided in Table 7.3. THMs may be classified as *instantaneous THMs* (the THM concentration at the moment of sampling), or as *terminal THMs,* which is the concentration of THM at the farthest end of the water distribution system. THM formation increases as the

- concentration of organic precursors in the finished water increases,

- concentration of free chlorine increases,

- water temperature increases,

- pH of water increases,

- chlorine contact time increases, and

- bromide concentration increases.

The concentration of bromide affects THM formation because bromide is oxidized to bromine by free chlorine, and bromine reacts very quickly with organic precursors to form THMs. Bromide is found in highest concentrations in water supplies that are currently influenced by seawater or have been influenced by seawater in the past.

Kawamura (2000) lists a three-step process for controlling THM formation:

Step 1: Find an alternative water source, especially groundwater

Step 2: Modify conventional treatment techniques:

- Improve flocculation and sedimentation to enhance removal of organic precursors

- Change the point of chlorination to decrease chlorine contact time

- Substitute potassium permanganate for prechlorination

- Use chloramines as an alternative disinfectant

- Add powdered activated carbon (PAC) to remove organic precursors

Step 3: Consider the following alternatives:

- Use preozonation to oxidize precursor

- Strip THMs via aeration

- Remove THMs with granular activated carbon (GAC) beds

- Utilize membrane processes to remove THMs

7.2.12 Taste and Odor Treatment

Although taste and odors are secondary contaminants, with no known negative impacts on human health, they can be a particularly troublesome contaminant to remove from the water and are the source of many customer complaints.

The most common sources of taste and odor problems are algae and *Actinomycetes*. The second most common cause is decaying vegetation. Thus, taste and odor problems are more likely to be found in drinking water that comes from surface water than that from groundwater. Other sources of tastes and odors are hydrogen sulfide, industrial chemical discharges, and agricultural chemicals, and these can affect both groundwater and surface water.

The detection and intensity of tastes and odors is often estimated from a flavor profile analysis (FPO), which is conducted by a panel of trained "taste testers." The current drinking water quality standard is that the threshold odor number (TON) must be less than 3. The TON analysis is obtained by starting with a 200 mL sample of water and diluting it until no odors can be detected by the panel. The volume of each dilution should be 200 mL. The TON is the inverse of the largest dilution that yielded a perceptible odor, as perceived by the taste-test panel. Thus, if 50 mL of sample is mixed with 150 mL of distilled water (a 4 to 1 dilution) and yields an odor, and any greater dilution does not yield an odor, the TON is 4. This analysis method is extremely sensitive, as some taste and odor compounds, such as geosmin and 2-methlyisoborneal (MIB) impart objectionable odors at concentrations as low as 20 ng/L ($20 \cdot 10^{-9}$ g/L).

Tastes and odors can be controlled by prevention at the source, removal at the treatment plant, and control in the distribution system (Kawamura, 2000).

Prevention at the source can be attained by controlling aquatic plant growth and reservoir mixing. Aquatic plant growth can be controlled by drastically altering reservoir water depths or using copper sulfate. Reservoir mixing works because when stratified (see section 9.2 on limnology) the hypolimnion of lakes becomes anaerobic. Under such conditions, hydrogen sulfide is formed and the degradation of organic substances occurs that results in objectionable odors. Reservoir mixing by mechanical mixers or pumps can effectively prevent the formation of some taste- and odor-causing compounds.

Methods for removal at the treatment plant include:

- Aeration via diffused air, cascading trays, or packed towers is effective for volatile compounds such as hydrogen sulfide.

■ Oxidation may be the most effective treatment technique and can be carried out with chlorine, ozone, potassium permanganate, etc.

■ Adsorption using PAC or GAC is also effective. PAC can be added to the water as a slurry, and GAC is used in adsorption beds. Section 11.4.2 describes the use of GAC in more detail.

■ Control in the distribution system is possible if the system has a minimum number of dead ends and has a large number of clean-out assemblies.

7.2.13 Fluoridation

A fluoride concentration of 1 mg/L is known to have a significantly positive impact on preventing dental cavities. It is added to virtually all public drinking water sources in the United States. The most common forms of fluoride are sodium fluoride (NaF), sodium silicofluoride (Na_2SiF_6), and hydrofluosilicic acid (H_2SiF_6). Both NaF and Na_2SiF_6 are available in dry form, while hydrofluosilicic acid is a liquid; the latter has the benefit of ease of handling but is a strong acid.

Example **7.7**

Fluoridation of drinking water

Estimate the flow requirements for a metering pump for hydrofluosilicic acid for a 1.25 MGD water treatment plant. The water supply has a fluoride concentration of 0.25 mg/L. The hydrofluosilicic acid is available in a 23% solution with a unit weight of 76 lb/ft³.

Solution

Assume that the purpose is to increase the fluoride concentration to 1 mg/L. Thus, the quantity of fluoride that needs to be added to the water daily is:

$$(1 \text{ mg/L} - 0.25 \text{ mg/L}) \cdot 1.25 \text{ MGD} \cdot 8.34 = 7.8 \text{ lb/day}$$

Next, we determine the number of gallons per day of hydrofluosilicic acid corresponding to 7.8 lb/day of fluoride ion. The molecular weight of hydrofluosilicic acid is 144 g/mol, and the molecular weight of fluorine is 19 g/mol. Consequently, given the molecular formula of hydrofluosilicic acid, we find each mole (or gram or gallon for that matter) of hydrofluosilicic acid is 79.2% fluorine.

Thus, for a 23% solution of hydrofluosilicic acid, each pound of the acid contains

$$0.23 \cdot 0.792 \cdot 1 \text{ lb} = 0.18 \text{ lb fluoride}$$

For this example problem, 7.8 lb of fluoride are needed, which corresponds to 43.3 lb of hydrofluosilicic acid required each day as follows:

$$\frac{7.8 \text{ lb F}^-}{\dfrac{0.18 \text{ lb F}^-}{1 \text{ lb H}_2\text{SiF}_6}} = 43.3 \text{ lb}$$

Given the unit weight of hydrofluosilicic acid, the number of gallons of hydrofluosilicic acid required each day is:

$$\frac{43.3 \text{ lb/day}}{76 \text{ lb/ft}^3 \cdot (1 \text{ ft}^3 / 7.48 \text{ gal})} = 4.3 \text{ gal/day} = 0.18 \text{ gal/hr}$$

This flow rate of 0.18 gal/hr gives the engineer a starting point in selecting a metering pump to deliver the hydrofluosilicic acid to the drinking water.

High concentrations of fluoride can have adverse effects on human health, including unsightly mottling of the teeth and cancer. When fluoride is found in source waters at relatively high concentrations (above 2 mg/L), it needs to be removed. Removal methods include chemical precipitation (via lime softening or alum) and ion exchange.

7.3 RESIDUALS MANAGEMENT

There are four main categories of residuals produced from drinking water treatment plants:

1. *Sludge* is a mixture of solids and water. The solids in sludge consist of solids found in the source water and solids originating from chemicals added during treatment. Sludge solids originate from presedimentation, coagulation/flocculation, filter backwashing, lime softening, iron and manganese removal, etc.

2. *Concentrates* are wastewaters generated by membrane filtration and reverse osmosis reject water.

3. *Other solids* consist of spent GAC, ion exchange resins, and spent filter media.

4. *Air emissions* include off gases from air stripping, ozone contact tanks, etc.

One goal of residuals management is to minimize the quantity of residuals to be disposed of. Residuals management is challenging, given the huge quantities of residuals generated, and given the fact that the residuals contain all of the undesirable "leftovers" of water treatment: pathogens, solids, DBPs, taste- and odor-causing compounds, manganese and iron, arsenic and other toxic compounds, and salt.

7.3.1 Sludges

Sludges are produced by a variety of drinking water treatment processes, including lime softening and coagulation/flocculation. Solids in sludge are also generated during backwashing of filters and from PAC addition.

Sludge Properties
One of the most common sludge properties used by environmental engineers is the *total solids content,* or sometimes simply called the *solids content.* The total solids content of a sludge sample is defined as

$$TS = \frac{\text{mass of solids}}{\text{mass of sludge}}$$

Given that sludge consists of solids and water, the equation can be rewritten as

Total Solids

$$TS = \frac{\text{mass of solids}}{\text{mass of solids} + \text{mass of water}} \qquad (7.11)$$

Total solids content is generally expressed as a percentage. Sludge exiting a sedimentation basin may have a total solids content of 1% or less. The goal of thickening and dewatering processes (described later in this section) is to increase the total solids content. As a result, thickening decreases the volume of sludge and decreases storage requirements, transportation costs, and so on. The characteristics of sludge vary greatly depending on the type of sludge and the solids content as shown in Table 7.17.

Table 7.17 Characteristics of sludge

Alum/Iron Coagulant Sludge Characteristics (ASCE/AWWA, 1990)

Solids Content	Sludge Characteristic
0–5%	Liquid
8–12%	Spongy, semisolid
18–25%	Soft clay
40–50%	Stiff clay

Chemical Softening Sludge Characteristics (ASCE/AWWA, 1990)

Solids Content	Sludge Characteristic
0–10%	Liquid
25–35%	Viscous liquid
40–50%	Semisolid
60–70%	Crumbly cake

Source: U.S. Environmental Protection Agency, *Management of Water Treatment Plant Residuals*, 625/R-95/008.

The dewaterability of sludge depends on the raw water turbidity, type of coagulant used, dose of coagulant, extent of lime use, and mechanism of coagulation (see Figures 7.5 and 7.6). The pH also affects the solids content, as it controls the coagulation mechanism. Sludge produced by the charge neutralization mechanism (pH ~ 6.5) will produce a higher solids content than sludge produced by the sweep floc mechanism (pH ~ 8), as the former uses lower doses of alum. Moreover, alum sludges typically have a lower solids content than iron and lime sludges.

Specific resistance, *r*, measures the "dewaterability" of a sludge (that is, the sludge's ability to release water). The specific resistance depends on the permeability of the sludge, as well as how tightly the water is bound up within the sludge structure. Specific resistance is measured by removing water from a sample by applying a vacuum to a sample placed on a laboratory filtration apparatus. The volume of filtrate removed from the sample is measured as a function of time. The equation for specific resistance is

$$r = \frac{2PA^2b}{\mu c} \qquad (7.12)$$

$$r = \frac{2PA^2b}{\mu c} = \text{Specific resistance}$$

where

r = specific resistance ($L \cdot M^{-1}$)

P = pressure drop across sludge cake ($M \cdot T^{-2} \cdot L^{-3}$)

A = surface area of filter (L^2)

μ = filtrate viscosity ($M \cdot L \cdot T^{-1}$)

c = weight of dry solids deposited per volume of filtrate ($M \cdot L^{-3}$)

b = slope of a plot of t/V vs. V ($T \cdot L^{-6}$)

t = time of filtrate (T)

V = filtrate volume (L^3)

Specific resistance varies widely among sludges (from $2 \cdot 10^{10}$ m/kg to $2 \cdot 10^{12}$ m/kg). Sludges for which r is less than $10 \cdot 10^{10}$ are said to dewater readily.

The density of sludge varies based on the water content and the nature of the flocs. Floc density is related to floc size, the type and dose of coagulant, the mechanism of coagulation, the suspended solids content. Floc specific gravities have been reported to be in the range of 1.1 to 1.3.

The specific gravity of sludge, SG_{sludge}, can be estimated based on its solids content and the specific gravity of the solids:

$$SG_{sludge} = SG_{solids} \cdot TS + SG_{water} \cdot (1 - TS)$$

where

SG_{solids} = specific gravity of sludge solids

TS = total solids content (expressed as a decimal)

SG_{water} = specific gravity of water

Sludge Quantity

The mass of alum sludge can be predicted by the following equation (Davis and Cornwell, 2008):

$$S = 8.34Q \cdot (0.44Al + SS + A) \tag{7.13}$$

where

S = sludge solids produced (lb/day)

Q = plant flow rate (MGD)

Al = alum dose (mg/L)

SS = raw water suspended solids (mg/L)

A = net solids from additional chemicals added such as polymer or PAC (mg/L)

The mass of iron coagulant sludge can be predicted by the following equation:

$$S = 8.34Q \cdot (2.9Fe + SS + A) \tag{7.14}$$

where

Fe = iron dose (mg/L, as Fe)

Alternatively, the mass of sludge produced due to the coagulant alone can be estimated from Table 7.18. The total mass of sludge produced is this mass plus the mass of suspended solids and any additional chemicals required.

The mass of softening sludge can be predicted by the following equation (Davis and Cornwell, 2008):

$$S = 8.34Q(2\text{CaCH} + 2.6\text{MgCH} + \text{CaNCH} + 1.6\text{MgNCH} + CO_2) \quad \textbf{(7.15)}$$

where

CaCH   = calcium carbonate hardness removed as $CaCO_3$ (mg/L)

MgCH   = magnesium carbonate hardness removed as $CaCO_3$ (mg/L)

CaNCH = noncarbonated calcium hardness removed as $CaCO_3$ (mg/L)

MgNCH = noncarbonated magnesium hardness removed as $CaCO_3$ (mg/L)

CO_2 = carbon dioxide removed by lime addition as $CaCO_3$ (mg/L)

For units of flow in m³/s and sludge production in kg/day, the 8.34 factor is replaced with 86.4.

Table 7.18 Sludge generation rates

Process	Unit	Range	Typical Value
Coagulation			
Alum, $Al_2(SO_4)_3 \cdot 14H_2O$	kg dry sludge/kg coagulant	0.33–0.44	0.33
Ferric sulfate, $Fe_2(SO_4)_3$	kg dry sludge/kg coagulant	0.59–0.8	0.59
Ferric chloride, $FeCl_3$	kg dry sludge/kg coagulant	0.48–1.0	0.48
PACl	kg dry sludge/kg PACl	$(0.0372\text{–}0.0489) \times Al(\%)$	$(0.0489) \times (Al, \%)$
Polymer addition	kg dry sludge/kg coagulant	1.0	1.0
Turbidity removal	mg TSS/NTU removed	0.9–1.5	1.25
Softening			
Ca^{2+a}	kg dry sludge/kg Ca^{2+} removed	2.0	2.0
Mg^{2+b}	kg dry sludge/kg Mg^{2+} removed	2.6	2.6

[a] Sludge is expressed as $CaCO_3$.
[b] Sludge is expressed as $Mg(OH)_2$.

Source: MWH, *Water Treatment: Principles and Design*, 2nd ed., © 2005, John Wiley & Sons, Inc. Reprinted by permission.

The *volume* of sludge produced per day (*V*) is given by the following equation:

$$V = \frac{S}{\rho_{sludge} \cdot TS} \quad \textbf{(7.16)}$$

where

ρ_{sludge} = sludge density (M·L^{-3})

    = $S.G._{sludge} \cdot \rho_{water}$

ρ_{water} = 998.2 kg/m³ at 20°C

TS = total solids content, expressed as decimal

Example **7.8**

Sludge generation rates

Consider a 30,000 m³/day drinking water treatment plant. In the treatment process, 3.0 mg/L of PAC are utilized and 30 mg/L of ferric sulfate ($Fe_2(SO_4)_3$) are added. The total solids content of sludge leaving the settling basin is 1.5%. Estimate the daily volume of sludge exiting the settling basin. The specific gravity of the PAC is 0.55 and is 1.6 for all other solids.

The raw water has a turbidity of 5 NTU. A student worker has collected raw water data and created a graph (Exhibit 4) showing the correlation between suspended solids and turbidity.

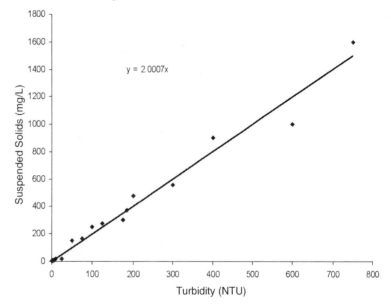

Exhibit 4 Suspended solids—turbidity correlation

Solution

First, express the ferric sulfate dose in terms of mg/L as Fe:

$$\left(\frac{30 \text{ mg Fe}_2(SO_4)_3}{L}\right)\left(\frac{1 \text{ mole Fe}_2(SO_4)_3}{400 \cdot 10^3 \text{mg}}\right)\left(\frac{2 \text{ mole Fe}}{\text{mole Fe}_2(SO_4)_3}\right)\left(\frac{55.8 \cdot 10^3 \text{ mg}}{\text{mole Fe}}\right)$$

$$= 8.4 \text{ mg/L as Fe}$$

The solids produced each day are obtained from Equation 7-14.

$$S = 8.34Q \cdot (2.9\text{Fe} + SS + A)$$

$$Q = \left(\frac{30,000 \text{ m}^3}{d}\right)\left(\frac{1 \text{ gal}}{3.785 \cdot 10^{-3} \text{ m}^3}\right)$$

$$= 7.9 \cdot 10^6 \text{ gal} = 7.9 \text{ MGD}$$

Fe $= 8.4$ mg/L

SS $= 10$ mg/L (from Exhibit 4)

$A = 3$ mg/L

Therefore, the total mass of solids produced is 2561 lb/day, or 1162 kg/day.

To determine the volume of residuals produced, the density of the sludge needs to be known. The specific gravity of the solids is a weighted average of the PAC (SG = 0.55) and the remaining solids (SG = 1.6). The mass of PAC, M_{PAC}, used daily is

$$M_{PAC} = 8.34 \cdot 3 \text{ mg/L} \cdot 1.5 \text{ MGD}$$

$$= 659 \text{ lb/day, or } 299 \text{ kg/day}$$

the mass of the remaining solids, M_{other}, is

$$M_{other} = 1162 \text{ kg/day} - 299 \text{ kg/day}$$

$$= 863 \text{ kg/day}$$

The specific gravity of the sludge solids is then calculated to be

$$SG_{solids} = \frac{\left(299 \text{ kg/day} \cdot 0.55\right) + \left(863 \text{ kg/day} \cdot 1.6\right)}{1162 \text{ kg/day}}$$

$$= 1.33$$

The specific gravity of the sludge can be determined by the following equation, with all values known:

$$SG_{sludge} = SG_{solids} \cdot TS + SG_{water} \cdot (1 - TS)$$

$$= (1.33 \cdot 0.015) + (1 \cdot 0.985)$$

$$= 1.005$$

Not surprisingly, this very low solids content sludge has a specific gravity very close to that of water.

Finally, the volume of sludge is determined from Equation 7.16 to be

$$V = \frac{S}{\rho_{sludge} \cdot TS}$$

$$= \frac{1162 \text{ kg/day}}{1.005 \cdot 998 \text{ kg/m}^3 \cdot 0.015}$$

$$= 77 \text{ m}^3/\text{day}$$

Sludge Dewatering

The purpose of sludge dewatering is to increase the total solids content of the sludge. This increase in solids content decreases the volume and mass of sludge produced, with concomitant decreases in storage volume required, transportation costs, disposal costs, and so on.

The decrease in volume as a result of a thickening or dewatering process is provided in the following equation:

$$\frac{V_2}{V_1} = \frac{P_1}{P_2} \tag{7.17}$$

where

V_1 = volume of sludge before thickening (or dewatering) (L^3)

V_2 = volume of sludge after thickening (L^3)

P_1 = total solids concentration of sludge before thickening

P_2 = total solids concentration of sludge after thickening

It is important to note that this equation assumes that the density of the sludge does not change as the total solids content changes. Also, the equation assumes that 100% of the solids are captured by the thickening and dewatering devices.

The primary processes associated with dewatering are shown in Figure 7.17. In practice, a water treatment facility does not need to include any or all of the solids management steps shown. Indeed, some utilities have the luxury of being able to send their sludge directly to the wastewater treatment plant.

Thickening, coagulant recovery, dewatering, drying, and disposal are further discussed in this section.

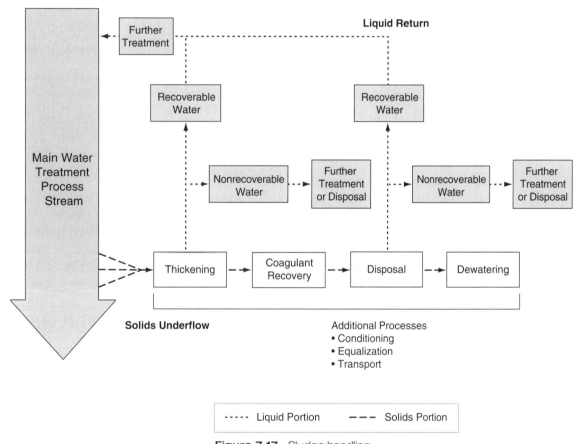

Figure 7.17 Sludge handling

Source: U.S. Environmental Protection Agency, *Management of Water Treatment Plant Residuals*, 625/R-95/008.

Thickening

Thickening increases the solids content of sludges. It usually follows immediately after sedimentation, filtration, or water softening and is often used as "predewatering" step. Thickening is accomplished most commonly by gravity thickening, in which the solids enter the thickener in the center of the tank. The solids settle to the bottom where they are collected, while the effluent is collected by weirs at the periphery of the tank. Metal hydroxide residuals only thicken to between 1% and 6%, while softening sludges can thicken to between 15% and 30% total solids. Sludge can also be thickened by dissolved air flotation, lagoons, or drying beds.

Coagulant Recovery

The purpose of coagulant recovery is to conserve coagulant use and limit metal concentration in the sludge to be disposed. Coagulants are most often extracted by acidification, with the pH between 1.8 and 3 used for acid contact times between 10 and 20 minutes. Simplified reactions for coagulant recovery using sulfuric acid are provided as follows:

$$2Al(OH)_3 + 3H_2SO_4 \rightarrow Al_2(SO_4)_3 + 6H_2O$$

$$2Fe(OH)_3 + 3H_2SO_4 \rightarrow Fe_2(SO_4)_3 + 6H_2O$$

According to MWH (2005), coagulant recovery is limited in practice due to contaminants found in the recovered coagulant.

Conditioning

The purpose of sludge conditioning is to enhance the dewaterability of the sludge, typically by addition of chemicals. The most common chemical conditioners are ferric chloride, lime, or polymer.

Dewatering

Similar to thickening, dewatering processes increase the solids content; some have defined dewatering as increasing the solids content to greater than 8% (U.S. EPA, 1996). Dewatering is most often accomplished mechanically. Common means of mechanical dewatering include belt filter presses, centrifuges, and vacuum filters:

- **Centrifugation**. Centrifuges use the concept of centrifugal force to remove solids from the water. In effect, a centrifuge is a sedimentation device that increases the settling speed of sludge solids by centrifugal force. In operation, sludge is fed in through the center of the centrifuge. Solids are forced to the periphery, where they are removed by a rotating screw conveyer. Centrifugation is most effective for lime softening sludges, and solids contents between 35% and 50% are common. If properly conditioned, solids contents up to 25% may be obtained for alum sludges.

- **Vacuum filtration.** In vacuum filtration, a cylindrical drum, covered with a filter fabric, rotates through a partly submerged vat of sludge. A vacuum is applied that pulls water through the fabric and into the drum, where it is released. The solids adhere to the filter fabric, from which they are removed by scraping.

- **Plate and frame.** Pressure filters, such as a plate and frame press, use positive pressure (greater than 100 psi) to force water out of the sludge. In the plate and frame press, sludge enters the filter pack, which is a series of chambers held together by the press "skeleton." As the chamber is filled, water passes through a filter cloth medium that lines each chamber and is discharged. Once the chamber completely fills with solids, the chambers are opened and the filter cake is removed.

- **Belt filter press.** A belt filter press (Figure 7.18) dewaters sludge by a combination of gravity drainage and mechanical application of pressure. Conditioned sludge is placed on a moving belt; initially, filtrate drains out. The sludge is then "sandwiched" between the original belt and a second belt, which moves across a series of rollers, effectively squeezing out the water.

The sludge cake is removed at the end. Capital costs and space requirements are significantly less than for plate and frame presses.

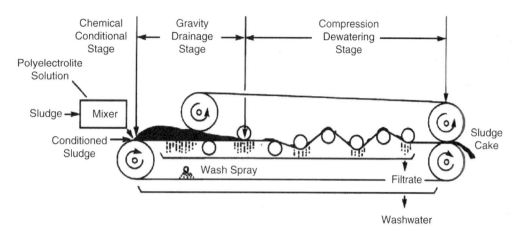

Figure 7.18 Belt filter press

Source: U.S. Environmental Protection Agency, *Process Design Manual, Sludge Treatment and Disposal,* 1979.

Drying

The purpose of drying is to remove water by evaporation and is defined as increasing the solids content to greater than 35%.

The most common method of drying uses sand drying beds (Figure 7.19), on which drying takes place by evaporation and drainage. Sludge is placed on the bed, the bottom of which is composed of sand or some other filter medium. Depending on the dryness of the climate, an underdrain system is installed to remove water that has drained from the sludge. Decanting also aids in managing the operation of a drying bed. The design goal may be to minimize the number of application and removal cycles or minimize land area. The designer must balance the need for drying speed (which increases with decreasing thickness of the applied sludge), space availability, and operations concerns. Thin layers require more cycles of loading and unloading and/or increase the land area required for the sand beds. Davis and Cornwell (1998) recommend a total of three or four beds for flexibility of operations.

Operation of a sand drying bed entails applying between 6 and 12 inches of sludge at a time, allowing the sludge to dry, and then removing the dried sludge. Chemical conditioning hastens drying time. Design requires either small-scale testing, or knowledge of *net evaporation rates* (net evaporation rate is the difference between evaporation rate and precipitation rate). In cold climates, freezing effectively conditions the sludge; upon thawing, water drains rapidly from the sludge.

In certain parts of the country, drying occurs in lagoons. The lagoons are operated as gravity settlers for a period of time, followed by a period of drying. Lagoons are equipped with decant capabilities and may be designed to allow drainage. Long-term settling in a lagoon produces solids contents for metal hydroxide sludges on the order of 10%. Without underdrain capabilities, drying of the sludge could take years, depending on the climate and the depth of sludge. Drying time is strongly affected by the fact that many sludges form a crust that dramatically decreases the sludge drying rate. Conversely, if cracking of the surface occurs, evaporation rates will increase significantly. Dewatering lagoons, as compared to storage lagoons without underdrain systems, offer the benefits of a sand drying bed in addition to the ability to store large volumes of sludge periodically.

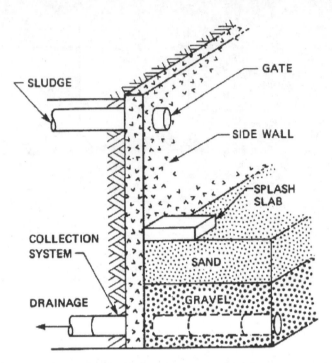

Figure 7.19   Sand drying bed section

Source: U.S. Environmental Protection Agency, *Management of Water Treatment Plant Residuals*, 625/R-95/008.

Example **7.9**

Preliminary design of a dewatering train

A 5 MGD water treatment utility produces 1250 lb of solids per day at a concentration of 1.5% total solids. In the past, the sludge has been sent to the wastewater treatment facility. However, reduced capacity at the wastewater treatment facility necessitates the design and construction of a new dewatering train. The proposed dewatering train will consist of a gravity thickener followed by a belt filter press. The gravity thickener produces a solids content of 3.5%, while the filter press produces a solids content of 22%. The filter cake produced by the belt filter press will be sent to drying beds. The climate permits that each bed can be cycled twice per year, and sludge will be spread in 6-inch layers on the drying beds.

Estimate the number of acres of drying bed required to accommodate a total of four beds, with one bed to be used as a spare.

Solution

A sketch of the system is provided in Exhibit 5.

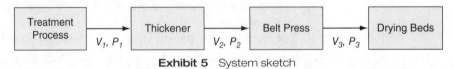

Exhibit 5   System sketch

For this problem, the density of the sludge will be assumed to remain constant at 8.34 lb/gal, irrespective of the solids content.

The daily volume of sludge entering the dewatering train is determined by first finding the daily mass of sludge produced and dividing this mass by the sludge density. The mass of sludge produced is determined using the definition of total solids:

$$\text{mass of sludge} = \frac{\text{mass of solids}}{\text{TS of sludge}} = \frac{1250 \text{ lb}}{0.015} = 83,000 \text{ lb of sludge per day}$$

$$\text{volume of sludge} = \frac{\text{mass of sludge}}{\text{sludge density}} = \frac{83,000 \text{ lb}}{8.34 \text{ lb/gal}}$$

$$= 10,000 \text{ gal of sludge per day}$$

The volume of sludge exiting the gravity thickener is determined from Equation 7.17.

$$\frac{V_2}{V_1} = \frac{P_1}{P_2}$$

$$V_1 = 10,000 \text{ gal}$$

$$P_1 = 1.5\%$$

$$P_2 = 3.5\%$$

Therefore, $V_2 = 4300$ gallons.

The volume of sludge exiting the belt filter press is determined using a similar analysis:

$$\frac{V_3}{V_2} = \frac{P_2}{P_3}$$

Given that $P_3 = 22\%$, the volume of sludge exiting the filter press and traveling to the drying beds, V_3, is 680 gallons. Thus, the sand drying beds must be sized to handle 680 gallons of sludge per day (or 250,000 gallons/year).

The total land area needed is estimated by dividing the volume of sludge generated per year by the depth of application:

$$\frac{\dfrac{250,000 \text{ gal}}{\text{yr}} \cdot \dfrac{\text{ft}^3}{7.5 \text{ gal}}}{0.5 \text{ ft}} = 67,000 \text{ ft}^2$$

Given that each bed can be filled twice per year, the land area required is 33,500 ft². This area must be distributed among three beds, so each bed is about 11,000 ft². The total land area that must be set aside for the four beds is:

$$\text{total area} = 11,000 \text{ ft}^2 \cdot 4 \text{ beds} \cdot 1.5$$

$$\underset{\text{spacing factor}}{}$$

$$= 66,000 \text{ ft}^2$$

$$= 1.5 \text{ acres}$$

The factor of 1.5 used in the above equation takes into account roadways and berms that will need to be constructed.

Recall that the problem statement assumed that density did not vary with solids content. The validity of this assumption can be examined by first determining the density of the 22% solids content sludge, assuming that the solids have a specific gravity of 2.5.

$$SG_{sludge} = SG_{solids} \cdot TS + SG_{water} \cdot (1 - TS)$$
$$SG_{sludge} = 2.5 \cdot 0.22 + 1 \cdot (0.78)$$
$$= 1.33$$

Clearly, the sludge has a specific gravity significantly greater than 1. To determine how this impacts the volume, consider that the mass of solids is conserved. Thus, the mass of solids entering the drying beds is equal to 1250 lb. The mass of sludge entering the drying beds is

$$TS = \frac{\text{mass of solids}}{\text{mass of sludge}}; \quad \text{mass of sludge} = \frac{\text{mass of solids}}{TS} = \frac{1250 \text{ lb}}{0.22} = 5700 \text{ lb}$$

The volume of this sludge is then

$$Vol = \frac{\text{mass}}{\text{density}} = \frac{5700 \text{ lb/day}}{1.33 \cdot 8.34 \text{ lb/gal}} = 514 \text{ gal/day}$$

This is significantly less than the estimated 680 gal/day.

Sludge Disposal and Reuse

The following disposal options are commonly used in the United States (U.S. EPA, 1996):

- Landfilling
  - Daily cover
  - Codisposal with municipal solid waste (MSW)
- Land application on…
  - Agricultural land
  - Forest land
  - Other designated disposal site
- Sewer disposal
- Direct discharge to receiving body of water
- Reuse
  - Coagulant recovery
  - Turf farming
  - Building materials
  - Fill material

The most common disposal option in the United States is landfilling in a municipal solid waste or hazardous waste landfill. The latter option is rare, as drinking water treatment sludges typically pass the TCLP test (see section 11.1 for a description of the TCLP test).

Land application has seen mixed results, as the benefits are much less obvious than the benefits of land application of wastewater treatment sludge. The spreading of coagulant sludges has raised concern about the increased concentrations of aluminum in the soil, and the ability of the sludges to bind plant-available phosphorus. Lime sludges can be beneficially used on agricultural fields in place of commercially available lime.

A growing concern with land application of water treatment sludges is increasing arsenic concentrations. As regulations require greater removal of arsenic from drinking water, the arsenic ends up in the sludge.

7.3.2 Liquid Residuals

In addition to the solid residuals (sludges) described above, water treatment facilities also must treat liquid residuals. The sources of liquid residuals include filter waste washwater (washwater from backwashing filters), *filter-to-waste* washwater (collected during the ripening stage of granular media filters), membrane concentrates (from nano filtration or reverse osmosis), and ion exchange brines. As such, the waste streams contain suspended and dissolved solids, pathogenic organisms, trace metals, and any other contaminants removed from the raw water in the treatment process.

The combined volume of these liquid wastes may comprise from roughly 5% of the total plant flow (for backwash and dewatering filtrates) to 50% for RO processes. Moreover, given that these liquid streams are highly concentrated, returning these waste streams to the head of the plant or discharging to local receiving waters is often infeasible.

Several techniques exist to treat the liquid residuals, including:

- Flow equalization (so that residual flows returned to the head of the plant is treated at a constant rate rather than in intermittent high flows)

- Lagoons

- Granular or membrane filtration

- Disinfection

Filter washwater has a low solids concentration, and has traditionally been returned to the plant headworks (either with or without flow equalization), discharged to lagoons, or discharged to surface water. However, there is increasing concern about recycling pathogenic microorganisms such as giardia and cryptosporidium to the plant headworks.

The liquid residuals from RO processes are typically high in dissolved solids and low in suspended solids. Typical recovery rates, *r,* are between 50% and 85%, which demonstrates that the liquid waste stream from RO processes could be equal to the flow rate of the permeate (that is, of the water that is sent to consumers).

The quantity and quality of liquid residuals from RO processes can be calculated using the definitions of recovery rate and the rate of rejection, Rej, discussed in section 7.2.7. This can be accomplished readily be writing a mass balance of the contaminant on an RO unit, and defining the following terms:

$$Q_c = \text{concentrate flow rate}$$

$$C_c = \text{concentrate concentration}$$

The mass balance equation at steady state and with no reaction term can be written by referring to Figure 7.20.

$$Q_f C_f = Q_p C_p + Q_c C_c$$

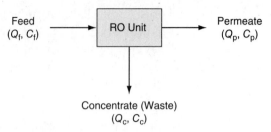

Figure 7.20   Mass balance on an RO unit

Also, knowing that the flow continuity equation is valid (that is, flow into the RO unit equals the flow out of the RO unit), the following can be stated:

$$Q_f = Q_c + Q_p$$

Combining these two equations with the definition of recovery rate yields this useful equation:

$$Q_c = \frac{Q_p(1-r)}{r}$$

This RO concentrate is regulated as an industrial waste in the United States. Moreover, treatment can be quite complicated and can include membrane concentration steps, such as membrane concentration, evaporation, distillation, and crystallization.

Example 7.10

Residuals from an RO process

A 12,000 m³/day RO facility has a recovery rate of 0.75 and a rejection rate of 0.95. The total dissolved solids concentration in the feed stream is 2100 mg/L. Determine:

(a) The daily volume of water that must be treated to produce 12,000 m³/day of finished water (permeate)

(b) The volume of residual wastewater

(c) The concentration of dissolved solids in the permeate

(d) The concentration of dissolved solids in the residual wastewater

Solution

(a) The daily volume of water to be processed can be determined from the definition of recovery rate (Equation 7.6):

$$r = \frac{Q_p}{Q_f}$$

where

$r\ \ = 0.75$

$Q_p = 12{,}000\ \text{m}^3/\text{day}$

Therefore, $Q_f = 16{,}000\ \text{m}^3/\text{day}$.

(b) The volume of residual wastewater is

$$Q_c = Q_f - Q_p$$

$$= 16,000 \text{ m}^3/\text{day} - 12,000 \text{ m}^3/\text{day}$$

$$= 4000 \text{ m}^3/\text{day}$$

Alternatively, Q_c can be determined from the following equation:

$$Q_c = \frac{Q_p(1-r)}{r} = \frac{\left(12,000 \text{ m}^3/\text{day}\right)\left(1-0.75\right)}{0.75} = 4000 \text{ m}^3/\text{day}$$

(c) The concentration of dissolved solids in the permeate can be calculated using the definition of rejection rate (Equation 7.7).

$$\text{Rej} = \frac{C_f - C_p}{C_f}$$

or $C_p = C_f(1 - \text{Rej})$

$$= 2100 \text{ mg/L}(1 - 0.95)$$

$$= 105 \text{ mg/L}$$

(d)   The concentration of dissolved solids in the residual wastewater can be determined from the mass balance relationship as follows:

$$C_c = \frac{Q_f C_f - Q_p C_p}{Q_c}$$

$$C_c = \frac{(16,000 \text{ m}^3/\text{day} \cdot 2100 \text{ mg/L}) - (12,000 \text{ m}^3/\text{day} \cdot 105 \text{ mg/L})}{4000 \text{ m}^3/\text{day}}$$

$$C_c = 8370 \text{ mg/L}$$

Ion exchange brines resulting from water softening processes will constitute 1.5% to 10% of the plant flow (MWH, 2005). The residual wastewater includes water used for regeneration, for rinsing of the resin following regeneration, and any water used to initially backflush the bed prior to regeneration. As such, the ion exchange residual waste stream is high in total dissolved solids, including the hardness ions removed from the water and the sodium ions used in regeneration. Typical values for the chemical constituents in ion exchange liquid residuals are provided in Table 7.19.

Table 7.19   Chemical composition of ion exchange wastewater

Constituents	Range of Averages (mg/L)
TDS	15,000–35,000
Ca^{++}	3000–6000
Mg^{++}	1000–2000
Hardness (as $CaCO_3$)	11,600–23,000
Na^+	2000–5000
Cl^-	9000–22,000

Source: U.S. Environmental Protection Agency, *Management of Water Treatment Plant Residuals*, 625/R-95/008.

7.4 WATER DISTRIBUTION SYSTEM DESIGN AND ANALYSIS

The purpose of a water distribution system is to deliver potable water to residential, industrial, and commercial users from the drinking water treatment facility. An effective system must deliver water at a sufficient flow and pressure. Typically, the goal is to provide pressures between 20 psi and 80 psi, although a maximum of 100 psi is often allowed. Flow requirements must be high enough to meet fire demands in addition to the daily water requirements of consumers. An effective system must provide this flow and pressure 24 hours a day, 365 days per year with minimal interruptions in service and variation in quality.

One conversion that is especially useful in analyzing and modeling water systems is the following:

$$1 \text{ psi} = 2.3 \text{ ft of water}$$

This relationship arises from the fundamental relationship of hydrostatics ($P = \gamma \cdot h$). In water system analysis and design, this relationship can readily provide estimates of system performance. For example, if the elevation of water in an elevated storage tank is 1180 feet above sea level, and the elevation of the lowest point in the service area is 1060 feet, the pressure available to customers at the low point can be estimated to be 52 psi based on the elevation difference. Although, in reality, many other factors will influence the pressure available at any point in the system, this simple relationship can be very helpful.

7.4.1 Flow Requirements

Three types of demands will be discussed in this section: Fire flow, customer demand, and unaccounted for water (UFW).

Fire Flow

The quantity of water needed to fight a fire depends on many factors, including structure size, construction materials, proximity of adjacent buildings, wind speed, temperature, and so on. Several methods have been suggested for estimating fire flows based on readily measured characteristics of the structure.

For one- and two-family dwellings not exceeding two stories in height, the Insurance Services Office (ISO) uses the flows shown in Table 7.20. The residential building lots common today (>0.25 acres) readily allow for spacing between houses greater than 100 feet, and thus a fire flow of 500 gpm is a common value used in design and analysis of water distribution systems in residential areas.

Table 7.20 Fire flow for residential construction

Distance between Buildings	Needed Fire Flow (gpm)
More than 100'	500 gpm
31'–100'	750 gpm
11'–30'	1000 gpm
10' or less	1500 gpm

For nonresidential construction, the ISO method is often used. The governing equation for the ISO method is:

$$NFF = (C) \cdot (O) \cdot (1 + (X + P)) \tag{7.18}$$

where

NFF = needed fire flow (gpm)

C = a factor related to the type of construction

O = a factor related to the type of occupancy

X = a factor related to the exposure to other buildings

P = a factor related to the communication between buildings

The following considerations apply to the ISO method:

- The maximum needed fire flow is 12,000 gpm.

- The minimum needed fire flow is 500 gpm.

- ISO rounds the final calculation of needed fire flow to the nearest 250 gpm if less than 2500 gpm and to the nearest 500 gpm if greater than 2500 gpm.

A simpler equation is the Iowa State University method which calculates the fire flow, G, in gpm as

$$G = \text{volume of space (in ft}^3)/100$$

The volume of water required to extinguish a fire is the product of fire flow rates and duration. Durations are provided in Table 7.21.

Table 7.21 Fire duration

Required Fire Flow (gpm)	Duration (h)
2500 or less	2
3000 to 3500	3
4000 to 4500	4
5000 to 5500	5
6000 to 7000	6
7000 to 8000	7
8000 to 9000	8
9000 to 10,000	9
10,000 to 12,000	10

Source: ISO, "Guide for Determination of Needed Fire Flow," *www.isomitigation.com/ downloads/ppc3001.pdf.*

Customer Demand

Customer demand includes residential, commercial, industrial, academic, and others. A rule of thumb for estimating residential water usage is 100 gallons/capita/ day. This value varies due to many factors, including:

- Cost of water

- Income level

- Water conservation efforts

- Climate

- Time of year

- Extent of metering

Water usage varies dramatically depending on the time of day. An example of a diurnal variation is shown in Figure 7.21.

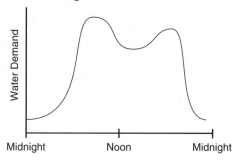

Figure 7.21   Example diurnal curve

The relative difference between peaks and valleys depends on the size of the community and the extent of nonresidential use. The peaks shown in Figure 7.21 correspond to times of high residential demands (for example, early morning and dinner time). As nonresidential uses increase, additional peaks may occur (for example, when an industry exerts a large demand at an off-peak time), or the peaks and valleys may be attenuated due to constant withdrawal by nonresidential use.

Various demands are used depending on the type of analysis being performed. *Average day demand* is the average demand for an average day. *Maximum day demand* is the average rate of use for the maximum day. The *peak hour demand* is the average rate of usage for the maximum hour of usage, which may or may not occur on the maximum day.

The demands can also be characterized by a *peaking factor* that relates average day demands to peak day demands. The definition of peaking factor (*PF*) is

$$PF = \frac{Q_{max}}{Q_{avg}} \qquad (7.19)$$

where

Q_{max} = maximum day demand ($L^3 \cdot T^{-1}$)

Q_{avg} = average day demand ($L^3 \cdot T^{-1}$)

Unaccounted for Water

Unaccounted for water, or *UFW,* typically comprises between 5% and 25% of a system's production and is the difference between the volume of water treated and the volume of water used by the utility's customers. Sources of UFW include leaking pipes and valves, poorly calibrated meters, illicit use, and hydrant flushing.

7.4.2 Water System Appurtenances

The water system appurtenances discussed in this section are pipes, hydrants, valves, pumps, and storage tanks.

Pipes

Pipes used in water distribution systems may be made of cast iron, PVC, galvanized iron, and copper. For modeling purposes, pipes are characterized by their lengths, roughness, and diameter. Generally, the pipe's nominal diameter is used, although this varies from the actual inner diameter for most pipes. Moreover, scaling (for example, accumulation of calcium carbonate) and tuberculation (formation of uneven lumps due to a buildup of corrosion products) can effectively decrease the inner diameter of pipes, in some cases to a dramatic extent.

AWWA provides guidelines to note whether a pipe segment is "deficient." (See Table 7.22.) These are not hard-and-fast numbers but are useful in system analysis.

Table 7.22   Criteria for deficient pipe segments

Criteria	Rationale
Velocities greater than 5 ft/s	Headloss is proportional to the square of velocity, and, as a result, maintaining relatively low velocity is important. Also, systems with high velocities are more susceptible to water hammer.
Headlosses greater than 10 ft/1000 ft	Pipes with these high headlosses are significant contributors to low system pressures.
Headlosses greater than 3 ft/1000 ft for large diameter pipes (> 16 in)	Since larger pipes generally are longer, the accumulation of headloss becomes significant.

Source: AWWA, "Distribution Network Analysis for Water Utilities," M32, 1989.

Hydrants

Hydrants have several purposes, primarily firefighting, system analysis, and system flushing. Hydrants are simply a valve mechanism. Hydrant spacing is typically 500 feet, or at every street intersection.

Valves

Valves are used to control the flow of water in a system. Three types of valves will be discussed in this subsection: isolation valves, directional valves, and control valves.

Isolation valves are the most commonly used valves in a water distribution system. By closing one or more of these valves, a section of the system can be isolated. Isolation is required most often to make repairs or otherwise modify a system. A well-designed system has a large number of valves, which minimizes the number of customers inconvenienced by such repairs. Regulatory agencies often require a minimum distance of 500 to 800 feet between valves and require that valves are located at every street intersection. Isolation valves are most often gate valves, although butterfly valves, globe valves, and plug valves are also used.

Directional valves, or check valves, allow the water to only flow in one direction. Check valves are often found on the discharge side of pumps to prevent potentially damaging backflow. Backflow preventers used on residential water hoses are a type of check valve that prevents the flow of pesticides or fertilizers back into the distribution system.

Control valves purposely add a headloss to the system to control the flow of water. Two examples of control valves are pressure sustaining valves (PSVs) and pressure reducing valves (PRVs). (See Figure 7.22.) When no flow is passing through a PRV, the spring acts to hold the linkage up against the seat. As water flows through the valve, it pushes down on the linkage, thereby opening up the system to flow. The tension in the spring regulates the flow rate exiting the PRV.

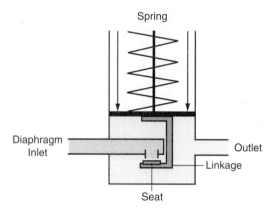

Figure 7.22 PRV valve operation

Source: Illustration courtesy of Reliance Water Controls.

PRVs are used when large changes of elevation would create extremely large pressures at the lowest elevations in a distribution system. For example, an operating range of 60 psi (between 20 psi and 80 psi) correlates to changes of elevation of approximately 140 feet. Thus, the maximum ground elevation differences in a distribution system that could be treated without a PRV would be 140 feet, assuming 20 psi was acceptable at the higher elevations and 80 psi was acceptable at the lowest elevations. The effect of a PRV is shown in Figure 7.23. Note the hydraulic grade line (HGL) sloping gradually in the direction of flow due to frictional headloss. Without a PRV, the HGL is too high above the service area at the lower elevation, but with the PRV, the HGL elevation is reduced to an acceptable level.

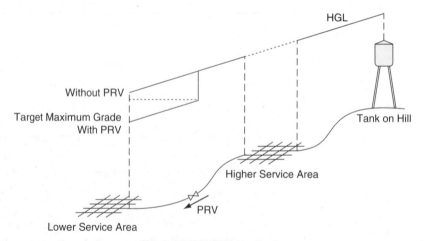

Figure 7.23 PRV impacts

Source: *Advanced Water Distribution Modeling and Management.* Reprinted with permission of the Bentley Institute Press.

Pumps

Pumps add energy to the water and are represented by a distinct jump in the HGL in the direction of flow. Water enters at the *suction* side of the pump and exits at the *discharge* side. As described in section 3.2, the performance of a pump is described by the pump characteristic curve. Pumps may be either *variable* speed or *fixed* speed. Variable-speed pumps, as the name implies, are able to provide different head-discharge characteristics depending on the speed.

As mentioned in section 3.2 selection of a pump is determined based on the intersection of the pump characteristic curve and the system curve. In practice, system curves are much more complicated than those discussed in Chapter 3. For example, a gridwork of pipes is much more complex to model than a single pipe. Also, water distribution pumps may be required to operate at several different operating points on the system curve, corresponding to average day demand, maximum day demand, and so on.

A third factor to be considered in reality that complicates "textbook" system curves is that the system curve varies as the elevation of water in the elevated storage tank varies. Consider a system curve for a system that contains a pump that delivers water in a transmission main to an elevated storage tank, with a change in elevation between the pump and water in the storage tank equal to Δz. A simplified system curve for this system is $h_s = \Delta z + CQ^2$, where CQ^2 represents the headloss in the system The Δz quantity will change for this system as the tank empties and fills, thus altering the system curve. This concept is illustrated in Figure 7.24 for the tank at empty and full stages.

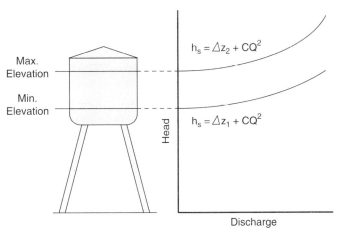

Figure 7.24 System curve variation with storage tank water level

Pump curves also give information about the pump efficiency and the electrical efficiency of the motor. Efficiency is a very important concept associated with pumps as it directly relates to the operations cost of the pump. (Pumps can account for up to 75% of the energy requirements for a water system.) Optimization of pump operations is very important and may take into account operating the system at off-peak times, when rate structures provide relatively inexpensive electricity rates.

Three terms associated with pump efficiency are shown in Figure 7.25, which depicts a pump as a combination of a motor and an impeller.

With respect to Figure 7.25, the types of efficiency related to pumps are

$$\text{motor efficiency} = \frac{\text{brake power}}{\text{input power}} \tag{7.20}$$

$$\text{hydraulic efficiency} = \frac{\text{water power}}{\text{brake power}} \qquad \textbf{(7.21)}$$

$$\text{overall efficiency} = \text{motor efficiency} \cdot \text{pump efficiency} = \frac{\text{water power}}{\text{input power}} \qquad \textbf{(7.22)}$$

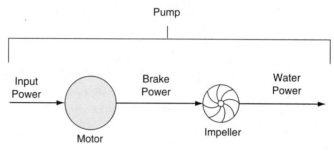

Figure 7.25 Pump efficiency terms

The terms *brake power* and *water power* are often termed *brake horsepower* (bhp) and *water horsepower* (whp), respectively. The more generic terms are used in this text to emphasize that horsepower are not the only units of power. Also, the term *overall efficiency* is sometimes referred to as *wire-to-water* efficiency.

Example 7.11

Estimating pump operating costs

A system curve is represented by this equation: $h_s = 1.32 \cdot 10^{-4} Q^2 + 0.022Q + 15$, where Q is in gpm and h_s is in feet. For the given pump curve (Exhibit 6), estimate the annual energy requirements, assuming the pump motor has an efficiency of 75% and that electricity costs \$0.08/kW-hr.

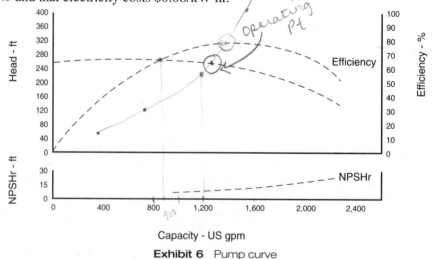

Exhibit 6 Pump curve

Solution

The operating point can be determined by plotting the system curve on this pump curve. A few points are provided in Exhibit 7 and plotted in Exhibit 8; in practice, a computer could be used to create a continuous smooth curve. The operating point corresponds to the point where the system curve and pump curve intersect.

h_s (ft)	Q (gpm)
45	400
117	800
232	1200
389	1600

Exhibit 7 System curve points

From the graph, the operating point is circled and corresponds to a flow of 1260 gpm at 247 ft. Also, the hydraulic efficiency of the pump is determined to be 78%.

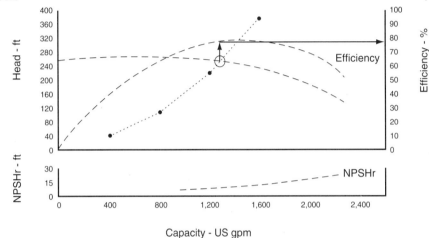

Capacity - US gpm

Exhibit 8 System curve plotted on pump curve

The pump efficiency can be determined from this equation: $\text{pump efficiency} = \dfrac{\text{water power}}{\text{brake power}}$. The water power is determined from the following relationship: $\text{water power} = \gamma \cdot Q \cdot h_s$, where $\gamma = 62.4$ lb/ft³, and Q and h_s correspond to the flow and head delivered at the operating point.

$$\text{water power} = 62.4\,\frac{\text{lb}}{\text{ft}^3}\ \cdot\ 1260\,\frac{\text{gal}}{\text{min}}\cdot 247\ \text{ft}\cdot\frac{1\ \text{ft}^3}{7.48\ \text{gal}}\ \frac{1\ \text{min}}{60\ \text{sec}}\ \frac{1\ \text{hp}}{550\ \text{ft lb/sec}}$$

$$= 78.7\ \text{hp}$$

Thus, the brake power equals $\dfrac{\text{water power}}{\text{pump efficiency}} = \dfrac{78.7\ \text{hp}}{0.78} = 100\ \text{hp}$.

The input power is determined from the definition of motor efficiency (Equation 7.20) and represents the power that must be purchased to operate the pump.

$$\text{input power} \ = \ \frac{\text{brake power}}{\text{motor efficiency}} = \frac{100\ \text{hp}}{0.75} = 133\ \text{hp}$$

The power purchased is estimated by assuming that the pump runs 18 hours per day.

$$133\ \text{hp}\,\frac{746\ \text{W}}{\text{hp}}\cdot\frac{1\ \text{kW}}{1000\ \text{W}}\cdot\frac{18\ \text{hr}}{\text{day}}\cdot\frac{365\ \text{day}}{\text{yr}}\cdot\frac{\$0.08}{\text{kW}\cdot\text{hr}} = \$52,300\,/\,\text{yr}$$

Storage Tanks

Storage tanks serve many purposes in a water distribution system. One purpose is to reduce the size of a pump that would otherwise be required; that is, without a storage tank, pumps would need to operate constantly and would need to be sized to deliver the peak hourly demand. With a storage tank in place, the pumping requirements are greatly reduced, to the point that the pumps only need to supply the maximum day demand rather than the maximum hourly demand. Other purposes of tanks are as follows:

■ They provide a reserve of treated water that will minimize interruptions of supply due to failures of mains, pumps, or plant equipment.

■ They help maintain uniform pressure throughout the day, in spite of diurnal variations in demand.

■ They provide a reserve of water for firefighting and other emergencies.

When designing a storage tank, the designer needs to size the tank, determine a location for the tank, consider using an elevated or ground level storage tank, and, often of utmost importance to community members, consider the tank aesthetics. Tanks may be constructed of steel or concrete and come in a variety of shapes. Tall cylindrical tanks are often called standpipes.

When choosing between an elevated storage tank or a ground storage tank, the following should be considered:

■ Construction costs (typically higher for elevated storage tanks)

■ Complexity of operation (elevated storage tanks do not require booster pumps and thus offer simpler operation)

■ Response of system to power failure (elevated storage tanks are not hindered by this)

■ Size required (ground storage tanks can be much larger than elevated storage tanks)

■ Flexibility of operation (operating characteristics of ground storage tanks can be readily modified through pump sizing and controls)

Some "elevated" storage tanks are simply ground storage tanks placed at high land elevation, if available. Another type of ground storage tank is a hydropneumatic tank, in which water is pressurized within the tank.

To size a tank, several different volumes must be considered (Figure 7.26):

■ Operational storage refers to the storage required to meet average daily demands when the pump supplying the tank is not running.

■ Equalizing storage is the storage required to satisfy high rate demands, such as peak hourly demand.

- Fire and emergency storage is the volume required to fight fires or satisfy demands in the advent of an unusual emergency, such as the shutdown of a plant in response to equipment failure or breakage of a transmission main. Fire storage can be determined by the ISO method described in section 7.4.1.

- Dead storage is water that cannot be used in the system because of elevation concerns; this water is typically found in the bottom of standpipes.

It is important not to oversize the tank. This may seem obvious from a cost standpoint, but it is also important in terms of minimizing water age.

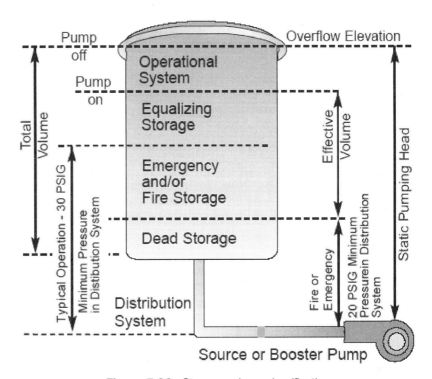

Figure 7.26 Storage volume classification

Source: Reprinted with permission of the National Environmental Services Center.

Figure 7.26 also illustrates another useful attribute of storage tanks, the ability to meet system pressures for different demand scenarios. Note that the elevation of the water level in the tank corresponding to the bottom of equalization storage is still adequate to generate an HGL that meets system pressures of 30 psi. In the advent of an emergency such that all of the emergency storage volume is depleted, the tank provides enough head to supply the entire system with water at 20 psi.

An elevated storage tank will also readily handle the diurnal variation, such as that shown in Figure 7.21. During periods of low demand, the tank can be filled by pumps, and during periods of high demand, the volume of water in the tank decreases. This cycling is depicted in Figure 7.27.

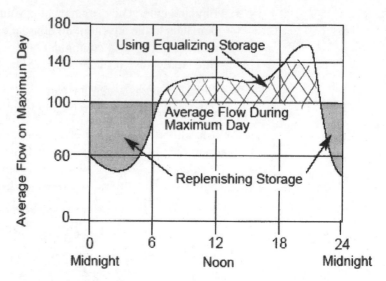

Figure 7.27   Response of storage tank to diurnal variation

Source: Reprinted with permission of the National Environmental Services Center.

Note the line in Figure 7.27 that represents the average flow during the maximum day, equal to 100 flow units. The shaded area below this line is equal to the crosshatched area above this line. The crosshatched area above the line represents the equalization volume. The shaded area is equal to the volume of water that is replenished, and the crosshatched area is equal to the volume of water that is emptied from the tank. A tank that is equal in size to either of these areas and filled at a constant rate of 100 flow units would adequately handle the maximum day demands. This example demonstrates the fact that a pump paired with such a storage tank would only need to be sized to meet the average flow for the maximum day, which would be much smaller (and less expensive to purchase and operate) than a pump that would have to deliver the peak hourly flow, in the case of no storage tank.

To save on piping costs, locating the storage tank in the center of the distribution system might be economical. Alternatively, tanks that are located at the extremities of the system will be better able to accommodate eventual growth of the service area and to provide a uniform pressure distribution throughout the entire service area. This latter characteristic is shown in the idealized drawing in Figure 7.28. In this figure, the distribution system is operating at a high demand time, such that the system is being served by the pump and by the tank. In Figure 7.28A, the system is being served from both sides, and the resulting HGL is shown. In Figure 7.28B, the tank is placed in an ineffective location directly next to the pumping station. Again, at times of peak flow, with the pump running, the HGL is shown. Note that the elevation of the HGL at the midpoint is equal in both cases as the slope of the HGL is assumed to be constant. (The slope will most likely be greater for the second case, given that the flow is much larger, which will produce a greater headloss.)

However, the customers at the far right of the drawing have a lower pressure, demonstrated by comparing the HGL between Figure 7.28A and B.

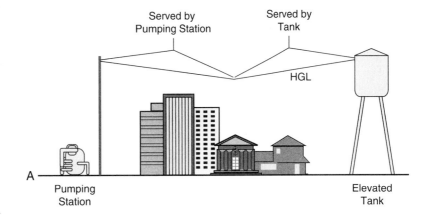

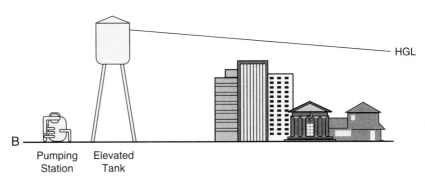

Figure 7.28   Effect of tank location on HGL

Example **7.12**

Estimating storage tank volume needs

An engineer designing a water system serving 5000 customers needs a preliminary estimate for the tank volume. Data has been collected and analyzed to create a typical diurnal demand pattern for this municipality and is provided in Exhibit 9 as a series of "pattern factors." The flow can be determined at any time as the product of the pattern factor and the average flow. For this analysis, the tank volume will be estimated for the peak day, which has a peaking factor of 1.5. On the peak day, the pump will run constantly and deliver a flow equal to the maximum daily flow. Equalization storage will be necessary to deliver water for the peak hourly flow, which is assumed to occur on the peak day.

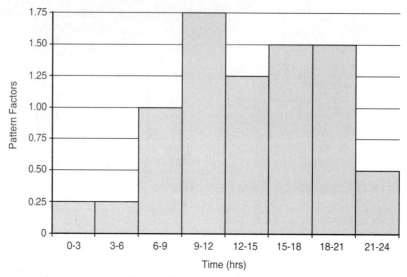

Exhibit 9   Diurnal demand variation

The tank will also need to provide fire storage to fight a fire at an industry that requires a fire flow of 2000 gpm.

Solution

This solution is based on the assumption that the tank needs to be sized to contain the sum of equalization volume and fire flow volume.

The first step is to determine the equalization volume required. This is accomplished by transforming the graph of pattern factors vs. time into a graph of flow vs. time. The flow by which the pattern factors will be multiplied is the maximum daily flow. Given the population of 5000 residents and assuming 100 gal/capita/day, an average daily flow rate of 500,000 gal/day is estimated. Given the peaking factor of 1.5, the maximum daily flow to be used for this problem is 750,000 gal/day. Using this information, a diurnal flow curve (Exhibit 10) can be prepared by multiplying each of the pattern factors by 750,000 gal/day. The equalization storage is equal to the darker shaded area shown in the graph.

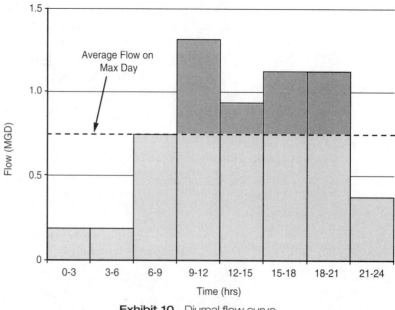

Exhibit 10   Diurnal flow curve

The calculated equalization volume is shown to be 187,500 gallons from Exhibit 11.

Time	Flow Rate	Equalization Volume
0–3	0.1875	
3–6	0.1875	
6–9	0.75	
9–12	1.3125	3 hr · (1.3125 MGD – 0.75 MGD) = 70,313 gal
12–15	0.9375	3 hr · (0.9375 MGD – 0.75 MGD) = 23,438 gal
15–18	1.125	3 hr · (1.125 MGD – 0.75 MGD) = 46,875 gal
18–21	1.125	3 hr · (1.125 MGD – 0.75 MGD) = 46,875 gal
21–24	0.375	
		Total = 187,500 gal

Exhibit 11 Equalization volume requirements

From Table 7.21, a fire duration of two hours is required. Thus, the fire flow storage volume is $2000 \frac{\text{gal}}{\text{min}} \cdot \frac{60 \text{ min}}{\text{hr}} \cdot 2 \text{ hr} = 240,000$ gal.

Therefore, an estimate of the volume required is 187,500 gal + 240,000 gal = 427,500 gal. Thus, if a fire at the cabinet making shop occurs on the peak day, the tower should hold enough water to fight the fire and meet all customer demands, assuming that the pump runs constantly. Also, this analysis does not guarantee that the fire flow and customer flows will be delivered at the proper pressure, as the pressure is governed by system characteristics such as pipe diameters, lengths, pipe layout, and so on.

7.4.3 Fundamentals of Modeling Water Systems

Water distribution systems are complex networks that are not conducive to analysis without the use of a computer. However, understanding the fundamentals is essential in order to carry out hand calculations to arrive at a "back of the envelope" estimate and to verify computer output.

Governing Principles

A water distribution network is a combination of pipes arranged in series and in parallel. Types of pipes in a system are as follows:

■ *Transmission mains*, also called trunk mains, are designed to carry large quantities of water over great distances, such as between a water treatment facility and a water tower.

■ *Distribution mains* are generally laid out to follow the street network and carry less flows than transmission mains. They are typically smaller in diameter and shorter in length.

■ *Service lines*, or sometimes called *services*, convey water from the distribution mains to individual homes

Table 7.23 Pipes in series and parallel

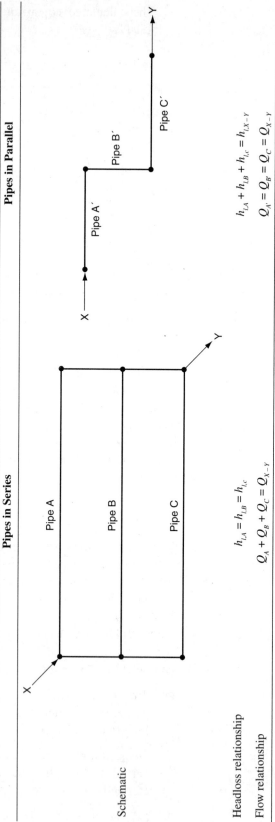

	Pipes in Series	Pipes in Parallel
Schematic	Pipe A, Pipe B, Pipe C	Pipe A´, Pipe B´, Pipe C´
Headloss relationship	$h_{LA} = h_{LB} = h_{Lc}$	$h_{LA} + h_{LB} + h_{Lc} = h_{LX-Y}$
Flow relationship	$Q_A + Q_B + Q_C = Q_{X-Y}$	$Q_{A'} = Q_{B'} = Q_{C'} = Q_{X-Y}$

The principles that apply to pipes in series and to pipes in parallel were discussed in section 3.1.2 and are reviewed in Table 7.23. These very simple concepts become the building blocks of assembling a complex model of an entire distribution system. In Figure 7.29, which shows a portion of an actual distribution system, you can see that the system consists of pipes in series and pipes in parallel.

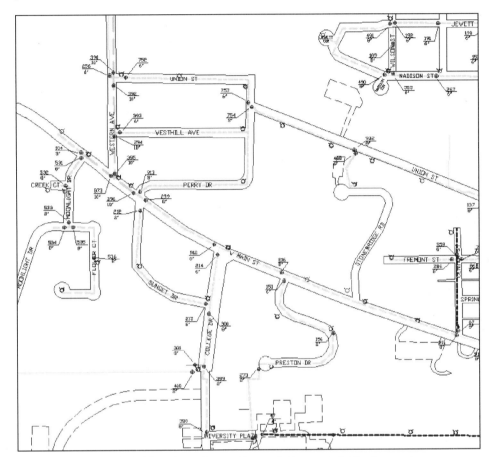

Figure 7.29   Pipes in series and in parallel in a real distribution system

Figure 7.29 also shows that pipe networks are often looped and that dead ends are avoided. Dead ends are avoided due to water quality issues, the need for additional maintenance in flushing of the pipe, and the fact that better pressure distributions are possible throughout a system that is composed of many loops, given that customers can be served from many directions. Perhaps most important, a branched system, compared to a looped system, causes the loss of service to many more customers given a breakage in a pipe, as shown in Figure 7.30. In Figure 7.30, two potential designs are provided to serve a new subdivision. In Figure 7.30A, dead ends occur at points A, C, G, H, and J. In Figure 7.30B, four pipes have been added (A'-B', C'-G', H'-I', and G'-J') such that there are no dead ends. Although the system in Figure 7.30B is more expensive due to the additional pipes, consider what happens if a break occurs in pipe E-F (or E'-F'). In the case of Figure 7.30A, all customers to the right of point F will be without water service while the repairs are taking place. In the system with more loops, only those customers located between points E' and F' will be without service during repairs.

Two other concepts of utmost importance in the modeling of water networks are the concepts of conservation of mass and conservation of energy. The conservation of mass principle states that the mass rate of water entering a control volume must be equal to the mass rate of water exiting the control volume. If the control volume is a simple set of pipes in parallel or pipes in series, such as those shown in Table 7.23, the conservation of mass equation says that the flow rate into the system at point X must equal the flow rate of water exiting the system at point Y.

The conservation of energy principle states that the energy loss (or gain) between two points must be the same, regardless of the path that the water travels between the two points. Given the fact that the loss in energy between two points is expressed as the headloss between the two points, the conservation of energy principle is restating the principle in Table 7.23, which states that the headloss in the parallel pipe network can be expressed as $h_{l,A} = h_{l,B} = h_{l,c}$.

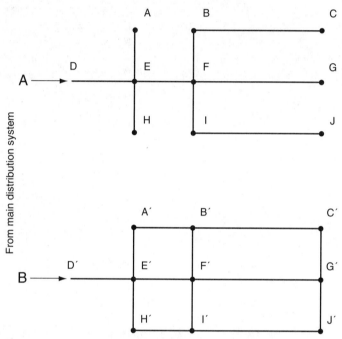

Figure 7.30 Importance of pipe looping

Headloss between two points in a distribution system can be found using any number of models, but the Darcy-Weisbach equation and the Hazen-Williams equation are most common for water distribution systems. The Darcy-Weisbach equation is

$$h_l = f \frac{L}{D} \frac{v^2}{2g}$$

where

h_l = head loss (L)

f = the friction factor, obtained from the Moody chart (see Figure 3.1)

L = length of pipe (L)

D = diameter of pipe (L)

v = velocity in pipe (L·T^{-1})

g = acceleration due to gravity (L·T^{-2})

The Hazen-Williams equation is

$$h_l = U \frac{L}{D^{4.87}} \left(\frac{Q}{C} \right)^{1.852}$$

(7.23)

where

U = unit conversion constant

Value of U	Units of Q	Units of D
10.7	m/s	m
4.73	cfs	ft
10.5	gpm	in

h_L = headloss (same units as L)

L = pipe length

D = pipe diameter

Q = pipe discharge

C = Hazen-Williams coefficient (see Table 7.24)

Table 7.24 Hazen-Williams values

Pipe material	C
Asbestos-cement	140
Cast iron	100
Concrete	100
Copper	130
Steel	120
Galvanized steel	120
Polyethylene	140
PVC	150

A nomograph for quickly estimating the flow due to the Hazen-Williams equation is provided in Figure 7.31.

Example 7.13

Comparison of empirical methods

Estimate the flow in a 1000-foot length of 12-inch cast iron pipe that increases 17 feet in elevation in the direction of flow. There are no other inflows or outflows to the pipe. The pressure at the upstream end is 75 psi, and the pressure at the downstream end is 62 psi. Determine the flow in this pipe using the following:

(a) Darcy-Weisbach equation

(b) Hazen-Williams equation

(c) Hazen-Williams nomograph

Solution

A schematic of the system is shown in Exhibit 12.

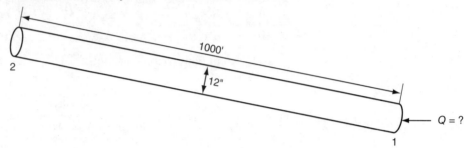

Exhibit 12 System schematic

Some preliminary values that will be used in this problem are computed first. P_1 and P_2 can be expressed as heads by multiplying the units of psi by 2.3, yielding $P_1 = 173$ ft of water and $P_2 = 143$ ft of water. Also, the cross-sectional area of the pipe is given by:

$$A = \pi \frac{D^2}{4} = \pi \frac{1\ \text{ft}^2}{4} = 0.785\ \text{ft}^2$$

Before using any of the methods, the headloss in the pipe can be computed by applying the energy equation between points 1 and 2.

$$\frac{P_1}{\gamma} + \frac{V_1^2}{2g} + z_1 = \frac{P_2}{\gamma} + \frac{V_2^2}{2g} + z_2 + h_l$$

By assuming that the datum runs through point 1 and by knowing that the velocities at points 1 and 2 are the same, the energy equation can be solved for headloss:

$$h_l = \left(\frac{P_1}{\gamma} - \frac{P_2}{\gamma} \right) + \left(z_1 - z_2 \right) = (173\ \text{ft} - 143\ \text{ft}) - (17\ \text{ft} - 0\ \text{ft}) = 13\ \text{ft}$$

Note that according to Table 7.22, a headloss of 13 ft/1000 ft is considered to be a high headloss.

(a) The Darcy-Weisbach equation will require the use of an iterative technique, as described in Example 3.1. The governing equation is:

$$h_l = f \frac{L}{D} \frac{v^2}{2g} = f \frac{L}{D} \frac{Q^2}{2g\ A^2}$$

Given values of h_l (13 ft), L (1000 ft), D (1 ft), g (32.2 ft/s²), and A (0.785 ft), this equation can be solved for Q.

$$Q = \left(\frac{0.516}{f} \right)^{0.5}$$

One other equation that will be used in the iterative technique is:

$$\text{Re} = \frac{D\,v}{v} = \frac{1\ \text{ft} \cdot v}{1.632 \cdot 10^{-4}\text{ft}^2/\text{s}} = \frac{v}{1.632 \cdot 10^{-4}\text{ft}^2/\text{s}}$$

The iterative technique can begin, knowing that $\varepsilon/D = 0.00085$ ft/1 ft = 0.00085.

Iteration 1: Guess $f = 0.03$

$$Q = \left(\frac{0.516}{f}\right)^{0.5} = \left(\frac{0.516}{0.03}\right)^{0.5} = 4.1 \text{ cfs}$$

$$v = \frac{Q}{A} = \frac{4.1 \text{ cfs}}{0.785 \text{ ft}^2} = 5.3 \text{ ft/s}$$

$$\text{Re} = \frac{5.3 \text{ ft/s}}{1.632 \cdot 10^{-4} \text{ft}^2/\text{s}} = 3.2 \cdot 10^4$$

$$f = 0.025$$

Iteration 2: $Q = 4.5$ cfs, $v = 5.8$ fps, Re $= 3.5 \cdot 10^4, f \sim 0.025$

Thus, the Darcy-Weisbach equation yields a flow rate of 4.5 cfs.

(b) The Hazen-Williams equation (Equation 7.23) is a bit more straightforward:

$$h_l = U \frac{L}{D^{4.87}} \left(\frac{Q}{C}\right)^{1.852}$$

$$13\text{ft} = 4.73 \frac{1000 \text{ ft}}{1^{4.87}} \left(\frac{Q}{100}\right)^{1.852}$$

This equation is solved to yield a flow rate of 4.14 cfs.

(c) The Hazen-Williams nomograph is the easiest method, but like all graphical techniques it only provides an estimate. However, in much engineering work such accuracy is acceptable.

Given a head loss of 13 ft/1000 ft, $C = 100$, and a diameter of 12 in, the nomograph yields a flow rate of 4.3 cfs.

Minor losses can be very important in modeling the hydraulics of a system and can consist of tees, elbows, valves, and a variety of other fittings. Minor losses are discussed in section 3.1.2.

A system can be modeled by applying the conservation of mass and conservation of energy equations to a pipe system. The conservation of mass equation is applied to individual nodes in the system, such that the sum of all flows entering the node must equal the sum of all flows exiting the node. The conservation of energy equation is applied around loops in the system. By traveling around a loop, the headlosses in each pipe segment sum up to zero. A sign convention is required, such that all loops are traveled clockwise, with decreases in head in the direction of flow being represented as positive headlosses, and increases in head in the direction of flow represented as negative headlosses. By applying these two equations to all nodes and loops within a system, a series of equations can be generated that form the basis of a computer model that can be used for a range of analysis tasks.

The process outlined above is the same process used when applying Kirchoff's rule to electric circuits. According to Kirchoff's rule, the sum of the currents coming into a junction is equal to the sum leaving the junction, and the sum of all the potential differences, or voltage changes, around a complete loop is equal to zero.

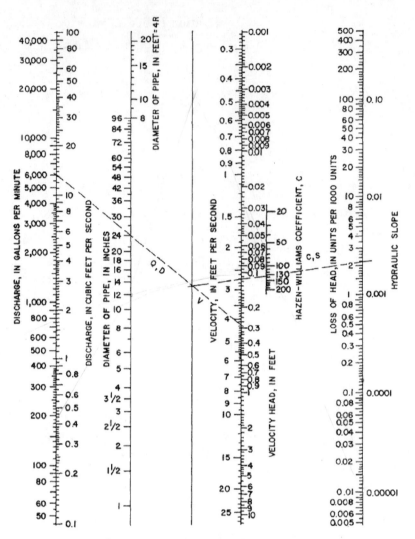

Figure 7.31 Hazen-Williams nomograph

Source: *Design and Construction of Sanitary and Storm Sewers (Manual and Report No. 37)*; 1986, American Society of Civil Engineers. Reprinted by permission.

Example 7.14

Examining a simple pipe system

For the simple schematic provided (Exhibit 13), develop a series of linear equations by applying the conservation of mass and conservation of energy equations. The schematic shows that the system is supplied at node 1, and demands are placed on the system and nodes 3, 4, and 5.

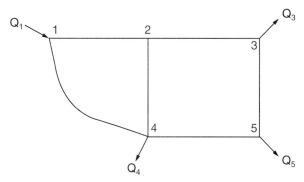

Exhibit 13   Simple pipe network

Solution

A direction of flow needs to be assumed for each pipe reach. This is shown in Exhibit 14. If the assumed flow directions are wrong, model results will simply return negative values for those flow directions.

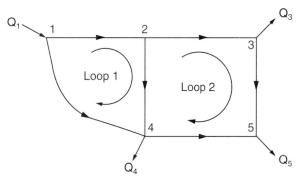

Exhibit 14   Pipe network with assumed flow directions

Using the notation for flow between node i and node j as $Q_{i\text{-}j}$, conservation of mass equations can be written for each node as follows:

Node 1:	$Q_1 - Q_{12} - Q_{14} = 0$
Node 2:	$Q_{12} - Q_{24} - Q_{23} = 0$
Node 3:	$Q_3 + Q_{35} - Q_{23} = 0$
Node 4:	$Q_4 + Q_{45} - Q_{14} - Q_{24} = 0$
Node 5:	$Q_5 - Q_{45} - Q_{35} = 0$

The conservation of energy equation can be written for both loops as

Loop 1:	$h_{L,12} + h_{L,24} - h_{L,14} = 0$
Loop 2:	$h_{L,23} + h_{L,35} - h_{L,45} - h_{L,42} = 0$

Each of these headlosses can be expressed in terms of the flow rate in each pipe using the Hazen-Williams equation. Thus, given values for the flow rate supplied to the system (Q_1), the demands on the system (Q_3, Q_4, and Q_5), and the pipe characteristics (such as length, pipe diameter, and material type), the flows in each pipe reach can be calculated by solving these equations. Note that the above analysis has produced seven equations and six unknowns.

This system of equations can be solved using a spreadsheet or a programmable calculator. Example input data and results are provided in Exhibit 15 and Exhibit 16, respectively.

	Length (ft)	D (in)		Flow (cfs)
Pipe$_{12}$	400	10	Q_1	5
Pipe$_{14}$	300	10	Q_3	3
Pipe$_{24}$	700	10	Q_4	1
Pipe$_{23}$	650	8	Q_5	1
Pipe$_{35}$	550	8		
Pipe$_{45}$	1000	10	$C = 130$ for all pipes	

Exhibit 15   Example input data

Nodes	Q (cfs)	H_1 (ft)
1 to 2	2.3	2.5
1 to 4	2.7	2.7
2 to 4	0.4	0.2
2 to 3	1.9	8.6
3 to 5	−1.1	2.8
4 to 5	2.1	5.6

Exhibit 16   Model output

Note that for this set of input data, the flow was assumed incorrectly to flow from node 3 toward node 5.

Computer Modeling

The hand calculation methods described above are rarely used in modern engineering practice due to the power of computer models. Modeling can serve many purposes, including the following (Walski et al., 2001):

■ Long-range master planning

■ Fire protection studies

■ Water quality studies

■ System design

■ Operator training

To have its results trusted, a computer model should be *calibrated*. Calibration is the process of comparing model results to real values that characterize the system's performance. For example, results from a model that predict the pressures at various nodes could be compared to pressure readings taken in the system at fire hydrants. If model results and reality do not agree to an acceptable level, the model can be modified such that the two values agree. Pipe diameters, demands, or roughness values can be changed within the model such that the model results and reality agree. However, if the modifications are so great that the model parameters are significantly different from physical reality (for example, pipes that were originally inputted as having a diameter of 10 inches need to be decreased to 4 inches in the calibration process), the model should be investigated closely to determine why such dramatic differences exist.

Skeletonization is a process used to simplify an existing system. In developing a model, the modeler has to determine the level of detail required. This level will vary depending on the model's intended use. For larger systems, it is a monumental task to include every distribution line. The system can be simplified by removing selected pipes, a process known as skeletonization. In the process of skeletonizing a system, demands associated with pipes that are removed are combined with other system demands. For example, Figure 7.32 shows how the skeletonization process has reduced a pipe network from 29 nodes to 3 nodes; however, the total demand to the system is unchanged.

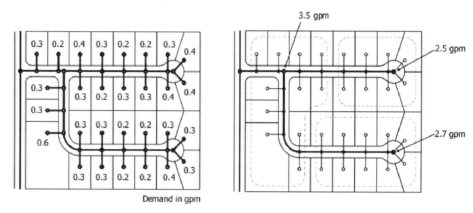

Figure 7.32 Skeletonization process

Source: *Advanced Water Distribution Modeling and Management.* Reprinted with permission of Bentley Institute Press.

Figure 7.33 is an example of a schematic used by a water distribution system model (in this case, WaterCAD by Bentley). This schematic is annotated to show the velocity and headloss in each pipe. This represents only a small fraction of the types of information that such a model can provide.

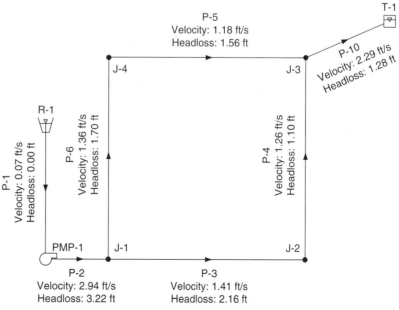

Figure 7.33 Sample model results

7.4.4 Applications

The following applications of water distribution system modeling and analysis will be presented in this subsection: hydrant tests, extending existing services, the lead and copper rule, and water hammer.

Hydrant Tests

There are several uses for hydrant tests, including:

■   Obtaining data for model calibration

■   Determining the amount of flow available for fire fighting purposes

■   Estimating the roughness in pipes

Hydrant tests can be run in any number of configurations. The simplest is to open a single hydrant (the *flowed hydrant*) and attach a pressure gage to a second hydrant, the *residual* hydrant. Such a hydrant test allows the technician to determine the flow rate and pressure at which water is available. Data from such a test can be used to estimate the pressure available at a given flow and compared to regulatory requirements (for example, 500 gpm at 20 psi). Given results from a hydrant test (P_s, P_t, and Q_t), the following equation can be used to estimate the flow Q_o at some other pressure P_o:

$$Q_o = Q_t \left(\frac{P_s - P_o}{P_s - P_t} \right)^{0.54} \tag{7.24}$$

where

P_s = static pressure (the pressure in the residual hydrant with no flow from the flowed hydrant)

P_t = test pressure (pressure in the residual hydrant while the flowed hydrant is flowing)

Q_t = flow rate from the flowed hydrant when opened

Example **7.14**

Using hydrant flow test data

Exhibit 17 shows a typical data sheet used to record hydrant flow test data. From this information, estimate the pressure at which a 500 gpm flow would be delivered.

Solution

$$Q_o = Q_t \left(\frac{P_s - P_o}{P_s - P_t} \right)^{0.54}$$

$$500 \text{ gpm} = 1405 \text{ gpm} \left(\frac{92 \text{ psi} - P_o}{92 \text{ psi} - 82 \text{ psi}} \right)^{0.54}$$

Therefore, $P_o = 101$ psi.

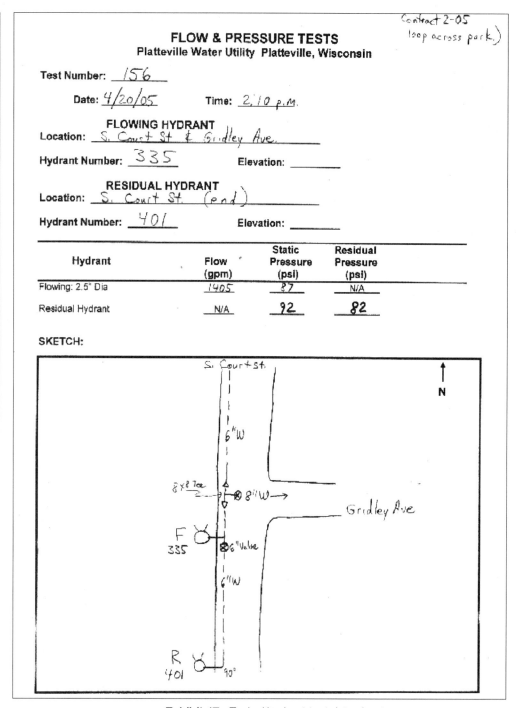

Exhibit 17 Typical hydrant test data sheet

Extending Existing Systems

The hydrant test results can be very helpful in predicting the impacts of adding new services to a system, such as expanding the system to serve a new residential subdivision or an extension of an industrial park. The test results can be entered into a computer model to predict the following:

- How well the existing system will serve the new services

- How the extension will impact the existing system

Perhaps the most important aspect to consider when a new area is being served is the elevation of the area. If the area is too high, the area will not receive enough flow for daily use and/or for emergency use. Such a problem is relatively expensive to rectify. Conversely, if the area to be served is too low, the pressures will be too high; however, this can be readily rectified by installation of a PRV (pressure reducing valve). These concepts are clearly illustrated in Figure 7.34, which shows the highest and lowest elevations that can be served given a system's minimum and maximum HGL.

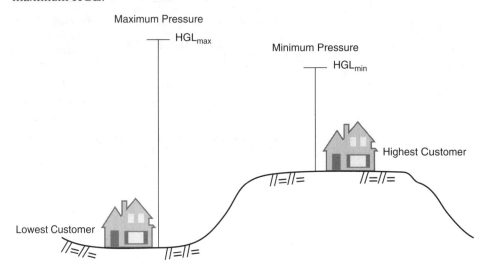

Figure 7.34 Limits of pressure zone

Source: *Advanced Water Distribution Modeling and Management.* Reprinted with permission of the Bentley Institute Press.

The minimum and maximum elevations shown in Figure 7.34 can be estimated from the following equations (Walski et al., 2001):

$$EL_{min} = HGL_{max} - P_{max}/\gamma$$

$$EL_{max} = HGL_{min} - P_{min}/\gamma$$

where

EL_{max} = maximum allowable elevation of customers in zone (L)

EL_{min} = minimum allowable elevation of customers in zone (L)

HGL_{max} = maximum expected HGL in pressure zone (L)

HGL_{min} = minimum expected HGL in pressure zone (L)

P_{max}/γ = maximum acceptable pressure head in pressure zone (L)

P_{min}/γ = minimum acceptable pressure head in pressure zone (L)

Water Hammer

Water hammer is a transient phenomenon that occurs in water distribution systems, including municipal drinking water systems and home plumbing networks. Water hammer is caused by a sudden change in water velocity, most commonly a sudden stop in water flow. A sudden decrease in water velocity may occur due to the sudden closing of a valve or the sudden shutting off of a pump during a power failure. This sudden change in velocity causes the water to be compressed and a pressure wave to move through the system. Water hammer can cause knocking in pipes, but more important, can burst pipes or separate joints.

Figure 7.35 shows the effect of suddenly closing a valve at $t = t_o$ and the resulting pressure wave. Once the pressure wave travels to the "end" of the pipeline, it will "bounce" back to the valve like a coiled spring. Thus, the pressure wave will oscillate back and forth until it is dampened out. The transient pressure wave is dampened out by pipe wall frictional effects and the flexibility of the pipe wall.

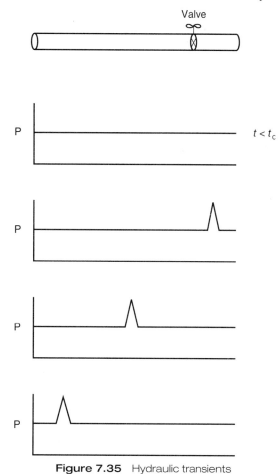

Figure 7.35   Hydraulic transients

The Joukowski equation describes the change in pressure resulting from water hammer effects:

$$\Delta h = \frac{c \cdot v}{g} \tag{7.25}$$

where

Δh = the rise in pressure (m)

c = the velocity of the shockwave (m/s)

v = the water velocity (m/s)

g = the acceleration due to gravity (m/s²)

Given a shockwave velocity on the order of 1000 m/s, Δh can be on the order of hundreds or thousands of psi! A rule of thumb for estimating the change in pressure is that a change of velocity of 1 ft/s can result in a change in head of 100 ft.

Water hammer can be minimized by one or more of the following practices:

■ Stop pumps slowly

■ Use slow-closing valves

■ Use an air vessel

■ Use a surge tank

■ Avoid designing long, straight lengths of pipe

■ Keep pipe velocities low

7.5 WATER SECURITY

Water systems are vulnerable to natural events (floods, earthquakes, natural contamination of the water source), purposeful events (terrorist activity or vandalism), and accidental events (cross connections, accidental discharge to source water, etc.). Drinking water may be contaminated at the source, in the treatment facility, or in the distribution system. Points of contamination within the distribution system include (Walski et al., 2003):

■ Water treatment plant, especially the clear well

■ Pump stations

■ Valves

■ Finished tanks and reservoirs

■ Hydrants

Walski et al. (2003) summarize some of the factors to consider when evaluating the threat of different contaminants:

■ Availability

■ Ability to be detected by monitoring equipment

■ Physical appearance

■ Taste and odor characteristics

■ Dosage required

■ Type of adverse health impact

■ Stability in water

■ Tolerance to chlorine

The Public Health Security and Bioterrorism Preparedness and Response Act of 2002 requires community drinking water systems serving populations of more than 3300 persons to conduct assessments of their vulnerabilities to terrorist attack or other intentional acts and to defend against adversarial actions that might substantially disrupt the ability of a system to provide a safe and reliable supply of drinking water. The requirements of the act assign the EPA and water utilities the responsibility to enhance water sector security and to develop response measures for potential threats to the nation's water supplies and systems.

ADDITIONAL RESOURCES

1. Davis, M. L., and D. A. Cornwell. *Introduction to Environmental Engineering,* 4th ed. McGraw-Hill, 2006.

2. Haestad, T. M. Walski, D. V. Chase, and D. A. Savic. *Water Distribution Modeling.* Haestad Press, 2001.

3. Kawamura, S. *Integrated Design and Operation of Water Treatment Facilities,* 2d ed. Wiley, 2000.

4. Kay, M. *Practical Hydraulics,* Spon Press, 1998.

5. MWH. *Water Treatment, Principles and Design,* 2d ed. Wiley, 2005.

6. U.S. Environmental Protection Agency. *Alternative Disinfectants and Oxidants Guidance Manual.* EPA 815-R-99-014, 1999.

7. U.S. Environmental Protection Agency, *Management of Water Treatment Plant Residuals,* EPA/625/R-95/008, 1996.

Wastewater

9/29/09

OUTLINE

Wastewater is generally defined as *liquid and liquid-carried solid wastes*. Wastewater that comes from residential, commercial, and institutional sources is called *sanitary* or *domestic*, while *municipal* wastewater may have some fraction of permitted industrial sources. *Industrial* wastewaters that contain high levels of contaminants that would inhibit biological activity in the treatment plant must be pretreated to permitted discharge levels on-site prior to discharge.

8.1 WASTEWATER CHARACTERISTICS

The composition of wastewater entering the wastewater treatment plant (WWTP) is usually described with respect to the constituents of primary concern. The primary water quality indicators (WQIs) that have discharge limits are TSS, BOD, fecal coliforms, oil and grease, and pH. Typically, TSS enter the WWTP at approximately 220–250 mg/L, approximately 75% of which are volatile. BOD_5 enters with an average concentration of 200 mg/L, an indication of the amount of organic matter (OM) present, while fecal coliforms can be on the order of 10^6–10^8 CFU/100 mL. Oil and grease are difficult to determine in the influent but typically are present at concentrations near 90 mg/L and are readily skimmed during processing. Values for pH are generally in the range of 6–8.

Treatment plant effluent standards may vary based on geographic region, and, often, more stringent standards may apply for a particular WWTP than those listed below. However, as a minimum, effluent wastewater characteristics for TSS and BOD are limited to 30 mg/L (monthly average), 45 mg/L (weekly average), or a minimum of 85% removal for influents below 200 mg/L. Fecal coliform discharge limits vary on end use; however, for discharge to recreational waters, concentrations should be below 200 CFU/100 mL (monthly average) and 400 CFU/100 mL (weekly average). For unrestricted use, fecal coliforms need to be below 2 CFU/100 mL, which requires considerable tertiary treatment. Oil and grease need to be below 10 mg/L (monthly average) and 20 mg/L (weekly average), while pH is maintained in the 6–9 range. You will note that, while monthly and weekly averages must be met, discharges that exceed the average may still occur, albeit on an infrequent basis. A summary of wastewater characteristics entering and leaving the WWTP can be found in Table 8.1.

Table 8.1 Average values for municipal wastewater

WQI	Influent	Effluent
pH	6–8	6–9
TSS	220 mg/L	< 30 mg/L
BOD	200 mg/L	< 30 mg/L
Total N	35 mg/L	25 mg/L
Total P	8 mg/L	5 mg/L
Coliforms	10^6–10^8 CFU/100 mL	200 CFU/100 mL

Wastewater flows are usually estimated using water use data and an assumed *return rate*, which is defined as *the percentage of water use directed toward the sanitary sewer*. Return rates are dependent on the types of human activities utilizing water and the relative proportion of domestic to industrial uses, but they are often higher in urban areas (~90%–95%) than in suburban (~70%–85%) or rural (~60%–80%) areas. The most common units used in water and wastewater system design are gallons per day (gpd), or for large systems million gallons per day (Mgal/day), often represented as mgd.

Wastewater flow estimation for new or replacement systems should use projected demand based on population forecasting. As in water demand, consideration must be made for the variation in flow that occurs throughout the day due to human schedules. This is often accomplished by estimating the maximum hourly flow

(Q_{max}) and minimum hourly flow (Q_{min}) using scaling factors derived from equations or graphical data, as provided in Figure 8.1.

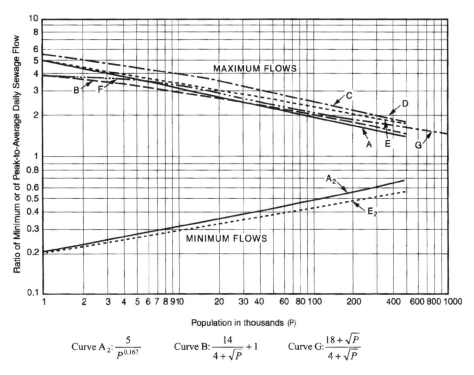

$$\text{Curve A}_2: \frac{5}{P^{0.167}} \qquad \text{Curve B:} \frac{14}{4+\sqrt{P}}+1 \qquad \text{Curve G:} \frac{18+\sqrt{P}}{4+\sqrt{P}}$$

Figure 8.1 Sewer flow ratio curves

Source: *Design and Construction of Sanitary and Storm Sewers*, 1970, American Society of Civil Engineers. Reprinted by permission.

In addition to direct flows, consideration must be made for extraneous waters that end up in the sanitary sewer line. This contribution is often called *infiltration and inflow* (*I&I*) and is due to groundwater entering through breeches in the pipe or stormwater runoff directed (often illegally) to the sanitary sewer from sump pumps and roof gutter downspouts. The amount of I&I is a function of infiltration waters present, age and diameter of the sewer line, and amount of inflow, for which data is rarely available. In the absence of any other data, the general practice is to assign a value of 30,000 gpd/mi (gallons per day per mile of sewer line), which is multiplied by the total length of sewer line in the service area.

$$I\&I \sim 30{,}000\ \frac{gpd}{mi}$$

Example **8.1**

Minimum/Maximum sewage flows

A city with an average sanitary sewage flow of 11.6 mgd has a maximum flow of 23.2 mgd. Determine the minimum anticipated flow. You may assume that the variations are described by the A and A_2 sewage flow ratio curves.

Solution

We start by finding the ratio of maximum to average flow as follows:

$$\frac{Q_{max}}{Q_{ave}} = \frac{23.2}{11.6} = 2$$

From curve A on the sewage flow ratio curves (Figure 8.1), a maximum-to-average ratio of 2 occurs at a population of 80,000. Using curve A_2 at this population gives a minimum-to-average ratio of 0.47. This allows us to estimate the minimum flow as:

$$\frac{Q_{min}}{Q_{ave}} = 0.47 \quad \Rightarrow \quad Q_{min} = (0.47)Q_{ave} = (0.47)\left(11.6 \cdot 10^6 \text{ mgd}\right) = 5.45 \cdot 10^6 \text{ mgd}$$

8.2 SEWER SYSTEM DESIGN

Design of sewer systems may be conducted using the nomographs for open channel flow through circular pipes. In general, flows from smaller lines, called *laterals*, are directed toward *submains*, which are collected in *mains* that run throughout the service area and end at the discharge point. All intersections for sewer lines occur at *manholes*, which are designed to provide service access to the sewer lines and which are usually added to long runs of line for maintenance purposes. The discharge is usually a river for *stormwater sewer* systems or the WWTP for *sanitary sewer* and *combined sewer* systems. Since all flow is gravity induced, the area topology, or ground surface profile, usually dictates layout and direction of flow. In order to ensure a continuous downward slope, it is necessary to evaluate the *invert elevation*, which is the elevation of the bottom of the pipe usually referenced to sea level, and to a lesser extent the *crown elevation*, which is the elevation of the top of the pipe.

Primary consideration for system design is to determine pipe diameter and slope at the estimated flows, given constraints for minimum and maximum velocity. Minimum velocities (V_{min}) are generally established at 2 ft/s and are necessary to ensure self-cleansing (that is, conveying all solids with the liquid flow). Maximum velocities (V_{max}) vary depending on pipe material but range from 10 ft/s to 20 ft/s to maintain pipe integrity by minimizing scour and pipe erosion. Another consideration is minimum slope, which varies depending on pipe diameter and anticipated depth of flow. In general, pipes 8 to 24 inches in diameter require minimum slopes of 0.4–0.1%, while larger pipes may have minimum slopes as small as 0.05% (0.5 ft per 1000 feet of pipe).

Since Q_{max} and Q_{min} can vary by as much as an order of magnitude, standard practice is to design for Q_{max} plus *I&I* near V_{max}, then check the velocity at Q_{min} plus *I&I* to ensure compliance with V_{min}. In order to provide sufficient capacity for extreme flow events and for the potential of population growth in the service area, standard practice for sanitary sewers is to design for Q_{max} plus *I&I* at 50% depth of flow for lateral lines and at 75% depth of flow for main lines. In general, this level of overdesign will still provide a sewer system that is capable of meeting V_{min} requirements.

The other design constraint is *depth of cover*, which is the amount of soil over the sewer line and is generally a design driving force due to the cost of trench excavation. Minimum depths are established to prevent freezing, while maximum depths are usually a function of the ability to excavate the trench, usually 25–35 feet. In flat areas, it is desired to keep the slope as small as possible to avoid *lift stations*, which are pump stations in a manhole designed to add energy to the water by pumping from the maximum depth of cover to the minimum depth of cover, where gravity flow is allowed to continue. Although highly discouraged in common practice, in high slope areas a *drop manhole* may be used to dissipate energy.

This is required where pipe slopes required to maintain a minimum depth of cover would be excessive and would cause extremely high velocities.

The design of a large system generally begins with a layout determined from topographic maps of the intended service area. Lateral lines are then designed with the upstream (inlet) side at the minimum ground cover. Downstream (outlet) invert elevation is determined by multiplying section length by installed slope, and depth of cover is determined by accounting for any change in surface elevation. Once all laterals for a given submain are designed, the submain is designed based on the lateral with the lowest invert elevation. Once the depth of the submain is determined, laterals are redesigned with the slope required by their outlet connection to the submain. This process is repeated for the main/submain connections. Sewer system design in this manner is labor intensive and cumbersome. Thankfully, new computer software has greatly enhanced the design process, and entire cities can be evaluated with relative ease.

Example **8.2**

Sewer pipe design

Determine the diameter and slope of a sewer line flowing at 50% depth at maximum daily flow for a population of 40,000 with a per capita flow of 140 gpcd. You may assume a velocity at 50% depth of 8.2 fps, I&I is negligible, n is 0.013 and varies with depth, and that curves A and A$_2$ on the sewage flow ratio curves apply.

Solution

We start by calculating the average flow rate as follows:

$$Q_{ave} = (40,000 \text{ people})\left(140 \frac{\text{gal}}{\text{person} \cdot \text{day}}\right) = 5.6 \frac{\text{Mgal}}{\text{day}} \times \frac{1.547 \text{ ft}^3/\text{s}}{\text{Mgal/day}} = 8.7 \frac{\text{ft}^3}{\text{s}}$$

Using curve A on the sewage flow ratio curves for a population of 40,000, $Q_{max}/Q_{ave} = 2.3$; therefore, Q_{max} may now be determined as

$$Q_{max} = (2.3)(8.7) = 20.0 \ cfs$$

Now we can determine the full pipe flow rate by using the partial flow nomograph, or plot of hydraulic elements for circular cross sections from Figure 3.7. From the nomograph at 50% depth of flow with variable n we see the following:

$$\frac{Q}{Q_f} = 0.4 \ @ \ 50\% \text{ depth of flow}$$

Therefore, the full pipe flow rate is

$$Q_f = \frac{Q_{max}}{0.4} = \frac{20.0}{0.4} = 50.0 \frac{\text{ft}^3}{\text{s}}$$

Also, from the partial flow nomograph at a depth of 50%, we can determine the full pipe velocity as follows:

$$\frac{V}{V_f} = 0.8 \quad \Rightarrow \quad V_f = \frac{V}{0.8} = \frac{8.2 \text{ fps}}{0.8} = 10.25 \text{ fps}$$

Finally, with full pipe flow rate and velocity, we can size the pipe as follows:

$$Q = VA = V\frac{\pi}{4}D^2 \quad \Rightarrow \quad D = \left[\frac{4Q}{\pi V}\right]^{1/2} = \left[\frac{(4)\left(50.0\,\dfrac{ft^3}{s}\right)}{(\pi)\left(10.25\,\dfrac{ft}{s}\right)}\right]^{1/2}$$

$$= 2.49\,ft \times \frac{12\,in}{ft} = 29.9\,in$$

Note that the final diameter will need to be increased to the next common pipe size, in this case, 30 inches. Now, with a known value for n and V for a full pipe, we can write the Manning's equation as

$$Q = \frac{1.486}{n}A R_H^{2/3} S^{1/2} \quad \Rightarrow \quad \frac{Q}{A} = \frac{1.486}{0.013}R_H^{2/3} S^{1/2} = 10.25\,\frac{ft}{s}$$

The hydraulic radius for a circular pipe flowing full is $D/4$, or equivalently,

$$R_H = \frac{D}{4} = \frac{30\,in}{4} = 7.5\,in \times \frac{1\,ft}{12\,in} = 0.625\,ft$$

Substituting this value into the expression above yields

$$S^{1/2} = \frac{\left(10.25\,\dfrac{ft}{s}\right)(0.013)}{(1.486)(0.625\,ft)^{2/3}} = 0.123 \quad \Rightarrow \quad S = 0.015 = 1.5\%$$

Example 8.3

Sewer pipe depth of cover

Pipe A is 18-in diameter and enters manhole 1 with a crown elevation of 672 ft. Pipe B is 12-in diameter and enters manhole 1 with a crown elevation of 674 ft. The exiting pipe (Pipe C) has a diameter of 24 in, a slope of 0.2%, and is 1000 ft from manhole 2. Assuming the exiting pipe crown must be at or below the inlet invert to keep from restricting flow, determine the minimum depth of cover for the exit pipe at manhole 2 if the ground elevation there is 680 ft.

Solution

Assuming the outlet pipe crown is at the elevation of the lowest inlet pipe invert, the exiting pipe crown elevation can be determined as

$$\text{elevation} = 672\,ft - \frac{18}{12}\,ft = 670.5\,ft$$

Next, we need to determine the elevation change of that pipe over its length as follows:

$$S = \frac{\Delta y}{\Delta x} \quad \Rightarrow \quad \Delta y = S\,\Delta x = \left(0.002\,\frac{ft}{ft}\right)(1000\,ft) = 2\,ft$$

This elevation change is subtracted from the initial elevation of the pipe to find the crown elevation at manhole as follows:

$$\text{elevation} = 670.5 - 2 = 668.5\,ft$$

Finally, cover at this location is determined by difference in crown and ground elevation as

$$\text{cover} = 680 - 668.5 = 11.5 \text{ ft}$$

INTRODUCTION TO WASTEWATER TREATMENT PROCESSES

Wastewater treatment processes are those steps taken to transform municipal wastewater resources into a sufficiently clean state so as to be able to be discharged into surface water receiving streams, or sometimes for unrestricted use (for example, irrigation water). Some of the fundamental concepts used in water treatment can be applied directly to wastewater, such as design of clarifiers and disinfection units; however, the focus in wastewater treatment is the removal of organic matter through *secondary treatment* processes, also called *biological aeration*. WWTPs are generally categorized into the following treatment categories: preliminary, primary settling, secondary (biological) aeration, secondary clarification, tertiary (advanced) treatment, disinfection, and sludge processing.

8.3 PRELIMINARY TREATMENT

Traditional preliminary treatment units serve to remove large particles prior to primary treatment. Historically, this is accomplished through the use of coarse or medium screens and grit removal. Additional processes that may occur prior to primary treatment include pumping, size reduction (shredding), and flow measurement. It is not uncommon to have these three combined through the use of metered grinding pumps that perform size reduction and flow logging while lifting raw wastewater to the grit chamber. Further, flow equalization may be required and either occurs just prior to or immediately following primary settling.

8.3.1 Screening

Screens are often referred to as bar racks, due to the physical configuration of the units. Often consisting of vertically oriented bars, classification of screens are based on the bar spacing.

Coarse screens are located just upstream of the main plant lift pumps and serve to protect the pumps from large debris. Since WWTP lift pumps can often accommodate 3-inch diameter solids, bar spacing is generally in the 2–2¾-inch range. Medium screens may be placed in open channels immediately following the pumps, often leading to the grit chamber. Medium screens typically have openings in the ½–1½-inch range and serve to remove additional solids prior to grit removal. Design considerations for screens typically require approach velocities of 2–2.5 fps and through velocities of 2.5–3 fps to minimize forcing solids through the screen and screen damage. Cleaning of screens is often by mechanical scraper; however, small facilities may still clean screens through manual raking.

8.3.2 Grit Removal

Grit is defined as particles (sand, gravel, cinders, bone chips, etc.) that have settling velocities much greater than the organic material typical of wastewater suspended solids. Its removal prior to primary settling is preferred due to the difference in nature of the solids and the potential for fouling (and increased maintenance) of downstream treatment units and pumps. Often, grit is washed and disposed of as solid waste, while the rinse water is returned to the WWTP flow stream. Grit removal is accomplished in horizontal-flow tanks with or without aeration, or in vortex-type units.

Often, raw wastewater is mixed in an *aerated grit chamber*, which provides enough mixing to keep organics suspended, while the reduced density of the bulk fluid allows heavier particles to settle. Sometimes the aerated grit chamber is used for the process of *preaeration*, where more dissolved oxygen is supplied to the biological community and, while little BOD is consumed, prepares the water for the subsequent biological aeration process.

8.3.2.1 Gravity Induced Settling

Although covered in section 7.2.4, the fundamentals of gravity-induced settling will be reviewed here as well. Examinees confident with this topic are encouraged to review Example 8.4 and then continue on to the next section. The rate at which particles move in the vertical direction is described by their *terminal settling velocity* (V_t), which can be expressed as

$$V_t = \left[\frac{4\,g\left(\rho_p - \rho_f\right)d}{3\,C_D\,\rho_f} \right]^{1/2}$$

where g is the gravity constant; ρ_p and ρ_f are the densities of the particle and the fluid, respectively; d is the diameter of the particle; and C_D is the drag coefficient. C_D is dependent on Reynolds number (Re) and can be determined as

$$C_D = \frac{24}{\text{Re}} \qquad\qquad \text{Re} \le 1$$

$$C_D = \frac{24}{\text{Re}} + \frac{3}{\sqrt{\text{Re}}} + 0.34 \qquad 1 \le \text{Re} \le 10^4$$

$$C_D = 0.4 \qquad\qquad \text{Re} > 10^4$$

where the Reynolds number is defined as

$$\text{Re} = \frac{\rho D V}{\mu} = \frac{DV}{v}$$

Since V_t and Re are dependent on each other, an iterative solution is required where a velocity is assumed and used to determine the value of Re. This number is then used to determine C_D, and this drag coefficient is used to calculate a corrected V_t. The Reynolds number is corrected using the new velocity, and the process is repeated until the error in V_t is small. To reduce the complexity of the solution method, it is often assumed that *Stoke's law* is applicable (Re $\le$ 1), which is used to describe external flow around spherical bodies at low values for Re. Low Re values are possible for very small particles (on the order of micrometers in diameter) and low settling velocities (on the order of centimeters per second) and simplifies the expression for V_t by setting $C_D = 24/\text{Re}$. In these cases, V_t can be expressed as

$$\text{Re} < 1$$
very small particles $\qquad V_t = \frac{g\left(\rho_p - \rho_f\right)d^2}{18\,\mu}$

where g, ρ_p, ρ_f, and d are defined as previously, and μ is the absolute viscosity of the fluie—in this case, water.

Example **8.4**

Settling velocity

Determine the distance a particle will settle in 1 hour if the particle has a SG = 1.01 and a diameter of 0.5 mm. Assume the water temperature is 15°C and the Re < 1.

Solution

Since we know Re is less than 1, Stoke's law applies, and we may use the Stoke's form of the settling equation as follows:

$$V_t = \frac{g(\rho_p - \rho_f)d^2}{18\mu}$$

Converting the diameter of the particle to meters and interpolating a value for the viscosity of water at 15°C from Table 3.2 in Chapter 3, we may write

$$V_t = \frac{\left(9.81\,\frac{m}{s^2}\right)\left[(1.01)(999) - (999)\,\frac{kg}{m^3}\right](0.0005\,m)^2}{(18)\left(1.1545 \times 10^{-3}\,\frac{kg}{m \cdot s}\right)} = 0.0012\,\frac{m}{s}$$

The distance traveled by a particle can be found by multiplying the velocity by the time period, which may be calculated as follows:

$$d = V_t \times t = \left(0.0012\,\frac{m}{s}\right)(1\,hr)\left(\frac{3600\,s}{hr}\right) = 4.32\,m$$

It is often a good idea to check the Stoke's law assumption by calculating Re as

$$Re = \frac{\rho D V}{\mu} = \frac{\left(999\,\frac{kg}{m^3}\right)(0.0005\,m)\left(0.0012\,\frac{m}{s}\right)}{1.1545 \times 10^{-3}\,\frac{kg}{m \cdot s}} = 0.52$$

Since this is less than 1, Stoke's law is valid.

8.3.2.2 Grit Chamber Design

Horizontal-flow grit chambers are often rectangular or square in geometry. Typical detention times are 1 minute with horizontal velocities of 1 fps. Detention time is sometimes called retention time or residence time (t_R) and has been defined previously as

$$t_R = \frac{V}{Q}$$

where V is the volume of the basin and Q is the volumetric flow through the basin. Note that the units on V and Q need to be equal, and the time units for t_r will be the same as the time units on Q.

Detention times in an aerated grit chamber range from 2–5 minutes at peak flow (typically 3 minutes) for mixing purposes to 20 minutes for a preaeration basin. Air is usually supplied at a rate of 3–8 ft³/min per foot of basin length. Width-to-depth (W:D) ratios are 1:1 to 2:1, and length-to-width (L:W) ratios are typically 3:1 to 5:1.

Example **8.5**

Aerated grit chamber design

Determine the depth of an aerated grit chamber that treats an average flow of 6.2 mgd if the design detention time is 3 minutes at Q_{peak}. You may assume a L:W:D of 5:1.5:1.

Solution

Following Example 8.2 and using the sewage flow ratio curves, we can estimate the peaking factor for this flow at approximately 2.2. Using the expression for detention time and the data given in the problem statement, the tank volume can be determined as follows:

$$t_R = \frac{V}{Q} \Rightarrow V = Q\, t_R = \frac{\left(3\,\text{min}\right)\left(6.2\,\text{mgd}\right)\left(2.2\right)}{\left(60\right)\left(24\right)\dfrac{\text{min}}{\text{day}}}$$

$$= 28{,}417\,\text{gal} \times \frac{1\,\text{ft}^3}{7.48\,\text{gal}} = 3799\,\text{ft}^3$$

With a known volume and L:W:D ratio, a rectangular tank can be dimensioned as

$$V = LWD = \left(5\,D\right)\left(1.5\,D\right)\left(D\right) = 7.5\,D^3 = 3799\,\text{ft}^3 \Rightarrow D = \left(\frac{3799}{7.5}\right)^{1/3}$$

$$= 7.97\,\text{ft} \approx 8\,\text{ft}$$

8.3.3 Flow Equalization

Due to the diurnal flow pattern of typical wastewater flows, it is anticipated that flow into the WWTP will vary by a factor of 4–10 or more. As treatment units are designed based on detention time, which is a function of this flow, variations in influent rates may have a negative impact on plant performance. One method to deal with this variation is to design all units for peak flow to ensure removal percentages at the highest flows. While this ensures all flows will be sufficiently treated, it does require all process units to be sized larger than they might have otherwise. Another method is to provide flow equalization, where some of the influent is diverted to a holding tank during the highest flows and later released back into the process stream at times when the flow is below the daily average. The equalization basin is placed after grit removal, most often prior to primary settling. This would require the basin to be mixed and aerated to maintain organic solids in suspension and to ensure the wastewater does not become anaerobic.

Required storage capacity is determined through the evaluation of inflows of a WWTP over the course of several seasons, if the data is available. To use the *graphical technique*, a plot is made of cumulative inflow as a function of time for a 24-hour period. Assuming the process flow is at a constant rate over the 24 hours, the average inflow curve will be a straight line. To determine the required size of the basin, a line parallel to the average inflow curve is drawn tangent to the points where the two curves are at a maximum and minimum difference, as shown in Figure 8.2. If the flow pattern is such that no maximum exists, the average flow curve would be used. The maximum vertical distance between the cumulative inflow curve and any of the parallel lines is the required storage.

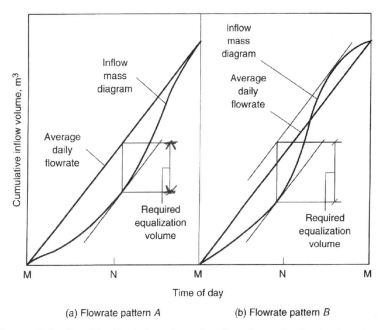

Figure 8.2 Graphical technique for estimation of equalization basin volume

Source: Metcalf and Eddy, *Wastewater Engineering: Treatment and Reuse*, 4th ed., © 2003, McGraw-Hill Education. Reprinted by permission of the McGraw-Hill Companies.

Two spreadsheet methods calculate both cumulative inflows and outflows and the disparity between the two for each date. In the first method, the difference is calculated as inflow minus outflow, which are summed over the time period to determine the cumulative difference. The required storage is determined by finding the maximum value from the cumulative difference column and subtracting the subsequent minimum value. This is called the *difference technique*. In the second method, the deficiency is calculated as outflow minus inflow. In the cumulative deficiency column, any negative values are set equal to zero, and the summation is continued. The required storage is taken directly from the maximum value in the cumulative deficiency column. This is called the *deficiency technique*.

Example 8.6

Flow equalization

Using the WWTP cumulative inflow curve provided in Exhibit 1, determine the volume of a flow equalization basin.

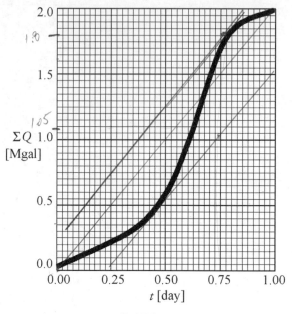

Exhibit 1

Solution

Start by drawing the Q_{ave} cumulative curve as shown in Exhibit 2, which is the 45° line from (0,0) to (1,2). Then draw a line tangent to the cumulative inflow curve given, parallel to Q_{ave} at the maximum and minimum points on the curve. Now, find Δy by drawing horizontal lines where the tangent lines cross any value of t. Note the value shown in the solution is $t = 0.75$.

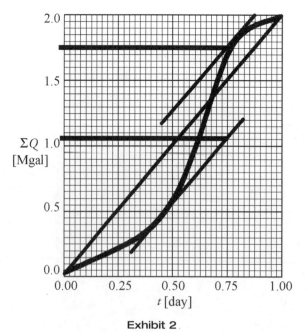

Exhibit 2

Δy = equalization basin volume = 1.75 − 1.05 = 0.7 Mgal

8.4   PRIMARY TREATMENT

Design of sedimentation units was previously covered in section 7.2.4. Examinees who are confident with this material are encouraged to review Examples 8.7 and 8.8 and then proceed to the next section. Primary treatment is accomplished through gravity-induced settling, usually called *sedimentation* or *clarification*, and is the process in which solid particles are allowed to fall out of suspension in relatively quiescent water. Tanks are designed to greatly reduce flow velocities such that particles are not being agitated by movement of fluid and are allowed to collect at the bottom of the tank. Sedimentation process fundamentals for wastewater treatment are identical to those in water treatment, where the presence of biological solids is a primary concern, and clarifiers are differentiated by task.

8.4.1 Clarifier Design Parameters

In addition to the detention time defined in section 8.3.2.2, the flow of water in the clarifier can be further characterized by the fluid velocity in the vertical direction, called the *surface settling rate*, and is expressed as

overflow rate [handwritten annotation]

$$V_0 = \frac{Q}{A}$$

[handwritten annotation: $V_t > V_0$ collected in clar. $V_t < V_0$ likely not collect.]

where V_0 has units ft/s or m/s, Q is the volumetric flow rate into the clarifier (cfs or m³/s), and A is the surface area of the clarifier (ft² or m²). Any particles possessing $V_t > V_0$ are collected in the clarifier, while particles with $V_t < V_0$ are collected only if they enter the clarifier at a height less than the water depth in the basin.

For design purposes, the surface settling rate is more commonly expressed in the common English units for Q, gallons per day (gpd), and is called the *overflow rate*, with units of gpd/ft². Another design parameter is the *side-water depth*, which usually describes the depth of water not including the sludge blanket that forms on the bottom of the tank. Because t_R, V_0, and side-water depth are not independent, care must be taken when using them in clarifier design to keep from overspecifying the design.

To ensure the wastewater does not cause mixing due to high velocity, another design consideration is the *horizontal* (or *flow-through*) *velocity* (V_h), which may be defined as

$$V_h = \frac{Q}{W H}$$

[handwritten annotation: $\frac{flow}{(width)(depth)}$ = flow through velo.]

where Q is volumetric flowrate, W is the tank width, and H is the tank depth.

An additional design parameter is the *weir loading* (*wl*), sometimes called the *weir overflow rate* (*WOR*), which is a measure of the rate of effluent flow over the discharge weir and can be expressed as

$$wl = \frac{Q}{L_w}$$

where *wl* has common units of gpd/ft, L_w is the length of the outlet weir (ft), and Q is as above.

Clarifier shape is usually rectangular, although some circular tanks exist, with the size of the tanks determined through the use of empirical design criteria. These criteria are based on operating conditions that have been found to be most effective in practice. Rectangular tanks generally have a L:W ratio of from 3:1 to 5:1, and weirs are set inside the tank at one end in a box or finger arrangement. For circular tanks, influent is in the center of the tank, and water flows radially toward the weir, located at the tank perimeter, with weir length equal to the tank perimeter.

8.4.2 Clarifier Design Criteria

Primary clarifiers are designed to treat raw wastewater, and removal rates are approximately 50% for suspended solids and 30%–40% for BOD. Primary wastewater clarifiers also allow for the removal of grease and oil by means of skimming at the air-water interface. Design criteria for clarifiers are often listed in tables and are provided for different wastewater treatment applications in Table 8.2. In general, V_0 for primary clarifiers ranges from 800–1200 gpd/ft² (approximately equal to 32–48 m³ per m² per day, or equivalently 32–48 m/d), and *horizontal* (or *flow-through*) *velocities* (V_h), should be kept below 0.5 ft/min. The design criterion for detention time is typically 2 hours, side-water depths should be maintained greater than 10 ft, and weir loadings are typically 20,000 gpd/ft.

Table 8.2 Design criteria for wastewater clarifiers based on average flows

	V_0		wl		t_R
Type of Basin	**(gpd/ft²)**	**m³/(m² · day)**	**gpd/ft**	**m³/(m · day)**	**hr**
Primary clarifiers	800–1200	32–48	20,000	250	2
Fixed film reactors	400–800	16–32	20,000	250	2
Air-activated sludge	400–800	16–32	20,000	250	2
Extended aeration	200–400	8–16	10,000	125	4

Example 8.7

Rectangular clarifier design

Determine the length of a rectangular clarifier for a flow of 3.1 mgd if the design detention time is 2 hours, the tank has a depth of 10 ft, and the L:W = 4:1.

Solution

Volume of a tank can be found given detention time and volumetric flow rate as

$$t_R = \frac{V}{Q} \quad \Rightarrow \quad V = Q\,t_R$$

$$V = \left(3.1 \times 10^6 \text{ gpd}\right)\left(2 \text{ hr}\right)\left(\frac{1 \text{ day}}{24 \text{ hr}}\right) = 258,333 \text{ gal} \times \left(\frac{\text{ft}^3}{7.48 \text{ gal}}\right) = 34,537 \text{ ft}^3$$

The area of the tank can be found from the volume and depth as follows:

$$A = \frac{V}{\text{depth}} = \frac{34,537 \text{ ft}^3}{10 \text{ ft}} = 3454 \text{ ft}^2$$

Further, for a L:W ratio of 4:1, we can find the tank dimensions as

$$A = LW = \left(4W\right)\left(W\right) = 4W^2 = 3454 \text{ ft}^2 \quad \Rightarrow \quad W = 29.4 \text{ ft} \approx 30 \text{ ft}$$

$$\text{and} \quad L = 4W = 120 \text{ ft}$$

Example **8.8**

Circular clarifier design

Find the diameter of a circular clarifier designed to treat 3.1 mgd. You may assume a detention time of 2 hours and a depth of 10 feet.

Solution

Tank volume can be found using the detention time as follows:

$$V = Q \cdot t_R = \left(3.1 \times 10^6 \text{ gpd}\right)\left(2 \text{ hr}\right)\left(\frac{1 \text{ day}}{24 \text{ hr}}\right) = 258{,}333 \text{ gal} \times \left(\frac{\text{ft}^3}{7.48 \text{ gal}}\right)$$

$$= 34{,}537 \text{ ft}^3$$

Area is calculated by dividing volume by depth, and the diameter of a circular tank is found using the area as follows:

$$A = \frac{V}{\text{depth}} = \frac{34{,}537 \text{ ft}^3}{10 \text{ ft}} = 3454 \text{ ft}^2 = \frac{\pi}{4} D^2 \quad \Rightarrow \quad D = 52.1 \text{ ft}$$

For construction purposes, it should be noted that tank diameter would be rounded to 50 ft or 55 ft.

8.5 SECONDARY TREATMENT

Secondary treatment is also called biological treatment or biological aeration due to the fact that microorganisms (primarily bacteria) are employed to reduce wastewater BOD to acceptable effluent levels. Optimization of this naturally occurring process requires the knowledge of biology covered in Chapter 2, coupled with process unit experience based on reactor characteristics. Most biological treatment process can be categorized as either suspended or attached growth processes. In suspended growth, individual organisms interact in a well-mixed environment that contains the food, oxygen, and nutrients required for removal of BOD. Attached growth processes use a biofilm surface and pass the food/nutrient/oxygen-containing wastewater over the film. Three types of biological treatment are discussed in this review: activated sludge, fixed film reactors, and treatment ponds and lagoons.

8.5.1 Activated Sludge

One method of encouraging interaction between wastewater organic matter, the bacteria that consume that organic matter as substrate, and the oxygen required to serve as terminal electron acceptor is a process known as *activated sludge*. As seen in Figure 8.3, a highly mixed reactor is designed to distribute the needed oxygen and keep the biological mass in suspension.

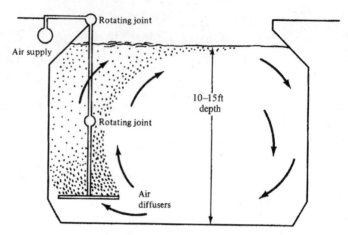

Figure 8.3 Cross section of a typical channel demonstrating mixing through aeration

Source: Warren Viessman and Mark J. Hammer, *Pollution Control*, 7th ed., © 2005. Reprinted by permission of Pearson Education, Inc., Upper Saddle River, N.J.

As bacteria consume the organic matter, the numbers increase significantly, converting BOD to TSS that still requires removal prior to final effluent discharge. However, these "biosolids" are rich with active bacteria that could be employed to remove additional organic matter from the wastewater. It is the return of the "activated" mass of organisms that provides the name for the activated sludge process. There are several typical process layouts as presented in Figure 8.4, but, in all cases, significant numbers of the bacteria removed in the final clarifier are subsequently returned to the aeration tank to continue the consumption of organic matter.

Continuous accumulation of biosolids is avoided through the natural process of cell lysis, through the wasting of a small amount of the returned sludge, and from the fraction that is discharged in the effluent as TSS. Modifications to the processes shown include the use of multiple feeds to the tank and direct aeration without the use of a primary clarifier.

System design considerations may be based on historical operation parameters or may be determined through kinetic evaluation of the biological process. The process of designing a system through operational parameters relies on a range of values that have been shown to be effective in the reduction of BOD through biological activity. Once the type of process is selected, aeration tank volume (V_A) may be calculated using a known wastewater flow rate (Q_0) and the *hydraulic residence time* (θ), also called the *aeration period*, which can be expressed as

$$\theta = \frac{V_A}{Q_0}$$

It is worthwhile to note that this is equivalent to the mean residence time (t_R) used previously, and that the volumetric flow rate used does not include any flow into the tank attributed to recirculated biosolids from the final clarifier. Typical values for θ are based on reactor configuration and are presented in Table 8.3, along with several additional activated sludge design parameters. It should be noted that definitions for each of the design criteria listed in Table 8.3 will be provided in the following portions of this section.

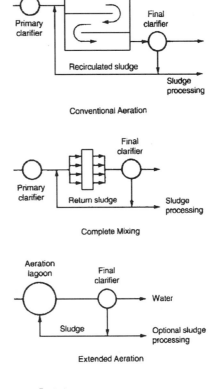

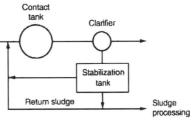

Figure 8.4 Variations of the activated sludge process

Table 8.3 Design criteria for activated sludge treatment systems

Process	SRT days	F/M (kg BOD/d)/ kg MLVSS	Volumetric Loading lb BOD/ 1000 ft³ · d	Volumetric Loading kg BOD/ m³ · d	MLSS mg/L	θ hr	RAS % of Q
High-rate aeration	0.5–2	1.5–2.0	75–150	1.2–2.4	200–1000	1.5–3	100–150
Contact stabilization							
Contact basin	5–10	0.2–0.6	60–75	1.0–1.3	1000–3000	0.5–1	50–150
Stabilization basin					6000–10,000	2–4	
High-purity oxygen	1–4	0.5–1.0	80–200	1.3–3.2	2000–5000	1–3	25–50
Conventional plug flow	3–15	0.2–0.4	20–40	0.3–0.7	1000–3000	4–8	25–75
Step feed	3–15	0.2–0.4	40–60	0.7–1.0	1500–4000	3–5	25–75
Complete mix	3–15	0.2–0.6	20–100	0.3–1.6	1500–4000	3–5	25–100
Extended aeration	20–40	0.04–0.10	5–15	0.1–0.3	2000–5000	20–30	50–150
Oxidation ditch	15–30	0.04–0.10	5–15	0.1–0.3	3000–5000	15–30	75–150
Sequencing batch reactor	10–30	0.04–0.10	5–15	0.1–0.3	2000–5000	15–40	NA

Source: adapted from Metcalf & Eddy, *Wastewater Engineering: Treatment and Reuse,* 4th ed. (McGraw-Hill, 2002), Table 8.16, p. 747.

Example **8.9**

Activated sludge tank design

A conventional activated sludge system treats 6.4 mgd. Based on a minimum hydraulic residence time, what is the length of aeration channel required if the cross section has dimensions of 15 × 12 feet?

Solution

From Table 8.3, the hydraulic residence time (also called aeration period) for a conventional activated sludge system is from 4 to 8 hours. Using the minimum of 4 hours, tank volume can be calculated as follows:

$$\theta = \frac{V}{Q} \quad \Rightarrow \quad V = \theta Q = \left(4 \text{ hr}\right)\left(\frac{1 \text{ day}}{24 \text{ hr}}\right)\left(6.4 \frac{\text{Mgal}}{\text{day}}\right)$$

$$= 1.07 \text{ Mgal} \times \frac{1 \text{ ft}^3}{7.485 \text{ gal}} = 142{,}500 \text{ ft}^3$$

Given the tank volume, the channel length can be determined as

$$V = L \cdot W \cdot H \quad \Rightarrow \quad L = \frac{V}{H \cdot W} = \frac{142{,}500 \text{ ft}^3}{(15 \text{ ft})(12 \text{ ft})} = 792 \text{ ft}$$

Tank volume calculations using hydraulic loadings must also satisfy the loading rate of the substrate (organic matter), which is approximated analytically using a value of BOD (or COD). The *volumetric substrate loading rate* (V_L) is defined as the *mass rate of BOD (or COD) fed per unit volume of aeration tank* and can be expressed as

$$V_L = \frac{Q_0 S_0}{V_A} = \frac{S_0}{\theta}$$

where Q_0 is the flow rate into the aeration tank and S_0 is the BOD (or COD) concentration fed to the aeration tank, not including any recirculated flow. Typical values for V_L are also based on reactor configuration and are given in Table 8.3. It should be noted that primary clarifiers remove 30%–40% of the BOD_5 that enters the plant, and activated sludge systems that are located after a primary clarifier should have their influent BOD_5 values reduced as such.

Example **8.10**

BOD loading criteria

Determine the BOD loading in units of $kg/m^3/day$ to the system in Example 8.9 if the BOD concentration into the activated sludge tank is 140 mg/L.

Solution

The units for the answer are in the metric system. We should start by converting the tank volume to metric units as follows:

$$V = 142{,}500 \text{ ft}^3 \times \left(\frac{1 \text{ m}}{3.28 \text{ ft}}\right)^3 = 4038 \text{ m}^3$$

We can calculate the BOD load by multiplying BOD concentration and flow rate as

$$\text{BOD load} = \left(140\,\frac{\text{mg}}{\text{L}}\right)\left(6.4\times10^6\,\frac{\text{gal}}{\text{day}}\right)\left(3.78\,\frac{\text{L}}{\text{gal}}\right)\left(10^{-6}\,\frac{\text{kg}}{\text{mg}}\right) = 3387\,\frac{\text{kg}}{\text{day}}$$

Now the BOD loading to the tank can be determined as

$$\text{BOD loading} = 3387\,\frac{\text{kg}}{\text{day}} \div 4038\,\text{m}^3 = 0.84\,\frac{\text{kg}}{\text{m}^3\cdot\text{day}}$$

Once tank volume is determined, geometric configuration can be determined depending on the type of activated sludge process unit. For conventional plug-flow systems, channels are assumed to have square or rectangular cross sections with depths of 10–15 feet. Total length of channel is determined by dividing tank volume by channel cross-sectional area. This channel length is divided into equal segments that will fit into the desired footprint, and overall tank dimensions are determined.

Example **8.11**

Aeration period

A flow of 2.6 mgd leaves a primary clarifier with a BOD of 131 mg/L. Determine the aeration period of an activated sludge tank that has a BOD loading of 35 lb of BOD per 1000 ft³ per day.

Solution

The BOD loading to the aeration tank can be determined using the BOD concentration and volumetric flow rate as follows:

$$\text{BOD load} = Q\times\text{BOD} = \left(2.6\,\text{Mgd}\right)\left(131\,\frac{\text{mg}}{\text{L}}\right)(8.34) = 2841\,\frac{\text{lb}}{\text{day}}$$

Notice the factor 8.34 used here. This is a great conversion factor to remember for use with wastewater treatment process unit design where flows are given in mgd and concentrations are given in mg/L. The factor 8.34 converts (mgd) × (mg/L) into the units of lb/day from the 2.205 lbs per kg and 3.785 L per gallon conversions. Notice the 10^6 on gallons and 10^{-6} on grams cancel each other out.

The volume of the tank can now be found by dividing the BOD load by the design loading rate as follows:

$$V = \frac{2841\,\dfrac{\text{lb}}{\text{day}}}{35\,\dfrac{\text{lb}}{1000\,\text{ft}^3\cdot\text{day}}} = 81{,}171\,\text{ft}^3$$

Now, the aeration period can be calculated using the given flow rate and volume as follows:

$$\theta = \frac{V}{Q} = \frac{\left(81{,}171\,\text{ft}^3\right)\left(7.48\,\dfrac{\text{gal}}{\text{ft}^3}\right)}{2.6\times10^6\,\dfrac{\text{gal}}{\text{day}}} = 0.233\,\text{day}\times\frac{24\,\text{hr}}{\text{day}} = 5.6\,\text{hr}$$

The concentration of bacteria in the aeration tank is called the *mixed liquor suspended solids* (MLSS) concentration. Typical values for MLSS are given in Table 8.3 for a variety of activated sludge processes. As a parameter in biological kinetic evaluations, MLSS is usually expressed by the variable X_A and is often expressed units of mg biomass (or COD) per L of tank volume. Another method of evaluating substrate loading is to regulate the food supply based on the amount of bacteria present in the aeration tank that consume the organic matter. This mass ratio is called the *food to microorganism (mass) ratio* (F/M) and can be calculated as

$$\frac{F}{M} = \frac{Q_0\,S_0}{V_A\,X_A} = MLSS$$

where F/M has units of mass of BOD per day per mass of biosolids. It should be noted that Table 8.3 uses the units of mass of BOD per day per unit mass of MLVSS, or mixed liquor volatile suspended solids. MLVSS is considered a better measure of the biomass fraction of the MLSS, as all bacterial would be considered volatile based on the definitions for solids analysis covered in section 1.3.5.

The F/M ratio is a powerful tool used to control the activated sludge process. As seen in Figure 8.5, high F/M ratios are good for high rates of consumption and, therefore, smaller aeration tanks, but the effluent from that tank exhibits poor settling characteristics. Further, the presence of large quantities of food to a relatively small bacteria population may result in the microbes preferentially consuming the simpler substrates, thus the potential for incomplete consumption of BOD. In order to achieve good solids removal in the final clarifier, the F/M ratio is lowered, and, while the substrate consumption rate is decreased, the consumption percentage is increased.

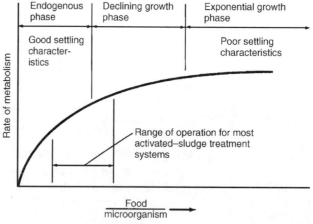

Figure 8.5   Effect of F/M ratio on degradation rate and settling characteristics

Source: Warren Viessman and Mark J. Hammer, *Pollution Control*, 7th ed., © 2005.
Reprinted by permission of Pearson Education, Inc., Upper Saddle River, N.J.

Since the majority of the solids that were removed from the wastewater in the final clarifier are returned to the aeration tank, an individual cell has the potential of cycling many times through the system. The average length of time that a single cell remains in the system is called the *mean solids (cell) residence time* (θ_c), often called the *sludge age* or *solids retention time* (SRT), and can be expressed as

Solids
Retention
Time

$$\theta_c = SRT = \frac{V_A\,X_A}{Q_e X_e + Q_w X_w} = \frac{V \times MLSS}{Q_e SS_e + Q_w SS_w}$$

where the subscript w indicates the waste stream, and the subscript e indicates the plant effluent stream. Notice that, while there may be substantial recirculation flow, the volumetric flow rate of the water in the effluent plus the wasted solids must equal the original flow into the aeration tank (that is, $Q_0 = Q_e + Q_w$) in order to avoid an accumulation of liquid volume in the system. Typical values for SRT are given in Table 8.3 for several different activated sludge treatment systems.

In the absence of other data, estimates for several terms in the sludge age equation can be made. Since there are regulatory limits on discharged suspended solids, X_e may be assumed to be < 30 mg/L and a value of 20 mg/L is typical. Further, the volumetric flow of the wasted solids is quite small compared to the effluent flow (Q_w is often ≤ 1% of Q_0); therefore, Q_e is often assumed to be equal to Q_0. Finally, the $Q_w X_w$ product represents the mass rate of wasted solids and can be estimated as a function of F/M ratio as

$$Q_w X_w = k\, Q_0\, S_0$$

(handwritten note: $k ≈ 0.43$, F/M = 0.2; $k ≈ 0.5$, F/M = 0.4)

where $k ≈ 0.43$ for F/M = 0.2 and $k ≈ 0.5$ for F/M = 0.4, with k estimated for other values of F/M through interpolation or extrapolation.

Example **8.12**

Sludge age

Determine the sludge age of the facility described in Example 8.11 given the following data: MLSS concentration is 2100 mg/L, effluent suspended solids equal to 20 mg/L, and the mass rate of wasted solids is equal to 44% of the BOD load.

Solution

The equation for sludge age is

$$\theta_c = \frac{\text{MLSS} \times V}{Q_e SS_e + Q_w SS_w}$$

The mass rate of wasted solids is related to BOD load in Example 8.11 as described in the problem statement and can be calculated as

$$Q_w SS_w = 44\% \times \text{BOD load} = (0.44)\left(2841\frac{\text{lb}}{\text{day}}\right) = 1250\frac{\text{lb}}{\text{day}}$$

We can simplify the sludge age expression if we assume that the effluent flow rate is approximately equal to the influent flow rate. Since Q_w is often ≤ 1% of Q_0, the assumption that $Q_e = 99\% Q_0$ or $Q_e = Q_0$ is good enough for estimations. Therefore, if we assume $Q_e ≈ 2.6$ mgd, we can estimate the sludge age as follows:

$$\theta_c = \frac{\left(2100\frac{\text{mg}}{\text{L}}\right)\left[\left(81{,}171\text{ ft}^3\right)\left(7.48\frac{\text{gal}}{\text{ft}^3}\right)\left(10^{-6}\frac{\text{Mgal}}{\text{gal}}\right)\right](8.34)}{1250\frac{\text{lb}}{\text{day}} + \left(2.6\text{ mgd}\right)\left(20\frac{\text{mg}}{\text{L}}\right)(8.34)} = 6.32\text{ days}$$

Notice the use of the 8.34 conversion factor in this problem as well. It should be noted that it works in the numerator with tank volume instead of volumetric flow rate, because it is really a conversion from milligrams to pounds and liters to gallons (that is, 8.34 ≈ 2.205 × 3.785). The time variable unit is not part of the conversion.

For completely mixed activated sludge systems, the microbial kinetics expressions developed in section 2.1.2 can be employed to design system parameters. Generally, laboratory data is generated that provides estimates for μ_{max}, K_s, k_d, and Y. Values for k_d range between 0.01 to 0.1 day^{-1} with a typical value of 0.05 day^{-1}, and Y ranges from 0.4 to 1.0 with a typical value of 0.7. The mean cell residence time, θ_c, as defined previously has units of mass of cells divided by the mass rate of change in the cells leaving the final clarifier, which may be expressed as

$$\theta_c = \frac{X_A}{\left(dX/dt\right)}$$

Substituting the biomass growth expression developed in section 2.1.2 allows θ_c to be related to the substrate utilization rate as follows:

$$\frac{1}{\theta_c} = \frac{\left(dX/dt\right)}{X_A} = \frac{Y\left(dS/dt\right)}{X_A} - k_d = \frac{\mu_{max}\, S}{K_s + S} - k_d = \mu - k_d$$

From this expression, the mean cell residence time can be estimated using the kinetic parameters μ_{max}, K_s, and k_d, along with a value for the substrate concentration inside of the aeration tank. Often, this substrate concentration is described as the soluble BOD, and since completely mixed reactors have the same effluent concentration as the concentration in the tank, the effluent soluble BOD concentration (S_e) can be determined from the above expression as

$$S_e = \frac{K_s\left(1 + k_d\,\theta_c\right)}{\theta_c\left(\mu_{max} - k_d\right) - 1}$$

Therefore, for a set of laboratory determined kinetic parameters, the effluent soluble BOD concentration can be estimated for a fixed value for sludge age.

Since substrate utilization is the change in substrate concentration that takes place during one unit of time in the aeration tank, we can express substrate utilization as

$$\frac{dS}{dt} = \frac{S_0 - S_e}{\theta} = \frac{Q_0\left(S_0 - S_e\right)}{V_A}$$

This may now be substituted into the sludge age expression above and written as

$$\frac{1}{\theta_c} = \frac{Y\, Q_0\left(S_0 - S_e\right)}{V_A\, X_A} - k_d = Y\left(\frac{F}{M}\right)\left(\frac{E}{100}\right) - k_d$$

where E is the BOD removal efficiency and may be expressed as

$$E = \frac{S_0 - S_e}{S_0} \times 100$$

Solving this expression for the tank volume (V_A) yields the following:

$$V_A = \frac{\theta_c\, Y\, Q_0\left(S_0 - S_e\right)}{X_A\left(1 + k_d\,\theta_c\right)}$$

or, if tank volume is known, the concentration of cells in the aeration tank (MLSS) can be estimated as

$$X_A = \text{MLSS} = \frac{\theta_c\, Y\, Q_0\left(S_0 - S_e\right)}{V_A\left(1 + k_d\,\theta_c\right)} = \frac{\theta_c\, Y\left(S_0 - S_e\right)}{\theta\left(1 + k_d\,\theta_c\right)}$$

Finally, the net mass of cell biomass produced (P_x) in g/d may be calculated as

$$P_x = \frac{Y\, Q_0 \left(S_0 - S_e\right)}{1 + k_d\, \theta_c}$$

Example 8.13

Biological yield

For the flow described in Example 8.11 and the activated sludge tank described in Example 8.12, determine the biological yield given a bacterial decay rate constant of 0.02 day^{-1}.

Solution

The expression for MLSS can be rearranged to express biological yield as a function of MLSS as follows:

$$\text{MLSS} = \frac{\theta_c\, Y (S_0 - S_e)}{\theta\,(1 + k_d \theta_c)} \quad \Rightarrow \quad Y = \frac{\text{MLSS}\,\theta\,(1 + k_d \theta_c)}{\theta_c\,(S_0 - S_e)}$$

Using the data given in the problem statements, biological yield can be calculated as

$$Y = \frac{\left(2100\ \dfrac{\text{mg biomass}}{\text{L}}\right)\left(\dfrac{5.6}{24}\ \text{days}\right)\left(1 + (0.02)(6.3)\right)}{(6.3\ \text{days})\left(131 - 20\ \dfrac{\text{mg BOD}}{\text{L}}\right)} = 0.789\ \frac{\text{mg biomass}}{\text{mg BOD}}$$

8.5.2 Fixed Film Treatment Units

Another type of biological treatment system is modeled from the natural tendency of microorganisms to attach to a surface and feed off of dissolved substrates that are carried past them. Because this biomass forms a thin film on the stationary surface, usually referred to as a biofilm, the process is called *fixed film* or *attached growth* biological treatment.

The most common type of fixed film treatment unit is the *trickling filter*, where the term *filter* does not refer to solids removal but the consumption of organic matter through biological utilization. The first trickling filters used rock as the biofilm surface, due to its ready availability anywhere a treatment system would be constructed. The surface area and void volume were low, however, and the weight of the media limited the depth of the units to 5–7 ft. Today, tricking filters consists of a tank that holds a high specific surface area (>100 m²/m³), high void fraction media (90%–95% voids), such as engineered plastic in the form of spheres, cylinders, or sheets. Additional characteristics of good filter media include durability, low unit weight (<100 kg/m³), low cost, and uniform flow properties.

Operation of a trickling filter requires that clarified wastewater is sprayed uniformly over the surface of the filter bed and allowed to trickle down through the bed and to collect in the underdrains below. Sufficient void space is required to maintain both a continuous flow of wastewater and a continuous flow of air, which provides oxygen to the wastewater as DO is needed by the bacteria for aerobic metabolism. Nearly all systems recirculate a portion of the treated wastewater at a ratio of 0.5 to 3.0 times the influent flow rate (that is, $Q_R = 0.5$ to $3 \times Q_0$) in order to reduce the strength of the wastewater that contacts the biofilm and provide a sufficient quantity of wastewater, regardless of flow variations that occur throughout the day. Several possible recirculation schemes may be applied, and some of the possible layouts are presented in Figure 8.6.

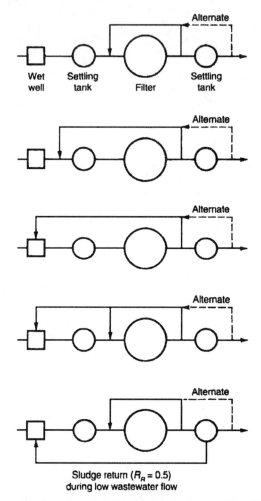

Figure 8.6   Process flow diagrams for some possible trickling filter recirculation scenarios

Design of a trickling filter unit is based on both hydraulic loading, which has units of volumetric flow rate per unit surface area of the tank in plan view (Q/A_s), and substrate loading, which has units of mass of BOD applied per day per unit volume of tank ($Q \cdot BOD/V$). While substrate loading is based upon the influent wastewater after primary sedimentation and ignores any BOD in the recirculation flow, hydraulic loading is based upon the combined flow (that is, $Q_0 + Q_R$).

The BOD removal efficiency for stone media trickling filters is calculated using the NRC equation, which can be expressed as

$$E_1 = \frac{100}{1 + 0.0561 \left(\dfrac{w}{VF} \right)^{1/2}}$$

where E_1 is the BOD removal efficiency at 20°C (%), w is the applied BOD load (lb/day), V is the volume of the filter (ft^3 × 10^{-3}), and F is the recirculation factor, which can be expressed as

$$F = \frac{1+R}{\left(1+0.1\,R\right)^2}$$

$$R = \frac{Q_R}{Q_o}$$

where R is the recirculation ratio, which is calculated as Q_R/Q_0. For the case where a high strength wastewater is being treated by trickling filter, it is common to require a second filter that is placed in series after an intermediate clarifier. The NRC efficiency equation for the second filter can be expressed as

$$E_2 = \frac{100}{1+\left(\dfrac{0.0561}{1-E_1}\right)\left(\dfrac{w_2}{V_2\,F_2}\right)^{1/2}}$$

Second filter

where V and F are as defined previously, the subscript 2 refers to the second filter, and w_2 is the BOD load applied to the second filter (lb/day) and can be calculated as

$$w_2 = \frac{\left(100-E_1\right)}{100}\,w_1$$

Since the NRC equation expresses efficiency at 20°C, correction for wastewater temperature must be made as follows:

$$E_T = E_{20}\left(1.035\right)^{T-20}$$

Temperature correction

where T is in units of °C. The efficiency as expressed by the NRC equation accounts for both the trickling filter and the final clarifier; therefore, the overall plant efficiency can be expressed as

$$E_{TOT} = 100 - 100\left(1-E_P\right)\left(1-E_{1T}\right)\left(1-E_{2T}\right)$$

Example 8.14

Trickling filter efficiency using the NRC equation

The flow in Example 8.11 is to be treated using a trickling filter that has the same BOD loading as the activated sludge tank. Find the efficiency of a stone media filter at 20°C if the recirculation flow is 5.2 mgd.

Solution

Remembering that the quantity w/V is the BOD loading in units of lb per 1000 ft^3 per day, we write

$$\frac{w}{V} = 35\,\frac{\text{lb}}{1000\ \text{ft}^3\cdot\text{day}}$$

Solving for the recirculation factor (F) requires the recirculation ratio (R) as follows:

$$R = \frac{Q_R}{Q_0} = \frac{5.2}{2.6} = 2$$

Now F may be calculated as

$$F = \frac{1+R}{(1+0.1R)^2} = \frac{3}{(1+0.2)^2} = 2.083$$

The efficiency of the trickling filter may now be determined as

$$E = \frac{100}{1 + 0.0561 \left(\dfrac{w}{VF} \right)^{1/2}} = \frac{100}{1 + 0.0561 \left(\dfrac{35}{2.083} \right)^{1/2}} = 81.3\%$$

In the case of deep beds of plastic media, this system is often called a *biological tower*, or simply a *biotower*. Biotowers are usually evaluated using biological kinetics to describe substrate utilization and, therefore, effluent quality and BOD removal efficiency. Laboratory data is generated to determine wastewater treatability, and BOD removal for single pass systems (that is, no recirculation flow) is expressed as

$$\frac{S_e}{S_0} = \exp\left[\frac{-k\,D}{q^n} \right]$$

where S_e and S_0 are the BOD concentration in the tower effluent and influent, respectively; k is the treatability constant (min^{-n}, often min$^{-1/2}$); D is the filter depth (m); q is the hydraulic loading (m^3/min per m^2 of filter surface area); and n is the empirical flow constant, generally assumed to be 0.5 for modular plastic media. The treatability constant can range from 0.01–0.1 min^{-1}, with a value of 0.06 min^{-1} generally used when specific kinetic data is unknown. This constant is often given at 20°C, and temperature compensation must be made using the equation given above for stone media trickling filters.

For high strength wastewaters, applied BOD concentration (S_a) can be reduced by recirculating treated water, and is calculated as follows:

$$S_a = \frac{S_0 + R\,S_e}{1 + R}$$

The effluent concentration can now be determined as follows:

$$\frac{S_e}{S_a} = \frac{\exp\left[\dfrac{-k\,D}{q^n} \right]}{(1+R) - R\exp\left[\dfrac{-k\,D}{q^n} \right]}$$

where q is the total hydraulic load to the tower that is, $(Q_0 + Q_R)$/tower surface area).

Example 8.15

Biotower design

Your boss asked you to redesign the trickling filter in Example 8.14 using high efficiency plastic media to fit into a 20-foot-diameter tower. If the treatability constant is 0.06 min^{-1} and the empirical flow constant is 0.5, determine the depth of filter needed when the recirculation flow is reduced to 2.6 mgd.

Solution

It is often useful to simplify the expression given for biotower efficiency as follows:

$$\frac{S_e}{S_a} = \frac{\exp\left[\dfrac{-kD}{q^n}\right]}{(1+R) - R\exp\left[\dfrac{-kD}{q^n}\right]} \quad \Rightarrow \quad \boxed{\frac{S_e}{S_a} = \frac{X}{(1+R) - RX}}$$

Biotower Eff.

Since S_e is known, S_a and R must be determined in order to solve for X. Since S_a is also a function of R, start by calculating the recirculation ratio as follows:

$$R = \frac{Q_R}{Q_0} = \frac{2.6}{2.6} = 1$$

Using the data above and in the problem statements for Examples 8.11 and 8.12, S_a may be determined as

$$S_a = \frac{S_0 + RS_e}{1+R} = \frac{131\,\dfrac{mg}{L} + (1)\left(20\,\dfrac{mg}{L}\right)}{1+1} = 75.5\,\frac{mg}{L}$$

Combining this with the results above, we may solve for X in the modified expression above as

$$\frac{S_e}{S_a} = \frac{X}{(1+R) - RX} \quad \Rightarrow \quad \frac{20\,\dfrac{mg}{L}}{75.5\,\dfrac{mg}{L}} = \frac{X}{2-X} = 0.265 \quad \Rightarrow \quad X = 0.42$$

Now, X was a substitute quantity that may be expressed as

$$X = \exp\left[\frac{-kD}{q^n}\right] = 0.42 \quad \Rightarrow \quad 0.866 = \frac{kD}{q^n}$$

The quantity q is measured in m/min but is actually expressed as m³ of flow per minute per m² of filter cross-section area, and may be calculated as

$$q = \frac{(5.2\,\text{mgd})\left(1.547\,\dfrac{cfs}{mgd}\right)\left(\dfrac{60\,s}{min}\right)}{\dfrac{\pi}{4}(20\,\text{ft})^2} = 1.54\,\frac{ft}{min} \times \frac{1\,m}{3.28\,ft} = 0.47\,\frac{m}{min}$$

Notice the factor 1.547 used here. This is another great conversion factor to remember for use with wastewater problems where flows are given in mgd and more standard English units of ft³/s are desired.

Finally, filter depth can be determined from q and given values of k and n as follows:

$$0.866 = \frac{kD}{q^n} \quad \Rightarrow \quad (0.866)(0.47)^{0.5} = (0.06)D \quad \Rightarrow \quad D = 9.9\,m \times \frac{3.28\,ft}{1\,m} = 32.4\,ft$$

Again, for construction purposes, the filter depth will be rounded to 30 ft or 35 ft.

Another form of the fixed film treatment system is called a *rotating biological contactor* (RBC). The RBC is a series of closely spaced circular plastic disks attached to a slowly rotating shaft, which are 10–15 feet in diameter and serve as the attachment surface for the biofilm. The disks are partially (~40%) submerged in the wastewater, which allows for contact between the organic matter and the bacteria while a section of the disk is submerged, and for contact with the air for replenishing depleted DO when rotated out of the wastewater. Figure 8.7 offers a schematic of an RBC along with the BOD removal efficiency expected as a function of hydraulic loading.

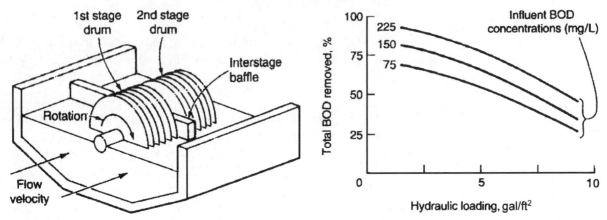

Figure 8.7   RBC schematic and BOD removal efficiency

8.5.3 Stabilization Ponds

Treatment of domestic wastewaters from small populations may be accomplished through the use of natural processes in shallow ponds. These ponds are often called *facultative ponds*, where symbiotic relationships between algae and aerobic bacteria cohabit with anaerobic activity in the sediment layer at the bottom of the pond. When stabilization ponds are used subsequent to secondary aeration, they are called *polishing* (*tertiary* or *maturation*) ponds. Because polishing ponds receive waters already treated, they generally possess detention times of only 10–15 days and maintain depths of 2–3 feet.

Volumetric BOD loads in facultative ponds generally range from 0.1–0.3 lb of BOD per 1000 ft³ of pond volume per day. Normal operating depths are kept greater than 3 feet to discourage rooted plant growth but less than 6 feet to maintain an aerobic aqueous environment. Detention times of 50–150 days (usually between 90 and 120 days) require substantial land area, and area-based BOD mass loadings are kept below 35 lb BOD per acre per day. Warmer climates can generally handle slightly larger BOD loads. However, freezing conditions essentially eliminate biological activity, and on-site storage volume should be sufficient to hold all wastewaters until late spring for locations that experience subfreezing temperatures.

Operation of the pond generally relies on controlling the depth as flow rates vary with the seasons. It is not uncommon for these ponds to have no discharge, as the evaporation rate is sufficient to maintain appropriate water depths. Often ponds are constructed in series or parallel configurations to provide additional control over flow distribution and to discourage short circuiting, and ponds may be lined with a compacted clay liner if groundwater contamination is a concern. Since odor is a minor problem during parts of the year, ponds are often located a sufficient distance downwind of any residential areas.

Example **8.16**

Stabilization ponds

Determine the minimum number of acres needed for a stabilization pond to treat the flow described in Example 8.11.

Solution

Stabilization ponds must meet several criteria. The design process requires an evaluation of all possibilities and selects the most conservative (that is, the one that meets all of the criteria). One criterion given in the narrative above is that the maximum BOD load is 35 lb of BOD per acre per day. Given the value for BOD load in solution 8.11, the minimum area required is

$$\text{Area} = \frac{2841 \dfrac{\text{lb BOD}}{\text{day}}}{35 \dfrac{\text{lb BOD}}{\text{acre} \cdot \text{day}}} = 81 \text{ acres}$$

Another design criteria noted in the text is 0.1–0.3 lb of BOD per day per 1000 ft³ of pond volume. Also, pond depth is limited to the range of 3–6 feet. Therefore, the minimum area would occur at the maximum loading rate and the maximum pond depth and can be calculated as

$$\text{Area} = \frac{2841 \dfrac{\text{lb BOD}}{\text{day}}}{0.3 \dfrac{\text{lb BOD}}{1000 \text{ ft}^3 \cdot \text{day}}} = 9.47 \times 10^6 \text{ ft}^3 \div 6 \text{ ft}$$

$$= 1.58 \times 10^6 \text{ ft}^2 \times \frac{1 \text{ acre}}{43{,}560 \text{ ft}^2} = 36 \text{ acres}$$

The third design criterion is a minimum detention time of 50 days. Using the same maximum pond depth of 6 feet to determine the minimum area, we can calculate area as follows:

$$t_R = 50 \, d = \frac{V}{Q} \Rightarrow V = (50 \text{ days})(2.6 \text{ mgd}) = 130 \text{ Mgal} \times \frac{1 \text{ ft}^3}{7.48 \text{ gal}} = 17.38 \times 10^6 \text{ ft}^3$$

$$\text{Area} = \frac{17.4 \times 10^6 \text{ ft}^3}{6 \text{ ft}} = 2.9 \times 10^6 \text{ ft}^2 \times \frac{1 \text{ acre}}{43{,}560 \text{ ft}^2} = 66.6 \text{ acres}$$

Therefore, to meet all of the criteria, the minimum pond size is 81 acres.

8.6 SECONDARY CLARIFICATION

Clarification following biological aeration may occur in two forms: *intermediate* and *final* clarifiers. Intermediate clarifiers are designed to treat the effluent of the first stage of a two-stage trickling filter. Final clarifiers treat the effluent from the secondary treatment aeration tanks, and their design depends on the type of biological treatment process used. One additional consideration for settling of activated sludge is the potential for *zone settling*, which may occur in waters of high biological cell concentrations. In these cases, a blanket of solids forms and settles as a mass. However, due to the particle density, fluid movement in the upwards direction is inhibited, and thus settling velocity is reduced. As water seeks a way back to the surface from the edges or breaks in the sludge blanket, high local fluid velocities and the potential for resuspension of particles exist.

The design of secondary clarifiers is identical to the clarifier design review previously. Design criteria were presented in Table 8.3, where values for overflow rate, weir loading, and detention time are given. Review of the criteria shows that clarification after secondary treatment has the same weir loadings and detention times as primary clarifiers; however, the overflow rate is about half of the value used in primary settling.

8.7 TERTIARY TREATMENT

Conventional wastewater treatment targets the water quality indicators identified in the effluent standards, namely, BOD, TSS, oil and grease, fecal coliforms, and pH. However, other contaminants may be present that have the potential to degrade the ecosystem upon discharge. Two chemical species identified in Table 8.1, N and P, are required for secondary treatment to be successful and usually are available in sufficient quantities. However, often these "nutrients" are present in amounts that may have a negative impact to receiving streams. For example, excess phosphorus may cause eutrophication of surface water bodies through the growth of algae and other aquatic plants, and ammonia nitrogen is toxic to fish at relatively low concentrations. Another possibility is that excess suspended solids may require additional processing for their removal to levels below discharge limits. Further, the potential for other chemicals, such as heavy metals or nonbiodegradable organic compounds from the improper disposal of hazardous chemicals into the municipal sewer system, may also pose a toxic threat to aquatic life forms and require additional processing for their removal.

8.7.1 Nitrification

Nitrification is the process of the biological oxidation of ammonia to nitrate through the following reaction:

$$NH_3 + 2O_2 = NO_3^- + H^+ + H_2O$$

In biological treatment, the aerobic degradation converts most of the organic nitrogen to ammonia, so nitrification systems are based upon the conversion of both ammonia and organic nitrogen to nitrate. This sum is known as the *total Kjeldahl nitrogen* (TKN). Theoretical oxygen requirement is 4.57 g O_2 per g N oxidized, although the actual amount is slightly less due to biological synthesis. Note that this process does not remove nitrogen from the system but instead converts it into the nontoxic nitrate (NO_3^-) form.

Nitrification can be achieved in a single-stage system (combined with activated sludge) or in a separate stage. The single-stage system requires no additional facilities but has limited control and results in increased oxygen utilization. A solids retention time of greater than 5 days is required at 20°C and greater than 10 days at 10°C to have effective nitrification. Nitrification can also be achieved in trickling filters and rotating biological contactors if the organic loading is limited. Organic loadings of less than about 5 lb BOD_5/1000 ft^3 · d (80 kg/m^3 · d) and 8 lb BOD_5/1000 ft^3 · d (130 kg/m^3 · d) for rock and plastic media, respectively, will result in greater than 90% nitrification. The kinetic coefficients are listed in Table 8.4.

Table 8.4   Kinetic coefficients for nitrification reactors

Coefficient	Value for Ammonia-N, 20°C
k	5 mg/mg·d
K_s	1 mg/L
k_d	0.05 d^{-1}
Y	0.2 mg/mg · d

Temperature corrections for k are

$$k_T = k_{20} e^{1.053(T-20)}$$

where k_T and k_{20} are the maximum N utilization rates (mg/mg · d) at temperatures T and 20°C, respectively. Nitrification is also strongly affected by pH below 7.2. The overall reaction produces a hydrogen ion, so adequate alkalinity must be present or added to maintain pH above 7.2.

8.7.2 Biological Denitrification

Biological denitrification is used to reduce nitrate to nitric oxide, nitrous oxide, and nitrogen gas. Coupled with nitrification, it is an integral part of biological nitrogen removal. Denitrification occurs in biological treatment under the presence of an organic substrate (BOD) and nitrate and the near absence of oxygen. A wide range of bacteria can use nitrate in the place of oxygen and are ubiquitous to wastewaters and wastewater treatment processes. The important parameter in designing denitrification systems is the ratio of BOD needed as an electron donor per NO_3 removed. Typical values range from about 2 to 5 mg BOD/mg NO_3-N.

Several different designs have been used to achieve denitrification for wastewaters. The complexity of the design results from having to provide BOD to reactors or zones of reactors that have high nitrate concentration. The BOD is commonly obtained from BOD in the influent or from endogenous decay of cells. Anoxic conditions can be easily obtained by eliminating aeration. The design of the systems in Figure 8.8 is beyond the scope of this review.

8.7.3 Phosphorus Removal

Phosphorus is removed from wastewater by precipitation with calcium (lime), aluminum (alum), or iron (ferric salt). Alum is the preferred process for a variety of reasons, including minimum sludge volumes and a greater removal at neutral pH. The metal salts can be added before the primary clarifier, aeration basin, or secondary clarifier as seen in Figure 8.9; however, alum is most effective when used downstream of the secondary treatment where a majority of the phosphate has been converted to orthophosphorus. Required alum dosages (mole Al:mole P) range from 1.4:1 for 75% phosphorus reduction, to 1.7:1 for 85% reduction, to 2.3:1 for 95% reduction.

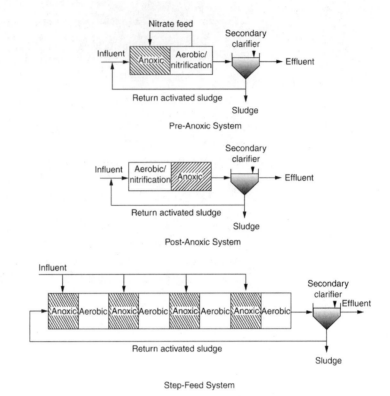

Figure 8.8 Examples of denitrification process systems

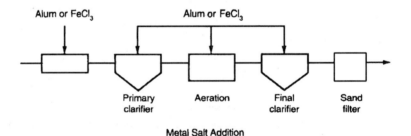

Figure 8.9 Phosphorus removal with alum or chloride

8.7.4 Adsorption

Removal of heavy metals and organic compounds that were not consumed in the secondary treatment unit may be removed through the transfer of mass from the aqueous phase to a suitable sorptive material through adsorption. For wastewater treatment, the adsorbent of choice is activated carbon (AC), either by passing the wastewater through a fixed bed of granular activated carbon (GAC) or through the mixing with powered activated carbon (PAC) in the aeration tank and its subsequent removal with the biosolids. The fine-grained nature of the PAC may require the addition of coagulants and the use of sand filters to achieve its complete removal from the process water prior to discharge.

Due to the common nature of the adsorption processes used throughout environmental engineering, the fundamentals of chemical sorption and the design of fixed-bed units will be covered in detail in section 11.4.2.

8.7.5 Filtration

There may be several reasons to consider suspended solids removal using a granular media (sand filtration), such as to meet TSS effluent limits, to remove particulate prior to disinfection to increase chlorination efficiency, and to prevent the fouling of a GAC adsorption column. Although care must be taken to account for the different nature of the particulate and the potential for substantial variations in flow rates (in the absence of flow equalization), a typical dual-media gravity filter is the process unit of choice. It is also possible to use the dual-media filter under pressure to overcome the potential headloss issues due to the nature of the solids removed. Filtration rates are similar to water treatment units at 3–6 gpm/ft^2 of filter area. The design of filtration units was covered in detail in section 7.2.5.

8.8 DISINFECTION

The primary consideration for disinfection of WWTP effluents is the removal of pathogens, as quantified through the fecal coliform test, although fecal viruses, protozoal cysts, and helminth eggs may need to be addressed. Two popular choices are chlorination and ultraviolet radiation.

8.8.1 Chlorination

Traditional chlorine disinfection as covered in section 7.2.10 is the most common form of treatment for wastewater effluents. However, the major difference between chlorination of water supplies and wastewater is the reaction of chlorine with ammonia to produce chloramines as

$$NH_3 + HOCl = NH_2Cl + H_2O$$

$$NH_2Cl + HOCl = NHCl_2 + H_2O$$

$$NHCl_2 + HOCl = NCl_3 + H_2O$$

Chloramines can undergo further oxidation to nitrogen gas with their subsequent removal.

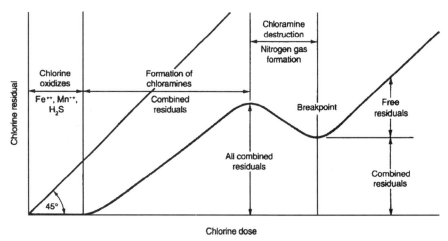

Figure 8.10 Chlorine dosage versus chlorine residual

This stepwise oxidation of various compounds is shown in Figure 8.10. As chlorine is initially added, it reacts with easily oxidized compounds such as reduced iron, sulfur, and manganese. Further addition results in the formation of chloramines, collectively termed combined-chlorine residual. This combined-chlorine residual results in the destruction of microorganisms, although much less effectively than with HOCI. Further addition of chlorine results in the destruction of the chloramines and ultimately leads to the formation of free-chlorine residuals such as HOCl and OCl⁻. The development of a free-chlorine residual is termed *breakpoint chlorination.*

The more contaminants present in the wastewater, the greater the chlorine demand before free chlorine is formed. Chlorination requirements vary by state; a typical requirements is a chlorine residual of 0.5 mg/L after 15 minutes of contact time. Typical dosages required to meet such a requirement are listed in Table 8.5. In practice, chlorine is introduced in a rapid-mix tank, followed by a minimum of 30 minutes of contact (designed at peak flow). Contact basins often are serpentine and/or baffled tanks to ensure adequate contact and limit short circuiting.

Table 8.5 Chlorine dosages for wastewaters

Wastewater	Dosage, mg/L
Untreated	6–25
Primary effluent	5–20
Activated sludge effluent	2–8
Trickling filter effluent	3–15
Sand filtered effluent	2–6

8.8.2 Ultraviolet Radiation

Gaining in popularity is the use of ultraviolet (UV) radiation for pathogen control in WWTP effluents. Similar to the $C \cdot t$ product used in chlorination of drinking water, UV disinfection is a product of the radiation intensity and time of exposure. Treatment times are determined based on the lamp intensity as well as the lamp construction. Low-pressure, low-intensity lamps emit light at 254 nm, which is nearly optimal for disinfection effectiveness. However, lamp efficiency is temperature dependent, with a maximum efficiency at 40°C, and power output is limited to approximately 25 W. Low-pressure, high-intensity lamps provide 2–20 times the power output, more emission stability over a greater temperature range, and a 25% life-span increase. Medium-pressure, high-intensity lamps possess a power output 50–100 greater than the low-pressure, low-intensity lamps and are used more commonly in WWTPs that have substantially higher flow rate.

A typical UV treatment unit consists of an open-channel filled with two or more banks of UV lamps in series, as shown in Figure 8.11. Orientation of the banks may be horizontal or vertical and are comprised of modules containing up to 16 lamps. Because of the short penetration distance of UV light into water, distance between lamp centers is on the order of 3 inches. Further, an additional channel is generally required to serve as backup for emergency or maintenance outages. Modules are usually easily removed, as regular cleaning is required to maintain optimum effectiveness.

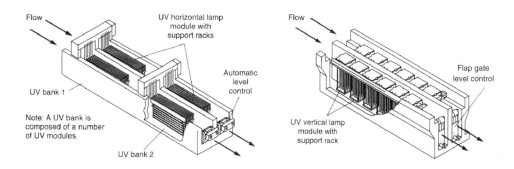

Figure 8.11 Open-channel UV disinfection systems

Source: Metcalf and Eddy, *Wastewater Engineering: Treatment and Reuse*, 4th ed., © 2003, McGraw-Hill Education. Reprinted by permission of the McGraw-Hill Companies.

8.9 SLUDGE TREATMENT PROCESSES

Water and wastewater treatment processes are generally designed to remove contaminants from solution, either through direct physical separation (sedimentation and filtration) or after chemical (coagulation) or biological (secondary aeration) treatment. While some sludge treatment processes differ based on the origin of the solids, many are the same regardless of source. The general progression of processing is (1) storage, (2) thickening, (3) conditioning, (4) dewatering, and (5) disposal.

8.9.1 Sludge Generation

Wastewater treatment results in a variety of sludge types, the vast majority from primary and secondary sedimentation. Typical quantities can be estimated based upon the type of processing unit generating the solids, as presented in Table 8.6.

Table 8.6 Sludge generation characteristics and quantities

	Specific Gravity		Quantity of Dry Solids (lb/1000 gal)
Source	**Solids**	**Sludge**	
Primary sedimentation	1.4	1.02	1.25
Activated sludge	1.25	1.005	0.7
Trickling filter	1.45	1.025	0.6
Extended aeration	1.3	1.015	0.8
Aerated lagoon	1.3	1.01	0.8

Sludge is composed primarily of water. The specific gravity of a particular sludge is given as

$$\frac{W_s}{S_s \gamma_w} = \frac{W_f}{S_f \gamma_w} + \frac{W_v}{S_v \gamma_w} \qquad \text{or} \qquad \frac{M_s}{S_s \rho_w} = \frac{M_f}{S_f \rho_w} + \frac{M_v}{S_v \rho_w}$$

[handwritten: S = solids; f = fixed solids; v = volatile; w = water]

where W is the weight, M is the mass, S is the specific gravity, γ is the specific weight, ρ is the density, and the subscripts s, f, v, and w indicate the solids, fixed solids, volatile solids, and water fractions, respectively. Typical values assumed for

$2.5 = SG \text{ fixed solids}$

$1.0 = SG \text{ volatile}$

the specific gravity of the fixed solids is 2.5 and 1.0 for the volatile solids. From this, the specific gravity of the sludge may be calculated as

$$\frac{M_{sl}}{S_{sl}} = \frac{M_s}{S_s} + \frac{M_w}{S_w} \qquad \text{or} \qquad \frac{1}{S_{sl}} = \frac{X_s}{S_s} + \frac{X_w}{S_w}$$

where S_{sl} is the specific gravity of the sludge, and X represents the mass fraction of the solids and water expressed as a decimal. The volume of sludge may be also calculated as

$$V = \frac{W_s}{\gamma_w S_{sl} P_{sl}} = \frac{M_s}{\rho_w S_{sl} P_{sl}}$$

where S_{sl} is the specific gravity of sludge, and P_{sl} is the percent solids of sludge expressed as a decimal. Solids content of several different sludges are provided in Table 8.7.

Table 8.7 Solids content of several sludge types

Sludge Type	Sludge Solids, % Dry Solids
Primary	5.0
Primary and activated sludge	4.0
Primary and trickling filter	5.0
Activated sludge	0.8
Pure oxygen-activated sludge	2.0
Trickling filter	1.5
Rotating biological contactor	1.5
Primary after gravity thickening	8.0
Activated sludge after air flotation	4.0
Activated sludge after centrifugation	5.0
Primary after anaerobic digestion	7.0
Primary and activated sludge after anaerobic digestion	3.5
Primary and trickling filter after anaerobic digestion	4.0
Primary after aerobic digestion	3.5
Primary and trickling filter after aerobic digestion	2.5
Activated sludge after aerobic digestion	1.3

Example 8.17

Sludge volume

Determine the volume of 1000 kg of primary sludge if 55% of the solids are volatile.

Solution

First determine the specific gravity of all of the solids in the primary sludge as

$$\frac{M_s}{S_s \rho_w} = \frac{M_f}{S_f \rho_w} + \frac{M_v}{S_v \rho_w} = \frac{(1000 \text{ kg})(0.45)}{(2.5)\left(1000 \dfrac{\text{kg}}{\text{m}^3}\right)} + \frac{(1000 \text{ kg})(0.55)}{(1.0)\left(1000 \dfrac{\text{kg}}{\text{m}^3}\right)}$$

$$= 0.73 = \frac{(1000 \text{ kg})}{S_s \left(1000 \dfrac{\text{kg}}{\text{m}^3}\right)}$$

or $S_s = 1.37$. Knowing that primary sludge is 5% solids (from Table 8.6), we can now determine the specific gravity of the sludge as

$$\frac{1}{S_{sl}} = \frac{X_s}{S_s} + \frac{X_w}{S_w} = \frac{0.05}{1.37} + \frac{0.95}{1.0} = 0.9865 \qquad \text{or} \qquad S_{sl} = 1.014$$

Now the volume of sludge may be calculated as

$$V = \frac{M_s}{\rho_w S_{sl} P_{sl}} = \frac{1000 \text{ kg}}{\left(1000 \dfrac{\text{kg}}{\text{m}^3}\right)(1.014)(0.05)} = 19.7 \text{ m}^3$$

That is equivalent to 5200 gallons of sludge.

Example **8.18**

Sludge production

Determine the solids loading to the wastewater treatment plant sludge digesters in kg/day given the following plant operating parameters: Q_{ave} = 8.2 mgd; plant inlet SS = 240 mg/L; a primary clarifier removes 50% of inlet SS; waste rate from final clarifier = 30%; final clarifier sludge concentration = 1% solids; sludge return to aeration tank = 105,000 gpd.

Solution

First, we can determine the solids loading from the primary clarifiers from the information given in the problem statement as follows:

$$\left(240 \frac{\text{mg}}{\text{L}}\right)\left(8.2 \frac{\text{Mgal}}{\text{day}}\right)\left(8.34 \frac{\dfrac{\text{lb}}{\text{day}}}{\dfrac{\text{mg}}{\text{L}} \cdot \dfrac{\text{Mgal}}{\text{day}}}\right)(0.5) = 8207 \frac{\text{lb}}{\text{day}}$$

For the final clarifier, we need to determine the waste mass flow rate. Since we were given a waste rate of 30%, we know that the returned sludge is at a 70% rate. Since this flow rate is given, we can find the waste flow rate as

$$Q_R = 0.7 \, Q_T \quad \Rightarrow \quad Q_T = \frac{Q_R}{0.7} = \frac{105,000 \text{ gpd}}{0.7} = 150,000 \text{ gpd}$$

Now the waste flow rate can be determined as

$$Q_W = 0.3 \, Q_T \quad \Rightarrow \quad Q_W = (0.3)(150,000 \text{ gpd}) = 45,000 \text{ gpd}$$

Now the mass rate of solids can be found noting that a concentration of 1% solids is equal to 10,000 mg/L (i.e., 1,000,000 mg/L × 0.01 = 10,000 mg/L) as follows:

$$\text{loading} = \left(45{,}000 \, \frac{\text{gal}}{\text{day}}\right)\left(\frac{1 \, \text{Mgal}}{10^6 \, \text{gal}}\right)\left(10{,}000 \, \frac{\text{mg}}{\text{L}}\right)\left(8.34 \, \frac{\frac{\text{lb}}{\text{day}}}{\frac{\text{mg}}{\text{L}} \cdot \frac{\text{Mgal}}{\text{day}}}\right) = 3753 \, \frac{\text{lb}}{\text{day}}$$

Finally, total solids can be found by adding the two sources and converting to kg as follows:

$$\text{total} = 8207 + 3753 = 11{,}960 \, \frac{\text{lb}}{\text{day}} \times \frac{1 \, \text{kg}}{2.205 \, \text{lb}} = 5424 \, \frac{\text{kg}}{\text{day}}$$

8.9.2 Preconditioning Processing

Storage often occurs in the processing units where the solid is collected (for example, at the bottom of the clarifier) and pumped daily to the sludge handling units. When necessary, a separate solids holding tank can be designed for temporary storage prior to processing. In some cases (as in aerobic digestion), sludges are pumped directly to the appropriate sludge processing unit as they are generated. *Thickening* is often accomplished using gravity settling, where the clarifier is designed with a greater depth, longer detention times, and lower overflow rates. Alternatives include *gravity belt thickening* for water treatments plant residues, where water drains from a conveyor belt that carries the sludge, to *air flotation thickening*, where fine bubbles lift the small diameter solids that are a result of activated sludge treatment to the liquid surface for subsequent removal via skimming.

Conditioning of sludges is dependent on the origin of the solid, which determines the composition and, as such, the treatment requirement. Water processing residues are basically inorganic slurries, and conditioning generally refers to additional lime or polymer additions to coagulate dissolved chemical species and increase overall solids removal in subsequent processes. Wastewater treatment sludges may be conditioned through *anaerobic digestion* or *aerobic digestion*, where biological activity is used to reduce sludge organic matter and produce a more stable product with no putrefaction.

8.9.3 Conditioning through Anaerobic Digestion

Anaerobic digestion has a long history of treating biomass due to the ability to recover an energy resource (methane) and biosolids possessing characteristics that are amenable to subsequent mechanical dewatering processes. This process occurs in two stages, the first being the hydrolysis of high molecular weight organic compounds to organic acids by specific bacterial strains. The second stage is the formation of methane and carbon dioxide by methanogenic bacteria. The volume of methane gas produced can be given by the following expression

$$V_{CH_4} = 0.35 \left[\frac{(S_0 - S_e)Q}{10^3} - 1.42 \, P_x \right]$$

where methane volume has units of m³/d measured at STP, S_0 and S_e are the influent and effluent BOD (or biodegradable COD) concentrations in mg/L, Q is the volumetric flow rate in m³/d, and P_x is the mass of cell production per day in kg/day. The expression for determination of P_x was presented previously in this chapter in section 8.5.1. Design parameters for anaerobic digesters are presented in Table 8.8.

Table 8.8 Design parameters for anaerobic sludge digestion

Parameter	Standard-Rate	High-Rate
Solids residence time [days]	30–90	10–20
Volatile solids loading [kg/m³/d]	0.5–1.6	1.6–6.4
Digested solids concentration [%]	4–6	4–6
Volatile solids reduction [%]	35–50	45–55
Gas production [m³/kg VSS added]	0.5–0.55	0.6–0.65
Methane content [%]	65	65

Source: *Fundamentals of Engineering Supplied-Reference Handbook*, 7th ed. (NCEES, 2005), p. 160.

The reactor volume required to anaerobically process wastewater sludge at the standard rate can be expressed as

$$V = \frac{V_1 + V_2}{2} t_r + V_2 t_s$$

where V_1 is the volumetric flow rate of influent raw sludge (volume/day), V_2 is the volumetric accumulation of digested sludge in the tank (volume/day), t_r is the digestion period (days), and t_s is the storage period (days). Digestion periods are generally 25–30 days at ~ 90°F, and storage periods range from 30–120 days, with 60 days a typical value.

For <u>high-rate digestion</u>, the process is typically split into two tanks, one for each stage of the digestion process. The volume of the first tank in high-rate anaerobic digestion can be determined as follows:

$$V_I = V_1 t_r$$

where V_1 and t_r are as defined previously. The second stage reactor volume can be calculated as

$$V_{II} = \frac{V_1 + V_2}{2} t_t + V_2 t_s$$

where V_1, V_2, and t_s are as defined previously, and t_t is the thickening period (days). While storage periods are similar for the two types of processes, the digestion period in the first stage of the high-rate system is reduced to 10–15 days.

| Example **8.19** | **Anaerobic sludge digestion** |

The wasted sludge from Example 8.12 is mixed with 1500 lb/day of sludge from the primary clarifier. Determine the volume of the anaerobic digester given the following sludge characteristics and design parameters: the combined raw sludge is 4% solids, and 70% of the solids are volatile; the digester has a volatile solids reduction of 40%; the digestion period is 25 days and the storage period is 60 days; and the digested sludge is 6% solids.

Solution

From Example 8.12, we know that $Q_w SS_w = 1250$ lb/day. Therefore, the total amount of sludge to be treated is

$$\text{total sludge} = 1250 + 1500 = 2750 \frac{\text{lb}}{\text{day}}$$

Given a solids concentration of 4%, we can calculate the mass rate of flow as follows:

$$\frac{2750 \dfrac{\text{lb}}{\text{day}}}{0.04} = 68{,}750 \frac{\text{lb}}{\text{day}}$$

If we assume this flow has a specific weight that is close to the specific weight of water, we may determine the volume of raw sludge entering the digester per day as follows:

$$V_1 = \frac{68{,}750 \dfrac{\text{lb}}{\text{day}}}{62.4 \dfrac{\text{lb}}{\text{ft}^3}} = 1102 \frac{\text{ft}^3}{\text{day}}$$

Next, we need to calculate the digested sludge accumulation. This is accomplished by noting that 30% of the solids are not volatile and will remain during treatment. The other 70% of the solids will degrade, but the reduction is only 40%, so 60% of the volatile solids remain. These solids are at a concentration of 6% solids and are converted to a volumetric flow rate using the specific weight of water as in computing V_1 above. The expression for V_2 may be expressed as follows:

$$V_2 = \frac{\left(2750 \dfrac{\text{lb}}{\text{day}}\right)(0.3)}{(0.06)\left(62.4 \dfrac{\text{lb}}{\text{ft}^3}\right)} + \frac{\left(2750 \dfrac{\text{lb}}{\text{day}}\right)(0.7)(0.6)}{(0.06)\left(62.4 \dfrac{\text{lb}}{\text{ft}^3}\right)} = 529 \frac{\text{ft}^3}{\text{day}}$$

Finally, we can calculate reactor volume as follows:

$$V_r = \frac{V_1 + V_2}{2} t_r + V_2 t_s = \left(\frac{1102 + 529 \dfrac{\text{ft}^3}{\text{day}}}{2}\right)(25 \text{ days}) + \left(529 \frac{\text{ft}^3}{\text{day}}\right)(60 \text{ days})$$

$$= 52{,}128 \text{ ft}^3$$

8.9.4 Conditioning through Aerobic Digestion

Aerobic digestion offers a stable end product with low capital cost and relative ease of operation, although operating costs may be greater due to oxygenation requirements. The process is similar to activated sludge treatment, where the slurry is maintained in the endogenous phase. After consuming the small amount of remaining BOD, the only food source available is from the biomass as cell lysis provides the remaining bacteria with the energy needed to maintain cell function. System design parameters are provided in Table 8.9.

Table 8.9 Design parameters for aerobic sludge digestion

Parameter	Value	Units
Hydraulic retention time, 20°C		
Waste activated sludge only	10–15	days
AS without primary settling	12–18	days
Primary plus waste activated or trickling-filter sludge	15–20	days
Solids loading	0.1–0.3	lb TVS/ft$^3 \cdot$ d
Oxygen requirements		
Cell tissue	~2.3	lb O$_2$/lb solids
BOD$_5$ in primary sludge	1.6–1.9	lb O$_2$/lb solids
Energy requirements for mixing		
Mechanical aerators	0.7–1.5	hp/1000 ft^3
Diffused-air mixing	20–40	ft^3/1000 ft$^3 \cdot$ min
Dissolved-oxygen residual in liquid	1–2	mg/L
Reduction in volatile suspended solids	40–50	%

Tank design is based primarily on the mean cell residence time (θ_c) required to reduce the volatile solids to the desired level. Tank volume (V) can be expressed as

$$V = \frac{Q_i\left(X_i + F\,S_i\right)}{X_d\left(K_d P_v + \dfrac{1}{\theta_c}\right)}$$

where Q_i is the volumetric flow rate to the digester, X_i is the suspended solids concentration in the influent, F is the fraction of BOD$_5$ from raw primary sludge, S_i is the influent BOD$_5$, X_d is the suspended solids concentration in the digester, K_d is the reaction rate constant, P_v is the fraction of the digester suspended solids that is volatile, and θ_c is the mean cell residence time in the digester. For sludge processing where primary sludge is not included, the FS_i term may be dropped from the above expression. A sludge age of 40 days is required for liquid temperatures of 20°C, increasing to 60 days as the temperature drops to 15°C. Oxygen requirements are approximately 2.3 pounds of oxygen per pound of cell mass, plus 1.6–1.9 pounds of oxygen per pound of BOD applied, where applicable.

Example **8.20**

Aerobic sludge digestion

An aerobic digester is being considered to treat only the wasted sludge from Example 8.12, which has a solids fraction of 1%. The volatile fraction of the sludge is 70%, and the reaction rate constant is 0.05 day^{-1}. Determine the digester volume if the digester solids concentration is 6% and the mean cell residence time is 50 days.

Solution

From Example 8.12, we know that $Q_w SS_w = 1250$ lb/day, and the problem statement gives a solids concentration of 1%, which is equivalent to 10,000 mg/L. We can therefore determine the volumetric flow rate as follows:

$$Q_w SS_w = 1250 \frac{lb}{day} = \left(10,000 \frac{mg}{L}\right) Q_w (8.34) \quad \Rightarrow \quad Q_w = 0.015 \text{ mgd}$$

$$Q_w = (0.015 \text{ mgd}) \left(\frac{10^6 \text{ gal}}{\text{Mgal}}\right) \left(\frac{1 \text{ ft}^3}{7.48 \text{ gal}}\right) = 2004 \frac{\text{ft}^3}{\text{day}}$$

Since there are no primary solids, F is equal to zero. Therefore, using the data given in the problem statement, we may write

$$V = \frac{Q X}{X_d \left(k_d P_v + \dfrac{1}{\theta_c}\right)} = \frac{\left(2004 \dfrac{\text{ft}^3}{\text{day}}\right)\left(10,000 \dfrac{mg}{L}\right)}{\left(60,000 \dfrac{mg}{L}\right)\left[(0.05 \text{ day}^{-1})(0.7) + \dfrac{1}{50 \text{ days}}\right]} = 6073 \text{ ft}^3$$

8.9.5 Postconditioning Processing

Dewatering has historically been completed in *sand beds*, where conditioned sludge was spread over a layer of sand containing drain pipe and allowed to dry until the desired solids content was obtained. However, more recent designs have attempted to reduce the manual labor with mechanized systems, including *vacuum filtration* and *pressure filtration* using belt filters or filter presses, or *centrifugation*, which is often used for large facilities. Final disposal of processed sludges is dependent on composition. Water treatment plant residues are often codisposed with municipal solid waste or directed into a waste monofill, whereas wastewater treatment plant sludges are often land-applied for agricultural purposes. Wastewater biosolids may also be composted, often with other organic waste matter, prior to use as a soil amendment.

Example 8.21

Comprehensive wastewater plant design

A small midwestern city produces a wastewater flow of 5.5 mgd and has an influent BOD of 170 mg/L. The WWTP consists of an aerated grit chamber (AGC), primary clarifier, activated sludge (AS) tank, final clarifier, and disinfection through UV radiation, with biosolids treated through anaerobic digestion. The primary removes 60% of the solids and 35% of the BOD. Laboratory data for the AS process determined a biological yield of 0.6 and a microbial decay constant of 0.06 day^{-1} in the 155,000 ft^3 tank when there was an MLSS concentration of 2100 mg/L. The complete-mix high-rate anaerobic digester has a biological yield of 0.08, a microbial decay constant of 0.03 day^{-1}, and a solids retention time (SRT) of 14 days. Final plant effluent has a BOD of 10 mg/L. Determine the following:

(a) The dimensions of the AGC for the flow above if the detention time is 3 minutes at peak flow, the width:depth ratio is 1.5:1, and the length:width ratio is 4:1

(b) The depth of the circular primary clarifier if the overflow rate is 1000 gpd/ft^2 and the detention time is 2 hours at Q_{AVE}

(c) The SRT in the AS tank

(d) Estimate of the monthly energy costs for disinfection if the electricity costs 5.5 cents per kW · hr. The UV system utilizes modules of 16 vertical lamps placed in rectangular flow channels. Each module can treat 0.5 mgd, and each lamp is rated at 100 W.

(e) Estimate of the daily volume of CH_4 produced at this WWTP. The anaerobic digesters were installed to provide CH_4 to produce power to run the WWTP operations.

Solutions

(a) Assuming an average per capita contribution of 100 gpd, the city population is approximately 55,000, which yields a peaking factor of approximately 3, or

$$Q_{peak} = 3\,Q_{ave} = (3)(5.5\ \text{Mgd})\left(1.547\frac{\text{cfs}}{\text{Mgd}}\right) = 25.53\ \text{cfs}$$

AGC tank volume can now be calculated from the design detention time as

$$t_R = \frac{V}{Q} \;\Rightarrow\; V = t_R \times Q = (3\ \text{min})\left(60\ \frac{\text{s}}{\text{min}}\right)\left(25.53\ \frac{\text{ft}^3}{\text{s}}\right) = 4595\ \text{ft}^3$$

Using the dimension ratios we know that

$$W = 1.5D \quad \text{and} \quad L = 4W = (4)(1.5D) = 6D$$

Therefore,

$$V = LWD = (6D)(1.5D)D = 9D^3 = 4595\ \text{ft}^3$$

Rearranging and solving for D gives us L and W also:

$$D = \left(\frac{4595\ \text{ft}^3}{9}\right)^{1/3} = 8.0\ \text{ft}$$

$$W = 1.5D = (1.5)(8\ \text{ft}) = 12\ \text{ft}$$

$$\text{and } L = (4)(12\ \text{ft}) = 48\ \text{ft}$$

(b) Using the design detention time to find clarifier volume we get

$$V = t_R \times Q = (2\ \text{hr})\left(\frac{1\ \text{day}}{24\ \text{hr}}\right)\left(5.5\times10^6\ \frac{\text{gal}}{\text{day}}\right)\left(\frac{1\ \text{ft}^3}{7.48\ \text{gal}}\right) = 61,275\ \text{ft}^3$$

Using the expression for overflow rate, we can find tank surface area as

$$V_O = \frac{Q}{A_S} \;\Rightarrow\; A_S = \frac{Q}{V_O} = \frac{5.5\times10^6\ \text{gpd}}{1000\ \dfrac{\text{gpd}}{\text{ft}^2}} = 5500\ \text{ft}^2$$

Now tank depth can be determined as

$$\text{depth} = \frac{V}{A_S} = \frac{61,275\ \text{ft}^3}{5500\ \text{ft}^2} = 11.14\ \text{ft}$$

(c) With the information provided in the problem statement, we can use the following expressions to find SRT:

$$\frac{1}{SRT} = Y\left(\frac{F}{M}\right)\left(\frac{E}{100}\right) - k_d \quad \text{where} \quad E = \frac{S_0 - S}{S_0} \times 100 \quad \text{and} \quad \left(\frac{F}{M}\right) = \frac{Q\,S_0}{V\,X}$$

For the AS process, S_0 is defined as the BOD concentration entering the tank, after the 35% is removed in the primary, therefore:

$$S_0 = \left(170\,\frac{mg}{L}\right)(1 - 0.35) = 110.5\,\frac{mg}{L}$$

Knowing the effluent BOD, we can calculate efficiency (E) as

$$E = \frac{110.5 - 10}{110.5} \times 100 = 90.95\%$$

Noting X is the MLSS concentration in the AS tank, we can calculate the F/M ratio as

$$\left(\frac{F}{M}\right) = \frac{\left(5.5 \times 10^6\,\frac{gal}{day}\right)\left(110.5\,\frac{mg}{L}\right)}{\left(155,000\,ft^3\right)\left(7.48\,\frac{gal}{ft^3}\right)\left(2100\,\frac{mg}{L}\right)} = 0.25\,day^{-1}$$

Combining expression allows for the calculation of SRT as

$$\frac{1}{SRT} = Y\left(\frac{F}{M}\right)\left(\frac{E}{100}\right) - k_d = (0.6)(0.25\,day^{-1})\left(\frac{90.95}{100}\right) - 0.06\,day^{-1}$$

$$= 0.0764\,day^{-1}$$

By taking the reciprocal of $0.0764\,day^{-1}$, the SRT is found to be 13.1 days.

(d) The number of modules required can be determined as

$$\frac{5.5 \times 10^6\,gpd}{500,000\,\dfrac{gpd}{module}} = 11\,\text{modules}$$

Total monthly power usage can now be determined as

$$(11\,\text{modules})\left(16\,\frac{lamps}{module}\right)\left(100\,\frac{W}{lamp}\right)\left(\frac{1\,kW}{1000\,W}\right)\left(30\,\frac{days}{mo}\right)\left(24\,\frac{hr}{day}\right)$$

$$= 12,672\,\frac{kW \cdot hr}{mo}$$

Power usage multiplied by the utility rate yield monthly cost as

$$\left(12,672\,\frac{kW \cdot hr}{mo}\right)\left(\frac{\$\,0.055}{kW \cdot hr}\right) = \$\,697\,\text{per month}$$

(e) With the information provided in the problem statement, we can use the following expressions to find the volume of methane produced as

$$V_{CH_4} = (0.35)\left[\frac{(S_0 - S)Q}{10^3} - 1.42\,P_X\right] \quad \text{where} \quad P_X = \frac{Y\,Q(S_0 - S)}{10^3\left[1 + (k_d)(SRT)\right]}$$

In these expressions, Q has units of m³/day, S has units of g/m³ (note that this is the same as mg/L), k_d has units of day⁻¹, SRT has units of days, and Y is unitless. It is also important to note that the S_0 here is the BOD entering the WWTP, not the wastewater entering the AS unit. This is because the BOD removed in the primary is included in the sludge to be treated anaerobically. This yields P_X in units of kg/day, which is converted to m³/day by the constant 0.35. First, Q must be converted into the correct units as follows:

$$Q = \left(5.5 \times 10^6 \frac{\text{gal}}{\text{day}}\right)\left(3.785 \frac{\text{L}}{\text{gal}}\right)\left(\frac{1\,\text{m}^3}{1000\,\text{L}}\right) = 20{,}818 \frac{\text{m}^3}{\text{day}}$$

Solving for P_X yields:

$$P_X = \frac{Y\,Q\,(S_0 - S)}{10^3\left[1 + (k_d)(SRT)\right]} = \frac{(0.08)\left(20{,}818\,\dfrac{\text{m}^3}{\text{day}}\right)\left(170 - 10\,\dfrac{\text{g}}{\text{m}^3}\right)}{\left(10^3\,\dfrac{\text{g}}{\text{kg}}\right)\left[1 + \left(0.03\,\text{day}^{-1}\right)(14\,\text{days})\right]} = 188 \frac{\text{kg}}{\text{day}}$$

Solving for the volume of methane produced yields:

$$V_{CH_4} = \left(0.35\,\frac{\text{m}^3\,CH_4}{\text{kg}\,CH_4}\right)\left[\frac{\left(170 - 10\,\dfrac{\text{g}}{\text{m}^3}\right)\left(20{,}818\,\dfrac{\text{m}^3}{\text{day}}\right)}{\left(10^3\,\dfrac{\text{g}}{\text{kg}}\right)} - (1.42)\left(188\,\frac{\text{kg}}{\text{day}}\right)\right]:$$

$$= 1072 \frac{\text{m}^3}{\text{day}}$$

Air Quality and Atmospheric Pollution Control

9.1 AIR QUALITY

Primary pollutants in air are those that are emitted by identifiable (albeit ubiquitous) sources, making it possible to address each specific discharge. *Secondary pollutants* are those that are formed in the atmosphere, often through reactions with primary pollutants. *Criteria pollutants* are those that have been identified as having a negative impact on the health of human populations and the health of ecosystems as a whole. The U.S. Environmental Protection Agency (U.S. EPA) has established criteria pollutant limits in ambient (outdoor) air called the *National Ambient Air Quality Standards* (NAAQS), as presented in Table 9.1. Large-scale (regional, national, and global) air pollution concerns include acid rain, global warming, and ozone depletion, each of which have identifiable causes with an opportunity for implementation of control technology.

Table 9.1   National Ambient Air Quality Standards for criteria pollutants

Criterion	Concentration		Averaging
Pollutant	[$\mu g/m^3$]	[ppm]	Period
CO	10,000	9	8-hour
CO	40,000	35	1-hour
Pb	1.5	—	3-month
NO_2	100	0.053	Annual
O_3	235	0.12	1-hour
O_3	157	0.08	8-hour
PM_{10}	150	—	24-hour
PM_{10}	50	—	Annual
$PM_{2.5}$	65	—	24-hour
$PM_{2.5}$	15	—	Annual
SO_2	80	0.03	Annual
SO_2	365	0.14	24-hour

Acid rain is primarily caused by the reaction of *sulfur dioxide* (SO_2) in the atmosphere, creating sulfuric acid, which is carried to the ground through precipitation events. SO_2 is a by-product from the combustion of fossil fuels containing sulfur compounds (which can be as high as 3%–4% by weight in some coals), and it is estimated that greater than 95% of all SO_2 is emitted from stationary sources such as power plants. Much has been done in the recent past to control SO_2 emissions, primarily through the implementation of scrubbing technologies, as covered in section 11.4.2. Other techniques to reduce sulfur emissions include the removal of sulfur from the fuel (for example, coal washing) or switching to the use of low-sulfur fuels.

Although there are still some scientists and politicians who question the link to global warming, the fact remains that the concentration of greenhouse gases in the atmosphere is on a measurable increase. Historically, *volatile organic compounds* (VOCs) such as CFCs (chlorofluorocarbons) received the primary attention of scientists, first from the *ozone depletion potential* (ODP) and later from a *global warming potential* (GWP). With the replacement of VOCs in most commercial products, other species have come under scrutiny. The most abundant *hydrocarbon* (HC), a class of compounds comprised solely of hydrogen and carbon, is *methane* (CH_4), and although the majority is naturally released, it is considered a primary greenhouse gas. Two gases from combustion processes, *carbon dioxide* (CO_2) and *carbon monoxide* (CO), are also on the greenhouse gas list, and efforts to control CO_2 emissions on a global scale are being discussed in the regulatory and industrial communities. Carbon monoxide also can have serious negative health impacts when respired at sufficient concentrations.

Ozone (O_3) is interesting in the fact that its appearance in the lower atmosphere causes smog and irritation of the mucous membranes, while the depletion of ozone in the upper atmosphere has been associated with a reduction in the ability to filter cosmic radiation. Ozone is a criteria pollutant, although it is also a secondary pollutant because it is derived from atmospheric reactions with primary pollutants such as NO_x and HCs. Besides being a respiratory irritant, it may also damage materials through enhanced oxidation as well as inhibit plant growth.

Particulate matter (PM) is emitted in urban areas, primarily through combustion processes, and is classified by the aerodynamic diameter of the particle in μm (10^{-6} m). PM is suspected of having a negative impact on human health. While the criteria pollutant in the recent past has been PM_{10} (PM with diameters less than 10 μm), new standards target $PM_{2.5}$ as the required capture limit. PM in the 2.5–10-μm range may become trapped in the lungs, minimally causing irritation and potentially causing more serious chronic health problems. Control of PM is covered in detail in section 9.4.

Oxides of nitrogen (NO_X), such as NO, NO_2, and N_2O, are also produced in the combustion process. However, only a fraction are fuel derived (oxidation of N in the fuel), even though some coals can have nitrogen concentrations up to 1% by weight. The majority of the NO_X species are due to the oxidation of N_2 that is part of the combustion air. NO_X contributes to smog and is also a respiratory irritant. Control of thermal NO_X, the portion of the NO_X that is created from N_2 in the combustion air, generally examines operating conditions in the combustion zone and attempts to inhibit creation of NO_X by maintaining the flame temperature within specified limits. After generation, or for controlling fuel-based NO_X, facilities often employ *selective catalytic reduction* (SCR) of the NO_X species after the combustion zone to convert NO_X back to N_2.

The final criteria pollutants to be discussed here are heavy metals, particularly *mercury* (Hg) and *lead* (Pb). While still of interest, the concern of atmospheric lead has been addressed by the regulatory and scientific communities in the recent past (for example, the removal of lead from gasoline). Lead may still be present in some industrial settings or as airborne particulate, but substantial efforts have been made to remove it from most commercial products. Mercury still remains a problem, due primarily to its high toxicity at low concentrations, its tendency to bioconcentrate and bioaccumulate, and its presence in many natural materials such as coal. However, advancements in air pollution control technology has done much to reduce the Hg emissions from power plant flue gases, and many scientists argue that the natural sources of mercury emissions such as global volcanic activity far exceed those from anthropogenic (those derived from human activity) sources.

9.2 METEOROLOGY

The atmosphere is in constant motion, driven by *mechanical* and *thermal* forces such as the Earth's rotation and solar energy. Wind and weather patterns are produced by these effects and also contribute to the distribution of all compounds that are discharged into the atmosphere. Once in the atmosphere, they are subject to a variety of chemical and physical transformations and are redeposited on the Earth's surface during precipitation events. Prediction of the distribution of pollutants from a point source is a valuable tool in evaluating the potential health impact an atmospheric discharge may have on any downwind populations.

The point source Gaussian dispersion model is used to evaluate contaminant concentrations downwind of a discharge, generally considered to be an industrial stack, as discussed in section 9.3. A primary consideration of the model is the *atmospheric stability*, which may be defined as the *tendency of the atmosphere to resist vertical motion*. Atmospheres are classified as *stable* when thermal effects *restrict* mechanical turbulence, *unstable* when thermal effects *enhance* mechanical turbulence, and *neutral* when thermal effects *neither enhance nor restrict* mechanical effects. An approximate stability classification was developed by Turner in which atmospheric stability increases from A to F, as defined in Table 9.2. These classifications were then related to prevailing environmental conditions that were based upon incident solar radiation and wind speed at an elevation of 10 m, as presented in Table 9.3.

Table 9.2 Atmospheric stability classifications

Degree of Stability	Classification
Extremely unstable	A
Unstable	B
Slightly unstable	C
Neutral	D
Slightly stable	E
Stable	F

Table 9.3 Atmospheric stability classifications as a function of environmental conditions

Wind speed at 10 m elevation (U_{10})	Day			Night	
	Incoming solar radiation			Thinly overcast or ≥ 1/2 low cloud	≤ 3/8 Cloud
(m/s)	Strong	Moderate	Slight		
<2	A	A–B	B	F	F
2–3	A–B	B	C	E	F
3–5	B	B–C	C	D	E
5–6	C	C–D	D	D	D
>6	C	D	D	D	D

Notes:

1. The neutral class (D) should be assumed for overcast conditions during day or night, regardless of wind speed.
2. "Strong" corresponds to a solar altitude greater than 60°F (summer) with clear skies.
3. "Moderate" corresponds to a summer day with a few broken clouds, or a clear day with a solar altitude between 35°F and 60°F (spring or fall).
4. "Slight" corresponds to a solar altitude from 15°F to 35°F (winter) with clear skies.
5. Night refers to the period from one hour before sunset to one hour after sunrise.
6. Cloud fractions refer to approximate amount of cloud cover.

Source: D. Bruce Turner, *Workbook of Atmospheric Dispersion Estimates: An Introduction to Dispersion Modeling*, 2nd ed. (Lewis Publishing, CRC Press, 1994).

The stability classification is used to predict the Gaussian (normal) dispersion distance as a function of distance downwind of a point source. As seen in the standard deviations of plumes graphs in Figures 9.1 and 9.2, dispersion in the vertical direction increases at variable rates with increased downstream distance, depending on the atmospheric stability classification. This is to be expected, because stability is defined as the tendency of the atmosphere to resist vertical motion. Horizontal dispersion still depends on stability classification; however, it does increase at a constant rate with downstream distance.

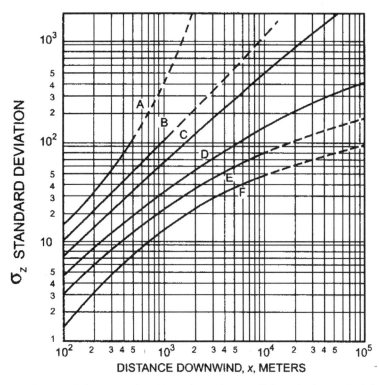

Figure 9.1 Standard deviations for plume dispersion coefficients in the vertical dimension

Source: D. Bruce Turner, *Workbook of Atmospheric Dispersion Estimates*, Washington, D.C., U.S. Department of Health, Education, and Welfare, 1970.

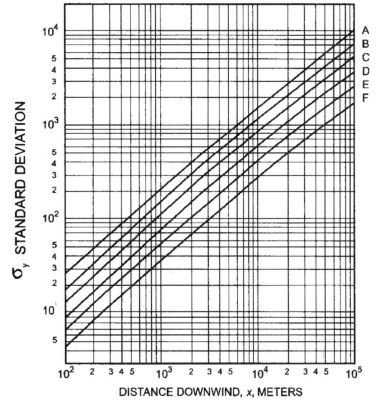

Figure 9.2 Standard deviations for plume dispersion coefficients in the horizontal dimension

Source: D. Bruce Turner, *Workbook of Atmospheric Dispersion Estimates*, Washington, D.C., U.S. Department of Health, Education, and Welfare, 1970.

9.3 ATMOSPHERIC DISPERSION MODELING

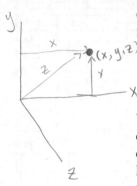

The Gaussian dispersion model as presented below assumes that material is discharged from a single point at elevation H, and any material that reaches the ground is reflected back into the air and can be expressed as

$$C_{(x,y,z)} = \frac{Q}{2\pi\,\sigma_y\,\sigma_z\,U} \exp\left[-\frac{1}{2}\left(\frac{y}{\sigma_y}\right)^2\right]\left\{\exp\left[-\frac{1}{2}\left(\frac{z-H}{\sigma_z}\right)^2\right]+\exp\left[-\frac{1}{2}\left(\frac{z+H}{\sigma_z}\right)^2\right]\right\}$$

where $C_{(x,y,z)}$ is the species concentration at location (x,y,z) downwind of the discharge (g/m³), x is the downwind or centerline distance (m), y is the horizontal distance that is perpendicular to the centerline (m), z is the elevation from ground level (m), Q is the species mass flow rate (g/s), σ_y and σ_z are the horizontal and vertical standard deviations (m) obtained from Figures 9.1 and 9.2, U is the wind speed at the point of discharge (m/s), and H is the *effective stack height* (m). Note that the variable used here for the emission rate (Q) has dimensions of mass per time, unlike in previous chapters where Q was defined as the volumetric flow rate (volume per time).

The effective stack height takes into account the elevation of the discharge, as well as the *plume rise*. Plume rise is a result of gas velocity exiting the stack, as well as the buoyancy effects due to the temperature differential between the emitted gas and the surrounding air, and can be estimated by the following relationships:

$$H = h + \Delta H$$

$$\Delta H = \frac{U_s\,d}{U}\left\{1.5+\left[0.0268\,P\left(\frac{T_s-T_a}{T_s}\right)d\right]\right\}$$

where h is the stack height (m), ΔH is the plume rise (m), U_s is the stack gas velocity (m/s), d is the stack diameter (m), U is the wind speed at the point of discharge (m/s), P is the prevailing atmospheric pressure (kPa), and T_s and T_a are the temperatures (K) of the stack gas and ambient air, respectively. It should be noted that the temperatures are expressed in absolute units, where K = °C + 273.

Often, it is important to evaluate ground level concentrations, where impact to human health would most likely occur. In this case, the elevation term z is set equal to zero, and the Gaussian model is simplified to

$$C_{(x,y,0)} = \frac{Q}{\pi\,\sigma_y\,\sigma_z\,U} \exp\left[-\frac{1}{2}\left(\frac{y}{\sigma_y}\right)^2\right]\exp\left[-\frac{1}{2}\left(\frac{H}{\sigma_z}\right)^2\right]$$

Further, based on the assumption of normally distributed dispersion, the maximum concentration at any downwind distance x would occur on the center line. In this case, the horizontal distance term y is set equal to zero, and the Gaussian model is simplified to

$$C_{(x,0,0)} = \frac{Q}{\pi\sigma_y\,\sigma_z\,U} \exp\left[-\frac{1}{2}\left(\frac{H}{\sigma_z}\right)^2\right]$$

Example 9.1

Plume dispersion modeling

An industrial stack with an effective stack height of 85 m emits SO_2 at a rate of 10 g/s. Estimate the maximum ground-level concentration at a distance 4 km downwind of the site if the wind is blowing at 5.5 m/s on a clear, sunny day.

Solution

First, use Table 9.3 to determine stability class given solar insolation and wind speed:

$U = 5.5$ m/s on a clear, sunny day; therefore, the stability class C

Now using Figures 9.1 and 9.2, the vertical and horizontal standard deviations of the plume may be determined as

class C at $x = 4$ km $\Rightarrow$ $\sigma_y = 400$ m and $\sigma_z = 220$ m

Now, use the $C_{(x,0,0)}$ equation to determine the maximum ground level concentration as follows:

$$C_{(x,0,0)} = \frac{Q}{\pi \, \sigma_y \, \sigma_z \, U} \exp\left[-\frac{1}{2}\left(\frac{H}{\sigma_z}\right)^2\right]$$

$$= \frac{10 \, \frac{g}{s}}{\pi \, (400 \text{ m})(220 \text{ m})(5.5 \, \frac{m}{s})} \exp\left[-\frac{1}{2}\left(\frac{85 \text{ m}}{220 \text{ m}}\right)^2\right]$$

$$C = 6.1 \times 10^{-6} \text{ g/m}^3 = 6.1 \text{ μg/m}^3$$

This expression may be rearranged to convey the maximum concentration of pollutant for a given wind speed and emission rate as a function of effective stack height and atmospheric dispersion as follows:

$$\left(\frac{CU}{Q}\right)_{max} = \frac{1}{\pi \, \sigma_y \, \sigma_z} \exp\left[-\frac{1}{2}\left(\frac{H}{\sigma_z}\right)^2\right]$$

The left-hand side of the expression is often plotted for various downwind distances and effective stack heights, and the curves for each atmospheric stability class have been regressed. The values for $(CU/Q)_{max}$ may be calculated from the following equation:

$$\left(\frac{CU}{Q}\right)_{max} = \exp\left[a + b\ln H + c\left(\ln H\right)^2 + d\left(\ln H\right)^3\right]$$

where C, U, Q, and H are as defined previously, and a, b, c, and d are constants based on atmospheric stability class as given in Table 9.4. The equation above is also solved for all stability classes as a function of effective stack height and presented graphically in Figure 9.3.

Table 9.4 Curve fitting constants for estimating $(CU/Q)_{max}$

Stability	Constants			
	a	b	c	d
A	−1.0563	−2.7153	0.1261	0
B	−1.8060	−2.1912	0.0389	0
C	−1.9748	−1.9980	0	0
D	−2.5302	−1.5610	−0.0934	0
E	−1.4496	−2.5910	0.2181	−0.0343
F	−1.0488	−3.2252	0.4977	−0.0765

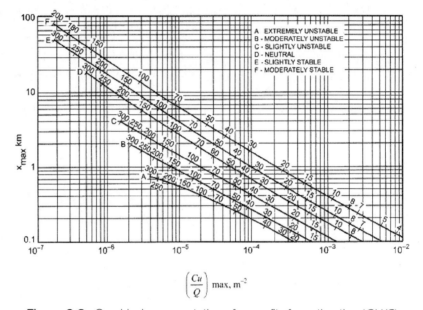

$$\left(\frac{Cu}{Q}\right) \text{max, m}^{-2}$$

Figure 9.3 Graphical representation of curve fits for estimating $(CU/Q)_{max}$

Source: D. Bruce Turner, *Workbook of Atmospheric Dispersion Estimates: An Introduction to Dispersion Modeling*, 2nd ed. (Lewis Publishing, CRC Press, 1994).

Example 9.2

Plume dispersion modeling II

Using the values for C, U, and H from Example 9.1, estimate the maximum ground-level concentration of SO_2 and the location that the maximum occurs.

Solution

Since the stability class was determined in Example 9.1, write the empirical expression and substitute the constants for atmospheric stability class C as

$$\left(\frac{CU}{Q}\right)_{max} = \exp\left[a + b\ln H + c\left(\ln H\right)^2 + d\left(\ln H\right)^3 \right]$$

$$= \exp\left[-1.9748 - 1.998\ln(85)\right] = 1.94 \times 10^{-5}$$

Now using the values for U and Q given in the problem statement, C may be estimated as

$$\left(\frac{CU}{Q}\right)_{max} = 1.94 \times 10^{-5} \quad \Rightarrow \quad C = \frac{\left(10\frac{g}{s}\right)\left(1.94 \times 10^{-5} \text{ m}^{-2}\right)}{5.5\frac{\text{m}}{\text{s}}}$$

$$= 3.524 \times 10^{-5} \frac{g}{\text{m}^3} = 35.2 \frac{\mu g}{\text{m}^3}$$

The location of the maximum ground-level concentration must occur on the centerline, so it is clear that $y = z = 0$. The value for x may be estimated from Figure 9.3 either by locating the intersection of $(CU/Q)_{max} \approx 2 \times 10^{-5}$ m^{-2} and the stability class C curve, or by estimating the location of a value of 85 for H on that same curve. Both yield a value of $x_{max} \approx 1.0$ km.

To check this result, use Figures 9.1 and 9.2 to find σ_y and σ_z at $x = 1$ km and class C stability as 110 m and 65 m, respectively, and substitute that back into the dispersion model as before:

$$C_{(x,0,0)} = \frac{Q}{\pi \sigma_y \sigma_z U} \exp\left[-\frac{1}{2}\left(\frac{H}{\sigma_z}\right)^2\right]$$

$$= \frac{10\frac{g}{s}}{\pi (110 \text{ m})(65 \text{ m})(5.5\frac{\text{m}}{\text{s}})} \exp\left[-\frac{1}{2}\left(\frac{85 \text{ m}}{65 \text{ m}}\right)^2\right]$$

$$C = 34.4 \times 10^{-5} \text{ g/m}^3 = 34.4 \ \mu g/\text{m}^3$$

Notice that this is in good agreement with the value of 35.2 $\mu g/\text{m}^3$ obtained previously. To determine if this is the maximum concentration, we can calculate the centerline concentration just upwind and downwind of $x = 1$ km. If we assume a value of x at 0.8 km, we find

$$\text{class } C \text{ at } x = 0.8 \text{ km} \quad \Rightarrow \quad \sigma_y = 90 \text{ m and } \sigma_z = 53 \text{ m}$$

$$C_{(x,0,0)} = \frac{10\frac{g}{s}}{\pi (90 \text{ m})(53 \text{ m})(5.5\frac{\text{m}}{\text{s}})} \exp\left[-\frac{1}{2}\left(\frac{85 \text{ m}}{53 \text{ m}}\right)^2\right] = 33.5 \frac{\mu g}{\text{m}^3}$$

If we assume a value of x at 1.2 km, we find

$$\text{class } C \text{ at } x = 1.2 \text{ km} \quad \Rightarrow \quad \sigma_y = 130 \text{ m and } \sigma_z = 80 \text{ m}$$

$$C_{(x,0,0)} = \frac{10\frac{g}{s}}{\pi (130 \text{ m})(80 \text{ m})(5.5\frac{\text{m}}{\text{s}})} \exp\left[-\frac{1}{2}\left(\frac{85 \text{ m}}{80 \text{ m}}\right)^2\right] = 31.6 \frac{\mu g}{\text{m}^3}$$

So it does appear that the maximum occurred at a distance of 1 km. This is true because, with an effective stack height of 85 m, at shorter distances the material has not had enough time to fully reach the ground, and most of the SO$_2$ is passing overhead. In addition, at distances greater than 1 km, the material continues to disperse in the vertical dimension and thus contaminate greater volumes of air but at lower concentrations.

Finally, for emissions that originate from ground level (for example, fires and fugitive emissions), the effective stack height H and elevation z may both be set equal to zero to yield an expression as follows:

$$C_{(x,y,0)} = \frac{Q}{\pi \, \sigma_y \, \sigma_z \, U} \exp\left[-\frac{1}{2}\left(\frac{y}{\sigma_y}\right)^2 \right]$$

emissions from ground level. ($H=0$)

to determine the ground level concentrations at a downwind location off the centerline, or as

$$C_{(x,0,0)} = \frac{Q}{\pi \, \sigma_y \, \sigma_z \, U}$$

to determine the maximum downwind concentration at ground level, which would occur on the centerline.

Example 9.3

Plume dispersion modeling III

A factory is on fire and is releasing smoke with a particle size less than 10 μm at a rate of 3 kg/s. What is a safe downwind distance if the NAAQS for PM_{10} is 150 μg/m³ for a 24-hour exposure? Assume a wind speed of 20 mph under neutral atmospheric conditions.

Solution

Noting that a factory fire would be considered a ground-level emission, the maximum downwind concentration can be estimated as

$$C_{(x,0,0)} = \frac{Q}{\pi \, \sigma_y \, \sigma_z \, U}$$

Plugging in the information given in the problem statement and converting units we get

$$C_{max} = 150\,\frac{\mu g}{m^3} = 0.00015\,\frac{g}{s} = \frac{\left(3000\,\frac{g}{s}\right)}{\pi \, \sigma_y \, \sigma_z \left(20\,\frac{mi}{hr}\right)\left(0.447\,\frac{m/s}{mi/hr}\right)}$$

$$\Rightarrow \quad \sigma_y \, \sigma_z = 712,10($$

Now we can use the vertical and horizontal dispersion plots provided in Figures 9.1 and 9.2 to estimate the required distance under neutral conditions (class D) to yield the value above for the product of σ_y and σ_z. Through trial and error we get

@ $X = 30$ km $\Rightarrow$ $\sigma_y = 1500$ m and $\sigma_z = 250$ m $\therefore$ $\sigma_y \sigma_z = 375,00$

@ $X = 50$ km $\Rightarrow$ $\sigma_y = 2200$ m and $\sigma_z = 310$ m $\therefore$ $\sigma_y \sigma_z = 682,00($

@ $X = 60$ km $\Rightarrow$ $\sigma_y = 2600$ m and $\sigma_z = 340$ m $\therefore$ $\sigma_y \sigma_z = 884,00$

Since 50 km is too small and 60 km is too large, we can interpolate to get an approximation of

$$\left(\frac{712,100 - 682,000}{884,000 - 682,000}\right)(60 - 50) + 50 = 51.5 \text{ km} \approx 32 \text{ miles}$$

INTRODUCTION TO ATMOSPHERIC POLLUTION CONTROL

Due primarily to the negative impact on human health, as well as to the deterioration of the flora and fauna in natural environments, much effort has been extended to control the airborne emissions generated from anthropogenic (man-made) sources. While several control technologies exist to address primary and secondary pollutants in air, the majority may be classified into a few main categories, as addressed in this chapter. The control of particulate matter is accomplished through the use of cyclones, electrostatic precipitators, and baghouses, as discussed in section 9.4. Certain chemical compounds may be removed through mass transfer operations such as absorption and adsorption (introduced in section 9.5 and covered in detail in section 11.4.2), or through the use of thermal controls (such as manipulation of combustion zone properties) and mass destruction through thermal oxidation processes (see section 9.6).

9.4   REMOVAL OF PARTICULATE MATTER

As one of the criteria pollutants, particulate matter has been the focus of much research regarding the continuous capture and processing of airborne solids. Although it is estimated that all anthropogenic sources combined account for less than 7% of primary particulate emissions per year, primarily due to the combustion of fossil fuels in industry, these sources are an identifiable point source amenable to environmental control. The three primary particulate control devices used in industry today are the cyclone, which is discussed in section 9.4.1, the electrostatic precipitator (ESP), which is covered in section 9.4.2, and the fabric filter (baghouse), as described in section 9.4.3.

9.4.1 Cyclones

A *cyclone* relies upon the inertial forces generated when a particle-laden gas is forced to experience spiral motion. The *centrifugal force* causes the higher density particulate matter to contact the outer wall of the cyclone and fall out of the gas flow to a collection tube at the bottom of the cyclone. Standard dimensions for cyclones are based on the diameter of the body, as shown in Figure 9.4 and Table 9.5. A cyclone can be quite effective for removing particles that possess diameters ≥ 10 μm (PM_{10}), with efficiencies dropping off substantially for $PM_{2.5}$ or smaller, limiting their use to coarse dusts or as a pretreatment unit.

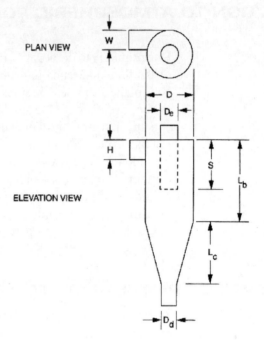

Figure 9.4 Plan and elevation view of cyclone dimensions

Source: *FE Supplied-Reference Handbook*, National Council of Examiners for Engineering and Surveying, 7th ed., 2005, *www.ncees.org*.

Table 9.5 Ratio of cyclone dimensions to body diameter (*D*)

Dimension		High Efficiency	Conventional	High Throughput
Inlet height	H	0.44	0.50	0.80
Inlet width	W	0.21	0.25	0.35
Body length	L_b	1.40	1.75	1.70
Cone length	L_c	2.50	2.00	2.00
Vortex finder length	S	0.50	0.60	0.85
Gas exit diameter	D_e	0.40	0.50	0.75
Dust outlet diameter	D_d	0.40	0.40	0.40

Ne   6.02

The estimation of particle capture efficiency is a function of the *approximate number of turns* (*N_e*) each particle will experience inside the cyclone and can be calculated as

$$N_e = \frac{1}{H}\left[L_b + \frac{L_c}{2}\right]$$

where H, L_b, and L_c are dimensions of the cyclone, as determined from the table of cyclone dimensions in Table 9.5, based on a known cyclone diameter. The *cut diameter* (*d_pc*) can be defined as the *particle diameter that is collected at 50% efficiency* and may be calculated as

$$d_{pc} = \left[\frac{9\mu W}{2\pi N_e v_i\left(\rho_p - \rho_g\right)}\right]^{1/2}$$

where μ is the gas viscosity, W is the inlet width, N_e is as defined above, v_i is the inlet gas velocity (calculated as volumetric flow rate divided by the inlet cross-sectional area), and ρ_p and ρ_g are the particle and gas densities, respectively. Often, the gas density is assumed negligible as compared to the particle density and is dropped from the calculation.

Once the cut diameter is known for a given cyclone configuration and flow rate, the *particle capture efficiency* (η) may be calculated as

$$\eta = \frac{1}{1 + \left(\dfrac{d_{pc}}{d_p}\right)^2}$$

where d_p is the diameter of the particle of interest. Since the particle capture efficiency is only a function of the ratio of d_{pc} and d_p, it is common to find the solution of the above expression solved for a range of ratios and expressed graphically, as shown in Figure 9.5. Cyclone design is generally based on sizing a cyclone to achieve a desired removal efficiency for a specified particle size.

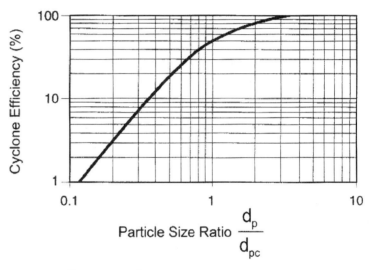

Figure 9.5 Collection efficiency for cyclones

Source: *FE Supplied-Reference Handbook*, National Council of Examiners for Engineering and Surveying, 7th ed., 2005, *www.ncees.org*.

Example **9.4**

High-efficiency cyclone

Determine the particle capture efficiency for PM_{10} of a high-efficiency cyclone that has a body diameter of 1 m if the gas flow rate is 2000 acfm. You may assume a specific gravity of 1.5 for the particles and a gas viscosity of 2.262×10^{-5} kg per m per s.

Solution

We need to start with the cyclone efficiency equation as follows:

$$\eta = \frac{1}{1 + \left(\dfrac{d_{pc}}{d_p}\right)^2}$$

Since d_p is given in the problem statement, we only need to calculate d_{pc}. This may be accomplished by using the equation for particle cut-diameter as

$$d_{pc} = \left[\frac{9 \mu W}{2 \pi N_e v_i (\rho_P - \rho_g)} \right]^{\frac{1}{2}}$$

We can simplify this equation if we assume that the particle density is much greater than the gas density as follows:

$$\text{assume } \rho_P \gg \rho_g \implies \rho_P - \rho_g \approx \rho_P$$

The particle density is found as

$$SG = 1.5 = \frac{\rho_p}{\rho_{H_2O}} \implies \rho_p = 1.5 \rho_{H_2O} = (1.5) \left(999 \frac{\text{kg}}{\text{m}^3} \right) = 1498.5 \frac{\text{kg}}{\text{m}^3}$$

The variable W is dependent on cyclone type and body diameter, and the inlet width to cyclone body ratio can be found in Table 9.5. W may be determined as

$$W = 0.21D = (0.21)(1 \text{ m}) = 0.21 \text{ m}$$

The variable N_e is dependent on H, L_b, and L_c, which are also dependent on cyclone type and body diameter. Again, using Table 9.5, N_e may be determined as follows:

$$N_e = \frac{1}{H} \left[L_b + \frac{L_c}{2} \right] = \frac{1}{0.44D} \left[1.4D + \frac{2.5D}{2} \right] = \frac{1}{0.44 \text{ m}} \left[1.4 \text{ m} + \frac{2.5 \text{ m}}{2} \right] = 6.02 \text{ turns}$$

Inlet gas velocity (v_i) is dependent on gas flow rate and inlet dimensions as follows:

$$v_i = \frac{Q}{A} = \frac{Q}{HW} = \frac{\left(2000 \frac{\text{ft}^3}{\text{min}} \right) \left(\frac{1 \text{ min}}{60 \text{ s}} \right) \left(\frac{1 \text{ m}^3}{35.3 \text{ ft}^3} \right)}{(0.44 \text{ m})(0.21 \text{ m})} = 10.22 \frac{\text{m}}{\text{s}}$$

Now, with a value for the viscosity of air given in the problem statement, we may substitute values for all variables into the expression for d_{pc} as follows:

$$\therefore d_{pc} = \left[\frac{(9)(2.262 \times 10^{-5})(0.21)}{(2)\pi(6.02)(10.22)(1498.5)} \right]^{1/2} = 8.59 \times 10^{-6} \text{m} = 8.6 \ \mu\text{m}$$

The units are left to the examinee to verify. Finally, the capture efficiency may be estimated as

$$\eta = \frac{1}{1 + \left(\frac{8.6}{10} \right)^2} = 0.575 = 57.5\%$$

As seen in Table 9.5, the primary difference between high-efficiency and high-throughput cyclones is the inlet and outlet dimensions. Inlet dimensions establish the gas velocity, and increased inlet velocity increases collection efficiency (that is, decreases particle cut-diameter). Unfortunately, as gas volumes increase, particle collection at high efficiency increases pressure drop in the cyclone, and thus increases the power costs required to maintain flow. Generally, this is addressed through the use of *multi-cyclones* (several cyclones operating in a parallel configuration), or by using other particulate capture devices in series to serve as polishing devices with lower mass loads after a high-throughput cyclone.

Example 9.5

High-efficiency vs. high-throughput cyclone efficiency

A high-throughput cyclone captures a 10-μm particle at an efficiency of 40%. What would the capture efficiency be for a high-efficiency cyclone assuming the inlet width and velocity remain unchanged?

Solution

From Figure 9.5, at an efficiency of 40% we see

$$\frac{d_p}{d_{pc}} = 0.8 \quad \Rightarrow \quad d_{pc} = \frac{d_p}{0.8} = \frac{10\,\mu m}{0.8} = 12.5\,\mu m$$

The particle cut-diameter may also be calculated using the following equation as

$$d_{pc} = \left[\frac{9\,W\mu}{2\pi\,N_e\,V_i\,(\rho_P - \rho_g)} \right]^{1/2}$$

Normally, values for W, N_e, and V_i are dependent upon the type of cyclone; however, the problem statement requires a constant value for W and V_i. Therefore, the difference in particle cut-diameter is only a function of N_e. Combining all other terms except N_e and representing them as a constant, we may write

$$d_{pc} = \left[\frac{C}{N_e} \right]^{1/2}$$

We can now determine the value of that constant term if we evaluate N_e for the high-throughput cyclone as follows:

$$N_e = \frac{1}{H}\left[L_b + \frac{L_C}{2} \right] = \frac{1}{0.8D}\left[1.7D + \frac{2D}{2} \right] = 3.4$$

Notice that the cyclone body diameter cancels out, so a value for diameter is not necessary. The result is that N_e is a constant that is only dependent on type of cyclone. With this value for N_e, we may now calculate the value for the lumped constant in the equation above as follows:

$$d_{pc} = \left[\frac{C}{N_e} \right]^{1/2} \quad \Rightarrow \quad C = d_{pc}^2\,N_e = (12.5)^2\,(3.4) = 531.25$$

To determine the particle cut-diameter for the high-efficiency cyclone, we will need to determine the value of N_e for the high-efficiency unit as follows:

$$N_e = \frac{1}{0.44D}\left[1.4D + \frac{2.5D}{2} \right] = 6.023$$

With the value for N_e and the constant determined previously, we may calculate the particle cut-diameter for the high-efficiency cyclone as follows:

$$d_{pc} = \left[\frac{531.25}{6.023} \right]^{1/2} = 9.4\,\mu m$$

From the particle cut-diameter, we can calculate the particle size ratio of d_p to d_{pc} as follows:

$$\frac{d_p}{d_{pc}} = \frac{10\,\mu m}{9.6\,\mu m} = 1.06$$

Now we can use this ratio with the cyclone collection efficiency plot as before to determine the efficiency of this unit. However, since the value of 1.06 is difficult to locate precisely on the plot, we can only estimate the efficiency as somewhere between 50% and 55%. An alternative method is to use the particle capture efficiency equation, noting that the ratio d_p to d_{pc} must be inverted as

$$\eta = \frac{1}{1+\left(\dfrac{d_{pc}}{d_p}\right)^2} = \frac{1}{1+\left(\dfrac{1}{1.06}\right)^2} = 0.529 = 52.9\%$$

9.4.2 Electrostatic Precipitation

The *electrostatic precipitator* (ESP) works by imparting a negative charge to particles entering the unit, which are then attracted to grounded collector plates. As particle mass is accumulated on the plates, they are tapped with hammers, causing the mass of solids to drop to a collection hopper below. Particle collection efficiency is dependent upon the velocity at which the particles travel toward the plates, called the *migration (drift) velocity (W)*, and can be estimated as

$$W = \frac{q E_p C}{6 \pi r \mu}$$

where q is the charge (C), E_p is the field intensity (V/m), r is the particle radius (m), μ is the gas viscosity (Pa · s), and C is the Cunningham correction factor, which may be calculated as

$$C = 1 + \frac{0.000621\,T}{d_p}$$

where T is the absolute temperature (K) and d_p is the particle diameter (μm). It should be noted that migration velocities are often provided for typical processes, with values of 0.015 to 0.018 m/s used for cement kilns and lime dust, and values of 0.08 to 0.17 m/s used for flyash from coal or solid waste incinerators.

Finally, the ESP collection efficiency (η) can be determined using the Deutsch-Anderson equation as follows:

$$\eta = 1 - \exp\left[\frac{-WA}{Q}\right]$$

where W is the migration velocity, A is the total plate collection area, and Q is the actual gas volumetric flow rate (at system temperature and pressure). Since many gas flows are given as *scfm (standard cubic feet per minute)*, values of *acfm (actual cubic feet per minute)* must be obtained through the use of the ideal gas law. Standard conditions assume a pressure of 1 atmosphere and a temperature of 25°C, or equivalently 77°F. Since the pressure in the ESP is not significantly different from atmospheric pressure, usually only the temperature needs to be accounted for in ESP problems.

Example 9.6

ESPs I

An existing ESP at an MSW energy recovery facility operates at 156°C and has plate dimensions of 4 m × 10 m. How many plates are required to obtain a 95% capture efficiency for flyash particles (average migration velocity of 13 cm/s) that are 1 μm in diameter if the flow rate is 20,000 scfm?

Solution

Start with the ESP efficiency equation as

$$\eta = 1 - \exp\left[\frac{-WA}{Q}\right]$$

Since η and W are given in the problem statement, and since we are solving for A, we are left with calculating the actual gas flow rate. Given gas flow rate in scfm (standard ft³/min) assumes standard temperature (25°C) and standard pressure (1 atm). Assuming the pressure in the ESP is near atmospheric pressure, the only factor to correct for is operating temperature. However, gas phase calculations require the use of absolute temperature. This is accomplished as follows:

$$Q_{actual} = Q_{scfm}\frac{T_{actual}}{25°C} = \left(20,000 \text{ scfm}\right)\frac{(273.15 + 156)}{(298.15 \ K)} = 28,789 \text{ acfm}$$

Next, we convert this into SI units as follows:

$$Q = 28,789 \ \frac{ft^3}{min} \times \frac{1 \ m^3}{35.3 \ ft^3} \times \frac{1 \ min}{60 \ s} = 13.6 \ \frac{m^3}{s}$$

Substituting all values into the efficiency equation, converting migration velocity to units of m/s and solving for A yields the required area as follows:

$$0.95 = 1 - \exp\left[\frac{-(0.13)(A)}{13.6}\right] \quad \Rightarrow \quad 0.05 = \exp\left[-0.00956 \ A\right]$$

$$-2.996 = -0.00956 \ A \quad \Rightarrow \quad A = 313.4 \ m^2$$

Calculating the available surface area for each plate, we get

$$A = (4 \ m)(10 \ m)(2 \text{ sides}) = 80 \ m^2$$

Therefore, the number of plates required is determined as

$$\frac{313.4 \ m^2}{80 \ m^2} = 3.92 \approx 4 \text{ plates}$$

Example 9.7

ESPs II

A small coal-fired power plant has an ESP unit that has four plates which are 4 m × 10 m each. If the efficiency must stay above 98%, what is the maximum actual gas flow rate allowed? You may assume an average migration velocity of 12 cm/s.

Solution

Starting with the ESP efficiency equation and substituting for the data given in the problem statement, converting cm/s to m/s, and recognizing that there are 4 total plates that are 4 × 10 meters with 2 sides each, we may write:

$$\eta = 1 - \exp\left[\frac{-WA}{Q}\right] = 0.98 \quad \Rightarrow \quad \ln(0.02) = \frac{-\left(0.12\,\frac{m}{s}\right)(4)(4\,m)(10\,m)(2)}{Q\,\frac{m^3}{s}} = \frac{-38.4}{Q}$$

Solving for volumetric flow rate we get:

$$Q = \frac{-38.4}{\ln(0.02)} = 9.8\,\frac{m^3}{s} \times \frac{35.3\,ft^3}{1\,m^3} \times \frac{60\,s}{1\,min} = 20{,}756\,\text{acfm}$$

9.4.3 Fabric Filters

Fabric filters are typified by the common filters used in most household air handling systems, where particle and gas flow rates are low. In industrial settings, the same principle is used on a large scale in a configuration often called a *baghouse*. A baghouse is a unit that holds several fabric tubes (bags), which are 8 to 12 inches in diameter and 10 to 20 feet long. The bags are open on one end, and flow is directed from inside the bag to the outside for particle collection in the bag (*shaker* or *reverse-flow* systems), or from outside the bag to inside for particle collection on the outside of the bag (*pulse-jet* systems).

Shaker systems have the advantage of using woven fabric, which generally has high tensile strength and therefore requires no additional support, and the flow from inside to outside keeps the bags inflated. Cleaning is dependent on mass loadings, which affect pressure drop and usually occur every 30 minutes to several hours. This is accomplished by isolating a compartment of bags by taking them offline and either shaking them or directing a flow of cleaned air in the reverse direction of flow, allowing the solids to drop into a hopper below. System design is based upon the *air-to-cloth ratio* to determine the total amount of fabric required for a specific application and are typically between 0.6 to 1.1 m³/min per m² of fabric for shaker units.

Pulse-jet systems use a felt fabric and require wire cages to maintain the cylindrical bag shape as flow goes from outside the bag to inside. Cleaning occurs quite frequently (every few minutes) through a pulse of high-pressure air inside the bag, which expands the bag slightly and causes accumulated dust on the outer surface to fall into the hopper below. Air-to-cloth ratios are typically larger for pulse-jet systems, usually in the range of 1.5 to 4.0 m³/min per m² of fabric.

Example **9.8**

Fabric filter design

A flue gas from a coal-fired power plant leaves the acid gas scrubbers at 85°C and at a flow rate of 50,000 scfm. Woven fabric filter bags are available from a manufacturer in units that contain 48 bags each, and each bag is 1 ft in diameter and 20 ft long. Determine the number of units required for the flow described if the air-to-cloth ratio is 0.8 m³ per minute per m² of fabric.

Solution

First, correct the volumetric flow rate to acfm based on gas temperature as in Example 9.6

$$Q_{actual} = Q_{scfm} \frac{T_{actual}}{25°C} = (50,000 \text{ scfm}) \frac{(273.15 + 85)}{(298.15)} = 60,062 \text{ acfm}$$

However, the air-to-cloth ratio given has units of m³ per minute per m² of fabric. Converting flow rate to m³/min, we get

$$Q = 60,062 \text{ acfm} \times \left(\frac{1 \text{ m}^3}{35.3 \text{ ft}^3} \right) = 1701 \frac{\text{m}^3}{\text{min}}$$

Therefore, the amount of fabric needed can be determined as

$$\frac{1701 \dfrac{\text{m}^3}{\text{min}}}{0.8 \dfrac{\text{m}^3}{\text{min} \cdot \text{m}^2}} = 2126 \text{ m}^2$$

Next, we need to determine the amount of fabric available in each unit as follows:

$$A_{unit} = \pi D l \times 48 = \pi (1 \text{ ft})(20 \text{ ft})(48) = 3016 \text{ ft}^2 \times \left(\frac{1 \text{ m}}{3.28 \text{ ft}} \right)^2 = 280 \text{ m}^2$$

Finally, the number of units may be estimated as

$$\frac{2126 \text{ m}^2}{280 \text{ m}^2} = 7.6 \text{ units} \quad \Rightarrow \quad 8 \text{ units}$$

9.5   ABSORPTION AND ADSORPTION

Two primary pollution control processes use preferential mass transfer for removing contaminants from gas streams: *absorption* and *adsorption*. These two processes are applied extensively throughout environmental engineering to a variety of process streams requiring remediation. This review text presents the fundamental equations and design applications in a single chapter with examples for several different scenarios (see section 11.4.2). Here, we will briefly overview the processes with respect to air pollution.

Absorption is a phenomenon through which contaminant mass is transferred from the gas phase to the liquid phase through a process known as *gas scrubbing*. The transfer of the pollution species occurs in a column where the scrubbing liquid is introduced at the top of the column in a spray pattern that contacts the gas, which is flowing in the upward (countercurrent) direction. Some columns may also be filled with a high-surface-area, high-void-volume packing material, often made of a plastic or ceramic compound, which is configured to provide uniform flow of both the liquid and gas phases. In either configuration, mass transfer occurs at the interface between the liquid and gas, and often a chemical is added to the liquid to further enhance the removal of the target species. One example is the spraying of hydrated lime (CaO) for SO_2 removal from flue gases that burn high-sulfur coal.

Adsorption is a process in which contaminant molecules (called the *adsorbate* or *solute*) that are present in the gas stream are removed through their attachment to a solid surface (called the *adsorbent*). The adsorbate preferentially attaches to the adsorbent due to chemical or physical attraction, and the attachment is generally stable as long as the gas properties (temperature, pressure, constituent compositions) remain constant. Activated carbon is a common adsorbent.

9.6 THERMAL CONTROLS AND DESTRUCTION

As discussed in section 9.1, oxides of nitrogen (NO_x) are produced in combustion processes, primarily due to the oxidation of N_2 that is part of the combustion air. Control of NO_x can be accomplished by preventing their formation or through removal technologies employed downstream of the combustion zone. Several prevention technologies focus on the fact that N_2 is oxidized by O_2 during combustion processes where the temperature exceeds 1600 K. Decreased flame temperatures may be accomplished through lowering the amount of excess air fed, lean combustion with sufficient air to keep flame temperatures down, combustion gas recirculation (lowers O_2 concentration), or injection of water or steam. Facilities may also employ low NO_x burners, staged combustion, or secondary combustion technologies, which are practices where some of the fuel is combusted at a point downstream of the initial (primary) combustion zone.

Additional thermal technologies are used for the oxidation of unwanted species, as typified by the combustion of municipal solid waste, and are often classified under the broad category of *incineration* processes. For air pollutants, incineration is especially useful for CO and organic compounds but may be employed for any species that do not form postcombustion by-products that are more hazardous than the original material (for example, chlorinated hydrocarbons may produce free chlorine or HCl gas). A general form of the combustion reaction may be represented as

$$C_aH_bO_c + \left(\frac{4a+b-2c}{4}\right)(O_2 + 3.76N_2) \rightarrow$$

$$aCO_2 + \left(\frac{b}{2}\right)H_2O + \left(\frac{4a+b-2c}{4}\right)(3.76N_2)$$

While the above equation may allow the calculation of the theoretical amount of air required for complete combustion, the rate of air actually fed to the combustion chamber is in excess of the stoichiometric amount. This excess amount can range from 5% to 100% (that is, multiply the stoichiometric coefficient for air by a value between 1.05 and 2.00), depending on the amount of temperature control required for reduction of other flue-gas constituents. Thermal oxidation is generally classified as either *direct-flame incineration*, where compounds are combusted with or without supplemental fuels, or *catalytic oxidation*, where a catalytic compound on a ceramic substrate allows oxidation to take place at much lower temperatures (300–800°F).

Two common means to express the ability to remove a particular contaminant species using thermal oxidation systems are the *destruction and removal efficiency* (DRE) and the *combustion efficiency* (CE). The DRE is a measure of the mass loss of the principal hazardous compound as a result of oxidation and can be expressed as

$$DRE = \frac{\dot{m}_{in} - \dot{m}_{out}}{\dot{m}_{in}} \times 100\%$$

where $\dot{m}_{in}$ and $\dot{m}_{out}$ are the mass flow rates of the principal compound into and out of the unit, respectively. Federal performance standards require a DRE of 99.99% on one or more selected Principal Organic Hazardous Constituents during a supervised trial burn.

The CE is a measure of the degree of completion of the oxidation and can be expressed as a ratio of the amount of carbon dioxide emitted as a percentage of the total oxidized carbon (that is, the sum of carbon dioxide and carbon monoxide). This may be represented as

$$CE = \frac{\left[CO_2\right]}{\left[CO_2\right]+\left[CO\right]} \times 100\%$$

where bracketed quantities represent the dry volumetric concentrations (ppmv) of the indicated species.

Example **9.9**

Thermal controls

The destruction and removal efficiency for a particular VOC in an incinerator is 97.5%. Determine the mass of VOC that enters the incinerator daily if the volumetric flow rate is 15 m³/s and the outlet concentration is 100 µg/m³.

Solution

Starting with the equation for DRE given and the value for DRE provided in the problem statement, we may write

$$DRE = \frac{\dot{m}_{in} - \dot{m}_{out}}{\dot{m}_{in}} = 0.975$$

Next, we can use the data in the problem statement to determine the mass flow rate out of the system as follows:

$$\dot{m}_{out} = Q\,C_{out} = \left(15\frac{m^3}{s}\right)\left(100\frac{\mu g}{m^3}\right) = 1500\frac{\mu g}{s}$$

Substituting this into the expression above, we can solve for mass flow rate into the system as

$$0.975 = \frac{\dot{m}_{in} - 1500}{\dot{m}_{in}} \quad \Rightarrow \quad 0.975\,\dot{m}_{in} = \dot{m}_{in} - 1500 \quad \Rightarrow \quad \dot{m}_{in} = 60{,}000\frac{\mu g}{s}$$

Now, convert this into a daily mass flow rate as

$$\dot{m}_{in} = \left(60{,}000\frac{\mu g}{s}\right)\left(\frac{kg}{10^9\,\mu g}\right)\left(\frac{(60)(60)(24)\,s}{day}\right) = 5.2\frac{kg}{day}$$

Solid Waste

Nearly every human activity generates some quantity of *solid waste,* which can be defined as *any unwanted and discarded materials that are solids or semisolids in their natural state.* Secure disposal is one of the least attractive options in the waste management hierarchy, just above direct release (see section 10.2 for a more detailed discussion of the waste management hierarchy). Often, other waste management options that are higher on the hierarchy may reduce the quantity of wastes. However, most of these options still generate a final waste stream (usually solid) that requires treatment and/or disposal.

10.1 WASTE QUANTITIES, CHARACTERIZATION, AND PROCESSING

There are two basic methods to determine total mass of municipal solid waste (MSW) and waste generation rates. *Direct measurement* techniques weigh the vehicles that carry wastes as they enter and exit the dumping station and are more accurate for the local community. However, extrapolation must be applied to very large, heterogeneous populations, which introduces the potential for substantial uncertainties. *Materials flow analysis* is based on mass balances performed at the generation source, prior to collection, and may offer good approximations for some sources but fail to give a good total picture. Based on several studies, it is estimated that the average person in the United States generates 3.5–4.0 pounds of residential waste per day, which works out to be around 1400 lb/yr. If the other sources are added, total MSW generation is over 6 pounds per person per day, or approximately 2200 lb/yr.

Since the residential component of MSW accounts for 60%–65% of the total, it is convenient to focus on this fraction when discussing composition. Typical components in residential MSW, with mass percentages in parentheses, include paper (34%), yard wastes (18.5%), food wastes (9%), glass (8%), plastics (7%), cardboard (6%), steel cans (6%), other metals (3%), dirt and ash (3%), textiles (2%), wood (2%), aluminum (0.5%), rubber (0.5%), and leather (0.5%). Chemical properties of MSW are generally based on *proximate analysis,* which includes moisture content, volatile (combustible) matter, fixed carbon, and ash, or based on *ultimate analysis,* which provides values for percentage of C, H, O, N, S, and ash on a dry-weight basis. For a typical MSW stream, ultimate analysis indicates that the waste is 47% C, 6% H, 40% O, 1% N, 0.2% S, and 6% ash. Finally, *energy content* is a function of the types of combustible materials present in the waste stream.

There are several physical properties of MSW that are of interest to the environmental engineer, including the MSW *specific weight (γ),* which is often specified in units of lb/yd^3 and requires descriptive qualifiers such as "loose," "as placed," "uncompacted," or "compacted." Typical values for MSW under various scenarios are 200–300 lb/yd^3 loose, 500 lb/yd^3 in a compaction truck, 700 lb/yd^3 normally compacted in a landfill, and up to 1000 lb/yd^3 well compacted in a landfill. Often, it is desirous to know the *particle size distribution,* especially for material recovery facilities that process waste streams for the purpose of resource recovery. The amount of moisture is generally described by the *moisture content (MC),* and while food and yard wastes can have MC of 60%–70%, the majority of the MSW mass has MC < 10%, and the overall MC averages 20%–25% for typical MSW. A related property is called *field capacity* and is a measure of the maximum amount of water that can be retained by the waste as placed in a landfill. While field capacity may be as high as 60% of the dry weight of the solids, this quantity is related to the overburden (waste placed above) pressure, which reduces capacity. Water present in excess of the field capacity generally contributes to the discharge of landfill leachate.

Once generated, collection and transfer of the waste is generally accomplished through municipal or private *curb-side collection* from residences or *container pick-up* from commercial establishments. The most common vehicle used for collection is the *compaction truck,* which allows for maximum time on route, minimizing lost time driving to and from the transfer station or landfill. *Inter-route transfer stations* can reduce the drive time by smaller collection trucks by consolidating MSW into larger carriers for transport to a distant processing facility or the landfill. Much planning goes into the collection and inter-route transfer of MSW, as 70%–85% of the total cost of MSW management is incurred in the collection phase. A *material recovery facility* (MRF) is used to segregate materials into fractions that are able to be recycled, composted, and/or combusted. Recycling efforts have been able to divert up to 20% of the U.S. waste stream, mostly in the paper and cardboard fractions. While much has been done in the aluminum, steel, glass, and plastic recovery sectors, the fact that they compose a small fraction of the total MSW stream means their removal does little to reduce the overall mass that requires final disposal.

10.2 WASTE MANAGEMENT HIERARCHY

As a concept, the waste management hierarchy is simply a prioritized list of possible management choices for any natural or industrial system that must contend with waste materials or energy. In its most basic form, the list would appear as follows: (1) *source reduction*, (2) *recycling*, (3) *waste treatment*, (4) *secure disposal*, and (5) *direct release* to the environment.

While definitions vary slightly by various federal or state agency, in general, *source reduction* can be defined as *any practice that reduces the amount of any hazardous substance entering the waste stream prior to recycling, treatment, or disposal*. The more recent version of this philosophy would be termed *pollution prevention*, as the basic tenets are the same between the two and will be explored in more detail in sections 10.3 and 11.2. Sometimes, the term *waste minimization* is also used synonymously; however, waste minimization will generally also include some of the recycling options discussed below. Regardless of the term used, source reduction is the idea that we do not have to worry about "what to do with the waste" if it is not generated in the first place.

Once a by-product stream is created, there are choices available on how to manage that material. *Recycling* is a common idea that has many applications, manifested through several subclassifications. Sometimes, the term *reuse* is incorporated as a separate entry in the waste management hierarchy before the recycling options; at other times, it is considered the highest level of recycling. In either case, reuse can be described as *using a discarded item for the original purpose* (for example, used cars or furniture). Often, materials can be *recirculated* within the same process, which is generally called *in-process* recycling. By definition, the U.S. Environmental Protection Agency (U.S. EPA) considers pollution prevention to include source reduction and in-process recycling. When a company chooses to use the waste stream from one process in another process within the plant boundaries, it is called *on-site* recycling. If that material is shipped to another facility to be incorporated into another process, it is called *off-site* recycling. Further, the discussion on life cycle assessments (section 10.4) will offer additional possible definitions for recycling. It is interesting to note that many corporations have become engaged in the system of waste trading, where employees identify a waste product stream from another participating corporation that meets their requirements as a process feed stream.

The United States recycles approximately 28% of its waste and has nearly doubled its recycling rate in the past 15 years. Industries are increasingly viewing many "wastes" as potential raw materials. The cost-effectiveness of recycling varies widely, depending on the efficiency of the collection system and the market for the recyclables. The cost obtained for recyclables depends on the supply available to users of recyclables, the type of material available (aluminum, steel, corrugated cardboard, etc.), the geographic locations of seller and buyer, the quality of the materials to be recycled, and so on. Moreover, the market prices vary drastically for some materials within a time frame of years or even months. In many cases, recycling of materials such as aluminum and steel does not require subsidies from the government. Other materials such as mixed glass and low-quality mixed paper are not in great demand and are typically subsidized. Subsidies are in place to take advantage of the many benefits of recycling for those materials for which immediate cost benefits are not seen (for example, the savings in landfill volume).

If it is determined that the generated waste material cannot be used in an economic or appropriate way, several waste treatment technologies are available to transform the material into a less hazardous, or low-volume, high-concentration state. Often, the treatment process still produces a small amount of waste material that requires *secure disposal,* which is generally assumed to be final disposal in a secure landfill. A secure landfill is one that was designed to comply with RCRA regulations for control of the material and will be discussed in section 10.5. It is unfortunately the case that *direct release* to the environment occurs with a high degree of regularity, both from the permitted (allowed) discharge standpoint and from a fugitive or even illegal discharge perspective. In the case of the permitted discharge, it is usually monitored to determine mass loadings to the environment and generally is followed up with an evaluation of the impact that the discharge has on flora and fauna sustained by the surrounding ecosystem.

10.3 POLLUTION PREVENTION AND WASTE MINIMIZATION

The fact that we utilize natural resources to maintain our existence and that natural processes cannot be 100% efficient means that waste streams will always be a part of human activity. In the United States alone, billions of tons of industrial waste are generated each year, and while technology exists to treat much of these streams and reduce the impact, the negative effects on the environment from insufficient treatment and questionable disposal practices are obvious. As defined previously, pollution prevention is any practice that reduces the amount of any hazardous substance entering the waste stream prior to recycling, treatment, or disposal, and waste minimization is pollution prevention with the inclusion of some recycling options. Pollution prevention would encompass the more efficient use of raw materials and energy, not only reducing material and energy waste but also in the process reducing costs and increasing yields. A more recent concept is the idea of *green engineering,* where the practice of pollution prevention becomes part of the design process from the very beginning of a project and is woven throughout in the evaluation of long-term cost and environmental benefit.

For existing manufacturing processes, pollution prevention begins with a *waste audit* (materials accounting in process lines) and *emission inventory* (direct releases through fugitive and secondary sources), where all material and energy flows are identified and evaluated for potential modification for each process unit in the facility. The engineer must also evaluate the interconnectedness of each separate process train in the facility, because a change in one shared material or energy flow could potentially impact several other flows. Special attention is placed on criteria pollutants, toxic chemicals, and other hazardous wastes, and a priority list is generated that provides realistic targets for the amounts of source reduction for those target compounds.

Identification of the sources is generally a simple task, and the real challenge comes in the cost justification to management. Often small changes are quite cost effective, such as valve or seal replacement for reduction of fugitive releases, and can be presented as appropriate maintenance practices. Substantial changes may require a more extensive *total cost accounting* to provide sufficient economic justification. Total cost accounting not only addresses the typical capital, operation and maintenance, and labor costs but also assesses less obvious costs, such as monitoring, permitting, and current or future liability, as well as relationships with employees, customers, and public perception of the corporation as an environmental steward.

Additional discussion on pollution prevention is provided in section 11.2 of this text as it applies to hazardous waste.

10.4 LIFE CYCLE ASSESSMENT

In the decision-making process, historically, corporations were nearsighted with respect to their cost-benefit analyses and rarely investigated the impact of production choices over the long term. However, more recently, it has become common practice to evaluate choices from the *cradle-to-grave* perspective. The formal description of this mindset is called *life cycle assessment (LCA)* and can be broken down into three components: (1) an inventory of all material, energy, waste, and emission flows associated with the life cycle of a product; (2) an environmental impact assessment of those flows; and (3) a feedback mechanism to assist in choosing between alternative products or processes or to address identified negative environmental impacts.

A conceptual framework for performing an LCA is provided in Figure 10.1. As seen in the graphical representation, the LCA is comprised of flow inventories during five stages of the product's life, namely, (1) *raw material acquisition* (extraction from nature), (2) *material manufacture* (raw material turned into feedstock), (3) *product manufacture*, (4) *product use*, and (5) *product disposal*. By summing all flows over the five stages, a direct comparison can be made between products or manufacturing processes to determine the one that minimizes the impact to the environment. Classical comparisons that can be made in this way include paper or plastic bags at the supermarket, wax-coated paper or polystyrene cups for take-out hot drinks, or cloth versus disposable diapers, and so on. Sometimes, the choice is clear, but more often it becomes a choice between high flows in one area such as waste compared to high flows in another such as energy, and it is left to the manager to determine the most appropriate choice.

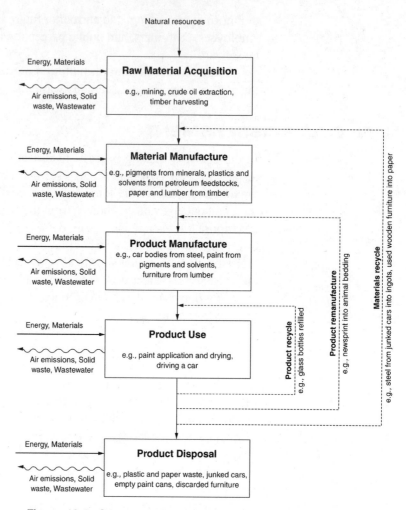

Figure 10.1 Conceptual framework for life cycle assessment (LCA)

Source: Allen and Rosselot, *Pollution Prevention for Chemical Processes*, © 2001, John Wiley & Sons, Inc. Reprinted by permission.

As seen in the figure, recycling has been divided into three categories. The first is termed product recycle and would be equivalent to the reuse term defined previously. The second process is termed product remanufacture and can be defined as *conversion of a product for reuse in a different manner*. The third term is called material recycle and is most similar to the concept of recycling most people hold.

10.5 LANDFILLS

Figure 10.2 shows a typical schematic of a MSW landfill. Such a landfill has been described as a "dry tomb," in that waste is placed in the landfill with every effort to keep water from entering the landfill and to quickly capture and remove any leachate from the landfill. Figure 10.2 depicts the major components of a landfill: the liner, the cap system, the leachate collection system, and the gas collection and recovery system. Each of these systems will be described in this section.

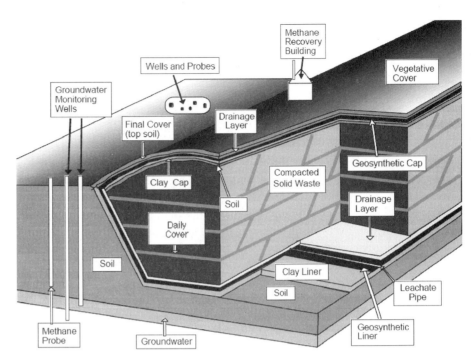

Figure 10.2 Schematic of a typical solid waste landfill

Source: P. Walsh and P. O'Leary, University of Wisconsin–Madison Solid and Hazardous Waste Education Center, reprinted from *Waste Age* Correspondence Course, 1991–1992.

10.5.1 Siting

The following restrictions apply when siting a landfill:

■ Airports: Landfills are not allowed within 10,000 feet of airports serving turbojets or within 5000 feet of airports serving piston-type aircraft, unless the owner can demonstrate that the design and operation of the landfill will not cause any bird hazards to the aircraft.

■ Floodplains: Landfills located on a 100-year flood plain must not restrict the flow of a 100-year flood, reduce the storage capacity of the flood plain, or result in the washout of solid waste.

■ Wetlands: The landfill must not cause or contribute to significant degradation of wetlands, taking into account impacts on fish and other wildlife. The potential effects of catastrophic failure should also be considered.

■ Fault areas: Landfills are not to be located within 200 feet (60 meters) of a fault that has had displacement in Holocene time unless it can be demonstrated that distances less than 200 feet will not cause damage to the landfill's structural integrity.

■ Seismic zones: Landfills are not allowed in seismic impact zones, unless the liners, leachate collection systems, and surface water control systems are designed to resist the maximum horizontal acceleration expected due to a seismic event.

■ Unstable areas: Landfills are not to be located in unstable areas, defined as locations that are susceptible to natural or human-induced events or forces capable of impairing the integrity of some or all of the landfill structural components responsible for preventing releases from a landfill. Unstable areas can include poor foundation conditions, areas susceptible to mass movements, and Karst terrains.[1]

10.5.2 Landfill Processes

Stabilization of wastes in a landfill undergoes five distinct phases. These phases differ in terms of the type of physical, chemical, and biological processes taking place within the landfill. Consequently, the rate and quality of the leachate and landfill gas generated also varies with time. These five phases are described following, and the variation in leachate and landfill gas quantity and quality as a function of the phase are illustrated in Figure 10.3.

1. *Phase I* is the initial adjustment phase, or *lag phase.* It is during this phase that moisture begins to accumulate and the oxygen entrained in freshly deposited solid waste begins to be consumed.

2. *Phase II* is the transition phase in which the moisture content of the waste has increased such that the field capacity is exceeded. Microbial processes change from an aerobic to anaerobic environment as oxygen is depleted. Detectable levels of volatile organic acids (VOAs) and an increase in the chemical oxygen demand (COD) are noted in the leachate.

3. *Phase III* is the acid forming stage. Acidogenic bacteria convert the VOAs from Phase II, resulting in lower pH. The lower pH solubilizes metal species from the waste into the leachate. This phase is also characterized by peak COD levels in leachate.

4. *Phase IV* is the methane fermentation phase in which methanogenic bacteria convert acid compounds produced in earlier phases to methane and carbon dioxide gas. This phase marks a return to more neutral pH conditions and a corresponding reduction in the solubilization of metals in the leachate.

5. *Phase V* is the maturation phase, and biodegradable matter and nutrients become limiting. Landfill gas production drops, and the leachate strength is much lower and less variable than in previous stages. Degradation of organics continues at a much slower rate.

1   U.S. EPA, *MSW Landfill Criteria Technical Manual*, 1998, accessed from *www.epa.gov/epaoswer/non-hw/muncpl/landfill/techman/*.

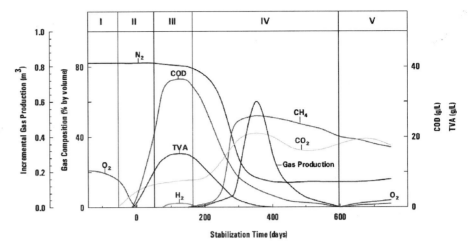

Figure 10.3 Progression of waste stabilization in landfills

Source: F. G. Pohland and S. R. Harper, *Critical Review and Summary of Leachate in Gas Production from Landfills*, EPA/600/2-86/073 (U.S. EPA, 1986).

10.5.3 Landfill Design

The following aspects of landfill design will be discussed in this section: liners, leachate collection, gas collection, and the landfill cap.

Liners

A number of geosynthetic materials are used in many aspects of landfill design, most important perhaps, in the liner system. The types of geosynthetics available include:

- Geonets, which provide a drainage conduit

- Geogrids, which provide strength to reinforce potentially unstable slopes

- Geomembranes, for isolation

- Geotextiles, which provide a means of filtration, drainage, and reinforcement

Geomembranes, or flexible membrane liners (FML), are required, in conjunction with clay, as a liner material for landfills. They have a very low permeability, but leaks are possible through seams, pinholes, and defects arising from the manufacturing process or due to accidental penetrations occurring during construction. Permeability rates can be estimated using Darcy's law (section 5.4) or by applying the Bernoulli equation to a hole and treating the hole as an orifice. This equation is also known as Torricelli's equation:

$$Q = C_d \cdot A \cdot (2 \cdot g \cdot h)^{0.5} \qquad \textbf{(10.1)}$$

where

C_d = discharge coefficient (typically 0.6)

A = area of hole, that is, of a defect in the membrane (L^2)

g = acceleration due to gravity ($L \cdot T^{-2}$)

h = head of water over hole (L)

Using this simple relationship, estimated flows through a composite liner as a function of hole size and hole density (holes/acre) can be calculated, and values are shown in Table 10.1. The number of holes per acre varies between 1 and 30, corresponding to high quality control and low quality control, respectively.

Table 10.1 Calculated leakage rate through a geomembrane

Size of Hole (cm²)	Number of Holes (hole/acre)	Flow Rate (gal/acre/day)
0.1	1	330
0.1	30	10,000
1	1	3300
1	30	100,000
10	1	33,000

In reality, the flow of leachate through a landfill liner will be more complicated than the flow through an orifice. The leakage through a composite system has been found to vary with the quality of contact between the geomembrane and the clay. Good contact conditions correspond to few waves or wrinkles in the geomembrane on a clay layer that is well compacted and has a smooth surface. The flow through the composite liner can be estimated from equations 10.2 and 10.3 (Quian et al., 2002):

geomembrane & clay layer contact conditions

$$\text{"Good" contact conditions: } Q = 0.21\, A^{0.1} \cdot h^{0.9} \cdot k_s^{0.74} \qquad \textbf{(10.2)}$$

$$\text{"Poor" contact conditions: } Q = 1.15\, A^{0.1} \cdot h^{0.9} \cdot k_s^{0.74} \qquad \textbf{(10.3)}$$

where

Q = leakage rate through membrane (m³/s)

A = area of hole in geomembrane (m²)

h = liquid head on top of the geomembrane (m)

k_s = hydraulic conductivity of the low-permeability soil component of the composite liner (m/s)

Typical values of k_s are 10^{-7} to 10^{-8} cm/s.

Example 10.1

Estimating landfill liner leakage rates

A 20-acre landfill liner is placed with very good quality control such that the FML and low permeability soil (10^{-8} cm/s) are in very good contact, and there are two 0.1-inch diameter holes per acre. Compare the leakage rate using Torricelli's equation to the leakage predicted by the equations above for composite liners.

Solution

From Torricelli's equation (Equation 10.1) and assuming a leachate head of 6 inches:

$$Q_{hole} = C_d \cdot A \cdot (2 \cdot g \cdot h)^{0.5}$$

$$Q_{hole} = 0.6 \frac{\pi(0.1\ in\ /\ 12)^2}{4} \left[2 \cdot 32.2 \frac{ft}{s^2} \cdot 0.5\ ft \right]^{0.5} = 1.86 \cdot 10^{-4}\ cfs = 120\ gal/day$$

$$Q_{total} = 120\ gal/day/hole \cdot 2\ hole/acre \cdot 20\ acre$$

$$= 4800\ gal/day$$

Considering the composite liner with good contact conditions, we will use Equation 10.2.

$$Q_{hole} = 0.21\ A^{0.1} \cdot h^{0.9} \cdot k_s^{0.74}$$

This empirical relationship must use the units specified in the equation definition. Thus,

$$A = \frac{\pi(0.1\ in\ /\ 12)^2}{4} = 5.5 \cdot 10^{-5}\ ft^2 = 5.1 \cdot 10^{-6}\ m^2$$

$$h = 6\ in = 0.152\ m$$

$$k_s = 10^{-8}\ cm/s = 10^{-10}\ m/s$$

Inserting these values into the above equation yields:

$$Q_{hole} = 4.5 \cdot 10^{-10}\ m^3/s$$

$$= 0.01\ gal/day$$

$$Q_{total} = 0.01\ gal/day/hole \cdot 2\ hole/acre \cdot 20\ acre$$

$$= 0.4\ gal/day$$

— flexible membrane liner

An alternative to using an FML is to use a GCL (*geosynthetic clay liner*). GCLs can be used in the liner and in the cap and are an effective replacement for the compacted clay. A GCL (Figure 10.4) is typically configured as a layer of bentonite fixed between two sheets of geotextile (it may also be configured as bentonite glued to a FML). Hydraulic conductivity on the order of 10^{-9} cm/s is possible.

GCLs offer advantages as compared to conventional bottom liners. Specifically, GCLs

- are fast and easy to install,

- have low hydraulic conductivity,

- have the ability to self-repair any rips or holes,

- are cost effective in regions where clay is not readily available, and

- consume less landfill volume, thus providing more space for MSW.

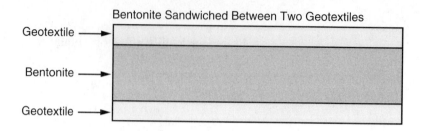

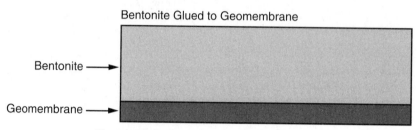

Figure 10.4 Typical configurations of GCL

Source: U.S. Environmental Protection Agency, *Geosynthetic Clay Liners Used in Municipal Solid Waste Landfills*, EPA/530-F-97-002, 2001.

Leachate Collection

The EPA defines leachate as the liquid that has passed through or emerged from solid waste and contains dissolved, suspended, or immiscible materials removed from the solid waste. The leachate generated in a landfill must be collected to minimize leakage from the landfill. Recall that the rate of leakage is directly related to the head of leachate over the landfill liner, and an effective leachate collection system minimizes this head. The goal of the system is to keep the leachate head less than 12 inches (30 cm).

A schematic of a typical leachate collection system is provided in Figure 10.5.

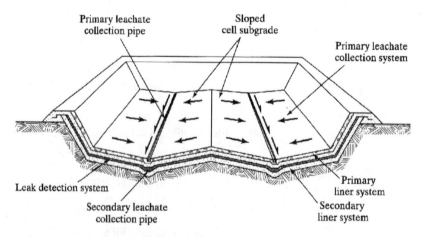

Figure 10.5 Leachate collection system

Source: X. Qian, R. M. Koerner, and D. H. Gray, *Geotechnical Aspects of Landfill Design and Construction*, © 2001. Reprinted by permission of Pearson Education, Inc., Upper Saddle River, N.J.

A leachate collection system consists of the following layers:

■ A composite liner, as described above

■ A high permeability drainage layer (hydraulic conductivity of 10^{-2} cm/s or higher) placed directly on the FML or on a protective bedding layer overlying the FML. A geosynthetic drainage net (also called a *geonet*) may be used in place of a layer of granular material.

■ Perforated leachate collection systems within the high permeability layer to collect and convey leachate quickly to the collection area. Pipes must be of adequate structural strength to withstand the pressure of the overlying waste.

■ A protective filter layer over the high permeability layer to prevent clogging. This may be a soil layer or a geotextile.

■ Leachate collection sump or header system for removal of leachate

A cross section of the area surrounding a typical leachate collection pipe is provided in Figure 10.6. An alternative is shown in Figure 10.7, which offers a leak detection option. For this latter option, if a leak occurs in the upper drainage system, the leakage will travel down to the geonet layer, which will convey the leaked leachate to a central collection location.

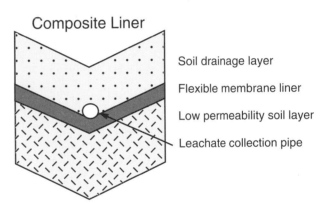

Figure 10.6 Leachate collection cross section

Source: Environmental Protection Agency, *Decision-Maker's Guide to Solid Waste Management*, 2d ed., EPA/530-R-95-023, 1995, p. 9-38.

When designing a leachate collection system, the liner should be sloped at a minimum slope of 2%, and collection pipes should be sloped at a minimum slope of 1%. Increasing the slopes prevents accumulation in localized depression area but leads to more excavation and possibly loss of volume for placement of waste.

When leachate flows across a liner, it forms a *leachate mound*, as shown in Figure 10.8. The goal of the leachate collection system is to keep h_{max} below 12 inches (30 cm). Equation 10.4 estimates the pipe spacing based on the value of h_{max}. Equation 10.4 is most commonly used to find the pipe spacing L for a given leachate collection system configuration.

$$h_{max} = \frac{L}{2}\left(\frac{q}{K}\right)\left[\frac{K\tan^2\alpha}{q} + 1 - \frac{K\tan\alpha}{q}\left(\tan^2\alpha + \frac{q}{K}\right)^{1/2}\right] \qquad \textbf{(10.4)}$$

where

L = distance between collection pipes (L)

q = vertical inflow per unit horizontal area (L/T)

K = hydraulic conductivity of drainage layer (L/T)

α = slope of liner (L/L)

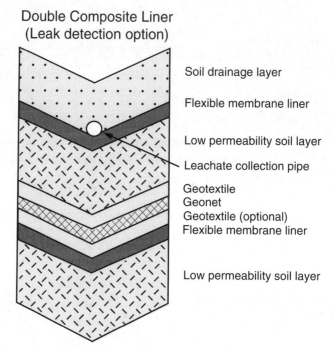

Figure 10.7 Leachate collection cross section with leak detection option

Source: Environmental Protection Agency, *Decision-Maker's Guide to Solid Waste Management*, 2nd ed., EPA/530-R-95-023, 1995, p. 9-38.

The flow into the drainage layer can be estimated using a model such as the HELP (Hydrologic Evaluation of Landfill Performance) model and can also be approximated using the water budget technique. (See section 5.1.1.)

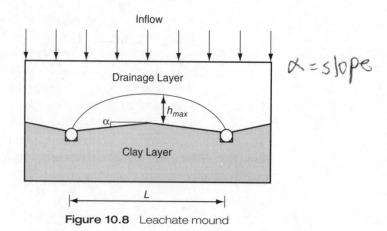

Figure 10.8 Leachate mound

Source: U.S. Environmental Protection Agency, *MSW Landfill Criteria Technical Manual*, 1993.

Example **10.2**

Analysis of a leachate collection system

The preliminary layout for a leachate collection system is provided in Exhibit 1. (Exhibit 1 is simply a schematic, as the number of collection pipes are not known until the spacing is determined.)

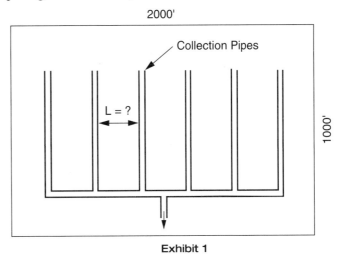

Exhibit 1

Compare the length and size of laterals required for a liner cross slope of 2% slope vs. a cross slope of 4%, given the following information:

Drainage layer hydraulic conductivity = 10^{-2} cm/s $= K$

Worst-case percolation rate = 3 in/day $= q$

Pipe slope = 1%

Solution

The pipe spacing, length, and diameter will first be estimated for a 2% cross slope of the liner using Equation 10.4.

$$h_{max} = \frac{L}{2}\left(\frac{q}{K}\right)\left[\frac{K\tan^2\alpha}{q}+1-\frac{K\tan\alpha}{q}\left(\tan^2\alpha+\frac{q}{K}\right)^{1/2}\right]$$

where

$L = ?$

h_{max} = 30 cm ~ 1 ft.

q = 3 in/day = $8.8 \cdot 10^{-5}$ cm/s

K = 10^{-2} cm/s

α = 0.02

$\frac{2000\ ft}{270\ ft} \sim 8$ pipes

These values yield a pipe spacing of 82 m, or 270 ft. Thus, a total of eight collection pipes will be required, separated by 250 ft spacing. The pipes on the ends will be placed 125 ft from the outer face of the landfill. A total pipe length of approximately 8000 ft will be required.

The flow in each pipe can be estimated based on the contributing land area. For each collection pipe, the contributing area is 1000 ft × 250 ft, or $2.5 \cdot 10^5$ ft². The flow is the product of the area and the percolation rate:

$$Q_{max} = 3 \text{ in/day} \cdot (2.5 \cdot 10^5 \text{ ft}^2) \cdot (1 \text{ ft/12 in}) \cdot (1 \text{ day/86,400 s})$$

$$= 0.7 \text{ cfs}$$

From Manning's nomograph (Figure 3.6), the diameter corresponding to a 1% pipe slope and conveying 0.7 cfs is between 6 in and 8 in, so a diameter of 8 in will be selected.

The same process applies when analyzing the impact of a 4% cross slope on the liner. The data is summarized for comparison's sake in the following table:

	2% cross slope	**4% cross slope**
Spacing (ft)	270 (use 250)	311 (use 300)
Number of collection pipes	8	7
Contributing area (ft^2)	$2.5 \cdot 10^5$	$3 \cdot 10^5$
Length of pipe (ft)	8000	7000
Q_{max} (cfs)	0.7	0.87
Diameter (in)	8	8

Gas Collection

The composition of landfill gas varies with time but can be approximated by values provided in Table 10.2.

Table 10.2   Typical landfill gas composition

Component	Percent
Methane	47.4
Carbon dioxide	47.0
Nitrogen	3.7
Oxygen	0.8
Paraffin hydrocarbons	0.1
Aromatic-cyclic hydrocarbons	0.2
Hydrogen	0.1
Hydrogen sulfide	0.01
Carbon monoxide	0.1
Trace compounds	0.5

Source: R. Ham, *Recovery Processing and Utilization of Gas from Sanitary Landfills* (U.S. EPA, 1979).

The EPA LandGEM model (Equation 10.5) may be used to estimate the volume of landfill gas generated by landfilled MSW. The equation is:

$$Q_T = \sum_{i=1}^{n} 2kL_o M_i e^{-kt_i} \qquad \textbf{(10.5)}$$

where

Q_T = total gas emission rate from landfill (L^3/T)

n = total time periods of waste placement

k = landfill gas emission constant (T^{-1})

L_o = methane generation potential per mass of waste (L^3/M)

t_i = age of the ith section of waste (T)

M_i = mass of waste placed at time i (M)

Use of this equation is best illustrated with Example 10.3.

Example **10.3**

Estimating landfill gas generation rates

A landfill accepts 100,000 tons of waste per year. Show how the landfill gas generation varies with time for this landfill, assuming that it accepts waste for five years. Use a landfill gas emission constant of 0.03 yr^{-1} and an L_o of $5 \cdot 10^3$ ft^3/ton.

Solution

Assuming that the entire waste is placed on the first day of the year, the landfill gas generated in year 1 from the first year's waste is

$$Q_{1,1} = 2(0.03 \text{ yr}^{-1})(5 \cdot 10^3 \text{ ft}^3/\text{ton}) \cdot (100,000 \text{ ton/yr})e^{-0.03 \text{yr}^{-1} \cdot 1 \text{yr}} = 2.9 \cdot 10^7 \text{ ft}^3/\text{yr}$$

The gas generated the second year by this waste placed in the first year is given by

$$Q_{1,2} = 2(0.03 \text{ yr}^{-1})(5 \cdot 10^3 \text{ ft}^3/\text{ton}) \cdot (100,000 \text{ ton/yr})e^{-0.03 \text{yr}^{-1} \cdot 2 \text{yr}} = 2.8 \cdot 10^7 \text{ ft}^3/\text{yr}$$

This gas generated from the waste placed during the first year will continue to decrease with time, as shown in Exhibit 2.

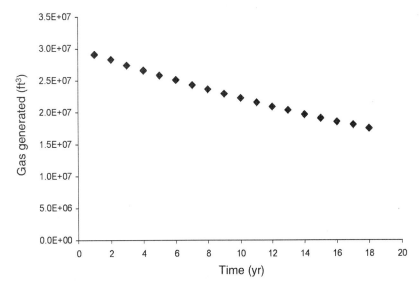

Exhibit 2   Gas generation from waste placed during year 1

However, during the second year, additional gas is also generated by the waste placed during the second year of operation. The gas produced in the second year by the waste placed during the second year ($Q_{2,2}$) is

$$Q_{2,2} = 2(0.03 \text{ yr}^{-1})(5 \cdot 10^3 \text{ ft}^3/\text{ton}) \cdot (100,000 \text{ ton/yr})e^{-0.03\text{yr}^{-1} \cdot 1\text{yr}} = 2.9 \cdot 10^7 \text{ ft}^3/\text{yr}$$

Thus, the total gas produced in the second year is

$$= 2.8 \cdot 10^7 \text{ ft}^3/\text{yr} + 2.9 \cdot 10^7 \text{ ft}^3/\text{yr} = 5.7 \cdot 10^7 \text{ ft}^3/\text{yr}$$

In the third year, gas will be generated due to the waste placed in years 1, 2, and 3, and the process will continue. Summing up the generation curve for each year results in the graph shown in Exhibit 3.

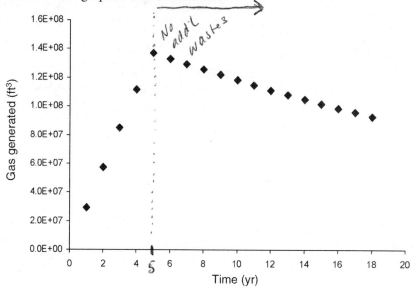

Exhibit 3   Cumulative gas generation for waste placed in years 1–3

A more generic view of generation rates and collection efficiency is shown in Figure 10.9.

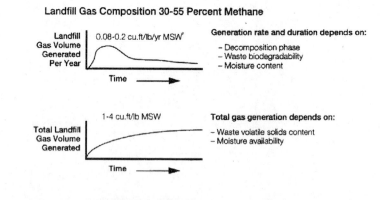

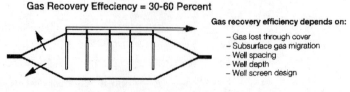

Figure 10.9   Landfill gas generation and recovery

Source: Source: P. Walsh and P. O'Leary, University of Wisconsin–Madison Solid and Hazardous Waste Education Center, reprinted from *Waste Age* Correspondence Course, 1991–1992.

Example **10.4**

Estimating gas production rates

Compare the gas production per pound of MSW, given an MSW composition of $C_{20}H_{30}O_{10}N$, to production values provided in Figure 10.9.

Solution

The governing equation is

$$C_{20}H_{30}O_{10}N + 33/4H_2O \rightarrow 87/8 \, CH_4 + 73/8 \, CO_2 + NH_3$$

The molecular weight of the MSW is 444 g/mol [(20 · 12) + (30 · 1) + (10 · 16) + (1 · 14)]. Consequently, 1 lb of MSW is equivalent to approximately 1 mole:

$$1 \, lb \frac{1 \, kg}{2.205 \, lb} \frac{1000 \, g}{1 \, kg} \frac{1 \, mol}{444 \, g} = 1.02 \, mole$$

Thus, every pound (or mole) of MSW produces 10.9 moles of methane and 9.1 moles of carbon dioxide. Assuming that methane and carbon dioxide act as ideal gases (22.4 L/mol), the volume of each gas can be determined:

$$10.9 \, mol \, CH_4 \frac{22.4 \, L}{mol} \frac{1 \, ft^3}{28.3 \, L} = 8.6 \, ft^3$$

$$9.1 \, mol \, CO_2 \frac{22.4 \, L}{mol} \frac{1 \, ft^3}{28.3 \, L} = 7.2 \, ft^3$$

Therefore, the total amount of gas generated per pound of MSW is 15.8 ft³. This is the maximum amount possible theoretically and is 4–16 times larger than the amount produced in reality, according to Figure 10.9.

Cap

The following design criteria apply to the final cover system (U.S. EPA, 1995):

- Minimize infiltration of precipitation into the waste

- Promote good surface drainage

- Resist erosion

- Prevent slope failure

- Restrict off-site landfill gas migration

- Enhance recovery of landfill gas

- Prevent penetration by vectors (for example, rodents)

- Improve aesthetics

- Minimize long-term maintenance

- Support the final intended use of the landfill site

Slopes should be kept to 5:1 or flatter, although 4:1 may be allowed. Vegetation should be selected that is native to the area and thus able to withstand annual weather conditions. Some plants are available that are able to evapotranspire large amounts of water, but these must be selected with care as they may require extensive care and maintenance.

10.5.4 Daily Operations

The daily operation of a landfill is illustrated in Figure 10.10. MSW is spread in thin layers and continuously compacted by the spreading machinery. At the end of each day, *daily cover* is spread over the waste placed that day. The purposes of landfill daily cover are as follows:

■ Prevent blowing of waste off of site

■ Control vectors and thus carrying of disease

■ Reduce odors

■ Minimize infiltration of water into the waste

Materials that are used for daily cover include soil excavated and stockpiled during the construction of the landfill; "waste" materials such as sludges and foundry sand; textiles; and foam or other proprietary materials.

The volume of waste placed and its corresponding cover material is termed a *cell.* The completed layer of adjacent cells is termed a *lift.* The *working face* is the portion of the cell that is uncovered and to which new waste is being added. Minimizing this area, while taking into account the need for efficient vehicle unloading, is necessary to minimize bird activity, unsightliness, and off-site blowing of litter.

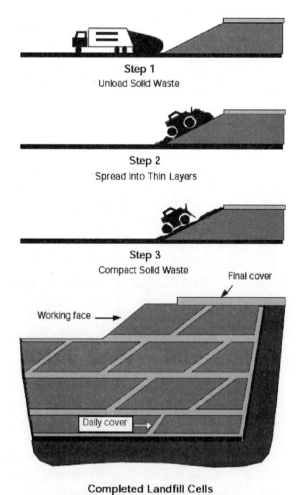

Step 1
Unload Solid Waste

Step 2
Spread Into Thin Layers

Step 3
Compact Solid Waste

Final cover

Working face

Daily cover

Completed Landfill Cells

Figure 10.10 Landfill daily operations

Source: Source: P. Walsh and P. O'Leary, University of Wisconsin–Madison Solid and Hazardous Waste Education Center, reprinted from *Waste Age* Correspondence Course, 1991–1992.

The volume of a landfill may be approximated as that of a truncated pyramid (Figure 10.11). The volume of such a pyramid is shown in Equation 10.6:

$$V = \frac{h}{3}(a^2 + ab + b^2) \tag{10.6}$$

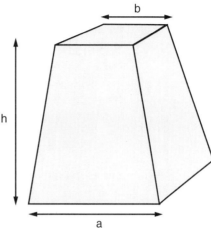

Figure 10.11 Truncated pyramid

Landfill volume estimate. (handwritten)

Example 10.5

Preliminary landfill sizing

P = 100,000 (handwritten)

A landfill serves a single city of 100,000 residents. Waste is delivered five days per week to the landfill. The city has an effective recycling program that results in 30% of the waste being recycled. Waste is delivered at a density of 300 lb/yd³, and its density is expected to double due to compaction by equipment. Daily cells are to be approximately 3 ft thick. The area of the landfill that will accept waste is 12 acres. The landfill is able to contain a total depth of 36 ft of waste, including daily cover. Daily cover will consist of 6 in of soil. Final cover slopes will be 5:1.

(a) Estimate the time for the landfill to be filled   *MSW/d ≈ 4 lb/d/pp* (handwritten)

(b) Investigate the impact of using a removable textile product as daily cover

(c) Investigate the impact of increasing recycling rates to 40%

Solution

(a) Given a 30% recycling rate and assuming a MSW generation rate of 4 lb/day and that one-fifth of the town is serviced by waste pickup on each weekday, the daily volume of waste accepted at the landfill is

$$\frac{20{,}000 \text{ people}}{\text{day}} \cdot \frac{4 \text{ lb}}{\text{person}}(1-0.30)\frac{1 \text{ yd}^3}{300 \text{ lb}} = 187 \text{ yd}^3/\text{day}$$

Consequently, the volume of waste in one cell after compaction is 93.5 yd³/day, given the doubling of the waste's density.

The volume of daily cover added is the product of the area corresponding to the daily cell and the thickness of the daily cover layer. The area of the landfill corresponding to a 3-ft lift is $\dfrac{93.5 \text{ yd}^3/\text{day} \cdot 27 \text{ ft}^3/\text{yd}^3}{3 \text{ ft}} = \dfrac{842 \text{ ft}^2}{\text{day}}$.

Thus, the volume of daily cover is 0.5 ft · 842 ft²/day = 421 ft³/day, or 15.6 yd³/day. Alternatively, this number could be arrived at by dividing the volume of compacted waste by six, since the daily cover amounts to one-sixth of the thickness of the daily lift.

The total volume of landfill airspace consumed in one day is 109 yd³ (93.5 yd³ + 15.6 yd³). The total volume of landfill available is found by determining the volume of a truncated pyramid. We will assume that the 12-acre site is a perfect square (723 feet on a side). Thus, the value a in Equation 10.6 is equal to 723 feet. Given the 5:1 side slopes, the value b in Equation 10.6 can be obtained by geometry to be equal to 468 ft. Thus, the volume of the truncated pyramid is

$$V = \frac{36 \text{ ft}}{3}(723 \text{ ft}^2 + 723 \text{ ft} \cdot 468 \text{ ft} + 468 \text{ ft}^2) = 480,000 \text{ yd}^3$$

Therefore, the landfill can accept waste for

$$\frac{480,000 \text{ yd}^3}{109 \text{ yd}^3/\text{day}} \frac{1 \text{ week}}{5 \text{ days}} \frac{1 \text{ year}}{52 \text{ weeks}} = 17 \text{ years}$$

(b) Using the removable cover will decrease the daily consumption of air space from 109 yd³/day to 93.5 yd³/day. As a result, the landfill will be able to accept waste for about 20 years.

(c) Using higher recycling rates will reduce the daily compacted cell volume to 80 yd³/day. In combination with 6 in of daily cover (or 13.3 yd³), the daily consumption of air space is 93.3 yd³/day, which is equivalent to the increase in available volume resulting from using a removable daily cover.

10.5.5 Closure and Postclosure Activity

For 30 years after closure, the owner/operator must continue to maintain the final cover, monitor groundwater to ensure the unit is not leaking, collect and monitor landfill gas, and perform other maintenance activities.

Financial assurance must be provided to ensure that monies are available to correct possible environmental problems. That is, landfill owners/operators are required to show that they have the financial means to pay for site closure, postclosure maintenance, and cleanups. Financial assurance can be provided with surety bonds, insurance, and letters of credit among other means.

Monitoring is a very important postclosure task. The following should be monitored: leachate head on liner, leakage through the liner, groundwater quality, air quality, gas in surrounding soil, leachate quality and quantity, and final cover stability (Vesilind et al., 2001).

Leakage can be determined using a liner system with leak detection (Figure 10.7) or a lysimeter. A lysimeter is an instrument used to collect water from soil (or in this case, a landfill) to determine the quantity and quality of water moving in a soil matrix. A lysimeter consists of a tube connected to a porous ceramic cup through which water is withdrawn. Lysimeters should be located in a landfill near a point on the liner corresponding to the occurrence of h_{max}.

Groundwater monitoring is accomplished with a series of monitoring wells. A means of collecting water at various depths is desirable, perhaps through the use of a cluster, or nest, of wells. Wells should be placed up-gradient and down-gradient of the landfill to ensure that the background groundwater quality is measured as well as the groundwater impacted by the landfill.

A closed landfill can serve many beneficial uses. Such uses have included model airplane flying parks, golf courses, and sledding hills, among other types of recreational uses. It is important that these activities are compatible with the design and construction of the landfill cover. Typically, building construction is not desirable due to settling issues, and care must be taken to prevent puncturing of the cover system (for example, through planting of deep-rooted plants such as trees).

10.6 BIOREACTOR LANDFILLS AND LEACHATE RECIRCULATION

The RCRA Subtitle D landfill described above has often been called a "dry tomb" landfill, and such a landfill inhibits biological activity. By keeping water out of the landfill, the volume of potential leakage is decreased to the detriment of biological activity. One means of correcting this problem is through leachate recirculation, which is possible given the increase in quality of landfill liner systems. By recirculating leachate, the waste can be stabilized much more rapidly, even as soon as closure.

In practice, leachate is recirculated by spraying leachate on the landfill surface, wetting the waste as it is applied or applying the waste into trenches or vertical columns. Design of other aspects of the landfill must be accomplished such that they are compatible with leachate recirculation (Vesilind et al., 2002).

Anaerobic and aerobic bioreactors are becoming increasingly popular alternatives to the "dry tomb" approach:

■ *Anaerobic bioreactor:* Although anaerobic conditions develop in landfills without any intervention, optimizing landfill conditions for the benefit of anaerobes can be accomplished by increasing the moisture content. The waste as accepted may have moisture contents of only 10%–20%, while optimal conditions occur in the moisture range between 45% and 60%. Additional moisture can be added via leachate recirculation, addition of sewage sludge, use of on-site stormwater, or other sources of nonhazardous liquids. Gas production is greatly accelerated in an anaerobic bioreactor landfill as compared to a conventional landfill, although the total volume of gas collected may not differ appreciably. As a consequence, the gas collection has to be sized to handle the larger peak gas generation rates.

■ *Aerobic bioreactor:* An aerobic bioreactor is operated via the addition of air, either by blowing air into the landfill through vertical or horizontal wells, or by creating a suction to pull air in. The aerobic process does not create methane, but degradation rates are increased significantly, such that stabilization occurs much more quickly than can be obtained in an anaerobic bioreactor or a conventional landfill. Anaerobic bioreactors require a higher level of management and the use of large quantities of water to ensure that conditions are not conducive to fires.

The benefits of a bioreactor landfill include the following[2]:

■ Total degradation of the waste mass can produce settlements approaching 40%, resulting in additional capacity that can be captured during the active

---

2   Source: *The Bioreactor Landfill: The Next Generation of Waste Management*, ©2000 Waste Management, Inc., Houston, Texas.

phase of the landfill, and substantially slowing the cost and demand for new landfill capacity.

■ Recirculation of leachate, an option provided for in current regulation and a key component of the bioreactor process, substantially lowers the cost of leachate management without having to change current liner and leachate head requirements, and eliminates the need for off-site management at municipal wastewater treatment plants.

■ Methane gas generation is expanded and concentrated during the active life of the landfill, which makes it easier to manage the gas and take advantage of beneficial gas use projects and, in turn, lowers the total mass of climate change gases and other potential emissions.

■ The bioreactor landfill may be used as a treatment center for liquid wastes otherwise destined for wastewater treatment plants or disposal without treatment.

■ Stabilization of the leachate and reduction in gas generation levels shortly after landfill closure can significantly reduce postclosure care costs and could ultimately result in a reduction in required final cap details and a shorter postclosure period.

■ The closed landfill that has been through the complete bioreactor process is no longer an environmental liability and can be integrated back into the conventional land use system within the host community.

Causes of concern include the following:

■ Increased gas emissions

■ Increased odors

■ Physical instability of waste mass due to increased moisture and density

■ Instability of liner systems

■ Surface seeps

■ Landfill fires

10.7   COMPOSTING

There are many benefits of composting, including keeping waste out of landfills. Compost is also an excellent soil conditioner with the ability to

■ improve water drainage,

■ increase water-holding capacity,

■ improve nutrient-holding capacity,

■ buffer pH,

■ increase soil organic content, and

■ reduce bulk density.

However, there are many challenges facing the implementation of commingled MSW recycling, including:

- The ability to find markets

- Odor problems

- The ability to create a uniformly high quality end product

- Inadequate design information

Composting is mediated by microorganisms, which need a carbon source, nutrients, moisture, oxygen, the proper particle size, and effective operating temperatures:

- *Carbon source.* Most MSW sources have adequate carbon sources, and when the MSW contains yard waste, the availability of carbon is not a problem. Most yard wastes (and food wastes) contain carbon that is highly bioavailable to the microorganisms, as compared to the carbon in lignins (from wood fibers), plastics, leather, and so on.

- *Nutrients.* Nitrogen is the most important nutrient of concern, and the ratio of carbon to nitrogen (C:N ratio) is considered critical to the success of the composting. C:N ratios of 20–30 are considered optimal. Nitrogen-rich feedstocks include food waste, yard trimmings, animal manures, and biosolids.

- *Moisture.* Moisture is also critical to the proper operation of a composting facility, and a moisture content between 50% and 60% is considered ideal. Lower moisture contents inhibit biological activity, while higher contents can lead to anaerobic conditions and the concomitant odors such as hydrogen sulfide.

- *Oxygen.* Oxygen is required for the aerobic degradation and, in practice, is supplied to the compost using a variety of means, as discussed below. The compost structure should have adequate void spaces to allow for proper transfer of oxygen through the pile.

- *Particle size.* the proper particle size allows for movement of oxygen through the compost. All other factors being equal, smaller particles tend to biodegrade much more rapidly than larger particles.

- *Temperature.* Figure 10.12 shows an idealized plot of temperature versus time in a large composting pile. As time progresses and the temperature varies, different types of microorganisms can flourish. In the mesophilic range, the most bioavailable organics are degraded and a corresponding rapid increase in temperature is found. The thermophilic range with temperatures between 40°C and 60°C provides near optimum conditions for a range of beneficial microorganisms. One important result of reaching the thermophilic range is that pathogen and weed seed destruction are achieved after three consecutive days at a temperature greater than 55°C. Following the thermophilic stage, the pile will enter the cooling/maturation phase. In reality, the actual temperature profile will vary significantly in response to turning of the pile.

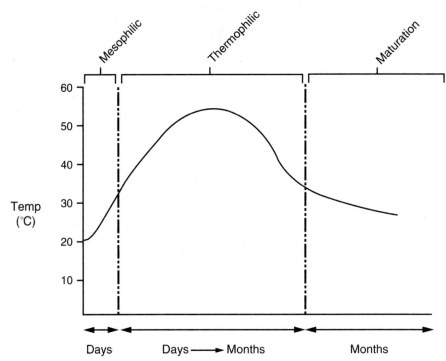

Figure 10.12 Temperature variation in a composting pile

Various means are available to create compost in practice and range from simple backyard composting piles to high-tech composting reactors. These composting methods vary one from another in terms of how the compost is mixed, how oxygen is provided, and how the proper moisture content and temperature are maintained.

The windrow method offers a relatively simple way of creating compost. The windrow width is usually twice the height, with widths between 14 feet and 16 feet common and heights between 4 feet and 8 feet. Optimal dimensions are large enough to allow for heat generation yet small enough for the transfer of oxygen. The piles are turned by specialized equipment. Turning adds oxygen to the pile and transfers material from the relatively cool outer portion of the pile to the relatively warm core. As a result of turning, the pile loses heat and may give off odors.

In the aerated static pile method, compost is placed in a large pile that is aerated. Aeration may be accomplished by a network of pipes underneath the pile hooked up to a blower or a vacuum. Air circulation provides relatively uniform oxygen concentrations throughout the pile, and the airflow can be controlled to maintain optimum temperatures. Since there is no turning, the outer layer of the pile may not reach proper temperatures to kill human pathogens and destroy weed seeds, and so a 6–12-inch "blanket" of finished compost may be added to help insulate the pile.

In-vessel composting systems are typically proprietary systems that allow for a high level of control of the composting system's temperature, moisture content, and oxygen level. Systems include drums, silos, digester bins, and tunnels. Some types of systems rotate to accomplish mixing, while other types are stationary with internal mixing mechanisms. Most in-vessel systems are continuous feed. Retention times are on the order of weeks. Minimal odors and, in some cases, no leachate are produced.

10.8   INCINERATION/COMBUSTION

The heat value of solid waste can be approximated using several methods. For example, the compositional analysis can be used to estimate the heat content as follows:

$$\text{Btu/lb} = 1238 + 15.6R + 4.4P + 2.7G \qquad (10.7)$$

where

R = plastics, % by weight on dry basis

P = paper, % by weight on dry basis

G = food waste, % by weight on dry basis

A second alternative is to estimate the heat value given the heat content of the various components of MSW. Sample data is provided in Table 10. 3.

Table 10.3   MSW proximate analysis and energy content

| Type of Waste | Proximate Analysis (% by weight) | | | | Energy Content (Btu/lb) | |
	Moisture	Volatiles	Fixed Carbon	Noncombustible	As Collected	Dry
Mixed food waste	70.0	21.4	3.6	5.0	1797	5983
Cardboard	5.2	77.5	12.3	5.0	7042	7428
Mixed paper	10.2	75.9	8.4	5.4	6799	7571
Newsprint	6.0	81.1	11.5	1.4	7975	8484
Mixed plastics	0.2	95.8	2.0	2.0	14,101	14,390
Textiles	10.0	66.0	17.5	6.5	7960	8844
Rubber	1.2	83.9	4.9	9.9	10,890	11,022
Yard wastes	60	30.0	9.5	0.5	2601	6503
Mixed wood	20.0	68.1	11.3	0.6	6640	8316
Residential MSW	21.0	52.0	7.0	20.0	5000	6250

Source: Adapted from Table 4-2, G. Tchobanoglous, H. Theisen, and S. A. Vigil, *Integrated Solid Waste Management* (McGraw-Hill, 1993).

Example 10.6

Calculating dry energy content

Confirm the value of the dry energy content of cardboard based on other information provided in Table 10. 3.

Solution

The heat value of cardboard on an as-collected basis is 7042 Btu/lb. In other words, 1 lb of cardboard, including water and dry cardboard solids, is equivalent to 7042 Btu. From Table 10.3, cardboard has a moisture content of 5.2%; thus, every pound of as-collected waste contains 0.052 lb of water and 0.948 lb of dry cardboard. The heat value on a dry basis is simply:

$$\text{heat value}_{dry} = \frac{7042 \text{ Btu/lb}}{0.948 \text{ lb dry MSW}} = 7428 \text{ Btu/lb on a dry basis}$$

Example 10.7

Heat value of MSW

Consider a waste (20% moisture content) comprised of the following, with the percentages given on a dry weight basis:

Mixed paper	42%
Yard wastes	20%
Food wastes	9%
Glass	7%
Plastics	7%
Steel cans	5%
Other	10%

Use two methods to estimate the heat value of this waste per pound of as collected waste.

Solution

Determine the heat content for 1 lb of dry MSW by using a weighted average of each constituent's heat value.

$$\text{heat content} = (0.42 \text{ lb} \cdot 6799 \text{ Btu/lb}) + (0.2 \text{lb} \cdot 2601 \text{ Btu/lb}) + (0.09 \text{ lb} \cdot 1797 \text{ Btu/lb}) + (0.07 \text{ lb} \cdot 14{,}101 \text{ Btu/lb})$$

$$= 4525 \text{ Btu/lb}$$

The heat content per as collected pound of MSW is 4525 Btu/lb · (1 − 0.2), or 3620 Btu/lb.

Alternatively, the heat content can be determined from the Equation 10.7:

$$\text{Btu/lb} = 1238 + 15.6R + 4.4P + 2.7G$$

$$= 1238 + 15.6(7) + 4.4(42) + 2.7(9)$$

$$= 1556 \text{ lb/Btu}$$

On a wet basis, the heat value is 1245 lb/Btu, which appears to be a severe underestimate.

If more precise estimates of the heat content of waste is desired, an *oxygen bomb calorimeter* (Figure 10.13) can be used. This calorimeter is used by placing a small sample of waste in a sealed sample holder and immersing the holder in a water bath. The sample is completely combusted within the sample holder, and the resulting release in energy causes the water bath temperature to increase.

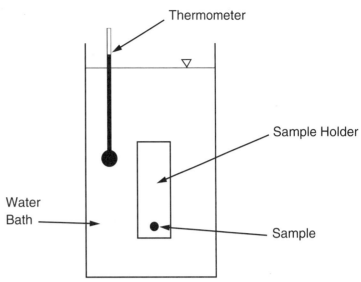

Figure 10.13 Oxygen bomb calorimeter

The change in temperature in the water bath can be related to the heat value (in units of cal/g) of the sample by Equation 10.8:

$$\text{heat value} = \frac{C_v \Delta T}{M} \tag{10.8}$$

where

C_v = heat capacity of the calorimeter (cal/°C)

ΔT = rise in temperature (°C) of water

M = mass of the unknown material (g)

Volume reduction is one significant benefit of MSW combustion. Typical values obtained are on the order of 90%.

Example 10.8

The effect of incineration on volume reduction

For the data of Example 10.7, estimate the volume reduction as a result of incineration, assuming an initial density of 300 lb/yd³ and a residue density of 950 lb/yd³.

Solution

This problem can be solved by referring to the noncombustible values of Table 10.3. The noncombustible fraction of the "other" portion of the MSW is assumed to be the same as for generic "residential MSW," and it is assumed that all water is removed. Thus, for every pound of waste, the mass of noncombustibles is:

$\text{mass}_{\text{noncombustibles}}$ = (0.42 lb · 0.054) + (0.20 lb · 0.005) + (0.09 lb · 0.05) + (0.07 lb · 0.98) + (0.07 lb · 0.02) + (0.05 · 0.98) + (0.1 lb · 0.2)

= 0.17 lb

This represents an 83% reduction in mass.

The reduction in volume is estimated by considering the densities of the waste; that is, the volume of waste entering the facility corresponding to 1 lb of MSW is:

$$V_{initial} = 1 \text{ lb}/(350 \text{ lb/yd}^3)$$

$$= 2.9 \cdot 10^{-3} \text{ yd}^3$$

The final volume of ash is:

$$V_{final} = 0.17 \text{ lb}/(950 \text{ lb/yd}^3)$$

$$= 1.7 \cdot 10^{-4} \text{ yd}^3$$

The reduction in volume is $\dfrac{V_{initial} - V_{final}}{V_{initial}} = 94\%$.

Incinerators require air to combust the organics, and the quantity of air required can be determined from stoichiometry, as illustrated in Example 10.9.

Example 10.9

Estimating air requirements for MSW combustion I

Assume that solid waste can be expressed by the chemical formula C_5H_{12}. Estimate the air requirements required to combust 10 tons per day of this waste.

Solution

Preliminary calculations include calculating the molecular weight of the waste (= 72 g/mol) and the total number of moles of waste:

$$\frac{10 \text{ ton}}{\text{day}} \frac{2000 \text{ lb}}{1 \text{ ton}} \frac{\text{kg}}{2.205 \text{ lb}} \frac{1000 \text{ g}}{\text{kg}} \frac{1 \text{ mol}}{72 \text{ g}} = 126{,}000 \text{ moles/day}$$

A balanced equation for the combustion of this MSW is:

$$C_5H_{12} + 8O_2 \rightarrow 5CO_2 + 6H_2O$$

Thus, every mole of waste requires eight moles of oxygen. The daily requirement of oxygen is 8 × 126,000 moles, or $1.0 \cdot 10^6$ moles/day.

Given that air is 21% oxygen, the daily requirements of air are

$$\frac{\left(1.0 \cdot 10^6 \text{ mol/day}\right) \cdot \left(32 \text{ g/mol}\right)}{0.21} = 1.5 \cdot 10^8 \text{ g/day} = 339{,}000 \text{ lb/day}$$

Using a specific weight of air equal to 0.075 lb/ft³ yields a daily volume of $4.1 \cdot 10^6$ ft³/day.

Another method of determining the oxygen requirements requires knowledge of the chemical composition of the solid waste (typical values were provided near the beginning of section 10.8). By combining the data describing the chemical composition with stoichiometric requirements, the oxygen requirements can be estimated. Table 10.4 provides helpful information.

Table 10.4 Stoichiometric oxygen requirements

Governing Equation	1 gram of	...requires ___ g of O_2
$C + O_2 \rightarrow CO_2$	C	32/12
$2H_2 + O_2 \rightarrow 2H_2O$	H_2	32/4
$S + O_2 \rightarrow SO_2$	S	32/32.1

In practice, *excess air* is required to ensure complete combustion. Excess air also affects the composition and temperature of the flue gas.

| Example **10.10** |

Estimating air requirements for MSW combustion II

A solid waste sample is characterized as follows:

Component	C	H	O	N	S
Weight percent	27	4	23	0.5	0.1

The remainder of the waste is comprised equally of water and inerts. Determine the air requirements, assuming 25% excess air.

Solution

Based on Table 10.4, the oxygen requirements, assuming 1 lb of waste, are:

Carbon: 0.27 lb · 32/12 = 0.72 lb O_2

Hydrogen: 0.04 lb · 32/4 = 0.32 lb O_2

Sulfur: 0.01 lb · 32/32.1 = 0.01 lb O_2

The total oxygen demand for combustion of these three elements is 1.05 lb. However, the pound of waste provides 0.23 lb of O_2, so the total requirement is 0.82 lb of O_2 per pound of waste.

Given that air is 21% oxygen, the air requirement per pound of waste is 0.82 lb/0.21, or 3.9 lb/lb of waste. At 25% excess air, the air requirements are 4.9 lb air/lb waste.

The efficiency of an MSW incinerator can be assessed by using an approach similar to the mass balance approach (section 1.4). However, instead of auditing the mass entering and exiting the control volume, the energy entering and exiting the control volume will be tracked. The rate of energy entering the control volume must equal the rate of energy exiting the control volume (since energy cannot be created or destroyed). The energy leaving the control volume, which may be the physical boundaries of the incineration process, is either usable energy or wasted energy (for example, energy lost as heat to the surroundings).

The heat energy Q required to raise the temperature of a substance is related to the mass of the substance and the *specific heat* (sometimes referred to as the *heat capacity*):

$$Q = m \cdot c \cdot \Delta T \tag{10.9}$$

where

M = mass of the substance (kg)

c = heat capacity (e.g., kJ/kg $\cdot$ K)

ΔT = change in temperature (K)

The specific heat of water is 1.0 Btu/lb/m $\cdot$ °F, or 4.19 kJ/kg $\cdot$ K.

The heat of vaporization is the energy required to transform a substance from its liquid form to its vapor form (or vice versa). The heat of fusion is the energy required to transform a substance from its solid form to its liquid form (or vice versa). The heat of fusion of water is 79.72 cal $\cdot$ g^{-1} or 334.5 kJ $\cdot$ kg^{-1}. The heat of vaporization of water is 2260 kJ $\cdot$ kg^{-1} or 539.8 cal $\cdot$ g^{-1}.

Example **10.11**

Energy generated by MSW incineration

Estimate the energy available due to incinerating MSW after analyzing for losses due to heat loss in flue gas, heat loss in bottom ash, heat loss to radiation, and heat loss to vaporization of water. Assume that 20 tons of MSW are incinerated per day, and that the MSW has a composition similar to that for "residential MSW" in Table 10.3. Other information is provided as follows:

Flue gas	mass flow rate = 0.25 kg/s
	temperature = 300°C
	specific heat, C_{flue} = 1.0 kJ/kg/K
Ash	temperature = 800°C
	specific heat, C_{ash} = 0.84 kJ/kg/K
Radiation	assume 5% loss of energy
Water vaporization	heat of vaporization = 2575 kJ/kg

Solution

From Table 10.3, the moisture content is 21%, the noncombustible fraction is 20%, and the waste has an energy content of 6250 Btu/lb ($1.454 \cdot 10^7$ J/kg) on a dry weight basis (5000 Btu/lb on an as-collected basis). From this information, the total energy available is

$$(5000 \text{ Btu/lb}) \cdot (20 \text{ ton/day}) \cdot (2000 \text{ lb/ton}) = 2 \cdot 10^8 \text{ Btu/day}$$

$$= 2.4 \cdot 10^6 \text{ J/s}$$

The heat lost in the flue gas can be determined as

$$\frac{0.25 \text{kg}}{\text{s}} \frac{1 \text{kJ}}{\text{kg} \cdot \text{°C}} 300°C = 75,000 \text{ J/s}$$

To determine the energy lost in the ash, the generation rate of ash needs to be determined. Given 20 tons/day of MSW with 20% of the waste being noncombustible, the ash production is estimated to be 8000 lb/day, or 0.042 kg/s.

$$\frac{0.042 \text{ kg}}{\text{s}} \frac{0.84 \text{ kJ}}{\text{kg} \cdot \text{°C}} 800°C = 28,000 \text{ J/s}$$

The heat lost to radiation is

$$0.05 \cdot (2.4 \cdot 10^6 \text{ J/s}) = 1.2 \cdot 10^5 \text{ J/s}$$

The energy required to vaporize the water from the solids is the product of the heat of vaporization of the water and the mass of water. The daily mass of water in the MSW is 20% of 20 tons, or 8000 lb. The energy required is thus

$$(8000 \text{ lb/day}) \cdot (2575 \text{ kJ/kg}) \cdot (1 \text{ kg}/2.205 \text{ lb}) = 9.3 \cdot 10^6 \text{ kJ/day}$$

$$= 1.1 \cdot 10^5 \text{ J/sec}$$

These energy flows are shown in Exhibit 4. From this diagram, the energy available for extracting useful energy is

$$(2.4 \cdot 10^6 \text{ J/s}) - [(7.5 \cdot 10^4 \text{ J/s}) + (1.2 \cdot 10^5 \text{ J/s}) + (2.8 \cdot 10^5 \text{ J/s}) + (1.1 \cdot 10^5 \text{ J/s})]$$
$$= 2.1 \cdot 10^6 \text{ J/s}.$$

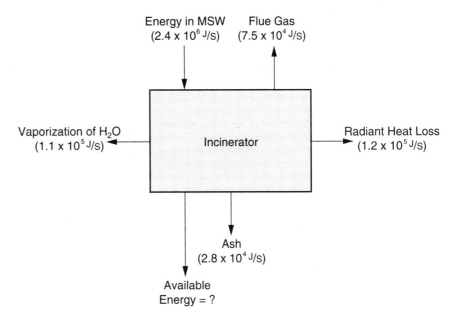

Exhibit 4 Energy flows in incinerator

RDF is created by processing MSW before combustion to remove noncombustible materials and to create a smaller, more uniform particle size. RDF may be compressed to make densified RDF, which decreases transportation and storage costs. RDF offers the advantages of a more uniform heat source and requires less excess air. Excess air requirements are 50%, as compared to 100% or higher for mass burn plants. However, RDF use is limited in practice due to difficulties related to processing the RDF.

Example 10.12

RDF vs. mass burning

Consider the impact of using RDF rather than mass burning the MSW of Example 10.7 and Example 10.8; that is, determine the heat value and volume reduction of the waste, assuming that all noncombustible materials have been removed. Assume all other values remain the same.

remove non-combustibles

Solution

The waste now consists of mixed paper, yard wastes, food wastes, and plastics. Thus, for one pound of the original waste, only 0.78 lb remains. The fraction of each of these materials in the RDF is:

Mixed paper	42% * 0.78 = 54%
Yard wastes	20% * 0.78 = 26%
Food wastes	9% * 0.78 = 12%
Plastics	7% * 0.78 = 9 %

The heat content of the RDF is

heat content = heat content of mixed paper + heat content of yard wastes + heat content of food wastes + heat content of plastic

$$= (0.54 \cdot 6799 \text{ Btu/lb}) + (0.26 \cdot 2601 \text{ Btu/lb}) + (0.12 \cdot 1797 \text{ Btu/lb}) + (0.09 \cdot 14{,}101 \text{ Btu/lb})$$

$$= 5832 \text{ Btu/lb (as compared to 4525 Btu/lb for Example 10.7)}$$

To determine the change in volume, the mass of noncombustible materials needs to be determined.

$$\text{mass}_{\text{noncombustibles}} = (0.54 \text{ lb} \cdot 0.054) + (0.26 \text{ lb} \cdot 0.005) + (0.12 \text{ lb} \cdot 0.05) + (0.09 \text{ lb} \cdot 0.02)$$

$$= 0.039 \text{ lb}$$

This represents a 96% reduction in mass, which is much larger than the 83% reduction found in Example 10.8.

The volume of RDF entering the facility, corresponding to 1 lb of MSW is

$$V_{\text{initial}} = 1 \text{ lb}/(350 \text{ lb/yd}^3)$$

$$= 2.9 \cdot 10^{-3} \text{ yd}^3$$

The final volume of waste is:

$$V_{\text{final}} = 0.039 \text{ lb}/(950 \text{ lb/yd}^3)$$

$$= 4.1 \cdot 10^{-5} \text{ yd}^3$$

The reduction in volume is $\dfrac{V_{\text{initial}} - V_{\text{final}}}{V_{\text{initial}}} = 98.5\%$, which is greater than the volume reduction of 94% determined in Example 10.8.

Building an MSW incinerator has to overcome a number of hurdles:

- The regulatory requirements are extensive.

- Public resistance is often active and well organized.

- The cost of actual implementation can be very high. One infamous example is that of Detroit's MSW incinerator, which charges tipping fees more than five times the average cost of landfill tipping fees in Michigan.

- A variety of solid, liquid, and gaseous waste streams must be considered.

- Contaminants include waste heat, particulates, sulfur emissions, mercury, and dioxins, to name a few.

The contaminants created by MSW combustion can all be treated to acceptable levels, but such treatment comes at a cost. Contaminant concentrations can be reduced given aggressive source reduction and sorting practices. Proper operation of an incinerator can also decrease contaminant creation. For example, dioxin is thought to be created due to incomplete combustion of the waste, although the exact mechanisms of formation are not presently understood.

10.9 PYROLYSIS AND GASIFICATION

Pyrolysis consists of heating the waste to high temperatures in the absence of oxygen. Thus it is not a combustion process. Pyrolysis is endothermic (that is, it requires an external heat source). The primary end products of pyrolysis are:

- Gases, including hydrogen, methane, carbon monoxide, carbon dioxide, and others

- A liquid fraction, which is a tar or oil stream that can be used as a synthetic fuel oil with further processing

- A solid fraction, or char

The oils have a heat value of about 9000 Btu/lb, and the gases can potentially have heat values of around 700 Btu/ft^3.

Pyrolysis is used widely for conversion of organic material to the useable end products and is used in the production of charcoal, but to date there has only been minimal application to the MSW field.

Gasification, unlike pyrolysis, is an exothermic reaction. Air or oxygen is used to partly combust the waste. The partial combustion, with stoichiometric proportions of air or slightly less (as compared to the excess air used in combustion), produces a combustible fuel gas that can be used to heat the remainder of the process to temperatures between 700°C and 900°C. Air emissions are typically lower of gasification processes than for combustion processes.

ADDITIONAL RESOURCES

1. Quian, X., R. M. Koerner, and D. H. Gray. *Geotechnical Aspects of Landfill Design and Construction*. Prentice Hall, 2002.

2. U.S. Environmental Protection Agency. *Decision-Maker's Guide to Solid Waste Management*, EPA 530-R-95-023, 1995.

3. Vesilind, P. A., W. A. Worrell, and D. R. Reinhart. *Solid Waste Engineering*, Thomson-Engineering, 2001.

Bellabridesmaid

Hazardous Waste

OUTLINE

This chapter will define hazardous wastes and discuss methods to minimize their production through "pollution prevention" practices. For those wastes that are still produced in spite of pollution prevention practices, a variety of treatment, storage, and disposal options exist and will be described.

The existence of hazardous waste in our environment has led to many deleterious health impacts. Assessing the risks associated with this contamination is covered in section 4.5 of this review manual.

11.1   CHARACTERISTICS AND TYPES OF HAZARDOUS WASTES

A waste may be defined as hazardous if it is a *listed* waste, or if it has one of the four following characteristics:

1. *Ignitability.* Ignitable wastes can create fires under certain conditions, are spontaneously combustible, or have a *flash point* less than 60°C (140°F). The flash point is the minimum temperature at which a liquid gives off a vapor in sufficient concentration to ignite when a source of ignition (sparks, open flames, cigarettes, etc.) is present.

2. *Corrosivity.* Corrosive wastes are acids or bases (pH less than or equal to 2 or greater than or equal to 12.5) that are capable of corroding metal containers, such as storage tanks, drums, and barrels.

3. *Reactivity.* Reactive wastes are unstable under normal conditions. They can cause explosions, toxic fumes, gases, or vapors when heated, compressed, or mixed with water.

4. *Toxicity.* Toxic wastes are harmful or fatal when ingested or absorbed (for example, containing mercury, lead, etc.). When toxic wastes are land disposed, contaminated liquid may leach from the waste and pollute groundwater. Toxicity is defined through a laboratory procedure called the Toxicity Characteristic Leaching Procedure (TCLP).

The TCLP test is a laboratory procedure designed to simulate the leaching a waste will undergo if disposed of in a sanitary landfill. In this procedure, a sample of a waste is extracted with an acetic acid solution. The extract obtained from the TCLP test (the *TCLP extract*) is analyzed to determine if any of the thresholds established for the 40 Toxicity Characteristic (TC) constituents (listed in Table 11.1) have been exceeded.

Table 11.1 TCLP leachate thresholds

Regulatory Level	Contaminant (mg/L)
Arsenic	5.0
Barium	100.0
Benzene	0.5
Cadmium	1.0
Carbon tetrachloride	0.5
Chlordane	0.03
Chlorobenzene	100.0
Chloroform	6.0
Chromium	5.0
o-Cresol	200.0[1]
m-Cresol	200.0[1]
p-Cresol	200.0[1]
Cresol	200.0[1]
2,4-D	10.0
1,4-Dichlorobenzene	7.5
1,2-Dichloroethane	0.5
1,1-Dichloroethylene	0.7
2,4-Dinitrotoluene	0.13[2]
Endrin	0.0[2]
Heptachlor (and its hydroxide)	0.008
Hexachlorobenzene	0.13[2]
Hexachloro-1,3-butadiene	0.5

(continued)

Regulatory Level	Contaminant (mg/L)
Hexachloroethane	3.0
Lead	5.0
Lindane	0.4
Mercury	0.2
Methoxychlor	10.0
Methyl ethyl ketone	200.0
Nitrobenzene	2.0
Pentachlorophenol	100.0
Pyridine	5.0^2
Selenium	1.0
Silver	5.0
Tetrachloroethylene	0.7
Toxaphene	0.5
Trichloroethylene	0.5
2,4,5-Trichlorophenol	400.0
2,4,6-Trichlorophenol	2.0
2,4,5-TP (Silvex)	1.0
Vinyl chloride	0.2

1. If o-, m-, and p-cresol concentrations cannot be differentiated, the total cresol (D026) concentration is used. The regulatory level of total cresol is 200 mg/L.

2. Quantitation limit is greater than the calculated regulatory level. The quantitation limit, therefore, becomes the regulatory level.

Source: *www.epa.gov/sw-846/pdfs/chap7up3b.pdf.*

Alternatively, a waste can be defined as a hazardous waste if it is *listed*. The U.S. Environmental Protection Agency (U.S. EPA) has defined four lists: F-list, K-list, P-list, and U-list.

■ *F-list (nonspecific source wastes).* This list identifies wastes from common manufacturing and industrial processes, such as solvents that have been used in cleaning or degreasing operations. Because the processes producing these wastes can occur in different sectors of industry, the F-listed wastes are known as wastes from nonspecific sources.

■ *K-list (source-specific wastes).* This list includes certain wastes from specific industries, such as petroleum refining or pesticide manufacturing. Certain sludges and wastewaters from treatment and production processes in these industries are examples of source-specific wastes.

■ *P-list and U-list (discarded commercial chemical products).* These lists include specific commercial chemical products in an unused form. Some pesticides and some pharmaceutical products become hazardous waste when discarded.

The U.S. EPA allows for a waste to be *delisted* and considers the following categories from petitioners: cost savings and aggregate economic impacts, impacts of delisting on the environment, and impacts of delisting on the RCRA hazardous waste management program.

Universal wastes are wastes that do meet the regulatory definition of hazardous waste but are managed under special, tailored regulations. These wastes include:

- Batteries
- Pesticides
- Lamps/fluorescent bulbs
- Mercury-containing equipment/thermostats

Leftover household products that contain corrosive, toxic, ignitable, or reactive ingredients are considered to be *household hazardous waste,* or HHW. Americans generate 1.6 million tons of HHW per year. Products such as paints, cleaners, oils, batteries, and pesticides that contain potentially hazardous ingredients require special care for disposal. However, household hazardous wastes are exempt from RCRA Subtitle C regulations.

Following is a list of common classes of contaminants found in hazardous wastes:

- VOCs (volatile organic compounds) are emitted as gases from liquids or solids. They have a high vapor pressure and a low solubility. VOCs are common groundwater contaminants. Paint thinners, dry cleaning solvents, and some constituents of petroleum fuels are examples of VOCs.

- Halogenated VOCs are compounds onto which a halogen (for example, fluorine, chlorine, bromine, or iodine) has been attached. The extent of halogenation can significantly affect performance of a treatment technology. For example, the resistance to biological treatment increases as a compound becomes more halogenated.

- Semi-volatile organic compounds (SVOCs) are compounds that do not readily volatilize at standard temperature and pressure. Two main classes of SVOCs are PAHs and pesticides. PAHs (polycyclic aromatic hydrocarbons) are hydrocarbon compounds with multiple benzene rings. PAHs comprise a group of over 100 different chemicals that are formed during the incomplete burning of coal, oil and gas, or other organic substances like tobacco or charbroiled meat. The lighter weight PAHs are more quickly degraded than the heavier PAHs, which has important implications for site cleanup as the heavier PAHs tend to be carcinogenic. Pesticides include insecticides, fungicides, and herbicides.

- Halogenated SVOCs are VOCs on which a halogen has attached. Of special concern are PCBs (polychlorinated biphenyls). Characteristics of PCBs include nonflammability, chemical stability, high boiling point, and electrical insulating properties, all of which made their use popular in hundreds of industrial and commercial applications. More than 1.5 billion pounds of PCBs were manufactured in the United States prior to cessation of production in 1977.

- Inorganics of concern are primarily metals. Metals may be found in the elemental form, but more often they are found as salts mixed in the soil.

■ Radionuclides are typically nonvolatile and less soluble in water than some other contaminants. Unlike organic contaminants (and similar to metals), radionuclides cannot be destroyed or degraded; therefore, remediation technologies applicable to radionuclides involve separation, concentration/volume reduction, and/or immobilization.

■ Explosives fall under the more general category of "energetic material." The various forms of energetic materials are shown in Figure 11.1. These materials are susceptible to "self-sustained energy release" when present in sufficient quantities and exposed to one of the following stimuli: heat, shock, friction, chemical incompatibility, or electrostatic discharge. Safety precautions are of utmost importance when investigating, sampling, treating, transporting, disposing, and otherwise managing explosives.

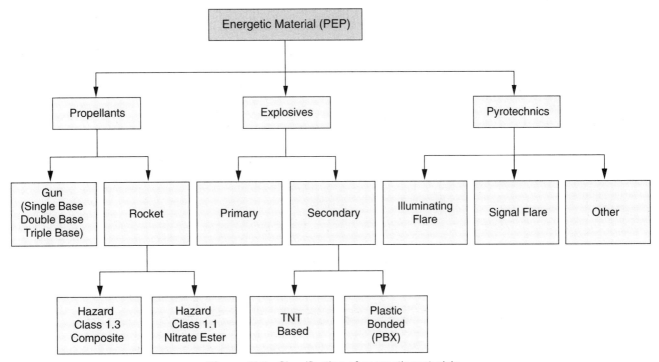

Figure 11.1 Classification of energetic materials

Source: *Remediation Technologies Screening Matrix and Reference Guide*, 4th ed., U.S. Army Environmental Command, SFIM-AEC-ET-CR-97053, January 2002.

11.2 POLLUTION PREVENTION

Pollution prevention, or P2, is a type of *source reduction*. The Pollution Prevention Act defines source reduction as "any practice which reduces the amount of any hazardous substance, pollutant, or contaminant entering any waste stream or otherwise released into the environment (including fugitive emissions) prior to recycling, treatment, or disposal; and reduces the hazards to public health and the environment associated with the release of such substances, pollutants, or contaminants."

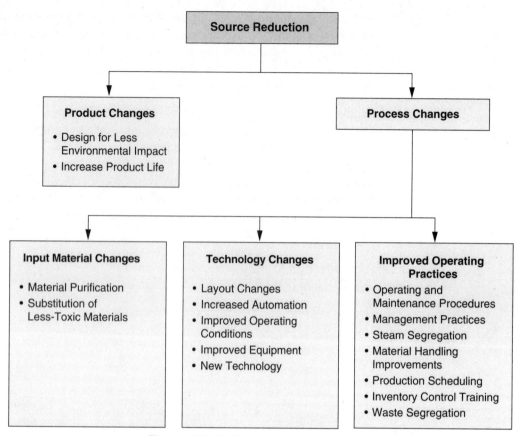

Figure 11.2 Pollution prevention methods

Source: U.S. EPA, *Facility Pollution Prevention Guide*, EPA/600/R-92/088, May 1992.

Pollution prevention has the following benefits:

■ The quantity and toxicity of hazardous and solid waste generation is reduced.

■ Raw material purchase costs are reduced.

■ Waste management record keeping and paperwork burden are minimized.

■ Waste management costs are minimized.

■ Workplace accidents and worker exposure decrease.

■ Environmental liability is lessened.

■ Profits can increase.

■ Public image is improved.

Methods of implementing P2 are shown in Figure 11.2.

The EPA's *Facility Pollution Prevention Guide* lists the following examples of process changes:

Examples of input material changes

■ Stop using heavy metal pigment

■ Use a less hazardous or toxic solvent for cleaning or as coating

- Purchase raw materials that are free of trace quantities of hazardous or toxic impurities

Examples of technology changes

- Redesign equipment and piping to reduce the volume of material contained, cutting losses during batch or color changes or when equipment is drained for maintenance or cleaning

- Change to mechanical stripping/cleaning devices to decrease solvent use

- Install a vapor recovery system to capture and return vaporous emissions

- Install speed control on pump motors to reduce energy consumption
  Examples of improved operating practices

- Train operators

- Segregate waste streams to avoid cross-contaminating hazardous and non-hazardous materials

- Improve control of operating conditions (for example, flow rate, temperature, pressure, residence time, and stoichiometry)

- Optimize purchasing and inventory maintenance methods for input materials

- Stop leaks, drips, and spills

The National Pollution Prevention Roundtable[1] lists the following barriers to P2:

- Lack of man-hours to devote to P2 implementation

- Perceived high cost of P2 implementation

- Low priority among some business owners

- Lack of awareness and interest of P2 success and programs in general

- Lack of regulatory enforcement

- Lack of strategic direction and organizational structure to help implement P2

- Misconception that all of the "low-hanging fruit" opportunities are already explored

11.3 TREATMENT, STORAGE, AND DISPOSAL FACILITIES[2]

RCRA Subtitle C established a *cradle-to-grave* concept, which controls hazardous waste from the time it is generated until its ultimate disposal. The latter stages of the cradle-to-grave time frame includes the *treatment, storage, and disposal (TSD)* of hazardous waste.

To track the hazardous waste from its cradle to grave, a *manifest* system is in place. The manifest system is a set of forms, reports, and procedures that tracks

---

1   *An Ounce of Pollution Prevention Is Worth 167 Billion Pounds of Cure: A Decade of Pollution Prevention Results, 1990-2000*, National Pollution Prevention Roundtable, Washington, D.C., 2003.

2   U.S. EPA, *www.epa.gov/epaoswer/osw/tsds.htm.*

hazardous waste from the time it leaves the generator facility until it reaches the TSD facility. Manifests allow the waste generator to verify that its waste has been properly delivered and that no waste has been lost or unaccounted for in the process. A manifest provides information about the generator of the waste, the facility that will receive the waste, a description and quantity of the waste (including the number and type of containers), and how the waste will be routed to the receiving facility.

The EPA requires the following of TSD facilities (TSDFs):

- *Air emissions.* TSDFs must control the emissions of volatile organic compounds (VOCs) from process vents from certain hazardous waste treatment processes; hazardous waste management equipment (for example, valves, pumps, and compressors); and containers, tanks, and surface impoundments.

- *Closure.* When a TSDF ends operations and stops managing hazardous waste, it must be closed such that it will not pose a future threat to human health and the environment.

- *Corrective action/hazardous waste cleanup.* The RCRA Corrective Action Program allows TSDFs to address the investigation and cleanup of any hazardous releases themselves. Cleanup at closed or abandoned RCRA sites can also take place under the Superfund program (section 12.1).

- *Financial assurance.* TSDFs must demonstrate that they will have the financial resources to properly close the facility or unit when its operational life is over, or provide the appropriate emergency response in the case of an accidental release.

- *Groundwater monitoring.* A TSDF must monitor the groundwater beneath its facility by installing groundwater monitoring wells and establishing a groundwater sampling regimen.

Building of TSDFs should be avoided in the following sensitive environments:

- 100-year floodplain (any land area that is subject to a 1 percent or greater chance of flooding in any given year from any source)

- Wetlands (see section 5.3)

- Recharge areas of high-quality groundwater

- Earthquake zones

- Karst terrain (rock such as limestone, dolomite, or gypsum that slowly dissolves when water passes through it and creates underground voids, tunnels, and caves, sometimes so large that their ceilings collapse forming large sinkholes[3])

- Unfavorable weather conditions (for example, siting an incinerator in an area prone to atmospheric inversion)

- Incompatible land use (for example, densely populated areas, areas near nursing homes, hospitals, day care facilities, etc.)

---

3 U.S. EPA, *Sensitive Environments and the Siting of Hazardous Waste Management Facilities,* www.epa.gov/epaoswer/hazwaste/tsds/site/sites.pdf.

Community resistance to siting of a hazardous waste site can be extremely high, exacerbated by the NIMBY (not in my backyard), NIMTOO (not in my term of office), and NOPE (not on planet earth) mindsets. Citizens are concerned primarily over potential increases in risk due to increased levels of toxins in the environment and due to sudden, catastrophic failure of the TSDF. Other concerns include increased truck traffic, increased noise, objectionable odors, decrease in property values, decreased potential for economic development, and so on.

11.4   TREATMENT TECHNOLOGIES

Treatment technologies for hazardous waste include the following:

■ Biopiles

■ Bioreactors

■ Constructed wetlands

■ Composting

■ Land farming

■ Slurry phase biological treatment

■ Chemical extraction

■ Adsorption/absorption

■ Advanced oxidation processes

■ Air stripping

■ Chemical reduction/oxidation

■ Granulated activated carbon (GAC)/liquid phase carbon adsorption

■ Ion exchange

■ Precipitation/coagulation/flocculation

■ Separation

■ Soil washing

■ Solidification/stabilization

■ Incineration

■ Pyrolysis

■ Thermal desorption

Some of these technologies have been discussed in sections 7.2.2, 7.2.3, 10.7, 10.8, and 10.9.

The following sections briefly describe the others.

11.4.1 Biological Treatment

Biological treatment involves treatment of waste by bacteria, fungi, or algae to remove and degrade hazardous constituents. Examples include bioreactors (the principles of which are discussed in section 8.5.1), constructed wetlands (section 5.3), biopiles, landfarming, and slurry phase biological reactors.

Biopiles

Biopiles (Figure 11.3) are a type of composting system in which contaminated soil is mixed with soil amendments (often proprietary) and aerated. The pile is placed on an impermeable membrane to protect groundwater quality and to allow leachate to be collected and recycled. VOCs may be released from a biopile, in which case further treatment of the VOCs may be necessary.

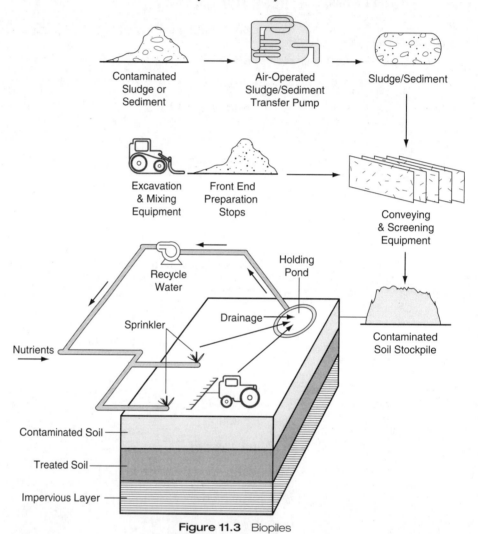

Figure 11.3 Biopiles

Source: *Remediation Technologies Screening Matrix and Reference Guide*, 4th ed., U.S. Army Environmental Command, SFIM-AEC-ET-CR-97053, January 2002.

Land farming

Land farming is a common means of hazardous waste treatment. Although the land is not farmed in the sense that edible crops are produced (due to concerns with plant uptake of contaminants), the process follows many agricultural practices. In land farming, contaminated soil, sediment, or sludge is applied to lined beds, and periodically turned over or tilled to aerate the waste. The process is illustrated in Figure 11.4.

In practice, contaminated soils are spread over or injected into the engineered soil medium. The soil/waste mixture is then mixed using a cultivator. This cultivation step serves two important functions: it brings the waste into contact with the soil and provides necessary aeration. Aeration occurs periodically, as often as weekly.

Waste is treated via land farming by a number of processes, including precipitation, ion exchange with soil particles, neutralization, volatilization, adsorption biological degradation chemical decay and plant uptake. The process has been used successfully for petroleum-based wastes and wood-preserving wastes and also is effective at removal of organics and metals.

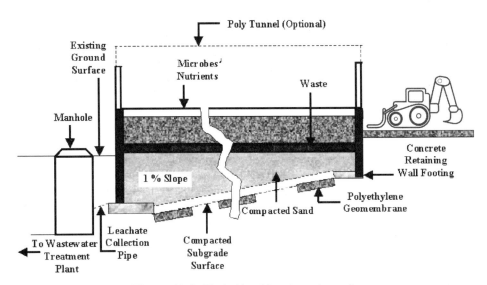

Figure 11.4 Typical land farming schematic

Source: *Remediation Technologies Screening Matrix and Reference Guide*, 4th ed., U.S. Army Environmental Command, SFIM-AEC-ET-CR-97053, January 2002.

Application of waste can continue as long as the assimilative capacity of the system is not exceeded. Once the assimilative capacity is reached, the site may be closed or the soil removed and disposed of. LaGrega et al. (2000) list three perspectives by which a site's assimilative capacity must be assessed:

1. *Capacity limiting.* Conservative substances applied to the site will accumulate with time. Once this cumulative amount has exceeded some threshold value, the site must be closed.

2. *Rate limiting.* Nonconservative substances will degrade with time. The amount added plus the amount remaining from previous applications must not sum up to be more than a threshold value.

3. *Application limiting.* The rate of application must take into account the migration of contaminants off-site (for example, by volatilization or runoff). The application rate must not be so large that the off-site migration exceeds some threshold value.

Limitations on the process include:

■ Specific amount of space required

■ Minimal control over the process

- Inorganic contaminants not degraded

- Must pretreat volatile contaminants to limit uncontrolled volatilization into the atmosphere

| Example **11.1** |

Assessing site assimilation

A liquid waste containing a single contaminant is applied on a pilot-scale test site such that the soil concentration after application is 15 mg/kg. After 60 days, the concentration of the contaminant in the soil is 3.9 mg/kg. Regulations prohibit the concentration of contaminant in the soil from exceeding 25 mg/kg. The producer would like to apply 5000 m^3 of the liquid waste on a full-scale plot every 90 days. Assess whether such an application rate would be appropriate given this additional information:

Full-scale site dimensions: 150 ft $\times$ 500 ft

Concentration of contaminant in the liquid waste: 50 mg/L

Treatment depth: 4 ft

Soil specific weight: 100 lb/ft^3

Assume first-order kinetics.

Solution

The first step is to use the relationship for first-order kinetics to estimate a decay coefficient. (First-order kinetics were reviewed in section 1.3.1.)

$$C = C_O e^{-k_1 t}$$

$$3.9 \text{ mg/kg} = 15 \text{ mg/kg} \cdot e^{-k_1 \ (60 \text{ days})}$$

The first-order decay constant is found to be 0.015 day^{-1}.

The loading of contaminants to the site for each application, in units of mg/kg, is determined next.

$$\frac{50 \text{ mg/L} \cdot 5000 \text{ m}^3 \cdot 1000 \text{ L/m}^3}{(150 \text{ ft} \cdot 500 \text{ ft}) \cdot 4 \text{ ft} \left(\dfrac{100 \text{ lb}}{\text{ft}^3} \right) \cdot \left(\dfrac{1 \text{ kg}}{2.205 \text{ lb}} \right)} = 18 \text{ mg/kg}$$

The initial loading rate will not increase the soil contaminant level above the regulatory limit of 25 mg/kg. The next step is to see how subsequent applications will raise the soil contaminant level.

The concentration in the soil after 90 days as a result of the initial application is 4.7 mg/kg. Thus, immediately after the second application, the concentration in the soil will be 22.7 mg/kg (18 mg/kg + 4.7 mg/kg). After 180 days, the original application will have decreased to 1.21 mg/kg, and the second application will have decreased to 4.7 mg/kg, giving a total concentration in the soil before the third application equal to 5.91 mg/kg. The graph in Exhibit 1 shows the progression of contamination in the soil.

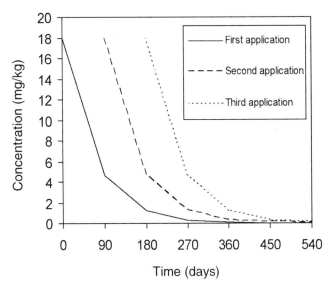

Exhibit 1 Soil concentration due to three applications

Summing up an infinite number of these curves demonstrates that the concentration in the soil will continue to increase until it asymptotically approaches a value of 24.3 mg/kg. This could also be determined using a tabular method. Both the graphical and tabular methods are provided in Exhibit 2 and Exhibit 3, respectively.

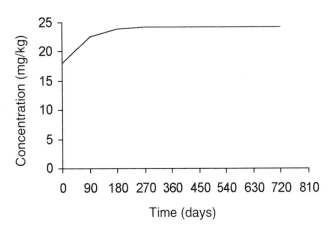

Exhibit 2 Soil concentration as a function of time

					Application #					
Times (days	1	2	3	4	5	6	7	8	9	SUM
0.0	18.0									18.0
90.0	4.7	18.0								22.7
180.0	1.2	4.7	18.0							23.9
270.0	0.3	1.2	4.7	18.0						24.2
360.0	0.1	0.3	1.2	4.7	18.0					24.3
450.0	0.0	0.1	0.3	1.2	4.7	18.0				24.3

(continued)

540.0	0.0	0.0	0.1	0.3	1.2	4.7	18.0			24.3
630.0	0.0	0.0	0.0	0.1	0.3	1.2	4.7	18.0		24.3
720.0	0.0	0.0	0.0	0.0	0.1	0.3	1.2	4.7	18.0	24.3

Exhibit 3   Tabular analysis of soil concentration

Slurry Phase Treatment

Bioremediation techniques have been successfully used to remediate soils, sludges, and sediments contaminated by explosives, petroleum hydrocarbons, solvents, pesticides, and a variety of other organic chemicals. The use of a bioreactor, such as slurry phase treatment, rather than *in situ* biological techniques (section 12.3.1) is preferred for heterogeneous soils, low-permeability soils, areas where underlying groundwater would be difficult to capture, or when faster treatment times are required.

Slurry phase biological treatment involves the controlled treatment of contaminated soil in a bioreactor. Slurry phase treatment is similar to conventional suspended growth except for the high concentration of solids (10% to 30% solids by weight) and the fact that many of the solids may be inert. Although biological degradation is the primary removal process, removal can also occur via volatilization and desorption. Figure 11.5 shows a slurry phase bioreactor that is closed for cases when volatilization is a concern; the closed reactor allows the contaminated air to be treated and released in a controlled manner. Desorption can be achieved through pretreatment by adding surfactants.

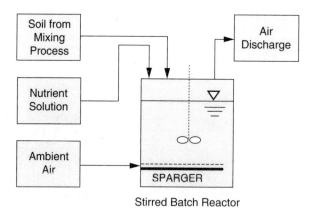

Figure 11.5   Slurry phase treatment

Source: *Remediation Technologies Screening Matrix and Reference Guide*, 4th ed., U.S. Army Environmental Command, SFIM-AEC-ET-CR-97053, January 2002.

The design of mixers is critical in the design of slurry phase treatment, as it is imperative to keep the solids in suspension and to supply sufficient agitation to break up particles. High energy mixers are required in conjunction with baffled tanks.

11.4.2 Physical-Chemical Treatment

The following physical-chemical treatment processes will be discussed in this section: chemical oxidation, advanced oxidation, extraction, microencapsulation, neutralization, separation, soil washing, granular activated carbon, air stripping, and solidification/stabilization.

Chemical Oxidation/Advanced Oxidation

Chemical oxidation uses strong oxidizing agents (for example, hypochlorite, peroxides, persulfates, percholorates, permanganates, etc.) to break down hazardous waste constituents to render them less toxic or mobile, while chemical reduction uses strong reducing agents (for example, sulfur dioxide, alkali salts, sulfides, iron salts, etc.). Oxidation may completely break down a contaminant, or at the least transform it into constituents that are more amenable to biodegradation. Some common oxidation/reduction reactions of environmental engineering concern are shown in Table 11.2.

Table 11.2 Typical oxidation/reduction reactions

Reagent	Contaminant	Reaction	Comments
Oxidation			
Ozone	Cyanide	$NaCN + O_3 \rightarrow NaCNO + O_2$	
Hydrogen peroxide	Cyanide Sulfide	$NaCN + H_2O_2 \rightarrow NaCNO + H_2O$ $H_2O_2 + H_2S \rightarrow 2H_2O + S$ $H_2O_2 + S^= \rightarrow SO_4 + 4H_2O$	pH 9.5–10.5
Chlorine	Ferrous to ferric	$2Fe^{++} + HOCl + 5H_2O \rightarrow 2Fe(OH)_3 + Cl^- + 5H^+$	
Chlorine	Cyanide to cyanate Cyanate to CO_2 & N_2	$NaCN + Cl_2 \rightarrow CNCl + NaCL$ $CNCl + 2NaOH \rightarrow NaCNO + H_2O + NaCl$ $2NaCNO + 3Cl_2 + 4NaOH \rightarrow N_2 + 2CO_2 + 6NaCl + 2H_2O$	CNCl is a toxic gas
Chlorine dioxide	Manganous ion to manganese dioxide	$Cl_2 + 2NaClO_2 \rightarrow 2ClO_2 + 2NaCl$ $Mn(NO_3)_2 + 2e \rightarrow MnO_2 + 2NO_2^-$	MnO_2 insoluble
Potassium permanganate	Sulfide	$4KMnO_4 + 3H_2S \rightarrow 2K_2SO_4 + S + 3MnO + MnO_2 + 3H_2O$	
Ozone/UV	Organics	$CH_3CHO + O_3 \rightarrow CH_3COOH + O_2$	Forms H_2O_2
UV/peroxide	Organics (CH_2Cl_2)	$CH_2Cl_2 + 2H_2O_2 \rightarrow CO_2 + 2H_2O + 2HCl$	
Oxygen	Organic matter (CH_2O)	$CH_2O + \frac{1}{2}O_2 \rightarrow CO_2 + H_2O$	
Oxygen	Cyanide	$2CN^- + O_2 \rightarrow 2CNO$ $CNO^- + 2H_2O + 2H^+ \rightarrow CO_2 + NH_4^- + H_2O$	Requires activated C and Cu catalyst
Reduction			
Sulfur dioxide	Chromium (VI)	$3SO_2 + 3H_2O \rightarrow 3H_2SO_3$ $2CrO_3 + 3H_2SO_3 \rightarrow Cr_2(SO_4)_3 + 3H_2O$	CR(III) is then precipitated as $Cr(OH)_3 \downarrow$ by adding lime
Ferrous sulfate	Chromium (VI)	$2CrO_3 + 6FeSO_4 + 6H_2SO_4 \rightarrow 3Fe_2(SO_4)_3 + Cr_2(SO_4)_3 + 6H_2O$	
Sodium borohydride	Copper	$NaBh_4 + 8Cu^+ + 2H_2O \rightarrow 8Cu + NaBO_2 + 8H^+$	

Source: *Remediation Technologies Screening Matrix and Reference Guide*, 4th ed., U.S. Army Environmental Command, SFIM-AEC-ET-CR-97053, January 2002.

Chlorine dioxide is a promising oxidizing agent to use to treat hazardous waste, and its use in drinking water treatment was discussed in section 7.2.10. Advantages of chlorine dioxide over chlorine include safety and the concern that chlorine can chlorinate organic compounds, possibly increasing their toxicity in the process. The following equation shows an example of how chlorine dioxide can transform an aldehyde, R-CHO, into the corresponding carboxylic acid (R-COOH):

$$R\text{-}CHO + H^+ + ClO_2 \rightarrow R\text{-}COOH + HOCL$$

With additional chlorine, the reaction can proceed to completion (O'Brien and Gere, 1995):

$$HOCl/CLO_2 + R\text{-}COOH \rightarrow CO_2 + H_2O, HCl$$

Chlorine dioxide can also treat organic wastes. The following reactions show the oxidation of metals to produce a precipitate that is less mobile and therefore less of a risk:

$$ClO_2 + M^{2+} \rightarrow ClO_2^- + M^{3+}$$

$$M^{3+} + 3OH^- \rightarrow M(OH)_{3(s)}, \text{ or}$$

$$M^{3+} + 3S^- \rightarrow MS_{3(s)}$$

Hydrogen peroxide is another strong oxidizing agent that can be used to treat hazardous wastes. It is readily available at a relatively low cost and does not produce any harmful by-products.

Example **11.2**

Peroxide requirements

Estimate the annual volume of hydrogen peroxide required to destroy cyanide in a 2500 m³/day waste stream. The waste stream contains 300 mg/L of cyanide as NaCN. Hydrogen peroxide has a density of 1400 kg/m³.

Solution

The annual volume will be approximated from stoichiometry (see Table 11.2), recognizing that more H_2O_2 will be required in practice due to short circuiting and other nonidealities.

The quantity of NaCN to be treated each day is:

$$\frac{300 \text{ mg}}{L} \frac{1 \text{ mol}}{49,000 \text{ mg}} = \frac{0.0061 \text{ mol}}{L} \frac{2.5 \cdot 10^6 \text{ L}}{\text{day}} = 1.53 \cdot 10^3 \text{ mol/day}$$

The quantity of hydrogen peroxide needed to oxidize this amount of NaCN can be found, recognizing that 1 mole of NaCN requires 1 mole of H_2O_2 according to the stoichiometry in Table 11.2:

$$\frac{1.53 \cdot 10^3 \text{ mol}}{d} \frac{34 \text{ g}}{\text{mol}} \frac{1 \text{ L}}{1400 \text{ g}} \frac{1 \text{ m}^3}{1000 \text{ L}} \frac{365 \text{ day}}{\text{yr}} = 136 \text{ m}^3/\text{day}$$

Photolysis is the breaking of chemical bonds by the action of UV (ultraviolet) radiation or visible light. UV radiation has been demonstrated to degrade

PCBs, dioxins, PAHs, BTEX, and so on. The energy of a photon E is governed by Planck's law:

$$E = h \cdot v$$

where

h = Planck's constant ($6.624 \cdot 10^{-27}$ erg $\cdot$ sec)

v = frequency (T^{-1})

The wavelength, frequency, and energy of various regions of the electromagnetic spectrum are shown in Table 11.3.

Table 11.3 Ranges of wavelengths, frequencies, and energies

	Wavelength (m)	Frequency (Hz)	Energy (J)
Radio	$> 1 \times 10^{-1}$	$< 3 \times 10^{9}$	$< 2 \times 10^{-24}$
Microwave	$1 \times 10^{-3} - 1 \times 10^{-1}$	$3 \times 10^{9} - 3 \times 10^{11}$	$2 \times 10^{-24} - 2 \times 10^{-22}$
Infrared	$7 \times 10^{-7} - 1 \times 10^{-3}$	$3 \times 10^{11} - 4 \times 10^{14}$	$2 \times 10^{-22} - 3 \times 10^{-19}$
Optical (visible)	$4 \times 10^{-7} - 7 \times 10^{-7}$	$4 \times 10^{14} - 7.5 \times 10^{14}$	$3 \times 10^{-19} - 5 \times 10^{-19}$
UV	$1 \times 10^{-8} - 4 \times 10^{-7}$	$7.5 \times 10^{14} - 3 \times 10^{16}$	$5 \times 10^{-19} - 2 \times 10^{-17}$
X-ray	$1 \times 10^{-11} - 1 \times 10^{-8}$	$3 \times 10^{16} - 3 \times 10^{19}$	$2 \times 10^{-17} - 2 \times 10^{-14}$
Gamma-ray	$< 1 \times 10^{-11}$	$> 3 \times 10^{19}$	$> 2 \times 10^{-14}$

Source: M. D. LaGrega, P. L. Buckingham, and J. C. Evans, *Hazardous Waste Management*, 2nd ed., © 2001, McGraw-Hill Education. Reprinted by permission of McGraw-Hill Companies.

The energy available in radiation is directly proportional to the strength of the bond that it can break. Radiation that strikes a molecule can be absorbed, transmitted, or scattered. The Grotthus-Draper law states that only the radiation that a molecule absorbs can cause chemical change.

Advanced oxidation systems generate the hydroxyl radical ($\bullet$OH), a powerful oxidant. The hydroxyl radical is typically generated through photochemical means. The hydroxyl radical can be generated via the following methods:

- VUV photolysis (photolysis of water using UV radiation)

- UV/Oxidation processes (UV photolysis of conventional oxidants such as hydrogen peroxide and ozone). Generation of the hydroxyl radical via UV photolysis of hydrogen peroxide is described by the following equation:

$$H_2O_2 + \text{light energy} \rightarrow 2\bullet OH$$

- Generation of the hydroxyl radical via UV photolysis of O_3 (ozone) is described by the following equations:

$$O_3 + \text{light energy} + H_2O \rightarrow H_2O_2 + O_2$$

$$H_2O_2 + \text{light energy} \rightarrow 2\bullet OH$$

$$2O_3 + H_2O_2 \rightarrow 2\bullet OH + 3O_2$$

- Photo-Fenton process (decomposition of hydrogen peroxide using iron-based reagents and irradiation with near-UV radiation). Fenton's reaction is

$$Fe(II) + H_2O_2 \rightarrow Fe(III) + OH^- + \bullet OH$$

■ Fenton's reaction can be enhanced with the use of near-UV radiation and visible light (the photo-Fenton reaction), as shown in the following equation:

$$Fe(III)(OH)^{2+} + \text{light energy} \rightarrow Fe(II) + \bullet OH$$

Advanced oxidation is more expensive than biological treatment; however, for those organic compounds that are biorefractory, advanced oxidation offers much shorter reaction times.

Extraction

Extraction (Figure 11.6) is a process that separates hazardous constituents by means of settling, filtration, adsorption, absorption, or other means. Chemical extraction does not destroy wastes but is a means of separating hazardous contaminants from soils, sludges, and sediments. As a result, the volume of the hazardous waste to be treated is reduced. Physical separation steps may be required to separate the soil into coarse and fine fractions, as the fine constituents may contain most of the contamination.

Acid extraction may be used to extract heavy metals from the contaminated matrix. The residence time varies depending on the soil type, types of heavy metal, and heavy metal concentrations but generally ranges between 10 and 40 minutes. When extraction is complete, solids are separated and transferred to the rinse system. The heavy metals are concentrated in a form potentially suitable for recovery. Finally, the soils are dewatered and mixed with lime to neutralize any residual acid.

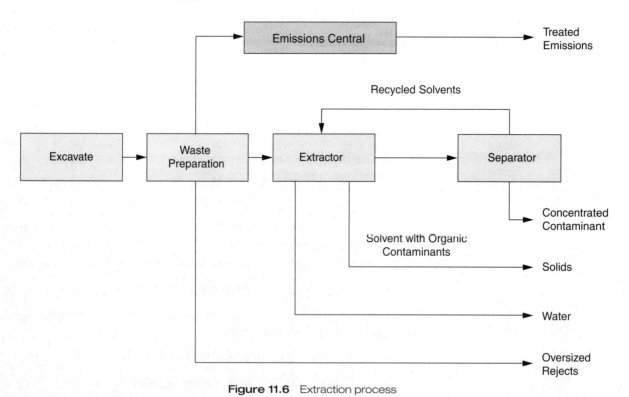

Figure 11.6 Extraction process

Source: *Remediation Technologies Screening Matrix and Reference Guide*, 4th ed., U.S. Army Environmental Command, SFIM-AEC-ET-CR-97053, January 2002.

Solvent extraction uses an organic solvent to remove organic contaminants and organically bound metals. Solvent extraction has been shown to be effective in treating sediments, sludges, and soils containing primarily organic contaminants, such as polychlorinated biphenyls (PCBs), volatile organic compounds (VOCs), and petroleum wastes. Traces of solvent may remain within the treated soil matrix; thus, the toxicity of the solvent is an important consideration.

Microencapsulation
Microencapsulation is a process that coats the surface of the waste material with a chemical coating (such as a thin layer of plastic or resin) to prevent the material from leaching hazardous waste constituents.

Neutralization
Neutralization is a process that is used to treat corrosive hazardous waste streams. Low pH acidic corrosive waste streams are usually neutralized by bases. High pH corrosive waste streams are usually neutralized by adding acids.

Separation
Separation processes seek to remove the contaminant from its liquid or solid matrix, effectively concentrating the contaminant in the process of purifying the original waste matrix.

One means of separation is distillation (Figure 11.7), whereby a liquid waste is heated and contaminants of varying volatilities are removed. In simple distillation, heat is applied to a liquid mixture in a *still*, causing a portion of the liquid to vaporize. The vapors are cooled and condensed producing a liquid product called *distillate*. The relative concentration of volatile contaminants is higher in the distillate than in the mixture remaining in the still. In practice, distillation can be very complex and expensive and may require extensive land area and tall heights (possibly exceeding 200 ft).

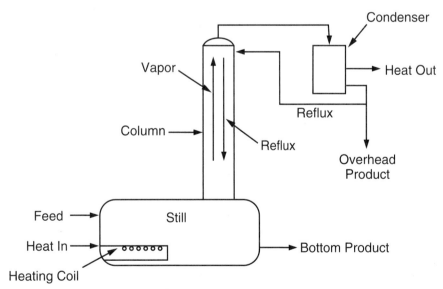

Figure 11.7 Distillation

Source: *Remediation Technologies Screening Matrix and Reference Guide*, 4th ed., U.S. Army Environmental Command, SFIM-AEC-ET-CR-97053, January 2002.

Freeze crystallization (Figure 11.8) is a separation process whereby a contaminated liquid is frozen. Ice is formed either by bringing the liquid in contact with a cooled surface, or by direct contact with a boiling refrigerant. As the ice crystals form, they are nearly pure, depending on freezing rate, contaminant concentration, and contaminant type. The ice crystals can be removed from the *mother liquid,* rinsed, and melted to form a nearly pure liquid. The mother liquid may be highly concentrated, depending on the freezing conditions, and as a result may be more amenable to further treatment.

High capital costs are associated with freeze crystallization. However, relatively low operating costs result, as the heat of fusion (6.013 kJ/mol) is much less than the heat of vaporization (40.63 kJ/mol) of water. Another advantage of the process is that pretreatment is not required.

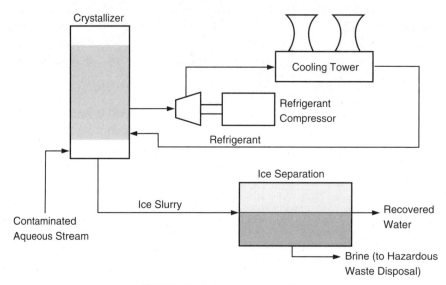

Figure 11.8 Freeze separation

Source: *Remediation Technologies Screening Matrix and Reference Guide*, 4th ed., U.S. Army Environmental Command, SFIM-AEC-ET-CR-97053, January 2002.

Membrane pervaporation (Figure 11.9) is another separation process. Pervaporation uses permeable membranes that preferentially adsorb volatile organic compounds (VOCs) from contaminated water. Water is heated before it enters the pervaporation module, in which VOCs diffuse by vacuum from the membrane-water interface through the membrane wall. Treated water exits the pervaporation module, while the organic vapors travel from the module to a condenser where they form a relatively high concentration liquid phase. Based on the affinity of the VOCs for the membrane, the relative composition of the gas may be much different from that of the original liquid. The name "pervaporation" arises from the fact that components in the liquid stream *permeate* through the membrane and *evaporate* into the vapor phase. Concentration factors greater than 1000 are possible.

Soil Washing

Soil washing (Figure 11.10) differs from extraction techniques in that it uses a water-based solution to separate contaminants from contaminated soil. The process removes contaminants from soils by

■ dissolving or suspending contaminants in the wash solution; or

■ concentrating them into a smaller volume of soil through particle size separation, gravity separation, and attrition scrubbing.

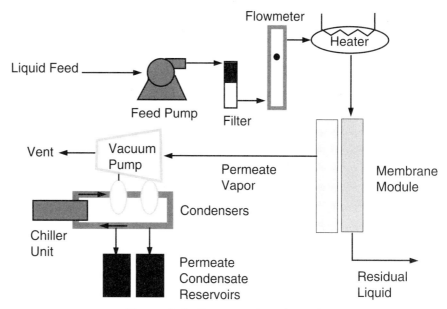

Figure 11.9 Pervaporation schematic

Source: U.S. EPA, *www.epa.gov/nrmrl/std/cppb/pervapor/pervaporationwhatis.htm.*

The latter process (attrition scrubbing) is necessary because many contaminants preferentially bind to the clay, silt, and organic portions of soil. The clays, silts, and organics are attached to the larger inert solids (that is, sand and gravel). Soil washing can separate these smaller particles from the sand and gravel, and collecting these smaller particles effectively concentrates the contaminants. The sand and gravel can be returned to the site, perhaps without any further processing. Like the acid and solvent extraction processes, soil washing does not destroy contaminants, but transfers them from one media to another.

Air Stripping

Absorption is a phenomenon by which contaminant mass is transferred from the liquid phase to the gas phase (*air stripping*), or from the gas phase to the liquid phase (*gas scrubbing*). Both scenarios involve a column filled with a high surface area, high void volume packing material, often made of a plastic or ceramic compound, which is configured to provide uniform flow of both the liquid and gas phases. Mass transfer occurs at the interface between the liquid and gas and often a chemical is added to the liquid to further enhance the removal of the target species. For example, lime is a common additive that increases pH and shifts the equilibrium between NH_4^+ and NH_3 to allow ammonia to be removed by air stripping, or is available to react with the gaseous SO_2 present in coal-fired power plant flue gasses to form $CaSO_3(s)$.

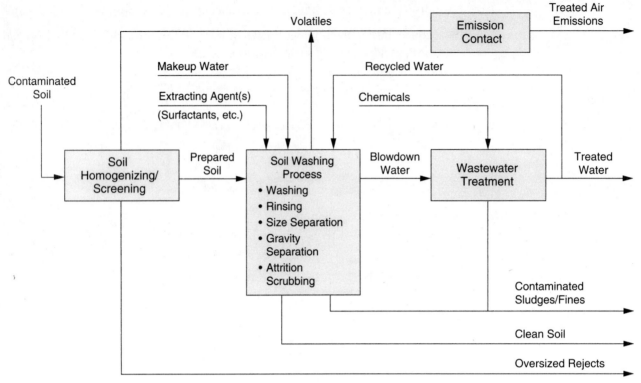

Figure 11.10   Soil washing

Source: *Remediation Technologies Screening Matrix and Reference Guide*, 4th ed., U.S. Army Environmental Command, SFIM-AEC-ET-CR-97053, January 2002.

The mass balance for an absorption system relies on the fact that environmental contaminants at low concentrations have a linear partitioning relationship between a liquid phase and a gas phase when the two phases are in equilibrium. This is known as *Henry's law* and may be expressed mathematically for systems operating at atmospheric pressure as

$$y_{out} = H'C_{in} = \frac{H}{RT}C_{in} \tag{11.1}$$

where y_{out} is the molar concentration of contaminant in the gas exiting the column, H' is the unitless form of the Henry's law constant, C_{in} is the molar concentration of contaminant in the liquid phase entering the column, H is the dimensioned form of the Henry's law constant (atm · m³/mol), R is the gas constant (8.026×10^{-5} atm · m³/mol · K), and T is the temperature (K). In general, the magnitude of H' is a good indicator of the direction of contaminant transfer, where values > 100 indicate species that have a strong preference for the air and values < 0.01 indicate species that have a strong preference for the aqueous phase.

For removing a pollutant species from water (air stripping), if we assume all of the contaminant entering in the liquid phase is removed by the gas, the steady state mass balance is written as

$$Q_w C_{in} = Q_a y_{out} = Q_a H'C_{in} \tag{11.2}$$

where Q_w and Q_a are the volumetric flow rates of the water and air, respectively (m³/s), and C_{in}, y_{out}, and H' are as defined previously. Canceling C_{in} from both sides of the above expression, we see that the ratio Q_w/Q_a is equal to the unitless Henry's law constant (H'), which is often referred to as the *liquid-to-gas ratio* (L/G). Thus, for H' values > 100 (air preference species), we see that Q_a is less than 1% of Q_w and contaminant species are easily removed, whereas H' values < 0.01 (water preference species) require an extremely high volumetric flow rate of air relative to the liquid flow rate making contaminant species removal substantially more difficult.

The design of a stripping column often focuses on the height of the column necessary to reduce contaminant concentrations in the water to acceptable levels. While several models have been developed for this purpose, one of the simplest may be expressed as

$$Z = HTU \times NTU \qquad (11.3)$$

where Z is the height of the stripping tower packing material (m), HTU is the *height of a transfer unit* (m), and NTU is the *number of transfer units* (unitless). The height of a transfer unit for an air stripping tower can be determined as follows:

$$HTU = \frac{Q_w}{K_L a A} = \frac{L}{M_w K_L a} \qquad (11.4)$$

where Q_w is defined previously, $K_L a$ is the overall mass transfer coefficient (s^{-1}), A is the column cross-sectional area (m^2), L is the molar liquid loading rate (kmol/s per m^2 of column cross-sectional area), and M_w is the molar density of water (equal to 55.6 kmol/m^3 or 3.47 lb mol/ft^3). The number of transfer units in the air stripping column may be determined from the following relationship:

$$NTU = \left(\frac{R}{R-1} \right) \ln \left[\frac{C_{in} / y_{out})(R-1) + 1}{R} \right] \qquad (11.5)$$

where C_{in} and C_{out} are the concentrations of the contaminant in the liquid phase in the influent and effluent, respectively, and R is the stripping factor, which may be expressed as

$$R = H' \frac{Q_a}{Q_w} \qquad (11.6)$$

For the air stripping column, since H' is equal to Q_w/Q_a for a value of $R = 1$, this corresponds to the minimum amount of air required for stripping of a particular contaminant. Values of $R < 1$ would provide an insufficient flow of air to remove the target species. Due to the limitations on interphase mass transfer within the column, excess air is always required, and values for R typically range from 1.5 to 5.0.

The overall mass transfer coefficient is a complicated function of a number of variables, including liquid density and viscosity, liquid mass loading rate, characteristics of the packing media, and contaminant volatility. The packing media is often described in terms of its *packing factor* (m^{-1}) and its *surface area:volume ratio* (m^2/m^3). Mass transfer coefficients can be obtained from mathematical functions (for example, the Onda Correlations); however, use of a pilot system is recommended.

The rate of mass transfer of a contaminant is described by the following equation:

$$J = -K_L a (C_s - C) \qquad (11.7)$$

where

J = rate of mass transfer ($M \cdot L^{-1} \cdot T^{-1}$)

C_s = equilibrium liquid-phase concentration (that is, the saturation concentration) ($M \cdot L^{-3}$)

C = operating liquid phase concentration ($M \cdot L^{-3}$)

$K_L a$ = overall mass transfer rate constant (T^{-1})

Water entering an air stripper has a much higher concentration than it would have if allowed to equilibrate with relatively pure air. It is this large concentration differential that drives the transfer from the liquid to the gas phase, and the equation above demonstrates that the greater this difference, the greater the rate at which the contaminant will transfer to the air phase.

One concern when designing air stripping towers is to make sure that the tower does not flood. There are several causes for *flooding,* which occurs when the selection of packing material and the air flow rate prohibit water from flowing freely through the column. To determine the potential for headloss in an air stripping tower, Figure 11.11 is used.

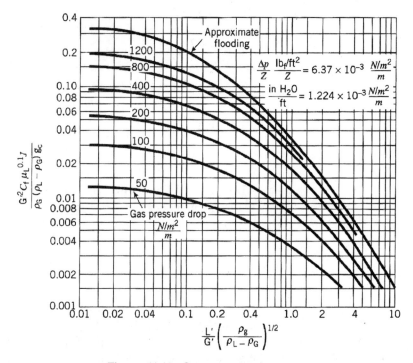

Figure 11.11   Gas pressure drop curve

To use Figure 11.11, values are obtained for the abscissa and ordinate, and the pressure drop is obtained. Note that the pressure drop is provided in $\dfrac{\text{N/m}^2}{\text{m}}$, or pascals per meter of packing depth. The terms in this graph are defined as follows:

L' = water mass loading rate (kg/m² · s)

G' = air mass loading rate (kg/m² · s)

ρ_g = density of air = 1.21 kg/m³

ρ_w = density of water (kg/m³)

μ_L = dynamic viscosity of water (kg/(m · s))

J = 1 for air-water systems

g_c = 1 for air-water systems

Reasonable gas pressure drops are between 20 and 100 N/m²/m.

Example **11.3**

Preliminary sizing of air stripper

Determine the stripping factor and the height of packing for removing ethylbenzene ($H' = 0.27$) from contaminated groundwater. The groundwater has an ethylbenzene concentration of 1.2 mg/L, and it must be decreased to 10 μg/L. Other data are

$K_L a = 0.020 \text{ s}^{-1}$

$Q_w = 18 \text{ m}^3/\text{hr}$

Column diameter = 0.75 m

Air-to-water ratio (Q_0/Q_w) = 25

Solution

The following values are found from some simple preliminary calculations:

Cross-sectional area = 0.442 m²

$C_{in}/y_{out} = 120$

$Q_w = 5 * 10^{-3} \text{ m}^3/\text{s}$

$Q_a = 1.25 * 10^{-4} \text{ m}^3/\text{s}$

R, *HTU*, and *NTU* are calculated next (Equations 11.6, 11.4, and 11.5, respectively):

$$R = H' \frac{Q_a}{Q_w} = 0.27 * 25 = 6.75$$

$$HTU = \frac{Q_w}{K_L a A} = \frac{5 \cdot 10^{-3} \text{ m}^3/\text{s}}{0.02 \text{ s}^{-1} \cdot 0.442 \text{ m}^2} = 0.57 \text{m}$$

$$NTU = \frac{R}{R-1} \ln\left[\frac{(C_{in}/y_{out})(R-1)+1}{R} \right] = \frac{6.75}{5.75} \ln\left[\frac{(120)(5.75)+1}{6.75} \right]$$

$$= 5.4 \text{ transfer units}$$

The height of the column packing is

$Z = HTU * NTU$

$= 0.57 \text{ m} \cdot 5.4 = 3.1 \text{ m}$

Example **11.4**

Assessing the potential for flooding

Determine whether the tower in Example 11.3 will flood. Assume a packing factor of 171 m⁻¹.

Solution

Initial parameters to calculate include:

$$L' = Q_w * \rho_w / A = (5 \text{ L/s} * 1 \text{ kg/L})/(0.442 \text{ m}^2)$$

$$= 11.3 \text{ kg/m}^2 \cdot \text{s}$$

$$G' = Q_a * \rho_a / A = (125 \text{ L/s} * 1.21 \text{ kg/m}^3)(1 \text{ m}^3/1000 \text{ L}) /(0.442 \text{ m}^2)$$

$$= 0.342 \text{ kg/m}^2 \cdot \text{s}$$

The *x*-axis term for Figure 11.11 is

$$\frac{L'}{G'}\left(\frac{\rho_g}{\rho_l - \rho_g}\right)^{1/2} = \frac{11.3}{0.342}\left[\frac{1.21 \text{ kg/m}^3}{(1000 \text{ kg/m}^3 - 1.21 \text{ kg/m}^3)}\right]^{1/2} = 1.15$$

The *y*-axis term for Figure 11.11 is

$$\frac{G'^2 C_f \mu_l^{0.1} J}{\rho_g(\rho_l - \rho_g)g_c} = \frac{0.3242^2 * 171 * (10^{-3})^{0.1} * 1}{121(1000 - 1.21) * 1} = 8.3 \cdot 10^{-3}$$

According to Figure 11.11, these operating conditions are not close to the flooding line, thus flooding should be avoided. The pressure drop is between 100 and 200 Pa/m.

The Federal Remediation Technologies Roundtable (*www.frtr.gov*) lists the following limitations to air stripping:

- The potential exists for inorganic (for example, iron greater than 5 ppm, hardness greater than 800 ppm) or biological fouling of the equipment, requiring pretreatment or periodic column cleaning.

- Air stripping is not effective for water contaminated with VOCs for which H′ is less than 0.01.

- Process energy costs are relatively high.

- Water sources containing compounds with low volatility at ambient temperature may require preheating.

- Off-gases may require treatment based on mass emission rate.

Gas Scrubbing

The design of a packed tower for gas scrubbing using a nonreactive solution (water) relies upon the concentration gradient as the mass transfer driving force and, therefore, is best operated in the countercurrent flow mode. This operation contacts the pure water inlet stream with the gas exiting the system, offering the potential to remove air contaminants to a very low level. The overall mass balance at steady state can be expressed as

$$(Q_g)_{in} y_{in} - (Q_g)_{out} y_{out} = (Q_w)_{out} C_{out} - (Q_w)_{in} C_{in} \tag{11.8}$$

where Q_g and Q_w are the volumetric flow rates of gas and water, respectively; y and C are gas phase and liquid phase contaminant concentrations, respectively; and the subscripts $_{in}$ and $_{out}$ refer to material entering and exiting the tower, respectively. For the specific case where the inlet water is pure ($C_{in} = 0$) and the contaminant is present in low concentrations (that is, $\Delta Q_g = 0$ and $\Delta Q_w = 0$), the mass balance may be written as

$$Q_g(y_{in} - y_{out}) = Q_w C_{out} \tag{11.9}$$

Design of the packed tower is based on the model presented previously as

$$Z = HTU \times NTU$$

where Z is the height of the stripping tower packing material (m), HTU is the *height of a transfer unit* (m), and NTU is the *number of transfer units* (unitless). For the gas scrubber, the height of a transfer unit is a function of the HTU for the gas (HTU_g) and water (HTU_w) phases, and may be expressed as

$$HTU = HTU_g + \left(\frac{H'Q_g}{Q_w}\right) HTU_w = \frac{Q_g}{(K_L a)_g A} + R'\left(\frac{Q_w}{(K_L a)_w A}\right) \tag{11.10}$$

where $(K_L a)_g$ and $(K_L a)_w$ are the overall mass transfer coefficients in the gas and water phase, respectively (s^{-1}); A is the stripping column cross-sectional area (m^2); H', Q_g, and Q_w are as defined previously; and R' is the scrubbing factor, which may be defined as

$$R' = H'\left(\frac{Q_a}{Q_w}\right) \tag{11.11}$$

Expressions for the overall mass transfer coefficients are quite complex and beyond the scope of this review; however, they can often be calculated based upon characteristic data supplied by the manufacturers of tower packing materials. The number of transfer units in the gas-scrubbing tower may be determined from the following relationship:

$$NTU = \frac{\ln\left[\left(\dfrac{y_{in} - H'C_{in}}{y_{out} - H'C_{in}}\right)(1-R) + R\right]}{1-R} \tag{11.12}$$

Often, the contaminant concentration of the inlet water is assumed to be zero, and the expression is simplified to

$$NTU = \frac{\ln\left[\left(y_{in} / y_{out}\right)(1-R) + R\right]}{1-R} \tag{11.13}$$

Example **11.5**

Gas scrubbing I

A 60-ft-tall gas scrubber has a HTU of 1.6 ft and a scrubbing factor of 0.95. Determine the gas contaminant concentration in the inlet gas if the outlet gas has a concentration of 50 µg/m^3.

Solution

Using the data given for Z and HTU in the problem statement, we first need to calculate the number of transfer units as follows:

$$Z = (HTU)(NTU) = 60 = 1.6\, NTU \Rightarrow NTU = 37.5$$

Next, using the equation for NTU (Equation 11.12) and the data given for y_{out} and R in the problem statement, we can solve for y_{in} as follows:

$$NTU = \frac{\ln\left[\left(\frac{y_{in}}{y_{out}}\right)(1-R)+R\right]}{1-R} = \frac{\ln\left[\left(\frac{y_{in}}{50}\right)(1-0.95)+0.95\right]}{1-0.95} = 37.5$$

$$1.875 = \ln\left[\left(\frac{y_{in}}{50}\right)(1-0.95)+0.95\right] \Rightarrow 6.52 = \left(\frac{y_{in}}{50}\right)(0.05)+0.95$$

$$\frac{6.52-0.95}{0.05}\left(\frac{y_{in}}{50}\right) \Rightarrow y_{in} = 5570\frac{\mu g}{m^3}$$

Example **11.6**

Gas scrubbing II

A 20-m-tall packed scrubber has an HTU of 2.1 m and scrubbing factor (R) of 1.5. Calculate the percent reduction in the contaminant concentration of the exiting gas.

Solution

The percent reduction in vapor phase concentration can be found in the expression above as y_{in}/y_{out}. To obtain a value for this ratio, we need the scrubbing factor given in the problem statement and the value for the number of transfer units (NTU). This may be calculated as

$$Z = HTU \times NTU \quad \Rightarrow \quad NTU = \frac{Z}{HTU} = \frac{20\ m}{2.1\ m} = 9.5$$

Now we can rearrange the expression given to solve for y_{in}/y_{out} as follows:

$$NTU = \frac{\ln\left[\left(\frac{y_{in}}{y_{out}}\right)(1-R)+R\right]}{1-R} \quad \Rightarrow \quad \frac{y_{in}}{y_{out}} = \frac{\exp\left[(NTU)(1-R)\right]-R}{1-R}$$

Substituting in the values for *NTU* and *R*, we can solve for y_{in}/y_{out} as follows:

$$\frac{y_{in}}{y_{out}} = \frac{\exp\left[(9.5)(1-1.5)\right]-1.5}{1-1.5} = \frac{\exp\left[(9.5)(-0.5)\right]-1.5}{-0.5} = 2.98$$

From this ratio, we may determine the percent reduction as follows:

$$\frac{y_{in}}{y_{out}} = 2.98 \Rightarrow y_{out} = \frac{y_{in}}{2.98} = 0.336\ y_{in} = 33.6\%\ of\ y_{in} \Rightarrow reduction = 66.4\%$$

Adsorption

Adsorption is a process in which contaminant molecules (called the *adsorbate* or *solute*) that are present either in the gas or liquid stream are removed from the fluid through their attachment to a solid surface (called the *adsorbent*). This technology is attractive for contaminants present in low concentrations that have a high monetary value or may impart a substantial negative impact to the environment upon release. The adsorbate preferentially attaches to the adsorbent due to chemical or physical attraction, and the attachment is generally stable as long as the fluid properties (temperature, pressure, constituent compositions) remain constant.

Regeneration of an adsorbent occurs when the majority of active sites have been saturated with adsorbate and a change in the fluid properties shifts the equilibrium of attachment, thereby releasing the previously adsorbed molecules. These released contaminant molecules are usually in much higher concentrations in the regeneration flow and may be further processed for recovery and reuse or for destruction and final, secure disposal.

Materials that make good adsorbents have large capacities due to extremely high specific surface areas that can be attributed to a highly porous internal structure. Commonly used adsorbents include *granular* or *powered activated carbon* (*GAC* or *PAC*), activated alumina, silica gel, and zeolites (aluminosilicate compounds often referred to as *molecular sieves*). Specific surface areas can range from 200–400 m^2/g for activated alumina, to 300–900 m^2/g for silica gel, to 700–1500 m^2/g for GAC and PAC.

The preference that a particular chemical contaminant has for either the fluid phase or the adsorbent is a function of the thermodynamic equilibrium of the system and is described by the *adsorption isotherm*. The isotherm data is obtained from laboratory tests. When testing a liquid medium, varying masses of GAC are placed in containers and each container is filled with the aqueous solution of interest. The containers are sealed and shaken for a period of time such that equilibrium conditions are achieved, which is typically on the order of days. The concentration of the contaminant in the liquid phase (C_e, where the subscript refers to the equilibrium conditions) is then measured. The mass of contaminant adsorbed, x, can be calculated using a mass balance.

These isotherms are often presented graphically as a plot of the ratio of mass of adsorbate adsorbed per mass of adsorbent versus the contaminant concentration in the fluid phase at equilibrium. Several researchers have mathematically described the curves that represent the data, the most common model suggested by Freundlich. The *Freundlich isotherm* is commonly applied to activated carbon adsorption in water and wastewater treatment processes and may be expressed as

$$\frac{x}{m} = X = KC_e^{1/n} \tag{11.14}$$

where x is the mass of adsorbate adsorbed, m is the mass of adsorbent, X is the mass ratio of the adsorbate adsorbed per mass of adsorbent, K is the Freundlich capacity factor, C_e is the equilibrium concentration of the solute in the aqueous phase after adsorption (mg/L), and $1/n$ is the Freundlich intensity factor. The constants K and $1/n$ in the Freundlich isotherm may be determined by plotting the adsorption data as $\ln(X)$ vs. $\ln(C_e)$ and performing a regression using the linearized form of the equation, which may be expressed as

$$\ln X = \frac{1}{n} \ln C_e + \ln K \tag{11.15}$$

Values for K and $1/n$ for many compounds of industrial and environmental significance have been determined and may be found in tables in many reference texts.

Another model frequently used to describe the behavior of contaminants in the gas phase is the *Langmuir isotherm,* which may be expressed as

$$\frac{x}{m} = X = \frac{abC_e}{1 + bC_e} \tag{11.16}$$

where values for the empirical constants a and b may be regressed from the linearized form of the equation as follows:

$$\frac{m}{x} = \frac{1}{X} = \frac{1}{a} + \frac{1}{ab}\frac{1}{C_e} \qquad \textbf{(11.17)}$$

In this form, isotherm data may be plotted as $1/X$ vs. $1/C_e$, which will yield a y-intercept of $1/a$ and a slope of $1/ab$.

Example 11.7

Estimating daily GAC requirements

Isotherm tests have been performed to determine the suitability of using GAC for removal of xylenes from groundwater. The initial contaminant concentration is 300 mg/L in groundwater, and the vials used for the analysis contain 100 mL. The laboratory results are provided in Exhibit 4. Given this data, estimate the Freundlich isotherm values and use the resulting Freundlich isotherm equation to estimate the daily GAC requirements for treating 1800 gal/day of the groundwater. An effluent concentration of 20 mg/L is required.

Container	Mass of GAC (mg)	C_e (mg/L)
1	50	250
2	200	150
3	300	105
4	400	45
5	600	10
6	700	5

Exhibit 4   Isotherm data

Solution

The mass of contaminant adsorbed to GAC in the first vial, x_1, is calculated by

$$x_1 = \Delta C * \text{sample volume} = (300 \text{ mg/L} - 250 \text{ mg/L}) * 100\text{mL}\ \frac{1 \text{ L}}{1000 \text{ ml}} = 5 \text{ mg}$$

Completing this calculation for each of the sample vials yields these results:

Container	1	2	3	4	5	6
x (mg)	5	15	19.5	25.5	29	29.5

Given that $X = x/m$, the linearized Freundlich plot can be readily generated and is shown in Exhibit 5, along with the equation of the best-fit linear regression line.

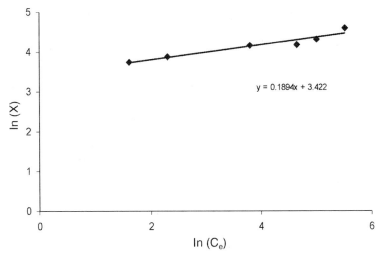

Exhibit 5 Linearization for Freundlich isotherm

From this plot, the value of the Freundlich constants are

$1/n$ = slope of line = 0.19

$\ln(K)$ = intercept of line

K = $e^{3.422}$ = 30.6 mg/g

This data produces in an isotherm equation of $x/m = 30.6\ C_e^{0.19}$ for x in mg, m in g, and C_e in mg/L. To determine the quantity of GAC required, the isotherm can be used, assuming that in practice the xylenes and GAC will reach equilibrium. To use the isotherm to solve for m, the mass of GAC required, the values of x and C_e are required. For this scale-up problem, the equilibrium concentration is 20 mg/L, and the value of x corresponds to the quantity of xylenes removed in one day:

$$\Delta C * Q * 1\ \text{day} = (300\ \text{mg/L} - 20\ \text{mg/L}) * \frac{1800\ \text{gal}}{\text{day}} * 1\ \text{day} * \frac{3.785\ \text{L}}{\text{gal}}$$

$$= 1.91 * 10^6\ \text{mg} = 1.91\ \text{kg}$$

The mass of GAC required can be obtained by solving the isotherm equation for x.

x = $m(30.6 \cdot C_e^{0.19})$

= $1.91 \cdot 10^6$ mg $[30.6(20\ \text{mg/L})^{0.19}]$

= 103 kg

Example **11.8**

Freundlich isotherms

A serum bottle is filled with 100 mL of an aqueous solution containing 100 mg/L of toluene and an unknown quantity of activated carbon. After 48 hours, the equilibrium concentration of the aqueous phase is 20 mg/L of toluene. Assuming the Freundlich isotherm parameters for toluene and activated carbon are $K = 45$ mg/g and $1/n = 0.2$, respectively, determine the amount of carbon in the bottle.

Solution

Substituting the values for K, C, and $1/n$ given in the problem statement into the Freundlich isotherm model allows us to solve for the concentration of toluene on the activated carbon as follows:

$$X = KC^{1/n} = \left(45 \frac{\text{mg}}{\text{g}} \right) \left(20 \frac{\text{mg}}{\text{L}} \right) 0.2 = 81.9 \frac{\text{mg}}{\text{g}}$$

To determine the amount of toluene associated with the carbon, we need to calculate how much toluene was in the aqueous phase initially and subtract the amount remaining in the aqueous phase at the end. This can be accomplished as follows:

$$\text{total toluene at beginning:} \quad 100 \frac{\text{mg}}{\text{L}} \times 0.1\,\text{L} = 10 \text{ mg toluene}$$

$$\text{total toluene at end:} \quad 20 \frac{\text{mg}}{\text{L}} \times 0.1\ \text{L} = 2 \text{ mg toluene}$$

$$\text{therefore, the carbon has } 10 - 2 = 8 \text{ mg toluene}$$

Now, we can estimate the amount of activated carbon in the bottle as follows:

$$\frac{8 \text{ mg toluene}}{81.9 \left(\dfrac{\text{mg toluene}}{\text{g carbon}} \right)} = 0.0977 \text{ g carbon} = 97.7 \text{ mg carbon}$$

Adsorption treatment units generally consist of a mass of adsorbent as a fixed bed in a column, sized to provide contaminant removal for a specified amount of time, with regularly scheduled regeneration cycles. A schematic representing the depth and travel of the adsorption zone in a column is presented in Figure 11.12. It would be useful to the student to refer to that figure during the discussion that follows. During the early stages of operation, contaminant mass is adsorbed onto the adsorbent closest to the fluid entrance region. As this adsorbent becomes saturated with the adsorbate, the contaminant travels further down the column until adsorbent with open active sites is encountered. In this way, the effluent fluid has a near-zero concentration of contaminant while this *adsorption zone* (Z_S) travels down the length of the column.

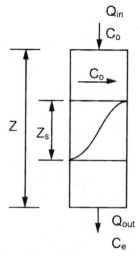

Figure 11.12   Schematic of adsorption column

When the majority of the column has been saturated with adsorbate (that is, the adsorption zone reaches the outlet of the column), low concentrations of contaminant begin to exit the adsorption column in a phenomenon called *breakthrough*. This concept can be visualized if we plot the effluent concentration of the contaminant species as a function of time (or throughput volume) as presented in Figure 11.13. The breakthrough point is typically defined as the time (or throughput volume) at which the contaminant concentration in the effluent stream reaches a value that is 5% of the inlet concentration. The column is considered *exhausted* when the effluent concentration of the contaminant reaches a value that is 95% of the influent concentration. Throughput volume is related to time by the volumetric flow rate of the fluid as $t = V/Q$.

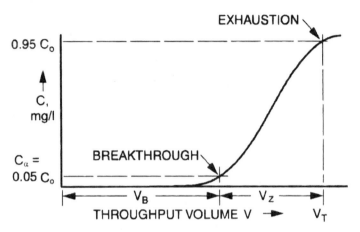

Figure 11.13   Effluent concentration profile (breakthrough curve) for adsorption column

Because it is undesirable to have any contaminant discharged from the column, it is common practice to run two or more columns in series. This allows for operation of the first column to the point of exhaustion, thereby taking advantage of all of the adsorbent capacity. After the first column is saturated with adsorbate, it is taken off-line, regencrated, and usually placed back in service as the last column in the series. Another configuration uses several columns in parallel, operated so as to stagger the regeneration cycles. The regeneration cycle may be triggered by the breakthrough of adsorbate or may be scheduled to precede breakthrough by some small time. In either configuration, it is beneficial to be able to predict the breakthrough time.

The breakthrough time t_B can be determined if the distance the adsorption zone travels is divided by the velocity of the adsorption zone V_S as it progresses down the column. Since the distance traveled is equal to the depth of the column Z minus the depth of the adsorption zone Z_S, this can be represented mathematically as

$$t_B = \frac{Z - Z_S}{V_S} \tag{11.18}$$

The depth of the adsorption zone may be calculated as

$$Z_S = Z\left[\frac{V_Z}{V_T - 0.5V_Z}\right] \tag{11.19}$$

$$V_Z = V_T - V_B$$

where V_T and V_B are the volumes of fluid treated at exhaustion and breakthrough, respectively. The velocity of the adsorption zone can be expressed as

$$V_S = \frac{QC_i}{X\rho_s A_c} = \frac{\dot{m}}{X\rho_s A_c} \tag{11.20}$$

where Q is the volumetric flow rate of the fluid (m³/s), C_i is the inlet concentration of the contaminant (g/m³), $\dot{m}$ is the mass flow rate of the contaminant species (g/s), X is the mass ratio of adsorbate to adsorbent (g/kg), ρ_s is the bulk density of the adsorbent as packed (kg/m³), and A_c is the cross-sectional area of the column (m²).

Example **11.9**

Adsorption column breakthrough I

A gas stream flowing at 1 m³/s with a VOC contaminant at a concentration of 100 mg/m³ is to be treated in an activated carbon adsorption column that is 0.5 m in diameter and 2 m long. Determine the breakthrough time t_B if the sorbent capacity is 120 g VOC per kg carbon. You may assume the packed density of the carbon is 300 kg/m³ and the volumes treated at breakthrough and at capacity are 85% and 95% of the maximum sorbent capacity, respectively.

Solution

The expression for breakthrough time was given as

$$t_B = \frac{Z - Z_S}{V_S}$$

The variable Z was given in the problem statement. The variable Z_S needs to be evaluated and may be calculated as

$$Z_S = Z\left[\frac{V_Z}{V_T - 0.5V_Z}\right] \Rightarrow V_Z = V_T - V_B$$

The problem statement provided values for V_T and V_B as percentages of the system volume. Therefore, we may calculate V_T and V_B as follows:

$$V_{column} = \frac{\pi}{4}D^2 H = \frac{\pi}{4}(0.5 \text{ m})^2(2 \text{ m}) = 0.393 \text{ m}^3$$

$$V_T = (0.95)(0.393 \text{ m}^3) = 0.3734 \text{ m}^3 \quad \Rightarrow \quad V_B = (0.85)(0.393 \text{ m}^3) = 0.334 \text{ m}^3$$

These can now be substituted into the expression for V_Z above to yield

$$V_Z = 0.3734 \text{ m}^3 - 0.334 \text{ m}^3 = 0.0394 \text{ m}^3$$

We can now solve for Z_S using these data and the expression above as follows:

$$Z_S = (2 \text{ m})\left[\frac{0.0394 \text{ m}^3}{0.3734 \text{ m}^3 - (0.5)(0.0394 \text{ m}^3)}\right] = 0.223 \text{ m}$$

Now, using the expression for v_s, we may write

$$v_s = \frac{QC_e}{X\rho_s A_c} = \frac{\left(1\frac{\text{m}^3}{\text{s}}\right)\left(100\frac{\text{mg VOC}}{\text{m}^3}\right)\left(\frac{1 \text{ g VOC}}{1000 \text{ mg VOC}}\right)}{\left(120\frac{\text{g VOC}}{\text{kg AC}}\right)\left(300\frac{\text{kg AC}}{\text{m}^3}\right)\frac{\pi}{4}(0.5)^2 \text{ m}^2} = 1.415\times10^{-5}\frac{\text{m}}{\text{s}}$$

Finally, breakthrough time may be determined by substituting into the original expression as

$$t_B = \frac{2 - 0.223 \text{ m}}{1.415 \times 10^{-5} \frac{\text{m}}{\text{s}}} = 125{,}583 \text{ s} \times \frac{1 \text{ hr}}{3600 \text{ s}} = 34.9 \text{ hr}$$

Example 11.10

Adsorption column breakthrough II

A gas flows at 1 m³/s with a VOC contaminant at a concentration of 50 μg/m³. A 3-m-tall column is packed with granular activated carbon (GAC) at a bulk density of 300 kg/m³ and has a sorbent capacity of 442 mg VOC per kg AC. Determine the column diameter if breakthrough was observed in 11 days and the adsorption zone is 15 cm in length.

Solution

Starting with the equation given for breakthrough time and the data given in the problem statement, we can calculate the velocity of the sorption zone as follows:

$$t_B = \frac{Z - Z_S}{V_S} \Rightarrow V_S = \frac{Z - Z_S}{t_B} = \frac{3 - 0.15 \text{ m}}{11 \text{ days}} = 0.26 \frac{\text{m}}{\text{day}} \times \frac{1 \text{ day}}{(24)(60)(60) \text{ s}}$$

$$= 3.0 \times 10^{-6} \frac{\text{m}}{\text{s}}$$

Now we may use the definition for V_S to determine the diameter of the column by rearranging it and substituting $\pi D^2/4$ for cross-sectional area as

$$V_S = \frac{QC_i}{X\rho \frac{\pi}{4}D^2} \Rightarrow D = \left[\frac{4QC_i}{X\rho\pi V_S}\right]^{1/2}$$

Finally, the values given in the problem statement and calculated above may be substituted into this expression to determine column diameter as

$$D = \left[\frac{(4)\left(1\frac{\text{m}^3 \text{ gas}}{\text{s}}\right)\left(50\frac{\mu\text{g VOC}}{\text{m}^3 \text{ gas}}\right)}{\left(442\frac{\text{mg VOC}}{\text{kg AC}}\right)\left(\frac{1000 \mu\text{g VOC}}{\text{mg VOC}}\right)\left(300\frac{\text{kg AC}}{\text{m}^3}\right)\pi\left(3\times10^{-6}\frac{\text{m}}{\text{s}}\right)}\right]^{1/2} = 0.40 \text{ m}$$

Isotherm tests provide guidance on how well a contaminant will be removed by GAC, but care must be taken when using these results for the design of full-scale experiments. A better alternative is to conduct a series of column experiments on a pilot scale and use these results to design a full-scale GAC column.

The BDST (Bed Depth Service Time) procedure allows for the scale-up of pilot-scale data to full-scale applications. Figure 11.14 characterizes the effluent concentration from three GAC columns in series, where C_o is the concentration of the feed solution and C is the concentration samples from the bottom of the columns. In this figure, *exhaustion* is defined as occurring when $C/C_o = 0.9$, and *breakthrough* is defined as occurring when $C/C_o = 0.1$. Thus, the point where the $C/C_o = 0.1$ line crosses an effluent curve corresponds to the location of the "leading edge" of the adsorption zone, and the point where the $C/C_o = 0.9$ line crosses an effluent curve corresponds to the "trailing edge" of the adsorption zone. In other words, for the data shown in the Figure 11.14, the leading edge of the adsorption zone is at the bottom of the first column between day 10 and day 12, and the trailing edge reaches the same point at day 28. Data is summarized in Table 11.4.

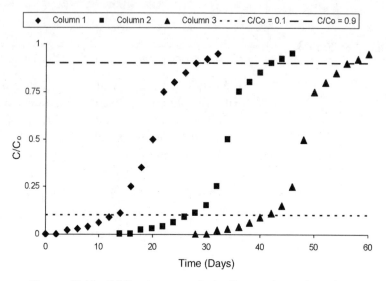

Figure 11.14 GAC column results for three columns in series

Table 11.4 Summary of GAC column results

Column #	Time at which $C/C_o = 0.1$	Time at which $C/C_o = 0.9$
1	11	28
2	27	40
3	41	56

Given a column length of Z, a BDST plot can be created (Figure 11.15). The abscissa of the graph represents the total bed depth, that is, the sum of the bed depths for the three columns in series. The line plotted on the graph is a least-squared regression linear trendline with equation $t = ax + b$.

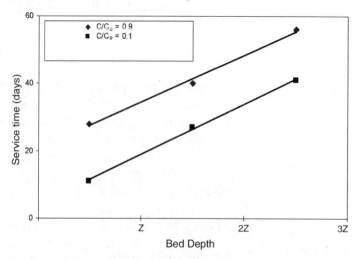

Figure 11.15 BDST curve

Useful information that can be extracted from the BDST curve is summarized in Table 11.5.

Table 11.5 Obtaining results from BDST plot

Height of the adsorption zone	=	Horizontal distance between the two best fit linear trendlines
Adsorption velocity	=	$\dfrac{1}{\text{Slope of trendline}} = \dfrac{1}{a}$
Exhaustion time	=	$\dfrac{\text{height of pilot column}}{\text{adsorption velocity}}$

Scale-up to full-scale columns consists of determining the number of columns required and their area. The area can be determined using the same loading rate from the pilot-scale tests (loading rate = flow rate / cross-sectional area) given the flow rate of the full-scale installation. The number of columns can be obtained from the following equation:

$$n = \frac{Z_s}{d} + 1 \tag{11.21}$$

where

n = number of columns in series required

Z_s = height of adsorption zone (L)

d = height of pilot columns (L)

Changes in flow rate (from the original design flow rate Q to some new flow rate Q') and influent concentration (from C to C') following design, construction, and operation can be accommodated by a procedure outlined by Watts (1998). For a change in flow rate:

$$a' = a(Q/Q') \tag{11.22}$$

where a and a' are the slopes of the BDST curves for the original and "new" conditions, respectively.

For a change in concentration:

$$a' = a \cdot \frac{C_0}{C_0'} \tag{11.23}$$

$$b' = b \cdot \frac{C_0}{C_0'} \cdot \frac{\ln\left[(C_0'/C')-1\right]}{\ln\left[(C_0/C)-1\right]} \tag{11.24}$$

where b and b' are the y-intercept of the BDST curves for the original and "new" conditions, respectively.

C_0' and C' are the influent and effluent concentration for the "new" conditions.

The FRTR (Federal Remediation Technologies Roundtable) lists the following limitations of contaminant removal via adsorption using GAC:

■ The presence of multiple contaminants can impact process performance. Single component isotherms may not be applicable for mixtures.

- Streams with high suspended solids and oil and grease concentrations (greater than 50 mg/L and greater than 10 mg/L, respectively) may necessitate pretreatment.

- Costs are high if used as the primary treatment on waste streams with high contaminant concentration levels.

- Temperature and GAC characteristics, such as pore size and quality, will impact process performance.

- Highly water-soluble compounds and small molecules are not adsorbed well.

- All spent carbon eventually needs to be properly disposed.

Solidification/Stabilization

Stabilization and solidification processes reduce the mobility of the hazardous constituents of a waste or make the waste easier to handle. Wastes may be solidified *in situ* or *ex situ*. Solidification processes are useful on inorganic and organic wastes. When used with a waste containing heavy metals, the metals are bound in place and unable to enter the environment. Such decrease in contaminant mobility can be characterized by the TCLP test.

Solidification and stabilization are not synonyms and are defined as follows (U.S. EPA Document EPA/542-R-00-010):

- *Solidification* refers to processes that encapsulate a waste to form a solid material and to restrict contaminant migration by decreasing the surface area exposed to leaching and/or by coating the waste with low-permeability materials. Solidification can be accomplished by a chemical reaction between a waste and binding (solidifying) reagents or by mechanical processes. Solidification of fine waste particles is referred to as microencapsulation, while solidification of a large block or container of waste (e.g., a 55-gallon drum) is referred to as *macroencapsulation*.

- *Stabilization* refers to processes that involve chemical reactions that reduce the leachability of a waste. Stabilization chemically immobilizes hazardous materials or reduces their solubility through a chemical reaction. The physical nature of the waste may or may not be changed by this process.

Advantages of solidification and stabilization include[4]:

- Ability to treat complex mixtures of different wastes

- Ability to restrict water access to waste contaminants by lowering waste permeability and raising waste density

- Cost-effectiveness

- Potential use as a building material

- Creation of a structurally sound material

---

4   Army Environmental Policy Institute, *Solidification Technologies for Restoration of Sites Contaminated with Hazardous Wastes* 1998 (*www.aepi.army.mil/internet/solidification-technologies-contam.pdf*).

Disadvantages of solidification and stabilization include[5]:

■ Inability to decrease contaminant toxicity

■ Possible increase in waste volume

■ Necessity of dealing with volatile air emissions

Table 11.6 shows how a variety of wastes can be stabilized using different solidification and stabilization processes.

Table 11.6 Compatibility* of wastes with various solidification techniques

Waste Component	Cement-based Treatment	Lime-based Treatment	Thermoplastic Solidification Treatment	Organic Polymer (UF— Urea-Formaldehyde Resin) Treatment
Organic solvents and oils	Many impede setting, may escape as vapor	Many impede setting, may escape as vapor	Organics may vaporize on heating	May retard set of polymers
Solid organics (for example, plastics, resins, tars)	Good—often increases durability	Good—often increases durability	Possible use as binding agent	May retard set of polymers
Acid wastes	Cement will neutralize acids	Compatible	Can be neutralized before incorporation	Compatible
Oxidizers	Compatible	Compatible	May cause matrix breakdown, fire	May cause matrix breakdown
Sulfates	May retard setting and cause spalling unless special cement is used	Compatible	May dehydrate and rehydrate causing splitting	Compatible
Halides	Easily leached from cement; may retard setting	May retard set; most are easily leached	May dehydrate	Compatible
Heavy metals	Compatible	Compatible	Compatible	Acid pH solubilizes metal hydroxides
Radioactive materials	Compatible	Compatible	Compatible	Compatible

* Compatible means that the solidification/stabilization process can generally be successfully applied to the indicated waste component. Exceptions may arise dependent upon regulatory and situation-specific factors.

Source: U.S. EPA Office of Water and Waste Management, Guide to the disposal of chemically stabilized and solidified waste, SW-872, 1982.

A brief description of some of the more common stabilization and solidification processes follows[6]:

■ *Bituminization.* Wastes are embedded in molten bitumen (or tar) and encapsulated when the bitumen cools. Heated bitumen and a waste slurry are mixed by a heated extruder. The final product is a homogenous mixture of extruded solids and bitumen.

---

5   Ibid.

6   *FRTR Remediation Technologies Screening Matrix and Reference Guide, Version 4.0,* Federal Remediation Technology Roundtable, January 2002 (*www.frtr.gov/matrix2/section4/4-21.html*).

- *Emulsified asphalt.* Asphalt emulsions are very fine droplets of asphalt dispersed in water. The small asphalt particles (5–10 μm) are stabilized by providing the droplets with a uniform electrical charge. Cationic or anionic emulsions are available. An emulsifying agent that provides a charge opposite to that of the contaminant to be removed is selected. Upon mixing, the overall charge of the waste/emulsion mixture is neutralized, which allows the particles to coalesce into a hydrophobic mass, leaving the higher quality water behind. Mixing breaks up the emulsion which releases water, leaving behind a matrix of asphalt encapsulating the waste solids.

- *Modified sulfur cement.* Modified sulfur cement is a thermoplastic material with a melting temperature between 127°C and 149°C. The molten thermoplastic and waste are mixed to form a slurry that is then allowed to cool and harden.

- *Polyethylene extrusion.* In this process, polyethylene binders and dry waste materials are mixed and heated. The heated, homogenous mixture exits the cylinder through an output die into a mold, where it cools and solidifies.

- *Portland cement.* The materials in Portland cement chemically react with water to form a solid cementitious matrix that improves the handling and physical characteristics of the waste. Mixing water and cement increases the pH of the water, which may help precipitate and immobilize some heavy metal contaminants. This type of solidification process is best suited to inorganic contaminants.

- *Vitrification/molten glass.* Vitrification processes are solidification methods that employ heat up to 1200°C to melt and convert waste materials into glasslike products. The high temperatures destroy organic constituents and create very few by-products. Materials such as heavy metals and radionuclides are incorporated into the glass structure, which is generally a relatively strong, durable material that is resistant to leaching. It is hoped that such materials are permanently leach-proof. In addition to solids, the waste materials can be liquids, wet or dry sludges, or combustible materials.

Many of these stabilization and solidification processes can be used *in situ*.

11.4.3 Thermal Treatment

In addition to the thermal methods used for stabilization and solidification described in section 11.4.2, hazardous wastes can also be treated by following thermal methods such as incineration and thermal desorption.

Incineration

Incineration of solid waste was discussed in section 10.8. Incineration is the high temperature burning (rapid oxidation) of a waste, usually at 1400°F to 2500°F. This process, unlike many of the other processes described in this chapter, destroys contaminants rather than stabilizes them or transfers them to another medium. Often auxiliary fuels are employed to initiate and sustain combustion. The destruction and removal efficiency (DRE) for properly operated incinerators exceeds the 99.99% requirement for hazardous waste and can be operated to meet the 99.9999% requirement for PCBs and dioxins. However, off-gases and combustion residuals require further treatment (for example, bottom ash may require stabilization).

Three critical factors ensure the completeness of combustion in an incinerator, also known as the "three Ts":

1. The *temperature* in the combustion chamber

2. The length of *time* wastes are maintained at high temperatures

3. The *turbulence,* or degree of mixing, of the wastes and the air

Thermal Desorption

Thermal desorption is a physical separation process and as such does not destroy organics. Rather, it transfers the contaminants from the liquid phase to the vapor phase. Wastes are heated to volatilize water and organic contaminants. The process has been used extensively to treat nonhazardous wastes, such as those from petroleum production. Such wastes only require temperatures below 400°C (*low temperature desorption*). Conversely, many hazardous wastes require higher temperatures for desorption (540°C–650°C), still much lower than incineration temperatures.

A carrier gas or vacuum system transports volatilized water and organics to the gas treatment system. The bed temperatures and residence times designed into these systems will volatilize selected contaminants but will typically not oxidize them. Units are modular and able to be moved from site to site. Temperature control is very important, especially given the wide array of contaminants that may be present at a site.

A typical schematic is shown in Figure 11.16. The gas treatment system may include afterburning of the gases, adsorption on GAC, or condensed and concentrated into a liquid that can be reused or disposed of.

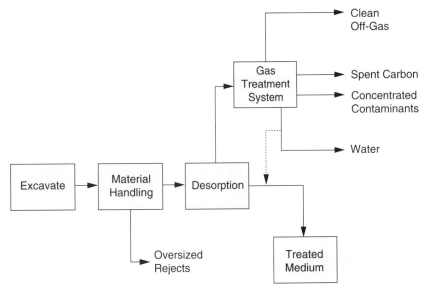

Figure 11.16 Thermal desorption

Source: *Remediation Technologies Screening Matrix and Reference Guide*, 4th ed., U.S. Army Environmental Command, SFIM-AEC-ET-CR-97053, January 2002.

The following limitations apply to thermal desorption[7]:

■ Pretreatment consisting of dewatering may be necessary to achieve acceptable soil moisture content levels.

■ Highly abrasive feed potentially can damage the processor unit.

■ Heavy metals in the feed may produce a treated solid residue that requires stabilization.

■ Clay and silty soils and high humic content soils increase reaction time as a result of binding of contaminants.

11.5 STORAGE

The EPA defines storage as the holding of waste for a temporary period of time prior to the waste being treated, disposed, or stored elsewhere. Hazardous waste is commonly stored prior to treatment or disposal and must be stored in one of the following, as defined by the U.S. EPA[8]:

■ *Containers.* A hazardous waste container is any portable device in which a hazardous waste is stored, transported, treated, disposed, or otherwise handled. Examples of containers include 55-gallon drums, tanker trucks, railroad cars, buckets, bags, and even test tubes.

■ *Tanks.* Tanks are stationary holding units constructed of nonearthen materials used to store or treat hazardous waste.

■ *Drip pads.* A drip pad is a wood drying structure used by the pressure-treated wood industry to collect excess wood preservative drippage. Drip pads are constructed of nonearthen materials with a curbed, free-draining base that is designed to convey wood preservative drippage to a collection system for proper management.

■ *Containment buildings.* Containment buildings are completely enclosed, self-supporting structures used to store or treat noncontainerized hazardous waste.

■ *Waste piles.* A waste pile is an open, uncontained pile used for treating or storing waste. Hazardous waste piles must be placed on top of a double liner system to ensure leachate from the waste does not contaminate surface or groundwater supplies.

■ *Surface impoundments.* A surface impoundment is a natural topographical depression, man-made excavation, or diked area such as a holding pond, storage pit, or settling lagoon. Surface impoundments are formed primarily of earthen materials and are lined with synthetic geomembranes to prevent liquids from escaping.

---

7   *FRTR Remediation Technologies Screening Matrix and Reference Guide, Version 4.0,* Federal Remediation Technology Roundtable, January 2002 (*www.frtr.gov/matrix2/section4/4-26.html*).

8   *www.epa.gov/epaoswer/osw/tsds.htm#store.*

11.6 DISPOSAL

Disposal may occur in one of the following: landfill, surface impoundment, waste pile, land treatment unit, injection well, salt dome formation, salt bed formation, underground mine, or underground cave. The latter four methods are geologic repositories. Such units vary greatly and are subject to environmental performance standards rather than prescribed technology-based standards.

11.6.1 Hazardous Waste Landfills

Hazardous waste landfills have many similarities to solid waste landfills; however, some additional considerations must be taken into account:

- A double composite liner is required, as is a leak detection system.

- Hazardous waste landfills must meet more stringent state requirements, which often include on-site state inspectors, additional groundwater monitoring wells, and restrictions on radioactive wastes.

- Liquid wastes are not allowed in hazardous waste landfills. Exceptions include very small containers (for example, laboratory ampules). Free liquids are eliminated by decanting or use of an absorbent material.

- All wastes delivered to a hazardous waste landfill must be manifested (to allow cradle-to-grave tracking).

- Care must be taken to ensure that incompatible wastes are not stored in close proximity to one another.

Operation of a hazardous waste landfill also differs from operation of a municipal solid waste landfill. Waste that is obtained as solid or semisolid form is spread in two- to three-foot layers and compacted. A 12-inch layer of daily cover must be used. Containers are placed upright in the cell. Space between the containers is filled with compatible bulk hazardous waste or soil.

11.6.2 Injection Wells

Injection wells have also been termed "deep-well disposal" or "subsurface injection." The wells must be deep enough to reach a porous, permeable, saline-water-bearing rock stratum that is confined by relatively impermeable layers above and beneath. Depths are on the order of several thousand feet. A schematic is shown in Figure 11.17. Some terms related to Figure 11.17 are defined as follows:

- *Surface casing.* This casing prevents contamination of any aquifers used for drinking water. It is cemented along its entire length.

- *Inner casing.* This casing is cemented along its entire length in order to seal off the injected wastes from any geologic formations above the injection zone.

- *Injection tubing.* Waste is injected into the injection zone through this tubing.

- *Annulus.* The area between the injection tubing and inner casing is filled with an inert, pressurized fluid and held in place by the *packer*. Thus, injected waste is prevented from entering the annulus.

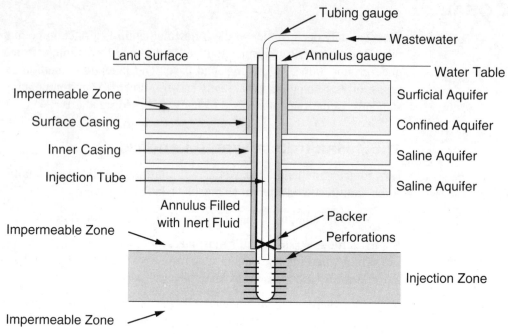

Figure 11.17 Deep-well injection

Source: *Remediation Technologies Screening Matrix and Reference Guide*, 4th ed., U.S. Army Environmental Command, SFIM-AEC-ET-CR-97053, January 2002.

The following limitations have been noted for deep-well injection by the Federal Remediation Technology Roundtable[9]:

- Injection is not feasible where seismic activity could occur.

- Injected wastes must be compatible with the mechanical components of the injection well system and the natural formation water. The waste generator may be required to perform pretreatment to ensure compatibility.

- High concentrations of suspended solids (typically > 2 ppm) can lead to plugging of the injection interval.

- Corrosive media may react with the injection well components, within the injection zone formation, or with confining strata with very undesirable results.

- High iron concentrations may result in fouling.

- Organic carbon may serve as an energy source for indigenous or injected bacteria resulting in rapid population growth and subsequent fouling.

- Extensive assessments are required before regulatory approval is obtained.

11.7 RADIOACTIVE WASTES

Radioactive wastes are not defined by RCRA as hazardous wastes. Radiation can be classified as nonionizing or ionizing, and the difference is illustrated in Fig-

---

9   *FRTR Remediation Technologies Screening Matrix and Reference Guide, Version 4.0*, Federal Remediation Technology Roundtable, January 2002 (*www.frtr.gov/matrix2/ section4/4-54.html*).

ure 11.18. Also, Table 11.3 provides useful information on the characteristics of the different types of radiation. Ionizing radiation has enough energy to break chemical bonds.

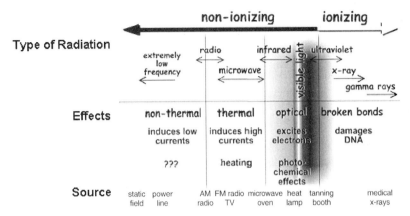

Figure 11.18 Ionizing vs nonionizing radiation

Source: U.S. EPA, *Understanding Radiation (www.epa.gov/radiation/understand/ionize_ nonionize.htm).*

Units of radioactivity include the curie (Ci) and the becquerel (Bq). A curie is defined as 37 billion disintegrations per second, while a becquerel is equal to 1 disintegration per second.

Radioactive wastes can be divided into six categories[10]:

1. Spent nuclear fuel from nuclear reactors

2. High-level radioactive waste from the reprocessing of spent nuclear fuel

3. Transuranic radioactive waste, resulting mainly from manufacture of nuclear weapons

4. Uranium mill tailings from the mining and milling of uranium ore

5. Low-level radioactive waste, generally in the form of radioactively contaminated industrial or research waste

6. Naturally occurring radioactive material

Remediation technologies include solidification (typically, cement ash or through vitrification) and stabilization. Stabilization involves the addition of chemical binders, such as cement, silicates, or pozzolans, which limit the solubility or mobility of radionuclides. Groundwater can be remediated by removal of radionuclides via coagulation and flocculation. Precipitation with hydroxides, carbonates, or sulfides is possible, as is ion exchange. Such technologies are generally applicable for low-level radioactive waste (LLW), transuranic waste (TRU), and/or uranium mill tailings. The technologies are not applicable to spent nuclear fuel and, for the most part, are not applicable for high-level radioactive waste.

Some special considerations when remediating sites contaminated with radionuclides include the following[11]:

---

10 U.S. EPA, *www.epa.gov/radiation/manage.htm.*

11 *Remediation Technologies Screening Matrix and Reference Guide, Version 4.0,* Federal Remediation Technology Roundtable, January 2002; *www.frtr.gov/matrix2/section2/2_9_1.html.*

■ Implementation of remediation technologies should consider the potential for radiological exposure to workers (internal and external). The degree of hazard is based on the radionuclide(s) present and the type and energy of radiation emitted (that is, alpha particles, beta particles, gamma radiation, and neutron radiation).

■ Because radionuclides are not destroyed, ex situ techniques will require eventual disposal of residual radioactive wastes. These waste forms must meet disposal site waste acceptance criteria.

■ Some remediation technologies result in the concentration of radionuclides. By concentrating radionuclides, it is possible to change the classification of the waste, which impacts requirements for disposal. For example, concentrating radionuclides could result in LLW becoming TRU waste (if TRU radionuclides were concentrated to greater than 100 nanocuries/gm with half-lives greater than 20 years per gram of waste). Also, LLW classifications (for example, Class A, B, or C for commercial LLW) could change due to the concentration of radionuclides. Waste classification requirements for disposal of residual waste (if applicable) should be considered when evaluating remediation technologies.

■ Disposal capacity for radioactive and mixed waste is extremely limited. Mixed waste is waste comprised of radioactive waste and hazardous waste.

ADDITIONAL RESOURCES

1. Freeman, H. M. *Standard Handbook of Hazardous Waste Treatment and Disposal*. McGraw-Hill, 1998.

2. LaGrega, M. D., P. L. Buckingham, and J. C. Evans. *Hazardous Waste Management*. McGraw-Hill, 2000.

3. O'Brien and Gere Engineers, Inc. *Innovative Engineering Technologies for Hazardous Waste Remediation*. Van Nostrand Reinhold, 1995.

4. Watts, R. J. *Hazardous Wastes: Sources, Pathways, Receptors*. Wiley, 1998.

Site Remediation

OUTLINE

12.1 SITE REMEDIATION UNDER SUPERFUND

The Comprehensive Environmental Response, Compensation, and Liability Act (CERCLA), commonly known as Superfund, was enacted by Congress on December 11, 1980. CERCLA[1]

■ established prohibitions and requirements concerning closed and abandoned hazardous waste sites,

■ provided for liability of persons responsible for releases of hazardous waste at these sites, and

■ established a trust fund to provide for cleanup when no responsible party could be identified.

The law authorizes two kinds of response actions:

1. Short-term removals to promptly address releases

2. Long-term remedial response actions that permanently and significantly reduce the dangers associated with releases or threats of releases of hazardous substances

This chapter will focus on the cleanup process governed by Superfund. However, it is important to note that the steps followed to clean up a site under Super-

---

1   U.S. EPA, *www.epa.gov/superfund/action/law/cercla.htm.*

fund are similar to the steps required to clean up a site without being under the auspices of Superfund (that is, voluntary cleanup or cleanup regulated by individual states).

The Superfund cleanup process consists of the following steps.[2]

Step 1: Site Discovery

Site discovery may be made by any number of parties, including local and state agencies, businesses, the U.S. Environmental Protection Agency (U.S. EPA), or by members of the public.

Step 2: Preliminary Assessment/Site Inspection (PA/SI)

Preliminary assessment evaluates the level of threat to human health and the environment. Sites requiring emergency response may be identified in the PA. If the PA recommends that additional investigations be made, a site inspection will be conducted. During the SI, soil, water, and air samples are obtained and analyzed. A vast amount of information is collected that serves as input to the Hazard Ranking System (HRS).

Step 3: HRS Scoring

The HRS is a numerical screening mechanism used to place sites on the National Priorities List (NPL). A high HRS score does not infer a higher priority for funding. The HRS score is the combination of scores from each of four pathways:

1. Groundwater migration

2. Surface water migration

3. Soil exposure

4. Air migration

The HRS site score S is calculated as follows:

$$S = \sqrt{\frac{S_{gw}^{\ 2} + S_{sw}^{\ 2} + S_{s}^{\ 2} + S_{a}^{\ 2}}{4}}$$

where

S_{gw} = groundwater migration pathway score

S_{sw} = surface water migration pathway score

S_{s} = soil exposure pathway score

S_{a} = air exposure pathway score

The mathematical nature of the root-mean-square equation is such that higher-scoring pathways exert a proportionately greater influence on the site score than lower-scoring pathways.

The groundwater pathway score depends on the following:

■ Likelihood of release as noted by direct observation or based on the possibility of release due to site characteristics such as depth to aquifer, soil properties, and net precipitation.

---

2   U.S. EPA, *www.epa.gov/superfund/action/process/sfproces.htm.*

- Waste characteristics, including contaminant toxicity and mobility and quantity

- Targets, characterized by the distance to the nearest well, existence of well-head protection program, and population served by groundwater

The surface water score is based on waste characteristics and targets also but with a specific focus on three threats: drinking water threat, human food chain threat, and environmental threat.

The soil pathway threat assesses two threats: the resident population threat and the nearby population threat. The score for the former depends on the size of the resident population, while the latter is based on the population within one mile of the site. As with the other pathways, consideration is given to the likelihood of exposure and waste characteristics.

The air pathway score depends on the likelihood of release, contaminant concentrations in the air, and other contaminant characteristics.

Step 4: NPL Site Listing Process

A site is placed on the NPL by one of three mechanisms:

1. Magnitude of the HRS score

2. Designation by states or territories regardless of score

3. Ability to meet each of the three following requirements:

   - The Agency for Toxic Substances and Disease Registry (ATSDR) of the U.S. Public Health Service has issued a health advisory that recommends removing people from the site.

   - The U.S. EPA determines the site poses a significant threat to public health.

   - The U.S. EPA anticipates it will be more cost-effective to use its remedial authority (available only at NPL sites) than to use its emergency removal authority to respond to the site.

Step 5: Remedial Investigation/Feasibility Study (RI/FS)

The remedial investigation serves as the mechanism for collecting data in addition to that data collected for the HRS score in order to

- characterize site conditions,

- determine the nature of the waste,

- assess risk to human health and the environment, and

- conduct pilot- or laboratory-scale testing to evaluate the potential performance and cost of the treatment technologies that are being considered.

The feasibility study is the mechanism for the development, screening, and detailed evaluation of alternative remedial actions.

The RI and FS are conducted concurrently. Data collected in the RI influence the development of remedial alternatives in the FS, which in turn affects the data needs and scope of treatability studies and additional field investigations.

Step 6: Record of Decision (ROD)

The Record of Decision (ROD) is a public document that explains which cleanup alternatives will be used to clean up a Superfund site.

Step 7: Remedial Design/Remedial Action (RD/RA)

Based on the specifications outlined in the ROD, full-scale remediation technologies are designed (remedial design) and constructed and implemented (remedial action).

Step 8: Construction Completion

Sites qualify for construction completion when

- all necessary physical construction is complete, regardless of whether final cleanup levels have been achieved; or

- the U.S. EPA has determined that the response action should be limited to measures that do not involve construction; or

- the site qualifies for deletion from the NPL.

Inclusion of a site on the Construction Completions List (CCL) has no legal significance.

Step 9: Postconstruction Completion

The goal of postconstruction tasks is to ensure the long-term protection of human health and the environment. Postconstruction completion steps may include:

- Operations and maintenance to ensure the effectiveness of the remedy

- Five-year reviews

- Assurance of proper institutional controls such as administrative and/or legal controls to minimize the potential for human exposure to contamination and/or protect the integrity of the remedy by limiting land or resource use

- Optimization of the remedy to decrease annual operating costs

- Deletion from the NPL if no further response action is appropriate

- Assurance that reuse activities do not adversely affect the implemented remedy

12.2   CONTAMINANT TRANSPORT

Contaminants may be transported in air, water, and soil. Exposure to water-borne contaminants will occur if people drink contaminated groundwater or surface water or accidentally ingest it while swimming, or if it comes into contact with their skin by any means. Humans will be exposed to hazardous substances in soil, sediment, or dust if they accidentally ingest it, inhale it (in the form of dust), or by direct dermal contact. Children are highly susceptible to exposure through soil pathways. Airborne contaminants can be inhaled or absorbed by the skin.

The movement of contaminants in the air has been discussed in section 9.3. Transport through surface water can be analyzed using concepts such as PFR and CSTR modeling (section 1.5) and the concepts of bioaccumulation.

Transport via groundwater occurs according to the principles of groundwater flow described in section 5.4. However, several important concepts related to contaminant flow in groundwater must be described in this chapter.

Groundwater resources are relatively easy to contaminate as compared to surface water, and given the large quantity of groundwater extracted for drinking water, groundwater contamination is one of the most pressing issues facing environmental engineers. The problem is further intensified by recognizing that very low concentrations of these contaminants can have significant health risks. Moreover, removal or destruction of these contaminants is a complicated and costly undertaking, further complicated by the fact that modeling contaminant movement in the groundwater is not nearly as straightforward as modeling the movement *of* the groundwater.

Contaminants can take one or more forms in the subsurface environment. They may be found in the gas phase, adsorbed to soil particles, dissolved in solution, or present in an immiscible phase. Transport of the contaminant in the groundwater system varies significantly depending on which form the contaminants take.

12.2.1 NAPLs in the Environment

Insoluble organic contaminants may be present as NAPLs (non-aqueous phase liquids); that is, they are sparingly soluble in water. Although they have low solubility, they often are soluble enough such that the maximum contaminant levels (MCLs) are violated. In addition to being in solution, NAPLs may be found in bulk (*free product*) or attached to soil particles as *residual* NAPL.

LNAPLs (light non-aqueous phase liquids) are NAPLs that are lighter than water. Upon release to the environment, LNAPLs will migrate downward until they encounter a physical barrier (for example, low permeability strata) or are affected by buoyancy forces near the water table. Once the capillary fringe is reached, LNAPLs may move laterally as a continuous, free-phase layer. Note the irregular shape of the LNAPL in Figure 12. 1 and the fact that a vapor plume of the LNAPL has formed. Examples of LNAPLs include gasoline and various types of oils. A common source of LNAPLs is a leaking underground storage tank.

Dense NAPLs (DNAPLs) have a specific gravity greater than 1 and will tend to sink to the bottom of surface waters and groundwater aquifers. The movement of DNAPLs in the subsurface is extremely complicated. DNAPLs will form a vapor phase in the unsaturated zone, will form pools based on soil heterogeneities in the saturated zone, will dissolve (albeit sparingly) and form an aqueous plume, and will eventually form a pool at the "bottom" of the aquifer but will continue to move from there into fractures in the confining layer and move en masse in the direction of groundwater flow. Clearly, modeling the movement of this contaminant in the subsurface and removing it for treatment is a tremendously complicated undertaking. Many chlorinated solvents, such as those used in dry cleaning operations, are DNAPLs, as are creosote, coal tar, and PCB oils. Sources include accidental spills or improper disposal practices (for example, unlined evaporation ponds or lagoons) in industries such as metal degreasing, pharmaceutical production, and pesticide formulation.

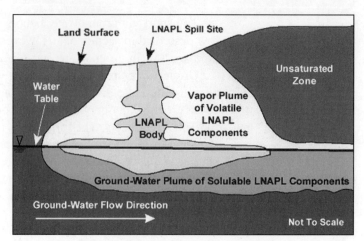

Figure 12.1 LNAPL in the subsurface

Source: U.S. Geological Survey, *http://toxics.usgs.gov/definitions/lnapls.html.*

12.2.2 Solute Movement in Groundwater

The movement of solutes in groundwater is governed by two processes: *advection* and *hydrodynamic dispersion*.

Advection is the movement of the contaminant with the bulk fluid. In other words, advection describes the plug flow behavior of contaminants moving at a velocity equal to the groundwater *pore velocity* (section 5.4). Movement of a contaminant due to advection for a continuous input and for a slug input are shown in Figures 12.2 and 12.3, respectively.

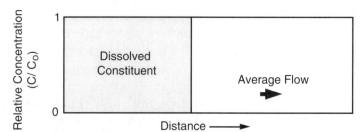

Figure 12.2 Movement of concentration front of a continuous input due to advection

Source: U.S. EPA, *Transport and Fate of Contaminants in the Subsurface*, EPA/625/4-89/019, 1989.

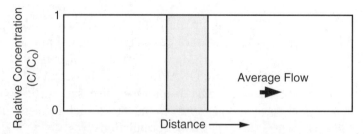

Figure 12.3 Movement of concentration front of a slug input due to advection

Source: U.S. EPA, *Transport and Fate of Contaminants in the Subsurface*, EPA/625/4-89/019, 1989.

Hydrodynamic dispersion is the spread of solute in the groundwater due to molecular diffusion and mechanical dispersion. The term *diffusion* is often used incorrectly to describe the overall movement of contaminants in the groundwater. Diffusion refers specifically to the movement of a dissolved solute in response to a *concentration gradient* and is governed by Fick's first law:

$$J = -D \frac{dc}{dx} \qquad (12.1)$$

where

J = diffusion flux $(M \cdot L^{-2} \cdot T^{-1})$

D = diffusion coefficient, or *diffusivity* $(L^2 \cdot T^{-1})$

C = concentration $(M \cdot L^3)$

x = distance (L)

Mechanical dispersion occurs due to physical mixing during fluid movement between the solid grains in the soil matrix. Specifically, water in the center of the pore moves faster than water at the soil particle surface. Also, the crossing of flow paths as the fluid flows around the solid grains in the rock further mixes the water. Additionally, the different paths that water takes through a soil matrix cause the contaminant to move at varying velocities.

The impact that dispersion has on contaminant movement is shown in Figure 12.4. Note that this is the same phenomenon that would occur in a plug flow reactor when dispersion is taken into account.

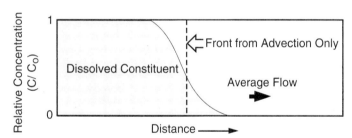

Figure 12.4 Movement of concentration front of a continuous input due to advection and dispersion

Source: U.S. EPA, *Transport and Fate of Contaminants in the Subsurface*, EPA/625/4-89/019, 1989.

The additional impacts of mechanical dispersion are illustrated in Figure 12.5. Between time A and time B, the effects of dispersion "flatten" out the bell-shaped curve of relative concentration versus distance. In reality, the bell-shaped curve would likely be skewed. It is important to note that the area under each curve is identical, corresponding to the mass of the contaminant.

As a result of the dispersion process, the contaminant is diluted as it moves through the subsurface. This process also leads to the formation of a *plume*. A plume is a body of contaminated groundwater in which the concentration of contaminant varies. The shape and movement of the plume is affected by the geologic conditions, contaminant properties, and groundwater flow characteristics. The contaminant concentration at various locations in the plume is characterized by extensive subsurface monitoring, typically through the use of a series of monitoring wells. The concentration of contaminant in the plume is often characterized by using concentration contours, as shown in Figure 12.6.

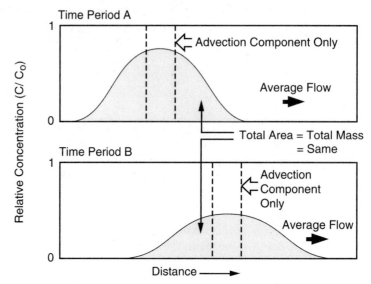

Figure 12.5   Advective vs. dispersive effects

Source: U.S. EPA, *Transport and Fate of Contaminants in the Subsurface*, EPA/625/4-89/019, 1989.

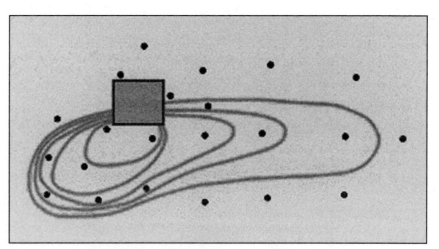

Figure 12.6   Contaminant concentration contours

Source: U.S. EPA, *www.epa.gov/OUST/graphics/cadnapl.htm.*

The movement of contaminants is further complicated by the fact that in addition to advection and hydrodynamic dispersion, contaminants can be "removed" as they travel through the subsurface. Such removal may be due to adsorption onto soil particles or due to a variety of chemical and biological transformations. The movement of the contaminant as affected by these various phenomena is shown in Figure 12.7. In this figure, the following notation is used: A = advection; D = dispersion; S = sorption; and B = biotransformation. The area under curve A is the same as the area under curve A+D; however, due to removal of mass via adsorption and biotransformation, the areas under the remaining curves are less.

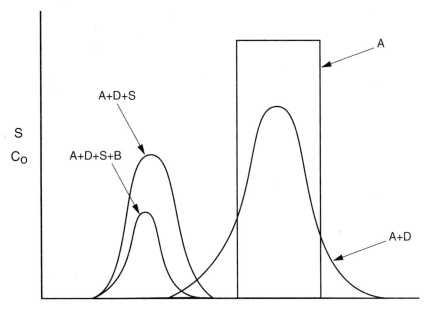

Distance from Slug-Release Contaminant Source

Figure 12.7 Impact of natural processes on contaminant movement

Source: U.S. EPA, *Transport and Fate of Contaminants in the Subsurface*, EPA/625/4-89/019, 1989.

The *retardation factor* quantifies the behavior of the A + D + S curve relative to the A + D curve of Figure 12.7. Specifically, the retardation factor R is defined as

$$R = \frac{\text{average linear groundwater velocity}}{\text{velocity of contaminant due to adsorption effects}}$$

The value of the retardation coefficient can be calculated from the following equation:

$$R = 1 + \frac{\rho_b}{\eta} K \tag{12.2}$$

where

ρ_b = soil bulk density $(M \cdot L^{-3})$

η = porosity

K = soil–water partition coefficient

The *soil–water partition coefficient* is the ratio of contaminant found on the soil to the ratio of contaminant found in the water phase and can be expressed as follows:

$$K = K_{oc} \cdot f_{oc} \tag{12.3}$$

where

K_{oc} = soil–water partition coefficient normalized to organic carbon

f_{oc} = fraction of organic carbon

The soil–water partition coefficient is related to the *octanol–water partition coefficient, K_{ow}*. To determine the value of K_{ow} for a compound, the compound is placed in a sealed container containing equal parts water and octanol ($C_8H_{17}OH$). After equilibrium is reached, the concentration of the contaminant in the water and in the octanol is measured. K_{ow} is defined as:

$$K_{ow} = \frac{[\text{contaminant}]_{octanol}}{[\text{contaminant}]_{water}} \qquad (12.4)$$

If a compound has a high K_{ow} value, it is said to by *hydrophobic;* that is, it is "water fearing," and the compound partitions into the octanol phase. There is a strong correlation between K_{ow} and K_{oc}, and the following equation may be used to estimate K_{oc} based on K_{ow}[3]:

$$\log K_{oc}\,\frac{cm^3}{g} = 0.903\,\log(K_{ow}) + 0.094 \qquad (12.5)$$

A solute may be classified as *conservative* or *reactive.* A conservative solute, or *tracer,* is one that does not react with soil and/or groundwater either through adsorption or chemical, biological, or radioactive decay. Conversely, a reactive solute reacts with the soil and/or groundwater.

Example 12.1

Retardation of contaminants II

Determine the retardation factor for the movement of benzene (log K_{oc} = 2.01) and pyrene (log K_{oc} = 4.69) through this soil matrix:

bulk density = 150 lb/ft³

porosity = 0.35

f_{oc} = 0.5%

Solution

The partition coefficients for both contaminants may be calculated first:

$$K = K_{oc} \cdot f_{oc}$$

$$K_{benzene} = 10^{2.01} \cdot 0.005 = 0.51$$

$$K_{pyren} = 10^{4.69} \cdot 0.005 = 250$$

The retardation coefficients can be calculated now using Equation 12.2:

$$R_{benzene} = 1 + \frac{150\ lb/ft^3}{0.35} 0.51 \frac{cm^3}{g} \frac{1000\ g}{2.205\ lb} \frac{1\ ft^3}{2.8 \cdot 10^4 cm^3} = 4.5$$

$$R_{pyrene} = 1 + \frac{150\ lb/ft^3}{0.35} 250 \frac{cm^3}{g} \frac{1000\ g}{2.205\ lb} \frac{1\ ft^3}{2.8 \cdot 10^4\ cm^3} = 1736$$

Consequently, the pyrene would take nearly 400 times longer to travel the same distance as the benzene.

---

3   James R. Mihelcic, *Fundamentals of Environmental Engineering* (Wiley, 1999).

Example **12.2**

Retardation of contaminants I

This example problem is based on Exhibit 1. For this example, a study was conducted by injecting three contaminants into a well and measuring their concentrations 5 meters down gradient. The average groundwater velocity on the site was 30 m/yr. Results are provided in Exhibit 1. Estimate a value for the retardation factor for carbon tetrachloride and tetrachloroethene.

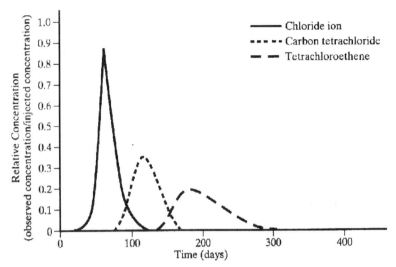

Exhibit 1 Concentration of contaminants in down-gradient well

Source: James R. Mihelcic, *Fundamentals of Environmental Engineering*, © 2001, John Wiley & Sons, Inc. Reprinted by permission.

Solution

The average velocity of each of the plumes is approximated by this relationship:

$$v = \frac{\text{distance}}{\text{time}} = \frac{5\text{ m}}{\text{time to peak}}$$

From the graphs, we estimate a time to peak for the carbon tetrachloride and tetrachloroethene to equal 116 days and 182 days, respectively. Therefore, the average velocities of these contaminants are 16 m/yr and 10 m/yr, respectively. Given the average groundwater velocity of 30 m/yr, the retardation factor for the carbon tetrachloride is 1.9 and 3 for the tetrachloroethene.

Exhibit 1 also demonstrates the use of a nonreactive tracer. The chloride peak occurs at approximately 70 days. This corresponds to a tracer velocity of 26 ft/s, demonstrating that the tracer moves nearly as fast as the groundwater.

This example problem further demonstrates that the relative concentration of contaminants from a source with multiple contaminants will vary with time and location.

12.3 TREATMENT PROCESSES

Hazardous waste sites can be remediated using a wide array of technologies. One way to classify the treatment methods is *ex situ* or *in situ*. For *in situ* (or in place) treatment, the contaminant is not moved from the subsurface, while *ex situ* treatment involves the excavation of soil (or pumping of contaminated groundwater from the aquifer) for treatment.

Advantages of *in situ* treatment include the following:

■ No costs for excavation or groundwater extraction

■ Ability to treat soils under buildings and other structures without affecting the structure

■ Avoidance of risks and costs associated with transportation

■ Decreasing likelihood of spreading contaminants off-site

Advantages of *ex situ* treatment include:

■ Shorter time periods for remediation

■ Greater uniformity of treatment due to homogenization of solid phase (for example, soil or sludge) or the ability to monitor and continuously mix the groundwater

The *ex situ* treatment of groundwater is often termed "pump and treat." Pump-and-treat technologies are used at approximately three-quarters of all Superfund sites where groundwater is contaminated.

Pump-and-treat systems are used for two main purposes: treatment of the groundwater and/or hydraulic containment of contamination. Figure 12.8 illustrates a pump-and-treat system for remediation of a site contaminated by DNAPLs from a dry cleaning facility. Treated groundwater may be injected directly back into the aquifer as shown or may be discharged to surface water.

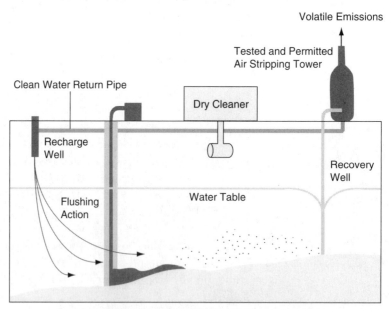

Groundwater Recovery, Treatment, and Recharge

Figure 12.8 Pump-and-treat system example

Source: U.S. EPA, *www.epa.gov/OUST/graphics/cadnapl.htm.*

Two phenomena that complicate pump-and-treat technology are *tailing* and *rebound*. Tailing refers to the progressively slower rate of decline in dissolved contaminant concentration as pump-and-treat remediation continues. As a result of tailing, the volume of groundwater to be pumped, and therefore the cleanup time, can be on the order of ten times greater than if tailing did not occur. Rebound is a relatively rapid increase in concentration following cessation of pumping. These two concepts are illustrated in Figure 12.9.

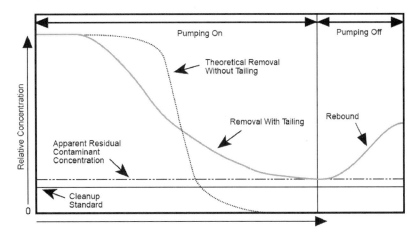

Figure 12.9 Tailing and rebound

Source: U.S. EPA, *Methods for Monitoring Pump-and-Treat Performance*, EPA/600/R-94/123, 1994.

A variety of *ex situ* treatment technologies for hazardous wastes, whether excavated soils or extracted groundwater, were discussed in section 11.4. This chapter focuses on the *in situ* treatment of waste during the site remediation process.

The process for selecting a remediation strategy as part of a feasibility study begins with identifying the objectives of the remediation and screening technologies based on their technical feasibility, effectiveness, and cost. These two steps are typically followed by treatability studies to assess the effectiveness of the alternatives and to provide information for eventual optimization of the process. Following these initial steps, a remediation process is selected using the following nine criteria suggested by the U.S. EPA:

1. Overall protection of human health and the environment

2. Compliance with appropriate regulations

3. Long-term effectiveness and permanence

4. Reduction of toxicity, mobility, or volume

5. Short-term effectiveness

6. Implementability

7. Cost

8. State acceptance

9. Community acceptance

Treatment technologies for hazardous waste include the following:

■ Bioventing

■ Phytoremediation

- Enhanced biodegradation

- Natural attenuation

- Air sparging

- Bioslurping

- Chemical oxidation

- Electrokinetic separation

- Passive/reactive treatment walls

- Soil flushing

- Soil vapor extraction

- Solidification/stabilization

- Thermal treatment

- Physical barriers

Several of these processes have been discussed in previous chapters; following sections briefly describe the rest.

12.3.1 Biological Treatment Technologies

Bioremediation techniques destroy contaminants by creating a favorable environment for the microorganisms such that the microorganisms are able to grow and use the contaminants as a food and energy source.

Biological processes are implemented at a relatively low cost. Bioremediation has been used to remediate groundwater contaminated by petroleum hydrocarbons, solvents, pesticides, wood preservatives, and other organic chemicals. Care must be taken when contemplating a remediation alternative to consider that some compounds may be broken down into more toxic by-products during the bioremediation process (for example, TCE to vinyl chloride).

The following must be considered when designing an *in situ* biological treatment process:

- Oxygen must be supplied at rates sufficient to maintain aerobic conditions and may be accomplished by forced air, liquid oxygen injection, or hydrogen peroxide injection. Alternatively, anaerobic conditions can be used to degrade highly chlorinated contaminants. The dechlorinated contaminants can be biodegraded by subsequent aerobic treatment.

- Nutrients required for cell growth include nitrogen, phosphorus, potassium, sulfur, magnesium, calcium, manganese, iron, zinc, and copper. Some or all of these may be added to the subsurface (for example, as ammonium for nitrogen and as phosphate for phosphorus).

- Control of pH is critical, as it affects the solubility of constituents that can affect biological activity. For example, many metals that are potentially toxic to microorganisms are insoluble at elevated pH; therefore, elevating the pH of the treatment system can reduce this risk.

- Temperature affects microbial activity, with biological activity increasing as temperature increases. Provisions for heating the bioremediation site, such as use of warm-air injection, may speed up the remediation process. Increasing temperature promotes volatilization of VOCs, increases the solubility of most contaminants, and decreases the amount of oxygen that can be dissolved in the water.

■ Bioaugmentation involves the use of microorganisms that have been specially bred for degradation of a specific contaminant or for survival under unusually severe environmental conditions. Bioaugmentation also takes the form of accelerating the growth of the natural microorganisms that preferentially feed on contaminants at the site.

■ *Cometabolism* is the transformation of an organic compound by a microorganism that does not use the compound as a source of energy or as one of its constituent elements. For example, microorganisms growing on one compound can produce an enzyme that chemically transforms another compound. A practical example is that of certain microorganisms that degrade methane; in the process of degrading the methane, the microorganisms produce enzymes that can initiate the oxidation of a variety of carbon compounds.

Bioventing

Bioventing involves the supply of relatively low airflow rates to sustain microbiological activity in the vadose (unsaturated) zone. The air (or oxygen) is supplied to the unsaturated zone. Compounds volatilized in the process are further degraded biologically as they travel upwards through the soil matrix. The process is relatively inexpensive and simple to operate and creates minimal disturbance to the area. Care must be taken to ensure that the soil permeability is high enough to encourage air movement and to ensure that the seasonally high groundwater table does not inhibit the transfer of air in the treatment area.

Phytoremediation

Phytoremediation is a process that uses plants to remove, transfer, stabilize, and destroy contaminants in soil and sediment. Upon uptake of contaminants (*phytoaccumulation*), the contaminants may be stored in the roots, stems, or leaves; changed into less harmful chemicals within the plant (*phyto-degradation*); or changed into gases that are released into the air as the plant transpires. For example, it is thought that poplar trees can degrade trichloroethylene in which the carbon is used for tissue growth and the chloride is expelled through the roots. Phytoaccumulation by itself does not destroy contaminants but, in effect, extracts the contaminants from the soil; the resulting mass is much smaller than the mass of soil that would otherwise require disposal. Alternatively, in *enhanced rhizosphere biodegradation,* plant roots can create an environment that encourages biodegradation through release of nutrients and by attracting water from deeper levels of the subsurface. Microbial counts in rhizosphere soils can be one or two orders of magnitude greater than in nonrhizosphere soils.

Phytoremediation is limited by the fact that in order to be successful the roots must come in contact with the contaminant. Root depths for plants commonly used for phytoremediation include Indian mustard (root depths to 12 inches), grasses (to 48 inches), and poplar trees (to 15 feet). One method to get the contaminated water to the roots is through irrigation, although this raises the concern of fugitive dust emissions.

Other concerns with phytoremediation include the potentially long cleanup times (on the order of years) and the fact that phytoremediation does not work for sites with high contaminant concentrations, as these may limit growth. In some cases, there is a concern with the disposal of biomass.

Enhanced Bioremediation

Enhanced bioremediation stimulates the natural biodegradation process by providing nutrients, electron acceptors, and even additional microorganisms. The process destroys contaminants in the process.

Oxygen is provided to the contaminated area typically by air sparging (Figure 12.10) or addition of hydrogen peroxide. Air sparging is the process by which air is injected into the groundwater. Care must be taken when using hydrogen peroxide as high concentrations (above 1000 ppm) are toxic to microorganisms.

Figure 12.10 shows the extraction of groundwater that receives additional treatment. (In some applications, extraction of the groundwater is not required.) In the case of groundwater recharge, the groundwater is mixed with an electron acceptor, nutrients, and other constituents, if required, and reinjected up-gradient of or within the contaminant plume. Infiltration galleries or injection wells may be used to reinject treated water. A "closed-loop" system could also be established in which all water extracted would be reinjected without treatment and all remediation would occur *in situ*. However, some states do not allow the direct reinjection of treated water into the aquifer.

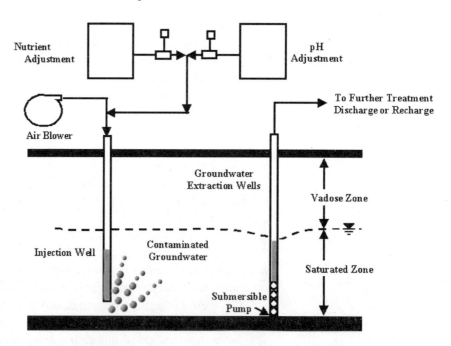

Figure 12.10 Enhanced bioremediation using air sparging

Source: *Remediation Technologies Screening Matrix and Reference Guide*, 4th ed., U.S. Army Environmental Command, SFIM-AEC-ET-CR-97053, January 2002.

According to the U.S. EPA,[4] the key parameters that determine the effectiveness of in situ groundwater bioremediation are as follows:

- Hydraulic conductivity of the aquifer, which controls the distribution of electron acceptors and nutrients in the subsurface

- Biodegradability of the contaminants, which determines both the rate and degree to which constituents will be degraded by microorganisms

- Location of contamination in the subsurface; contaminants must be dissolved in groundwater or adsorbed onto more permeable sediments within the aquifer

  Advantages of bioremediation include:

- The ability to remediate contaminants adsorbed and soluble contaminants in the groundwater

- Relatively simple infrastructure requirements

- Minimal disruption and/or disturbance to ongoing site activities

- Relatively short time requirements

- The ability to be combined with other technologies (for example, bioventing, and SVE)

- Absence of waste products requiring disposal

Challenges facing successful implementation of enhanced bioremediation include:

- Possible plugging of injection wells by microbial growth or mineral precipitation

- Difficulties associated with low-permeability aquifers

- Possible requirement for continuous monitoring and maintenance

- Heterogeneities in subsurface leading to inconsistent treatment throughout the aquifer

Monitored Natural Attenuation

Monitored natural attenuation is not a technology per se but a process whereby a number of fundamental processes are allowed to occur, thus reducing contaminant concentrations. These fundamental processes include dilution, dispersion, volatilization, biodegradation, radioactive decay, adsorption, and other chemical reactions.

Given the long-term time frame associated with natural attenuation, the U.S. EPA recommends this process only for those sites for which extensive off-site migration is not a concern. Also, the U.S. EPA encourages the use of natural attenuation in conjunction with more "active" techniques, perhaps as a follow-up step to more traditional active techniques.

---

4   U.S. EPA, *www.epa.gov/OUST/cat/insitbio.htm.*

At its best, natural attenuation offers the following advantages:

- Low cost and low complexity

- Minimal intrusion to the subsurface

- Generation of lesser volume of remediation wastes, reduced potential for cross-media transfer of contaminants commonly associated with ex situ treatment, and reduced risk of human exposure to contaminants, which are all common to all in situ techniques

Natural attenuation has been disparaged as a "do nothing" approach; however, such criticism is somewhat unfair. A great deal of data collection and analysis must occur prior to implementation, and extensive monitoring is required after implementation.

The Federal Remediation Technologies Roundtable (FRTR) lists the following limitations to monitored natural attenuation:

- Data collection requirements and long-term monitoring are significant.

- Intermediate degradation products may be more mobile and more toxic than the original contaminant.

- Natural attenuation is not appropriate where imminent site risks are present.

- Contaminants may migrate before they are degraded.

- If free product exists, it may have to be removed.

- Longer time frames may be required to achieve remediation objectives compared with active remediation.

- The hydrologic and geochemical conditions are likely to change over time, which will affect mobility of previously stabilized contaminants.

- More extensive outreach efforts may be required in order to gain public acceptance of natural attenuation.

12.3.2 Physical-Chemical Treatment

A variety of physical/chemical treatment processes are available that rely on such fundamental physical and chemical processes as volatilization, electrical separation, and a variety of chemical reactions.

Air Sparging

Air sparging was mentioned previously as a means of introducing air into the saturated portions of the aquifer for enhanced bioremediation. The incorporation of dissolved air into groundwater systems is common to many biological, physical, and chemical treatment processes, and the factors influencing the effectiveness of soil aeration systems are shown in Table 12.1.

Table 12.1 Factors affecting soil aeration systems

Soil Properties	Contaminant Properties	Environmenta Properties
Permeability	Henry's law constant	Temperature
Porosity	Solubility	Humidity
Grain size distribution	Adsorption coefficient	Wind speed
Moisture content	VOC concentration in soil	Solar radiation
pH	Polarity	Rainfall
Organic content	Vapor pressure	Terrain
Bulk density	Diffusion coefficient	Vegetation

Source: *Remediation Technologies Screening Matrix and Reference Guide*, 4th ed., U.S. Army Environmental Command, SFIM-AEC-ET-CR-97053, January 2002.

The goal of using air sparging for physical/chemical treatment is to remove contaminants by volatilization. In effect, air sparging acts as an in situ, underground air stripper (see section 11.4.2 for a review of air strippers). Contaminants that move from the water into the vapor phase bubble up through the aquifer, where perhaps a vapor extraction system is located to collect the gas-phase contaminants.

Spacing of injection wells such that their radius of influences (ROIs) overlap is a key design consideration. Well spacing of 20 feet is recommended. A 1–2-foot-long screen is used and must be placed entirely below the groundwater table. For LNAPLs, the screen should be below the area of contamination (approximately 5 feet). Wells placed lower than this may cause air to bypass the contaminated area, while shallower screens may not contact the lower portions of the contaminated area. For DNAPLs, the screen should be located just above the aquitard. A competent seal must be installed in the borehole to ensure that injected air does not escape out of the borehole.

Passive/Reactive Treatment Walls

Treatment walls, or *permeable reactive barriers,* are trenches dug into the ground perpendicular to groundwater flow. The trench may be large enough to capture the entire contaminant plume; alternatively, a *funnel and gate* system may be used. A funnel and gate system relies on low-permeability walls formed using slurry walls or sheet piles. The groundwater plume cannot flow through these walls and is thus directed to gaps in the walls, or gates, that contain the reactive material.

The most common reactive material is elemental iron granules, although chelators (ligands selected for their specificity for a given metal), sorbents, microbes, and surfactants can also be used. The iron walls are able to dechlorinate chlorinated compounds such as TCE by the mechanism of reductive dechlorination. Reductive dechlorination is described in the following reactions. These reactions illustrate the availability of electrons from the zero-valence iron and how these electrons can be used to dechlorinate a "generic" chlorinated compound (X-Cl) and specifically the oxidation of trichloroethylene (C_2HCl_3) to ethylene (C_2H_4).

$$Fe^0 \rightarrow Fe^{2+} + 2e^-$$

$$2H_2O \rightarrow 2H^+ + 2OH^-$$

$$X\text{-}CL + H^+ + 2e^- \rightarrow X\text{-}H + Cl^-$$

$$C_2HCl_3 + 3H^+ + 6e^- \rightarrow C_2H_4 + 3Cl^-$$

Soil Flushing

Soil flushing involves the injection of an extraction fluid such as water or a mixture of water and chemicals into a contaminated soil mass and recovery of the fluid at a series of down-gradient wells. Contaminants in the soil partition into the extraction fluid, which is treated before being recycled or discharged from the site.

The following factors may limit the applicability and effectiveness of soil flushing:

- Similar to a number of in situ technologies, low-permeability soils and sub-surface heterogeneities limit the effectiveness of the process.

- There is a risk of the extraction fluid, with its relatively high concentration of contaminants, from spreading to uncontaminated areas.

- Formulating an extraction fluid that can extract a variety of contaminants from a complex waste mixture may be difficult.

Soil Vapor Extraction

Soil vapor extraction (SVE) is used to remove contaminants from the unsaturated region (vadose zone) of the subsurface. Removal of VOCs from the vadose zone follows these two steps:

Step 1: Volatilization of VOCs to soil gas may occur directly from the residual contaminant adsorbed to the soil particle surface or via desorption from the soil particle surface; solubilization into soil water; and eventual volatilization into the soil gas. The transfer of the contaminant from the aqueous phase to the gas phase is governed by Henry's law, and the transfer from concentrated NAPL solutions to the gas phase is governed by Raoult's law.

Step 2: Once the contaminant has volatilized, movement of the contaminant through the soil media vapor phase occurs via advection (movement with bulk airflow) and diffusion (movement due to concentration gradient). The former controls in high permeability soil while the latter controls in low permeability soils.

The target for SVE is the residual contamination adhered to the soil particles in the vadose zone. (Although SVE is discussed in this chapter as an in situ treatment method, it may be used on stockpiled, excavated soils.) In practice, vapor extraction wells or perforated piping is installed in the zone of contamination, and a vacuum is placed on these systems. The extracted vapor typically requires further treatment, such as moisture extraction followed by adsorption or thermal destruction of the contaminants.

The design of the vapor extraction wells is critical when designing an SVE system. The well should be screened for the area above the groundwater table, not the entire well depth. As with air sparging wells, the ROIs for vapor extraction wells should overlap. In practice, SVE wells have an effective radius of between 20 and 150 feet, although spacing up to 300 feet has been used.

Enhancements to a basic system include the following:

- Lowering the groundwater table to increase the depth of the vadose zone, thus providing a greater subsurface volume that can be treated by SVE

- Covering the area with an impermeable barrier, such as a geomembrane, to prevent short circuiting of air from the surface to the extraction wells and to allow for increased extraction well spacing

- Adding air sparging to the saturated subsurface

- Heating the soil, known as *thermally enhanced SVE*. Advantages of heating include decreased viscosity of the residual contaminant, increased volatility, and increase solubility (for most compounds).

Although enhancing biodegradation is not the aim of SVE, some degree of such enhancement does occur. In a similar manner, a certain degree of volatilization occurs when bioventing is used.

SVE is a proven technology with relatively low implementation costs. Treatment times are relatively short, perhaps between six months and two years.

Factors that may limit the applicability and effectiveness of the process include the following[5]:

- Soil that has a high percentage of fines and a high degree of saturation will require higher vacuums (increasing costs) and/or hindering the operation of the in situ SVE system.

- Concentration reductions greater than 90% are difficult to achieve.

- SVE only works on the unsaturated region of the subsurface, requiring additional methods for the region below the groundwater table.

- Soil heterogeneities will affect local removal rates.

- High soil organic content has a high sorption capacity of VOCs, resulting in reduced removal rates.

Example 12.3

Soil vapor extraction of PCE

PCE has leaked from a dry cleaning facility such that 900 lb of PCE have contaminated the aquifer underlying the facility. Groundwater concentrations as high as 37 mg/L have been found. Treatment by SVE has commenced, using an extraction airflow of 300 ft³/min. The extracted air is being treated by a GAC bed. The average concentration of PCE in the air entering the GAC bed for each month following startup is shown in Exhibit 2.

Determine the mass of PCE that has been removed in the first 10 weeks and estimate the time by which 90% of the TCE will be removed.

Time (month)	1	2	3	4	5	6	7	8	9	10
Air concentration (mg/m³)	120	93	65	45	35	26	17	9	5	3

Exhibit 2

---

5   *Remediation Technologies Screening Matrix and Reference Guide, Version 4.0*, Federal Remediation Technology Roundtable, January 2002 (*www.frtr.gov/matrix2/section4/4-7.html*).

Solution

The mass of contaminant removed in the first month can be determined as follows:

$$\frac{120 \text{ mg}}{\text{m}^3} \cdot \frac{300 \text{ ft}^3}{\text{min}} \frac{1440 \text{ min}}{\text{day}} \frac{30 \text{ day}}{\text{month}} \frac{0.0283 \text{ m}^3}{1 \text{ ft}^3} \frac{2.205 \text{ lb}}{10^6 \text{ mg}} = 96 \text{ lb}$$

A similar procedure can be completed for each of the 10 months, and results are shown in Exhibit 3.

Time (month)	1	2	3	4	5	6	7	8	9	10
Mass (lb)	96.0	74.4	52.0	36.0	28.0	20.8	13.6	7.2	4.0	2.4

Exhibit 3 Mass of contaminant removed per month

The total mass removed is the sum of the monthly masses removed, or 334.5 lb. A first estimate for the time of cleanup would be to assume a linear model for removal; that is, if 334.5 lb were removed in the first 10-month period, an additional 334.5 lb could be removed in a second period of 10 months. However, a linear trend is clearly not the case, as shown in Exhibit 4. This decrease in the rate of removal occurs for many reasons, such as the decrease in diffusion rates as concentration gradients decrease. Moreover, removing the entire mass of PCE is not feasible, given the inability of extracted air to reach all of the contamination, the fact that some of the PCE is dissolved in the groundwater, and the fact that some of the PCE has most likely volatilized prior to implementation of SVE.

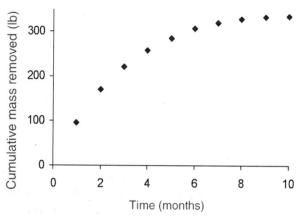

Exhibit 4 Cumulative mass of contaminant removed

Solidification /Stabilization

One of the most common means of in situ solidification and stabilization is vitrification. The concept behind in situ vitrification is the same as for ex situ vitrification (which is covered in section 11.4.2): high temperatures (1600°C to 2000°C or 2900°F to 3650°F) are employed to melt the soil, producing a glasslike final product.

In in situ vitrification, the high temperatures are obtained by passing an electric current through the soil matrix. The off-gases are collected in a large hood and treated before release. Organic compounds are destroyed by the process, while radionuclides and metals are retained within the vitrified mass in which they are resistant to leaching.

This process is not applicable to sites with large amounts of explosive or combustible materials. Also, if there are any buried electrical conduction paths (such as metal drums or cables), melting of the soil will not occur.

12.3.3 Containment

The purpose of containment is not to treat the waste but rather to isolate the waste. The isolated waste may be subjected to further treatment techniques or allowed to naturally attenuate.

Physical Barriers

Physical barriers such as *vertical cutoff walls* are used to contain or divert contaminated groundwater or divert uncontaminated groundwater flow from a contaminated region. These barriers can be installed quickly to provide additional time to design and implement a remediation process.

Physical barriers can be constructed using a variety of materials. *Sheet pile cutoff walls* can be constructed by driving steel or HDPE (high density polyethylene) piles into the soil and sealing the joints between the piles. Care must be taken when locating sheet pile walls to avoid underground structures and utilities.

Slurry walls are constructed by digging a 0.5–2-m-wide trench in the soil and backfilling with a slurry mixture of bentonite and native materials. Depths up to 50 m (160 ft) are possible. A third possibility is the use of *grout walls,* in which grouting compounds are forced under pressure into the ground. Grouting compounds include cement, bentonite, and silicate.

Cutoff walls can be designed as *hanging walls,* in which the bottom of the wall is "suspended" above the bottom of the aquifer; although water can pass under this wall, LNAPLs will be contained. Alternatively, the wall can be "keyed in" to the low permeable formation (for example, bedrock or clay) that serves as an aquitard. The keyed in depth should be 2–3 feet.

Vertical cutoff walls do not directly treat the contaminants, although such treatment may occur indirectly due to natural attenuation. Another limitation of the method is that at depths greater than 80 feet, sheet pile walls and slurry walls become economically infeasible.

12.4   BROWNFIELDS

The U.S. EPA defines a *brownfield* as "real property, the expansion, redevelopment, or reuse of which may be complicated by the presence or potential presence of a hazardous substance, pollutant, or contaminant." The U.S. EPA estimates that there are more than 450,000 brownfields in the United States.

Converting brownfields to usable properties has the following benefits:

- Increases local tax bases

- Facilitates job growth

- Utilizes existing infrastructure

- Takes development pressures off of undeveloped, open land

- Protects the environment

ADDITIONAL RESOURCES

1. LaGrega, M. D., P. L. Buckingham, and J. C. Evans. *Hazardous Waste Management.* McGraw-Hill, 2000.

2. Sellers, K. *Fundamentals of Hazardous Waste Site Remediation.* CRC Press, 1998.

3. Watts, R. J. *Hazardous Wastes: Sources, Pathways, Receptors.* Wiley, 1998.

Engineering Economics

OUTLINE

This is a review of the field known variously as *engineering economics, engineering economy*, or *engineering economic analysis*. Since engineering economics is straightforward and logical, even people who have not had a formal course should be able to gain sufficient knowledge from this chapter to successfully solve most engineering economics problems.

There are 32 example problems scattered throughout the chapter. These examples are an integral part of the review and should be examined as you come to them.

The field of engineering economics uses mathematical and economics techniques to systematically analyze situations which pose alternative courses of action. The initial step in engineering economics problems is to resolve a situation, or each alternative in a given situation, into its favorable and unfavorable consequences or factors. These are then measured in some common unit—usually money. Factors which cannot readily be equated to money are called intangible or irreducible factors. Such factors are considered in conjunction with the monetary analysis when making the final decision on proposed courses of action.

CASH FLOW

A cash flow table shows the "money consequences" of a situation and its timing. For example, a simple problem might be to list the year-by-year consequences of purchasing and owning a used car:

Year	Cash Flow ($)	
Beginning of first year 0	−4500	Car purchased "now" for $4500 cash. The minus sign indicates a disbursement.
End of year 1	−350	
End of year 2	−350	
End of year 3	−350	Maintenance costs are $350 per year.
End of year 4	−350	
	+2000	This car is sold at the end of the fourth year for $2000. The plus sign represents the receipt of money.

This same cash flow may be represented graphically, as shown in Figure A.1. The upward arrow represents a receipt of money, and the downward arrows represent disbursements. The horizontal axis represents the passage of time.

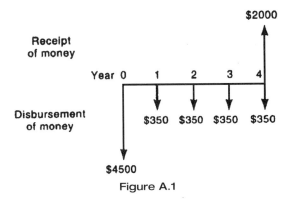

Figure A.1

Example **A.1**

In January 1993 a firm purchased a used typewriter for $500. Repairs cost nothing in 1993 or 1994. Repairs are $85 in 1995, $130 in 1996, and $140 in 1997. The machine is sold in 1997 for $300. Complete the cash flow table.

Solution

Unless otherwise stated, the customary assumption is a beginning-of-year purchase, followed by end-of-year receipts or disbursements, and an end-of-year resale or salvage value. Thus the typewriter repairs and the typewriter sale are assumed to occur at the end of the year. Letting a minus sign represent a disbursement of money and a plus sign a receipt of money, we are able to set up the cash flow table:

Year	Cash Flow ($)
Beginning of 1993	−500
End of 1993	0
End of 1994	0
End of 1995	−85
End of 1996	−130
End of 1997	+160

Notice that at the end of.1997 the cash flow table shows +160, which is the net sum of −140 and +300. If we define year 0 as the beginning of 1993, the cash flow table becomes

Year	Cash Flow ($)
0	−500
1	0
2	0
3	−85
4	−130
5	+160

From this cash flow table, the definitions of year 0 and year 1 become clear. Year 0 is defined as the *beginning* of year 1. Year 1 is the *end* of year 1, and so forth.

TIME VALUE OF MONEY

When the money consequences of an alternative occur in a short period of time—say, less than one year—we might simply add up the various sums of money and obtain the net result. But we cannot treat money this way over longer periods of time. This is because money today does not have the same value as money at some future time.

Consider this question: Which would you prefer, $100 today or the assurance of receiving $100 a year from now? Clearly, you would prefer the $100 today. If you had the money today, rather than a year from now, you could use it for the year. And if you had no use for it, you could lend it to someone who would pay interest for the privilege of using your money for the year.

Simple Interest

Simple interest is interest that is computed on the original sum. Thus if one were to lend a present sum P to someone at a simple annual interest rate i, the future amount F due at the end of n years would be

$$F = P + Pin$$

Example **A.2**

How much will you receive back from a $500 loan to a friend for three years at 10% simple annual interest?

Solution

$$F = P + Pin = 500 + 500 \times 0.10 \times 3 = \$650$$

In Example A.2 one observes that the amount owed, based on 10% simple interest at the end of one year, is $500 + 500 \times 0.10 \times 1 = \$550$. But at simple interest there is no interest charged on the $50 interest, even though it is not paid until the end of the third year. Thus simple interest is not realistic and is seldom used. *Compound interest* charges interest on the principal owed plus the interest earned to date. This produces a charge of interest on interest, or compound interest. Engineering economics uses compound interest computations.

EQUIVALENCE

In the preceding section we saw that money at different points in time (for example, $100 today or $100 one year hence) may be equal in the sense that they both are $100, but $100 a year hence is *not* an acceptable substitute for $100 today. When we have acceptable substitutes, we say they are *equivalent* to each other. Thus at 8% interest, $108 a year hence is equivalent to $100 today.

| Example **A.3** | At a 10% per year (compound) interest rate, $500 now is *equivalent* to how much three years hence? |

Solution

A value of $500 now will increase by 10% in each of the three years.

$$\text{Now} = \$500.00$$

$$\text{End of 1st year} = 500 + 10\%(500) = 550.00$$

$$\text{End of 2nd year} = 550 + 10\%(550) = 605.00$$

$$\text{End of 3rd year} = 605 + 10\%(605) = 665.50$$

Thus $500 now is *equivalent* to $665.50 at the end of three years. Note that interest is charged each year on the original $500 plus the unpaid interest. This compound interest computation gives an answer that is $15.50 higher than the simple-interest computation in Example A.2.

Equivalence is an essential factor in engineering economics. Suppose we wish to select the better of two alternatives. First, we must compute their cash flows. For example,

Year	Alternative	
	A	*B*
0	−2000	−2800
1	+800	+1100
2	+800	+1100
3	+800	+1100

The larger investment in alternative *B* results in larger subsequent benefits, but we have no direct way of knowing whether it is better than alternative *A*. So we do not know which to select. To make a decision, we must resolve the alternatives into *equivalent* sums so that they may be compared accurately.

COMPOUND INTEREST

To facilitate equivalence computations, a series of compound interest factors will be derived here, and their use will be illustrated in examples.

Symbols and Functional Notation

i = effective interest rate per interest period. In equations, the interest rate is stated as a decimal (that is, 8% interest is 0.08).

n = number of interest periods. Usually the interest period is one year, but it could be something else.

P = a present sum of money.

F = a future sum of money. The future sum F is an amount n interest periods from the present that is equivalent to P at interest rate i.

A = an end-of-period cash receipt or disbursement in a uniform series continuing for n periods. The entire series is equivalent to P or F at interest rate i.

G = uniform period-by-period increase in cash flows; the uniform gradient.

r = nominal annual interest rate.

Table A.1   Periodic compounding: Functional notation and formulas

Factor	Given	To Find	Functional Notation	Formula
Single payment				
Compound amount factor	P	F	$(F/P, i\%, n)$	$F = P(1 + i)^n$
Present worth factor	F	P	$(P/F, i\%, n)$	$P = F(1 + i)^{-n}$
Uniform payment series				
Sinking fund factor	F	A	$(A/F, i\%, n)$	$A = F\left[\dfrac{i}{(1+i)^n - 1}\right]$
Capital recovery factor	P	A	$(A/P, i\%, n)$	$A = P\left[\dfrac{i(1+i)^n}{(1+i)^n - 1}\right]$
Compound amount factor	A	F	$(F/A, i\%, n)$	$F = A\left[\dfrac{(1+i)^n - 1}{i}\right]$
Present worth factor	A	P	$(P/A, i\%, n)$	$P = A\left[\dfrac{(1+i)^n - 1}{i(1+i)^n}\right]$
Uniform gradient				
Gradient present worth	G	P	$(P/G, i\%, n)$	$P = G\left[\dfrac{(1+i)^n - 1}{i^2(1+i)^n} - \dfrac{n}{i(1-i)^n}\right]$
Gradient future worth	G	F	$(F/G, i\%, n)$	$F = G\left[\dfrac{(1+i)^n - 1}{i^2} - \dfrac{n}{1}\right]$
Gradient uniform series	G	A	$(A/G, i\%, n)$	$A = G\left[\dfrac{1}{i} - \dfrac{n}{(1+i)^n - 1}\right]$

From Table A.1 we can see that the functional notation scheme is based on writing (to find/given, i, n). Thus, if we wished to find the future sum F, given a uniform series of receipts A, the proper compound interest factor to use would be $(F/A, i, n)$.

Single-Payment Formulas

Suppose a present sum of money P is invested for one year at interest rate i. At the end of the year, the initial investment P is received together with interest equal to Pi, or a total amount $P + Pi$. Factoring P, the sum at the end of one year is $P(1+i)$. If the investment is allowed to remain for subsequent years, the progression is as follows:

Amount at Beginning of the Period	+	Interest for the Period	=	Amount at End of the Period
1st year, P	+	Pi	=	$P(1+i)$
2nd year, $P(1+i)$	+	$Pi(1+i)$	=	$P(1+i)^2$
3rd year, $P(1+i)^2$	+	$Pi(1+i)^2$	=	$P(1+i)^3$
nth year, $P(1+i)^{n-1}$	+	$Pi(1+i)^{n-1}$	=	$P(1+i)^n$

The present sum P increases in n periods to $P(1+i)^n$. This gives a relation between a present sum P and its equivalent future sum F:

$$\text{Future sum} = (\text{present sum})(1+i)^n$$

$$F = P(1+i)^n$$

This is the *single-payment compound amount formula*. In functional notation it is written

$$F = P(F/P, i, n)$$

The relationship may be rewritten as

$$\text{Present sum} = (\text{Future sum})(1+i)^{-n}$$

$$P = F(1+i)^{-n}$$

This is the *single-payment present worth formula*. It is written

$$P = F(P/F, i, n)$$

Example A.4

At a 10% per year interest rate, $500 now is *equivalent* to how much three years hence?

Solution

This problem was solved in Example A.3. Now it can be solved using a single-payment formula. $P = \$500$, $n = 3$ years, $i = 10\%$, and $F = $ unknown:

$$F = P(1+i)^n = 500(1+0.10)^3 = \$665.50.$$

This problem also may be solved using a compound interest table:

$$F = P(F/P, i, n) = 500(F/P, 10\%, 3)$$

From the 10% compound interest table, read $(F/P, 10\%, 3) = 1.331$.

$$F = 500(F/P, 10\%, 3) = 500(1.331) = \$665.50$$

Example A.5

To raise money for a new business, a man asks you to lend him some money. He offers to pay you $3000 at the end of four years. How much should you give him now if you want 12% interest per year?

Solution

P = unknown, F = \$3000, n = 4 years, and i = 12%:

$P = F(1 + i)^{-n} = 3000(1 + 0.12)^{-4} = \$1906.55$

Alternative computation using a compound interest table:

$P = F(P/F, i, n) = 3000(P/F, 12\%, 4) = 3000(0.6355) = \$1906.50$

Note that the solution based on the compound interest table is slightly different from the exact solution using a hand-held calculator. In engineering economics the compound interest tables are always considered to be sufficiently accurate.

Uniform Payment Series Formulas

Consider the situation shown in Figure A.2. Using the single-payment compound amount factor, we can write an equation for F in terms of A:

$$F = A + A(1 + i) + A(1 + i)^2 \qquad \textbf{(i)}$$

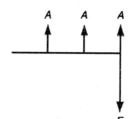

A = End-of-period cash receipt or disbursement in a uniform series continuing for n periods

F = A future sum of money

Figure A.2

In this situation, with n = 3, Eq. (i) may be written in a more general form:

$$F = A + A(1 + i) + A(1 + i)^{n-1} \qquad \textbf{(ii)}$$

Multiply Eq. (ii) by $(1 + i)$ $(1 + i)F = A(1 + i) + A(1 + i)^{n-1} + A(1 + i)^n$ **(iii)**

Subtract Eq. (ii): $-F = A + A(1 + i) + A(1 + i)^{n-1}$ **(iv)**

$$iF = -A + A(1 + i)^n$$

This produces the *uniform series compound amount formula:*

$$F = A\left[\frac{(1+i)^n - 1}{i}\right]$$

Solving this equation for A produces the *uniform series sinking fund formula:*

$$A = F\left[\frac{i}{(1+i)^n - 1}\right]$$

Since $F = P(1 + i)^n$, we can substitute this expression for F in the equation and obtain the *uniform series capital recovery formula:*

$$A = P\left[\frac{i(1+i)^n}{(1+i)^n - 1}\right]$$

Solving the equation for P produces the *uniform series present worth formula:*

$$P = A\left[\frac{(1+i)^n - 1}{i(1+i)^n}\right]$$

In functional notation, the uniform series factors are

Compound amount $(F/A, i, n)$

Sinking fund $(A/F, i, n)$

Capital recovery $(A/P, i, n)$

Present worth $(P/A, i, n)$

Example A.6

If $100 is deposited at the end of each year in a savings account that pays 6% interest per year, how much will be in the account at the end of five years?

Solution

$A = \$100$, F = unknown, $n = 5$ years, and $i = 6\%$:

$$F = A(F/A, i, n) = 100(F/A, 6\%, 5) = 100(5.637) = \$563.70$$

Example A.7

A fund established to produce a desired amount at the end of a given period, by means of a series of payments throughout the period, is called a *sinking fund*. A sinking fund is to be established to accumulate money to replace a $10,000 machine. If the machine is to be replaced at the end of 12 years, how much should be deposited in the sinking fund each year? Assume the fund earns 10% annual interest.

Solution

Annual sinking fund deposit $A = 10,000(A/F, 10\%, 12)$

$$= 10,000(0.0468) = \$468$$

Example A.8

An individual is considering the purchase of a used automobile. The total price is $6200. With $1240 as a down payment, and the balance paid in 48 equal monthly payments with interest at 1% per month, compute the monthly payment. The payments are due at the end of each month.

Solution

The amount to be repaid by the 48 monthly payments is the cost of the automobile *minus* the $1240 down payment.

$P = \$4960$, A = unknown, $n = 48$ monthly payments, and $i = 1\%$ per month:

$$A = P(A/P, 1\%, 48) = 4960(0.0263) = \$130.45$$

Example A.9

A couple sell their home. In addition to cash, they take a mortgage on the house. The mortgage will be paid off by monthly payments of $450 for 50 months. The couple decides to sell the mortgage to a local bank. The bank will buy the mortgage, but it requires a 1% per month interest rate on its investment. How much will the bank pay for the mortgage?

Solution

$A = \$450$, $n = 50$ months, $i = 1\%$ per month, and $P =$ unknown:

$$P = A(P/A, i, n) = 450(P/A, 1\%, 50) = 450(39.196) = \$17,638.20$$

Uniform Gradient

At times one will encounter a situation where the cash flow series is not a constant amount A. Instead, it is an increasing series. The cash flow shown in Figure A.3 may be resolved into two components (Figure A.4). We can compute the value of P^* as equal to P' plus P. And we already have the equation for $P':P' = A(P/A, i, n)$. The value for P in the right-hand diagram is

$$P = G\left[\frac{(1+i)^n - 1}{i^2(1+i)^n} - \frac{n}{i(1-i)^n}\right]$$

This is the *uniform gradient present worth formula.* In functional notation, the relationship is $P = G(P/G, i, n)$.

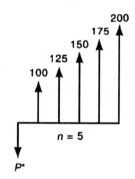

Figure A.3

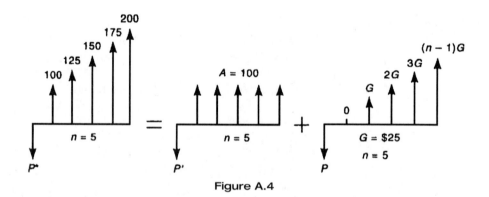

Figure A.4

Example A.10

The maintenance on a machine is expected to be $155 at the end of the first year, and it is expected to increase $35 each year for the following seven years (Exhibit 1). What sum of money should be set aside now to pay the maintenance for the eight-year period? Assume 6% interest.

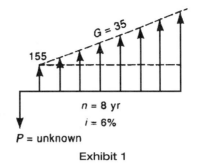

$n = 8$ yr

$i = 6\%$

P = unknown

Exhibit 1

Solution

$$P = 155(P/A, 6\%, 8) + 35(P/G, 6\%, 8)$$

$$= 155(6.210) + 35(19.841) = \$1656.99$$

In the gradient series, if—instead of the present sum, P—an equivalent uniform series A is desired, the problem might appear as shown in Figure A.5. The relationship between A' and G in the right-hand diagram is

$$A' = G\left[\frac{1}{i} - \frac{n}{(1+i)^n - 1}\right]$$

In functional notation, the uniform gradient (to) uniform series factor is: $A' = G(A/G, i, n)$.

The uniform gradient uniform series factor may be read from the compound interest tables directly, or computed as

$$(A/G, i, n) = \frac{1 - n(A/F, i, n)}{i}$$

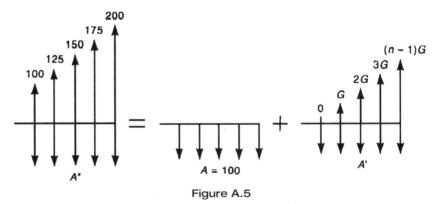

Figure A.5

Note carefully the diagrams for the uniform gradient factors. The first term in the uniform gradient is zero and the last term is $(n - 1)G$. But we use n in the equations and function notation. The derivations (not shown here) were done on this basis, and the uniform gradient compound interest tables are computed this way.

Example **A.11**

For the situation in Example A.10, we wish now to know the uniform annual maintenance cost. Compute an equivalent A for the maintenance costs.

Solution

Refer to Exhibit 2 The equivalent uniform annual maintenance cost is

$$A = 155 + 35(A/G, 6\%, 8) = 155 + 35(3.195) = \$266.83$$

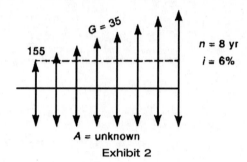

Exhibit 2

Standard compound interest tables give values for eight interest factors: two single payments, four uniform payment series, and two uniform gradients. The tables do *not* give the uniform gradient future worth factor, $(F/G,i,n)$. If it is needed, it may be computed from two tabulated factors:

$$(F/G, i, n) = (P/G, i, n)(F/P, i, n)$$

For example, if $i = 10\%$ and $n = 12$ years, then $(F/G, 10\%, 12) = (P/G, 10\%, 12) (F/P, 10\%, 12) = (29.901)(3.138) = 93.83$.

A second method of computing the uniform gradient future worth factor is

$$(A/G, i, n) = \frac{1 - n(A/F, i, n)}{i}$$

Using this equation for $i = 10\%$ and $n = 12$ years, $(F/G, 10\%, 12) = [(F/A, 10\%, 12) - 12]/0.10 = (21.384 - 12)/0.10 = 93.84$.

Continuous Compounding

Table A.2 Continuous compounding: Functional notation and formulas

Factor	Given	To Find	Functional Notation	Formula
Single payment				
Compound amount factor	P	F	$(F/P, r\%, n)$	$F = Pe^{rn}$
Present worth factor	F	P	$(P/F, r\%, n)$	$P = Fe^{-rn}$
Uniform payment series				
Sinking fund factor	F	A	$(A/F, r\%, n)$	$A = F\left[\dfrac{e^r - 1}{e^{rn} - 1}\right]$
Capital recovery factor	P	A	$(A/P, r\%, n)$	$A = P\left[\dfrac{e^r - 1}{1 - e^{-rn}}\right]$
Compound amount factor	A	F	$(F/A, r\%, n)$	$F = A\left[\dfrac{e^{rn} - 1}{e^r - 1}\right]$
Present worth factor	A	P	$(P/A, r\%, n)$	$P = A\left[\dfrac{1 - e^{-rn}}{e^r - 1}\right]$

r = nominal annual interest rate, n = number of years.

| Example **A.12** | Five hundred dollars is deposited each year into a savings bank account that pays 5% nominal interest, compounded continuously. How much will be in the account at the end of five years? |

Solution

$A = \$500$, $r = 0.05$, $n = 5$ years.

$$F = A(F/A, \ r\%, \ n) = A\left[\frac{e^{rn} - 1}{e^r - 1}\right] = 500\left[\frac{e^{0.05(5)} - 1}{e^{0.05} - 1}\right] = \$2769.84$$

NOMINAL AND EFFECTIVE INTEREST

Nominal interest is the annual interest rate without considering the effect of any compounding. *Effective interest* is the annual interest rate taking into account the effect of any compounding during the year.

Non-Annual Compounding

Frequently an interest rate is described as an annual rate, even though the interest period may be something other than one year. A bank may pay 1% interest on the amount in a savings account every three months. The *nominal* interest rate in this situation is $4 \times 1\% = 4\%$. But if you deposited $1000 in such an account, would you have $104\%(1000) = \$1040$ in the account at the end of one year? The answer is no, you would have more. The amount in the account would increase as follows:

Amount in Account

Beginning of year: 1000.00

End of three months: $1000.00 + 1\%(1000.00) = 1010.00$

End of six months: $1010.00 + 1\%(1010.00) = 1020.10$

End of nine months: $1020.10 + 1\%(1020.10) = 1030.30$

End of one year: $1030.30 + 1\%(1030.30) = 1040.60$

At the end of one year, the interest of $40.60, divided by the original $1000, gives a rate of 4.06%. This is the *effective* interest rate.

Effective interest rate per year: $i_{eff} = (1 + r/m)^m - 1$

where

r = nominal annual interest rate

m = number of compound periods per year

r/m = effective interest rate per period

Example A.13

A bank charges 1.5% interest per month on the unpaid balance for purchases made on its credit card. What nominal interest rate is it charging? What is the effective interest rate?

Solution

The nominal interest rate is simply the annual interest ignoring compounding, or 12(1.5%) = 18%.

$$\text{Effective interest rate} = (1 + 0.015)^{12} - 1 = 0.1956 = 19.56\%$$

Continuous Compounding

When m, the number of compound periods per year, becomes very large and approaches infinity, the duration of the interest period decreases from Δt to dt. For this condition of *continuous compounding*, the effective interest rate per year is

$$i_{\text{eff}} = e^r - 1$$

where r = nominal annual interest rate.

Example A.14

If the bank in Example A.13 changes its policy and charges 1.5% per month, compounded continuously, what nominal and what effective interest rate is it charging?

Solution

Nominal annual interest rate, $r = 12 \times 1.5\% = 18\%$

$$\text{Effective interest rate per year, } i_{\text{eff}} = e^{0.18} - 1 = 0.1972 = 19.72\%$$

SOLVING ENGINEERING ECONOMICS PROBLEMS

The techniques presented so far illustrate how to convert single amounts of money, and uniform or gradient series of money, into some equivalent sum at another point in time. These compound interest computations are an essential part of engineering economics problems.

The typical situation is that we have a number of alternatives; the question is, which alternative should we select? The customary method of solution is to express each alternative in some common form and then choose the best, taking both the monetary and intangible factors into account. In most computations an interest rate must be used. It is often called the minimum attractive rate of return (MARR), to indicate that this is the smallest interest rate, or rate of return, at which one is willing to invest money.

Criteria

Engineering economics problems inevitably fall into one of three categories:

1. *Fixed input.* The amount of money or other input resources is fixed.
   Example: A project engineer has a budget of $450,000 to overhaul a plant.

2. *Fixed output*. There is a fixed task or other output to be accomplished. *Example*: A mechanical contractor has been awarded a fixed-price contract to air-condition a building.

3. *Neither input nor output fixed*. This is the general situation, where neither the amount of money (or other inputs) nor the amount of benefits (or other outputs) is fixed.
   Example: A consulting engineering firm has more work available than it can handle. It is considering paying the staff to work evenings to increase the amount of design work it can perform.

There are five major methods of comparing alternatives: present worth, future worth, annual cost, rate of return, and cost-benefit analysis. These are presented in the sections that follow.

PRESENT WORTH

Present worth analysis converts all of the money consequences of an alternative into an equivalent present sum. The criteria are

Category	Present Worth Criterion
Fixed input	Maximize the present worth of benefits or other outputs
Fixed output	Minimize the present worth of costs or other inputs
Neither input nor output fixed	Maximize present worth of benefits minus present worth of costs, or maximize net present worth

Appropriate Problems

Present worth analysis is most frequently used to determine the present value of future money receipts and disbursements. We might want to know, for example, the present worth of an income-producing property, such as an oil well. This should provide an estimate of the price at which the property could be bought or sold.

An important restriction in the use of present worth calculation is that there must be a common analysis period for comparing alternatives. It would be incorrect, for example, to compare the present worth (PW) of cost of pump *A*, expected to last 6 years, with the PW of cost of pump *B*, expected to last 12 years (Figure A.6). In situations like this, the solution is either to use some other analysis technique (generally, the annual cost method is suitable in these situations) or to restructure the problem so that there is a common analysis period.

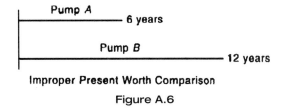

Improper Present Worth Comparison

Figure A.6

In this example, a customary assumption would be that a pump is needed for 12 years and that pump *A* will be replaced by an identical pump *A* at the end of 6 years. This gives a 12-year common analysis period (Figure A.7). This approach is easy to use when the different lives of the alternatives have a practical least-common- multiple life. When this is not true (for example, the life of *J* equals 7 years and the life of *K* equals 11 years), some assumptions must be made to select a suitable common analysis period, or the present worth method should not be used.

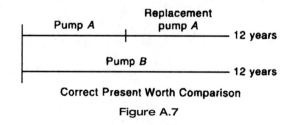

Correct Present Worth Comparison

Figure A.7

Example A.15

You have been tasked with providing an alkali dosing unit to treat a stream polluted with acid mine drainage. Chemical doser A has an initial cost of $10,000, an annual maintenance cost of $500 per year, and no salvage value at the end of its 4-year useful life. Doser B costs $20,000, and the first year there is no maintenance cost. Maintenance is $100 the second year, and it increases $100 per year thereafter. Doser B has an anticipated $5000 salvage value at the end of its 12-year useful life. If the minimum attractive rate of return (MARR) is 8%, which doser should be selected?

Solution

The analysis period is not stated in the problem. Therefore, we select the least common multiple of the lives, or 12 years, as the analysis period.

Present worth of cost of 12 years of doser A:

PW = 10,000 + 10,000(*P/F*, 8%, 4) + 10,000(*P/F*, 8%, 8) + 500(*P/A*, 8%, 12)

= 10,000 + 10,000(0.7350) + 10,000(0.5403) + 500(7.536) = $26,521

Present worth of cost of 12 years of doser B:

PW = 20,000 + 100(*P/G*, 8%, 12) − 5000(*P/F*, 8%, 12)

= 20,000 + 100(34.634) − 5000(0.3971) = $21,478

Choose doser B, with its smaller PW of cost.

Example A.16

Two alternatives have the following cash flows:

Year	Alternative	
	A	*B*
0	−2000	−2800
1	+800	+1100
2	+800	+1100
3	+800	+1100

At a 4% interest rate, which alternative should be selected?

Solution

The net present worth of each alternative is computed:

$$\text{Net present worth (NPW)} = \text{PW of benefit} - \text{PW of cost}$$

$$\text{NPW}_A = 800(P/A, 4\%, 3) - 2000 = 800(2.775) - 2000 = \$220.00$$

$$\text{NPW}_B = 1100(P/A, 4\%, 3) - 2800 = 1100(2.775) - 2800 = \$252.50$$

To maximize NPW, choose alternative *B*.

Infinite Life and Capitalized Cost

In the special situation where the analysis period is infinite ($n = \infty$), an analysis of the present worth of cost is called *capitalized cost.* There are a few public projects where the analysis period is infinity. Other examples are permanent endowments and cemetery perpetual care.

When n equals infinity, a present sum P will accrue interest of Pi for every future interest period. For the principal sum P to continue undiminished (an essential requirement for n equal to infinity), the end-of-period sum A that can be disbursed is Pi (Figure A.8). When $n = \infty$, the fundamental relationship is

$$A = Pi$$

Some form of this equation is used whenever there is a problem involving an infinite analysis period.

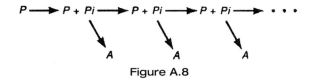

Figure A.8

Example **A.17**

In his will, a man wishes to establish a perpetual trust to provide for the maintenance of a small local park. If the annual maintenance is $7500 per year and the trust account can earn 5% interest, how much money must be set aside in the trust?

Solution

When $n = \infty$, $A = Pi$ or $P = A/i$. The capitalized cost is $P = A/i = \$7500/0.05 = \$150,000$.

FUTURE WORTH OR VALUE

In present worth analysis, the comparison is made in terms of the equivalent present costs and benefits. But the analysis need not be made in terms of the present—it can be made in terms of a past, present, or future time. Although the numerical calculations may look different, the decision is unaffected by the selected point in time. Often we do want to know what the future situation will be if we take some particular couse of action now. An analysis based on some future point in time is called *future worth analysis.*

Category	Present Worth Criterion
Fixed input	Maximize the present worth of benefits or other outputs
Fixed output	Minimize the present worth of costs or other inputs
Neither input nor output fixed	Maximize present worth of benefits minus present worth of costs, or maximize net present worth

Example **A.18**

Two alternatives have the following cash flows:

Year	Alternative	
	A	*B*
0	−2000	−2800
1	+800	+1100
2	+800	+1100
3	+800	+1100

At a 4% interest rate, which alternative should be selected?

Solution

In Example A.16, this problem was solved by present worth analysis at year 0. Here it will be solved by future worth analysis at the end of year 3.

$$\text{Net future worth (NFW)} = \text{FW of benefits} - \text{FW of cost}$$

$$NFW_A = 800(F/A, 4\%, 3) - 2000(F/P, 4\%, 3)$$

$$= 800(3.122) - 2000(1.125) = +\$247.60$$

$$NFW_B = 1100(F/A, 4\%, 3) - 2800(F/P, 4\%, 3)$$

$$= 1100(3.122) - 2800(1.125) = +\$284.20$$

To maximize NFW, choose alternative B.

ANNUAL COST

The annual cost method is more accurately described as the method of equivalent uniform annual cost (EUAC). Where the computation is of benefits, it is called the method of equivalent uniform annual benefits (EUAB).

Criteria

For each of the three possible categories of problems, there is an annual cost criterion for economic efficiency.

Category	Annual Cost Criterion
Fixed input	Maximize the equivalent uniform annual benefits (EUAB)
Fixed output	Minimize the equivalent uniform annual cost (EUAC)
Neither input nor output fixed	Maximize EUAB − EUAC

Application of Annual Cost Analysis

In the section on present worth, we pointed out that the present worth method requires a common analysis period for all alternatives. This restriction does not apply in all annual cost calculations, but it is important to understand the circumstances that justify comparing alternatives with different service lives.

Frequently, an analysis is done to provide for a more-or-less continuing requirement. For example, one might need to pump water from a well on a continuing basis. Regardless of whether each of two pumps has a useful service life of 6 years or 12 years, we would select the alternative whose annual cost is a minimum. And this still would be the case if the pumps' useful lives were the more troublesome 7 and 11 years. Thus, if we can assume a continuing need for an item, an annual cost comparison among alternatives of differing service lives is valid. The underlying assumption in these situations is that the shorter-lived alternative can be replaced with an identical item with identical costs, when it has reached the end of its useful life. This means that the EUAC of the initial alternative is equal to the EUAC for the continuing series of replacements.

On the other hand, if there is a specific requirement to pump water for 10 years, then each pump must be evaluated to see what costs will be incurred during the analysis period and what salvage value, if any, may be recovered at the end of the analysis period. The annual cost comparison needs to consider the actual circumstances of the situation.

Examination problems are often readily solved using the annual cost method. And the underlying "continuing requirement" is usually present, so an annual cost comparison of unequal-lived alternatives is an appropriate method of analysis.

Example A.19

Consider the following alternatives:

	A	*B*
First cost	$5000	$10,000
Annual maintenance	$500	$200
End-of-useful-life salvage value	$600	$1000
Useful life	5 years	15 years

Based on an 8% interest rate, which alternative should be selected?

Solution

Assuming both alternatives perform the same task and there is a continuing requirement, the goal is to minimize EUAC.

Alternative *A*:

$$\text{EUAC} = 5000(A/P, 8\%, 5) + 500 - 600(A/F, 8\%, 5)$$

$$= 5000(0.2505) + 500 - 600(0.1705) = \$1650$$

Alternative *B*:

$$\text{EUAC} = 10{,}000(A/P, 8\%, 15) + 200 - 1000(A/F, 8\%, 15)$$

$$= 10{,}000(0.1168) + 200 - 1000(0.0368) = \$1331$$

To minimize EUAC, select alternative *B*.

RATE OF RETURN ANALYSIS

A typical situation is a cash flow representing the costs and benefits. The rate of return may be defined as the interest rate where PW of cost = PW of benefits, EUAC = EUAB, or PW of cost – PW of benefits = 0.

Example **A.20**

Compute the rate of return for the investment represented by the following cash flow table.

Year:	0	1	2	3	4	5
Cash flow:	−595	+250	+200	+150	+100	+50

Solution

This declining uniform gradient series may be separated into two cash flows (Exhibit 3) for which compound interest factors are available.

Note that the gradient series factors are based on an increasing gradient. Here the declining cash flow is solved by subtracting an increasing uniform gradient, as indicated in the figure.

PW of cost – PW of benefits = 0

$$595 - [250(P/A, i, 5) - 50(P/G, i, 5) = 0$$

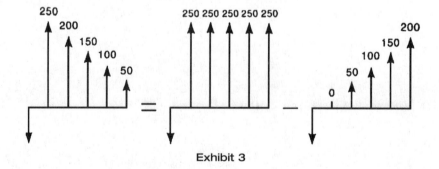

Exhibit 3

Try $i = 10\%$:

$$595 - [250(3.791) - 50(6.862)] = -9.65$$

Try $i = 12\%$:

$$595 - [250(3.605) - 50(6.397)] = +13.60$$

The rate of return is between 10% and 12%. It may be computed more accurately by linear interpolation:

$$\text{Rate of return} = 10\% + (2\%)\left(\frac{9.65 - 0}{13.60 + 9.65}\right) = 10.83\%$$

Two Alternatives

Compute the incremental rate of return on the cash flow representing the difference between the two alternatives. Since we want to look at increments of investment, the cash flow for the difference between the alternatives is computed by taking the higher initial-cost alternative minus the lower initial-cost alternative. If the incremental rate of return is greater than or equal to the predetermined minimum attractive rate of return (MARR), choose the higher-cost alternative; otherwise, choose the lower-cost alternative.

Example **A.21**

Two alternatives have the following cash flows:

Year	Alternative	
	A	*B*
0	−2000	−2800
1	+800	+1100
2	+800	+1100
3	+800	+1100

If 4% is considered the minimum attractive rate of return (MARR), which alternative should be selected?

Solution

These two alternatives were previously examined in Examples A.16 and A.18 by present worth and future worth analysis. This time, the alternatives will be resolved using a rate-of-return analysis.

Note that the problem statement specifies a 4% MARR, whereas Examples A.16 and A.18 referred to a 4% interest rate. These are really two different ways of saying the same thing: The minimum acceptable time value of money is 4%.

First, tabulate the cash flow that represents the increment of investment between the alternatives. This is done by taking the higher initial-cost alternative minus the lower initial-cost alternative:

Year	Alternative		Difference Between Alternatives
	A	*B*	*B – A*
0	−2000	−2800	−800
1	+800	+1100	+300
2	+800	+1100	+300
3	+800	+1100	+300

Then compute the rate of return on the increment of investment represented by the difference between the alternatives:

$$PW \text{ of cost} = PW \text{ of benefits}$$

$$800 = 300(P/A, i, 3)$$

$$(P/A, i, 3) = 800/300 = 2.67$$

$$i = 6.1\%$$

Since the incremental rate of return exceeds the 4% MARR, the increment of investment is desirable. Choose the higher-cost alternative B.

Before leaving this example, one should note something that relates to the rates of return on alternative A and on alternative B. These rates of return, if calculated, are

	Rate of Return
Alternative *A*	9.7%
Alternative *B*	8.7%

The correct answer to this problem has been shown to be alternative B, even though alternative A has a higher rate of return. The higher-cost alternative may be thought of as the lower-cost alternative plus the increment of investment between them. Viewed this way, the higher-cost alternative B is equal to the desirable lower-cost alternative A plus the difference between the alternatives.

The important conclusion is that computing the rate of return for each alternative does not provide the basis for choosing between alternatives. Instead, incremental analysis is required.

Example **A.22**

Consider the following:

Year	Alternative	
	A	*B*
0	−200.00	−131.00
1	+77.60	+48.10
2	+77.60	+48.10
3	+77.60	+48.10

If the MARR is 10%, which alternative should be selected?

Solution

To examine the increment of investment between the alternatives, we will examine the higher initial-cost alternative minus the lower initial-cost alternative, or A − B.

Year	Alternative		Increment
	A	*B*	*A − B*
0	−200.00	−131.00	−69.00
1	+77.60	+48.10	+29.50
2	+77.60	+48.10	+29.50
3	+77.60	+48.10	+29.50

Solve for the incremental rate of return:

$$PW \text{ of cost} = PW \text{ of benefits}$$

$$69.0 = 29.5(P/A, i, 3)$$

$$(P/A, i, 3) = 69.0/29.5 = 2.339$$

From compound interest tables, the incremental rate of return is between 12% and 18%. This is a desirable increment of investment; hence we select the higher-initial-cost alternative A.

Three or More Alternatives

When there are three or more mutually exclusive alternatives, proceed with the same logic presented for two alternatives. The components of incremental analysis are listed below.

Step 1. Compute the rate of return for each alternative. Reject any alternative where the rate of return is less than the desired MARR. (This step is not essential, but helps to immediately identify unacceptable alternatives.)

Step 2. Rank the remaining alternatives in order of increasing initial cost.

Step 3. Examine the increment of investment between the two lowest-cost alternatives as described for the two-alternative problem. Select the better of the two alternatives and reject the other one.

Step 4. Take the preferred alternative from step 3. Consider the next higher initial-cost alternative and proceed with another two-alternative comparison.

Step 5. Continue until all alternatives have been examined and the best of the multiple alternatives has been identified.

Example **A.23**

Consider the following:

Year	Alternative	
	A	B
0	−200.00	−131.00
1	+77.60	+48.10
2	+77.60	+48.10
3	+77.60	+48.10

If the MARR is 10%, which alternative, if any, should be selected?

Solution

One should carefully note that this is a three-alternative problem, where the alternatives are A, B, and DN. In this solution we will skip step 1. Reorganize the problem by placing the alternatives in order of increasing initial cost:

Year	Alternative		
	DN	A	B
0	0	−200.00	−131.00
1	0	+77.60	+48.10
2	0	+77.60	+48.10
3	0	+77.60	+48.10

Examine the B − DN increment of investment:

Year	B − DN
0	−131.00 − 0 = −131.00
1	+48.10 − 0 = +48.10
2	+48.10 − 0 = +48.10
3	+48.10 − 0 = +48.10

Solve for the incremental rate of return:

$$PW \text{ of cost} = PW \text{ of benefits}$$

$$131.0 = 48.1(P/A, i, 3)$$

$$(P/A, i, 3) = 131.0/48.1 = 2.723$$

From compound interest tables, the incremental rate of return is about 5%. Since the incremental rate of return is less than 10%, the $B − DN$ increment is not desirable. Reject alternative B.

Next, consider the increment of investment between the two remaining alternatives, $A - DN$.

Year	$A - DN$
0	$-200.00 - 0 = -200.00$
1	$+77.60 - 0 = +77.60$
2	$+77.60 - 0 = +77.60$
3	$+77.60 - 0 = +77.60$

Solve for the incremental rate of return:

$$\text{PW of cost} = \text{PW of benefits}$$

$$200.0 = 77.6(P/A, i, 3)$$

$$(P/A, i, 3) = 200.0/77.6 = 2.577$$

The incremental rate of return is 8%, less than the desired 10%. Reject the increment (alternative A) and select the remaining alternative: DN.

If you have not already done so, you should go back to Example A.22 and see how the slightly changed wording of the problem has radically altered it. Example A.22 required a choice between two undesirable alternatives. This example adds the DN (do nothing) alternative, which is superior to A and B.

BENEFIT-COST ANALYSIS

Generally, in public works and governmental economic analyses, the dominant method of analysis is the *benefit-cost ratio*. It is simply the ratio of benefits divided by costs, taking into account the time value of money.

$$B/C = \frac{\text{PW of benefits}}{\text{PW of cost}} = \frac{\text{Equivalent uniform annual benefits}}{\text{Equivalent uniform annual cost}}$$

For a given interest rate, a B/C ratio ≥ 1 reflects an acceptable project. The B/C analysis method is parallel to rate-of-return analysis. The same kind of incremental analysis is required.

Example **A.24**

Solve Example A.22 by benefit-cost analysis.

Solution

Year	Alternative		Increment
	A	B	$A - B$
0	-200.00	-131.00	-69.00
1	$+77.60$	$+48.10$	$+29.50$
2	$+77.60$	$+48.10$	$+29.50$
3	$+77.60$	$+48.10$	$+29.50$

The benefit-cost ratio for the A – B increment is

$$B/C = \frac{\text{PW of benefits}}{\text{PW of cost}} = \frac{29.5(P/A,10\%,3)}{69.0} = \frac{73.37}{69.0} = 1.06$$

Since the B/C ratio exceeds 1, the increment of investment is desirable. Select the higher-cost alternative A.

BREAKEVEN ANALYSIS

In business, "breakeven" is defined as the point where income just covers costs. In engineering economics, the breakeven point is defined as the point where two alternatives are equivalent.

Example A.25

An environmental engineering firm is considering purchasing $50,000 in analytical equipment for its in-house lab. The firm will save $600 per day compared to sending samples out for external analysis. You may assume the MARR is 12%, and the analyzer has no salvage value at the end of its ten year lifespan. How many days per year must the analyzer be in operation to justify the investment?

Solution

This breakeven problem may be readily solved by annual cost computations. We will set the equivalent uniform annual cost (EUAC) of the analyzer equal to its annual benefit and solve for the required annual utilization. Let X = breakeven point = days of operation per year.

$$\text{EUAC} = \text{EUAB}$$

$$50,000(A/P, 12\%, 10) = 600X$$

$$X = 50,000(0.1770)/600 = 14.8 \text{ days/year}$$

OPTIMIZATION

Optimization is the determination of the best or most favorable situation.

Minima-Maxima

In problems where the situation can be represented by a function, the customary approach is to set the first derivative of the function to zero and solve for the root(s) of this equation. If the second derivative is positive, the function is a minimum for the critical value; if it is *negative*, the function is a maximum.

Example A.26

A consulting engineering firm estimates that their net profit is given by the equation

$$P(x) = -0.03x^3 + 36x + 500 \qquad x \geq 0$$

where x = number of employees and $P(x)$ = net profit. What is the optimal number of employees?

Solution

$$P'(x) = -0.09x^2 + 36 = \qquad P''(x) = -0.18x$$

$$x^2 = 36/0.09 = 400$$

$$x = 20 \text{ employees.}$$

$$P''(20) = -0.18(20) = -3.6$$

Since $P''(20) < 0$, the net profit is maximized for 20 employees.

Economic Problem—Best Alternative

Since engineering economics problems seek to identify the best or most favorable situation, they are by definition optimization problems. Most use compound interest computations in their solution, but some do not. Consider the following example.

| Example **A.27** |

A firm must decide which of three alternatives to adopt to expand its capacity. It wants a minimum annual profit of 20% of the initial cost of each increment of investment. Any money not invested in capacity expansion can be invested elsewhere for an annual yield of 20% of the initial cost.

Alternative	Initial Cost	Annual Profit	Profit Rate
A	$100,000	$30,000	30%
B	$300,000	$66,000	22%
C	$500,000	$80,000	16%

Which alternative should be selected?

Solution

Notice that the profit rate was determined as the ratio of the annual profit to the initial cost. Since alternative C fails to produce the 20% minimum annual profit, it is rejected. To decide between alternatives A and B, examine the profit rate for the $B - A$ increment. Determine the incremental cost (cost of B – cost of A), the incremental profit (profit of B – profit of A), and the incremental profit rate as shown in the following table.

Alternative	Initial Cost	Annual Profit	Incremental Cost	Incremental Profit	Incremental Profit Rate
A	$100,000	$30,000			
B	300,000	66,000	$200,000	$36,000	18%

The $B - A$ incremental profit rate is less than the minimum 20%, so alternative B should be rejected. Thus the best investment of $300,000, for example, would be alternative A (annual profit = $30,000) plus $200,000 invested elsewhere at 20%

(annual profit = $40,000). This combination would yield a $70,000 annual profit, which is better than the alternative B profit of $66,000. Select A.

Economic Order Quantity

One special case of optimization occurs when an item is used continuously and is periodically purchased. Thus the inventory of the item fluctuates from zero (just prior to the receipt of the purchased quantity) to the purchased quantity (just after receipt). The simplest model for the economic order quantity (EOQ) is

$$EOQ = \sqrt{\frac{2BD}{E}}$$

where

B = ordering cost, $/order

D = demand per period, units

E = inventory holding cost, $/unit/period

EOQ = economic order quantity, units

Example A.28

An analytical laboratory uses 8000 high-capacity 0.45-μm filters per year for sample preparation. They are purchased at a cost of $5 each and are purchased from an outside vendor. The money invested in inventory costs 10% per year, and the warehousing costs amount to an additional 2% per year. It also costs $50 to process each purchase order. When an order is placed, how many filters should be ordered?

Solution

$$EOQ = \sqrt{\frac{2 \times \$50 \times 8000}{(10\% + 2\%)(5.00)}} = 1155 \text{ filters}$$

VALUATION AND DEPRECIATION

Depreciation of capital equipment is an important component of many after-tax economic analyses. For this reason, one must understand the fundamentals of depreciation accounting.

Notation

BV = book value

C = cost of the property (basis)

D_j = depreciation in year j

S_n = salvage value in year n

Depreciation is the systematic allocation of the cost of a capital asset over its useful life. *Book value* is the original cost of an asset, minus the accumulated depreciation of the asset.

$$\text{BV} = C - \Sigma(D_j)$$

In computing a schedule of depreciation charges, four items are considered.

1. Cost of the property, C (called the *basis* in tax law).

2. Type of property. Property is classified as either *tangible* (such as machinery) or *intangible* (such as a franchise or a copyright), and as either *real property* (real estate) or *personal property* (everything that is not real property).

3. Depreciable life in years, n.

4. Salvage value of the property at the end of its depreciable (useful) life, S_n.

Straight-Line Depreciation

The depreciation charge in any year is

$$D_j = \frac{C - S_n}{n}$$

An alternative computation is

$$D_j = \frac{C - \text{depreciation taken to beginning of year } j - S_n}{\text{Remaining useful life at beginning of year } j}$$

Sum-of-Years'-Digits Depreciation

Depreciation charge in any year,

$$D_j = \frac{\text{Remaining useful life at beginning of year } j}{\text{Sum of years' digits for total useful life}} \times (C - S_n)$$

Declining-Balance Depreciation

Double declining-balance depreciation charge in any year, $D_j = \dfrac{2C}{m}\left(1 - \dfrac{2}{n}\right)^{j-1}$

Total depreciation at the end of n years, $C = \left[1 - \left(1 - \dfrac{2}{n}\right)^n\right]$

Book value at the end of j years, $BV_j = C\left(1 - \dfrac{2}{n}\right)^j$

For 150% declining-balance depreciation, replace the 2 in the three equations above with 1.5.

Sinking-Fund Depreciation

Depreciation charge in any year, $D_j = (C - S_n)(A/F, i\%, n)(F/P, i\%, j - 1)$

Modified Accelerated Cost Recovery System Depreciation

The modified accelerated cost recovery system (MACRS) depreciation method generally applies to property placed in service after 1986. To compute the MACRS depreciation for an item, one must know

1. Cost (basis) of the item.

2. Property class. All tangible property is classified in one of six classes (3, 5, 7, 10, 15, and 20 years), which is the life over which it is depreciated (see Table A.3). Residential real estate and nonresidential real estate are in two separate real property classes of 27.5 years and 39 years, respectively.

3. Depreciation computation.

Table A.3 MACRS classes of depreciable property

Property Class	Personal Property (All Property Except Real Estate)
3-year property	Special handling devices for food and beverage manufacture Special tools for the manufacture of finished plastic products, fabricated metal products, and motor vehicles Property with an asset depreciation range (ADR) midpoint life of 4 years or less
5-year property	Automobiles* and trucks Aircraft (of nonair-transport companies) Equipment used in research and experimentation Computers Petroleum drilling equipment Property with an ADR midpoint life of more than 4 years and less than 10 years
7-year property	All other property not assigned to another class Office furniture, fixtures, and equipment Property with an ADR midpoint life of 10 years or more, and less than 16 years
10-year property	Assets used in petroleum refining and preparation of certain food products Vessels and water transportation equipment Property with an ADR midpoint life of 16 years or more, and less than 20 years
15-year property	Telephone distribution plants Municipal sewage treatment plants Property with an ADR midpoint life of 20 years or more, and less than 25 years
20-year property	Municipal sewers Property with an ADR midpoint life of 25 years or more

Property Class	Real Property (Real Estate)
27.5 years	Residential rental property (does not include hotels and motels)
39 years	Nonresidential real property

* The depreciation deduction for automobiles is limited to $2860 in the first tax year and is reduced in subsequent years.

■ Use double-declining-balance depreciation for 3-, 5-, 7-, and 10-year property classes with conversion to straight-line depreciation in the year that increases the deduction.

■ Use 150%-declining-balance depreciation for 15- and 20-year property classes with conversion to straight-line depreciation in the year that increases the deduction.

■ In MACRS, the salvage value is assumed to be zero.

Half-Year Convention

Except for real property, a half-year convention is used. Under this convention all property is considered to be placed in service in the middle of the tax year, and a half-year of depreciation is allowed in the first year. For each of the remaining years, one is allowed a full year of depreciation. If the property is disposed of prior to the end of the recovery period (property class life), a half-year of depreciation is allowed in that year. If the property is held for the entire recovery period, a half-year of depreciation is allowed for the year following the end of the recovery period (see Table A.4). Owing to the half-year convention, a general form of the double-declining-balance computation must be used to compute the year-by-year depreciation.

DDB depreciation in any year, $D_j = \dfrac{2}{n}(C - \text{depreciation in years prior to } j)$

Table A.4 MACRS* depreciation for personal property—half-year convention

If the Recovery Year Is	The Applicable %age for the Class of Property Is			
	3-Year Class	**5-Year Class**	**7-Year Class**	**10-Year Class**
1	33.33	20.00	14.29	10.00
2	44.45	32.00	24.49	18.00
3	14.81†	19.20	17.49	14.40
4	7.41	11.52†	12.49	11.52
5		11.52	8.93†	9.22
6		5.76	8.92	7.37
7			8.93	6.55†
8			4.46	6.55
9				6.56
10				6.55
11				3.28

* In the *Fundamentals of Engineering Reference Handbook*, this table is called "Modified ACRS Factors."

† Use straight-line depreciation for the year marked and all subsequent years.

Example **A.29**

A $5000 computer has an anticipated $500 salvage value at the end of its five-year depreciable life. Compute the depreciation schedule for the machinery by (a) sum-of-years'-digits depreciation and (b) MACRS depreciation. Do the MACRS computation by hand, and then compare the results with the values from Table A.4.

Solution

(a) Sum-of-years'-digits depreciation:

$$D_j = \frac{n-j+1}{\frac{n}{2}(n+1)}(C-S_n)$$

$$D_1 = \frac{5-1+1}{\frac{5}{2}(5+1)}(5000-500) = \$1500$$

$$D_2 = \frac{5-2+1}{\frac{5}{2}(5+1)}(5000-500) = \$1200$$

$$D_3 = \frac{5-3+1}{\frac{5}{2}(5+1)}(5000-500) = \$\,900$$

$$D_4 = \frac{5-4+1}{\frac{5}{2}(5+1)}(5000-500) = \$\,600$$

$$D_5 = \frac{5-5+1}{\frac{5}{2}(5+1)}(5000-500) = \$\,300$$

$$\overline{\$4500}$$

(b) MACRS depreciation. Double-declining-balance with conversion to straight-line. Five-year property class. Half-year convention. Salvage value S_n is assumed to be zero for MACRS. Using the general DDB computation,

Year

1 (half-year) $D_1 = \frac{1}{2} \times \frac{2}{5}(5000-0)$ $= \$1000$

2          $D_2 = \frac{2}{5}(5000-1000)$ $= \$1600$

3          $D_3 = \frac{2}{5}(5000-2600)$ $= \$\,960$

4          $D_4 = \frac{2}{5}(5000-3560)$ $= \$\,576$

5          $D_5 = \frac{2}{5}(5000-4136)$ $= \$\,346$

6 (half-year) $D_6 = \frac{1}{2} \times \frac{2}{5}(5000-4482)$ $= \$\,104$

$$\overline{\$4586}$$

The computation must now be modified to convert to straight-line depreciation at the point where the straight-line depreciation will be larger. Using the alternative straight-line computation,

$$D_5 = \frac{5000 - 4136 - 0}{1.5 \text{ year remaining}} = \$576$$

This is more than the $346 computed using DDB, hence switch to straight-line for year 5 and beyond.

$$D_6 \text{(half year)} = \frac{1}{2}(576) = \$288$$

Answers:

Year	Depreciation	
	SOYD	MACRS
1	$1500	$1000
2	1200	1600
3	900	960
4	600	576
5	300	576
6	0	288
	$4500	$5000

The computed MACRS depreciation is identical to the result obtained from Table A.4.

TAX CONSEQUENCES

Income taxes represent another of the various kinds of disbursements encountered in an economic analysis. The starting point in an after-tax computation is the before-tax cash flow. Generally, the before-tax cash flow contains three types of entries:

1. Disbursements of money to purchase capital assets. These expenditures create no direct tax consequence, for they are the exchange of one asset (money) for another (capital equipment).

2. Periodic receipts and/or disbursements representing operating income and/or expenses. These increase or decrease the year-by-year tax liability of the firm.

3. Receipts of money from the sale of capital assets, usually in the form of a salvage value when the equipment is removed. The tax consequences depend on the relationship between the book value (cost – depreciation taken) of the asset and its salvage value.

Situation	Tax Consequence
Salvage value > Book value	Capital gain on differences
Salvage value = Book value	No tax consequence
Salvage value < Book value	Capital loss on difference

After determining the before-tax cash flow, compute the depreciation schedule for any capital assets. Next, compute taxable income, the taxable component of the before-tax cash flow minus the depreciation. The income tax is the taxable income times the appropriate tax rate. Finally, the after-tax cash flow is the before-tax cash flow adjusted for income taxes.

To organize these data, it is customary to arrange them in the form of a cash flow table, as follows:

Year	Before-Tax Cash Flow	Depreciation	Taxable Income	Income Taxes	After-Tax Cash Flow
0	•				•
1	•	•	•	•	•

Example A.30

An environmental consulting firm expects to receive \$32,000 each year for 15 years from the sale of a patented treatment technology. There will be an initial investment of \$150,000. Manufacturing and sales expenses will be \$8067 per year. Assume straight-line depreciation, a 15-year useful life, and no salvage value. Use a 46% income tax rate. Determine the projected after-tax rate of return.

Solution

Straight-line depreciation, $D_j = \dfrac{C - S_n}{n} = \dfrac{\$150,000 - 0}{15} = \$10,000$ per year

Year	Before-Tax Cash Flow	Depreciation	Taxable Income	Income Taxes	After-Tax Cash Flow
0	−150,000				−150,000
1	+23,933	10,000	13,933	−6,409	+17,524
2	+23,933	10,000	13,933	−6,409	+17,524
•	•	•	•	•	•
•	•	•	•	•	•
•	•	•	•	•	•
15	+23,933	10,000	13,933	−6,409	+17,524

Take the after-tax cash flow and compute the rate of return at which the PW of cost equals the PW of benefits.

$$150,000 = 17,524(P/A, i\%, 15)$$

$$(P/A, i\%, 15) = \frac{150,000}{17,524} = 8.559$$

From the compound interest tables, the after-tax rate of return is $i = 8\%$.

INFLATION

Inflation is characterized by rising prices for goods and services, whereas deflation produces a fall in prices. An inflationary trend makes future dollars have less purchasing power than present dollars. This helps long-term borrowers of money, for they may repay a loan of present dollars in the future with dollars of reduced buying power. The help to borrowers is at the expense of lenders. Deflation has the opposite effect. Money borrowed at one point in time, followed by a deflationary period, subjects the borrower to loan repayment with dollars of greater purchasing power than those borrowed. This is to the lenders' advantage at the expense of borrowers.

Price changes occur in a variety of ways. One method of stating a price change is as a uniform rate of price change per year.

f = General inflation rate per interest period

i = Effective interest rate per interest period

The following situation will illustrate the computations. A mortgage is to be repaid in three equal payments of $5000 at the end of years 1, 2, and 3. If the annual inflation rate, f, is 8% during this period, and a 12% annual interest rate (i) is desired, what is the maximum amount the investor would be willing to pay for the mortgage?

The computation is a two-step process. First, the three future payments must be converted to dollars with the same purchasing power as today's (year 0) dollars.

Year	Actual Cash Flow		Multiplied by		Cash Flow Adjusted to Today's (yr. 0) Dollars
0	—		—		—
1	+5000	×	$(1 + 0.08)^{-1}$	=	+4630
2	+5000	×	$(1 + 0.08)^{-2}$	=	+4286
3	+5000	×	$(1 + 0.08)^{-3}$	=	+3969

The general form of the adjusting multiplier is

$$(1 + f)^{-n} = (P/F, f, n)$$

Now that the problem has been converted to dollars of the same purchasing power (today's dollars, in this example), we can proceed to compute the present worth of the future payments.

Year	Actual Cash Flow		Multiplied by		Present Worth
0	—		—		—
1	+4630	×	$(1 + 0.12)^{-1}$	=	+4134
2	+4286	×	$(1 + 0.12)^{-2}$	=	+3417
3	+3969	×	$(1 + 0.12)^{-3}$	=	+2825
					$10376

The general form of the discounting multiplier is

$$(1 + i)^{-n} = (P/F, i\%, n)$$

Alternative Solution

Instead of doing the inflation and interest rate computations separately, one can compute a combined equivalent interest rate, d.

$$d = (1 + f)(1 + i) - 1 = i + f + i(f)$$

For this cash flow, $d = 0.12 + 0.08 + 0.12(0.08) = 0.2096$. Since we do not have 20.96% interest tables, the problem has to be calculated using present worth equations.

$$PW = 5000(1 + 0.2096)^{-1} + 5000(1 + 0.2096)^{-2} + 5000(1 + 0.2096)^{-3}$$

$$= 4134 + 3417 + 2825 = \$10,376$$

Example A.31

One economist has predicted that there will be 7% per year inflation of prices during the next 10 years. If this proves to be correct, an item that presently sells for $10 would sell for what price 10 years hence?

Solution

$$f = 7\%, P = \$10$$

$$F = ?, n = 10 \text{ years}$$

Here the computation is to find the future worth F, rather than the present worth, P.

$$F = P(1 + f)^{10} = 10(1 + 0.07)^{10} = \$19.67$$

Effect of Inflation on Rate of Return

The effect of inflation on the computed rate of return for an investment depends on how future benefits respond to the inflation. If benefits produce constant dollars, which are not increased by inflation, the effect of inflation is to reduce the before-tax rate of return on the investment. If, on the other hand, the dollar benefits increase to keep up with the inflation, the before-tax rate of return will not be adversely affected by the inflation.

This is not true when an after-tax analysis is made. Even if the future benefits increase to match the inflation rate, the allowable depreciation schedule does not increase. The result will be increased taxable income and income tax payments. This reduces the available after-tax benefits and, therefore, the after-tax rate of return.

Example A.32

A man bought a 5% tax-free municipal bond. It cost $1000 and will pay $50 interest each year for 20 years. The bond will mature at the end of 20 years and return the original $1000. If there is 2% annual inflation during this period, what rate of return will the investor receive after considering the effect of inflation?

Solution

$$d = 0.05, \; i = \text{unknown}, \; j = 0.02$$

$$d = i + j + i(j)$$

$$0.05 = i + 0.02 + 0.02i$$

$$1.02i = 0.03, \; i = 0.294 = 2.94\%$$

REFERENCE

Newnan, Donald G. *Engineering Economic Analysis*, 5th ed. Engineering Press, San Jose, CA, 1995.

INTEREST TABLES

Compound interest factors

$\frac{1}{2}$ %

$\frac{1}{2}$ %

	Single Payment		Uniform Payment Series				Uniform Gradient		
	Compound Amount Factor	Present Worth Factor	Sinking Fund Factor	Capital Recovery Factor	Compound Amount Factor	Present Worth Factor	Gradient Uniform Series	Gradient Present Worth	
	Find F Given P	Find P Given F	Find A Given F	Find A Given P	Find F Given A	Find P Given A	Find A Given G	Find P Given G	
n	F/P	P/F	A/F	A/P	F/A	P/A	A/G	P/G	n
1	1.005	.9950	1.0000	1.0050	1.000	0.995	0	0	1
2	1.010	.9901	.4988	.5038	2.005	1.985	0.499	0.991	2
3	1.015	.9851	.3317	.3367	3.015	2.970	0.996	2.959	3
4	1.020	.9802	.2481	.2531	4.030	3.951	1.494	5.903	4
5	1.025	.9754	.1980	.2030	5.050	4.926	1.990	9.803	5
6	1.030	.9705	.1646	.1696	6.076	5.896	2.486	14.660	6
7	1.036	.9657	.1407	.1457	7.106	6.862	2.980	20.448	7
8	1.041	.9609	.1228	.1278	8.141	7.823	3.474	27.178	8
9	1.046	.9561	.1089	.1139	9.182	8.779	3.967	34.825	9
10	1.051	.9513	.0978	.1028	10.228	9.730	4.459	43.389	10
11	1.056	.9466	.0887	.0937	11.279	10.677	4.950	52.855	11
12	1.062	.9419	.0811	.0861	12.336	11.619	5.441	63.218	12
13	1.067	.9372	.0746	.0796	13.397	12.556	5.931	74.465	13
14	1.072	.9326	.0691	.0741	14.464	13.489	6.419	86.590	14
15	1.078	.9279	.0644	.0694	15.537	14.417	6.907	99.574	15
16	1.083	.9233	.0602	.0652	16.614	15.340	7.394	113.427	16
17	1.088	.9187	.0565	.0615	17.697	16.259	7.880	128.125	17
18	1.094	.9141	.0532	.0582	18.786	17.173	8.366	143.668	18
19	1.099	.9096	.0503	.0553	19.880	18.082	8.850	160.037	19
20	1.105	.9051	.0477	.0527	20.979	18.987	9.334	177.237	20
21	1.110	.9006	.0453	.0503	22.084	19.888	9.817	195.245	21
22	1.116	.8961	.0431	.0481	23.194	20.784	10.300	214.070	22
23	1.122	.8916	.0411	.0461	24.310	21.676	10.781	233.680	23
24	1.127	.8872	.0393	.0443	25.432	22.563	11.261	254.088	24
25	1.133	.8828	.0377	.0427	26.559	23.446	11.741	275.273	25
26	1.138	.8784	.0361	.0411	27.692	24.324	12.220	297.233	26
27	1.144	.8740	.0347	.0397	28.830	25.198	12.698	319.955	27
28	1.150	.8697	.0334	.0384	29.975	26.068	13.175	343.439	28
29	1.156	.8653	.0321	.0371	31.124	26.933	13.651	367.672	29
30	1.161	.8610	.0310	.0360	32.280	27.794	14.127	392.640	30
36	1.197	.8356	.0254	.0304	39.336	32.871	16.962	557.564	36
40	1.221	.8191	.0226	.0276	44.159	36.172	18.836	681.341	40
48	1.270	.7871	.0185	.0235	54.098	42.580	22.544	959.928	48
50	1.283	.7793	.0177	.0227	56.645	44.143	23.463	1035.70	50
52	1.296	.7716	.0169	.0219	59.218	45.690	24.378	1113.82	52
60	1.349	.7414	.0143	.0193	69.770	51.726	28.007	1448.65	60
70	1.418	.7053	.0120	.0170	83.566	58.939	32.468	1913.65	70
72	1.432	.6983	.0116	.0166	86.409	60.340	33.351	2012.35	72
80	1.490	.6710	.0102	.0152	98.068	65.802	36.848	2424.65	80
84	1.520	.6577	.00961	.0146	104.074	68.453	38.576	2640.67	84
90	1.567	.6383	.00883	.0138	113.311	72.331	41.145	2976.08	90
96	1.614	.6195	.00814	.0131	122.829	76.095	43.685	3324.19	96
100	1.647	.6073	.00773	.0127	129.334	78.543	45.361	3562.80	100
104	1.680	.5953	.00735	.0124	135.970	80.942	47.025	3806.29	104
120	1.819	.5496	.00610	.0111	163.880	90.074	53.551	4823.52	120
240	3.310	.3021	.00216	.00716	462.041	139.581	96.113	13415.56	240
360	6.023	.1660	.00100	.00600	1004.5	166.792	128.324	21403.32	360
480	10.957	.0913	.00050	.00550	1991.5	181.748	151.795	27588.37	480

Compound interest factors

	Single Payment		Uniform Payment Series				Uniform Gradient		
	Compound Amount Factor	Present Worth Factor	Sinking Fund Factor	Capital Recovery Factor	Compound Amount Factor	Present Worth Factor	Gradient Uniform Series	Gradient Present Worth	
	Find F Given P	Find P Given F	Find A Given F	Find A Given P	Find F Given A	Find P Given A	Find A Given G	Find P Given G	
n	F/P	P/F	A/F	A/P	F/A	P/A	A/G	P/G	n
1	1.010	.9901	1.0000	1.0100	1.000	0.990	0	0	1
2	1.020	.9803	.4975	.5075	2.010	1.970	0.498	0.980	2
3	1.030	.9706	.3300	.3400	3.030	2.941	0.993	2.921	3
4	1.041	.9610	.2463	.2563	4.060	3.902	1.488	5.804	4
5	1.051	.9515	.1960	.2060	5.101	4.853	1.980	9.610	5
6	1.062	.9420	.1625	.1725	6.152	5.795	2.471	14.320	6
7	1.072	.9327	.1386	.1486	7.214	6.728	2.960	19.917	7
8	1.083	.9235	.1207	.1307	8.286	7.652	3.448	26.381	8
9	1.094	.9143	.1067	.1167	9.369	8.566	3.934	33.695	9
10	1.105	.9053	.0956	.1056	10.462	9.471	4.418	41.843	10
11	1.116	.8963	.0865	.0965	11.567	10.368	4.900	50.806	11
12	1.127	.8874	.0788	.0888	12.682	11.255	5.381	60.568	12
13	1.138	.8787	.0724	.0824	13.809	12.134	5.861	71.112	13
14	1.149	.8700	.0669	.0769	14.947	13.004	6.338	82.422	14
15	1.161	.8613	.0621	.0721	16.097	13.865	6.814	94.481	15
16	1.173	.8528	.0579	.0679	17.258	14.718	7.289	107.273	16
17	1.184	.8444	.0543	.0643	18.430	15.562	7.761	120.783	17
18	1.196	.8360	.0510	.0610	19.615	16.398	8.232	134.995	18
19	1.208	.8277	.0481	.0581	20.811	17.226	8.702	149.895	19
20	1.220	.8195	.0454	.0554	22.019	18.046	9.169	165.465	20
21	1.232	.8114	.0430	.0530	23.239	18.857	9.635	181.694	21
22	1.245	.8034	.0409	.0509	24.472	19.660	10.100	198.565	22
23	1.257	.7954	.0389	.0489	25.716	20.456	10.563	216.065	23
24	1.270	.7876	.0371	.0471	26.973	21.243	11.024	234.179	24
25	1.282	.7798	.0354	.0454	28.243	22.023	11.483	252.892	25
26	1.295	.7720	.0339	.0439	29.526	22.795	11.941	272.195	26
27	1.308	.7644	.0324	.0424	30.821	23.560	12.397	292.069	27
28	1.321	.7568	.0311	.0411	32.129	24.316	12.852	312.504	28
29	1.335	.7493	.0299	.0399	33.450	25.066	13.304	333.486	29
30	1.348	.7419	.0287	.0387	34.785	25.808	13.756	355.001	30
36	1.431	.6989	.0232	.0332	43.077	30.107	16.428	494.620	36
40	1.489	.6717	.0205	.0305	48.886	32.835	18.178	596.854	40
48	1.612	.6203	.0163	.0263	61.223	37.974	21.598	820.144	48
50	1.645	.6080	.0155	.0255	64.463	39.196	22.436	879.417	50
52	1.678	.5961	.0148	.0248	67.769	40.394	23.269	939.916	52
60	1.817	.5504	.0122	.0222	81.670	44.955	26.533	1192.80	60
70	2.007	.4983	.00993	.0199	100.676	50.168	30.470	1528.64	70
72	2.047	.4885	.00955	.0196	104.710	51.150	31.239	1597.86	72
80	2.217	.4511	.00822	.0182	121.671	54.888	34.249	1879.87	80
84	2.307	.4335	.00765	.0177	130.672	56.648	35.717	2023.31	84
90	2.449	.4084	.00690	.0169	144.863	59.161	37.872	2240.56	90
96	2.599	.3847	.00625	.0163	159.927	61.528	39.973	2459.42	96
100	2.705	.3697	.00587	.0159	170.481	63.029	41.343	2605.77	100
104	2.815	.3553	.00551	.0155	181.464	64.471	42.688	2752.17	104
120	3.300	.3030	.00435	.0143	230.039	69.701	47.835	3334.11	120
240	10.893	.0918	.00101	.0110	989.254	90.819	75.739	6878.59	240
360	35.950	.0278	.00029	.0103	3495.0	97.218	89.699	8720.43	360
480	118.648	.00843	.00008	.0101	11764.8	99.157	95.920	9511.15	480

Compound interest factors

$1\frac{1}{2}\%$ $1\frac{1}{2}\%$

	Single Payment		Uniform Payment Series				Uniform Gradient		
	Compound Amount Factor	Present Worth Factor	Sinking Fund Factor	Capital Recovery Factor	Compound Amount Factor	Present Worth Factor	Gradient Uniform Series	Gradient Present Worth	
	Find F Given P	Find P Given F	Find A Given F	Find A Given P	Find F Given A	Find P Given A	Find A Given G	Find P Given G	
n	F/P	P/F	A/F	A/P	F/A	P/A	A/G	P/G	n
1	1.015	.9852	1.0000	1.0150	1.000	0.985	0	0	1
2	1.030	.9707	.4963	.5113	2.015	1.956	0.496	0.970	2
3	1.046	.9563	.3284	.3434	3.045	2.912	0.990	2.883	3
4	1.061	.9422	.2444	.2594	4.091	3.854	1.481	5.709	4
5	1.077	.9283	.1941	.2091	5.152	4.783	1.970	9.422	5
6	1.093	.9145	.1605	.1755	6.230	5.697	2.456	13.994	6
7	1.110	.9010	.1366	.1516	7.323	6.598	2.940	19.400	7
8	1.126	.8877	.1186	.1336	8.433	7.486	3.422	25.614	8
9	1.143	.8746	.1046	.1196	9.559	8.360	3.901	32.610	9
10	1.161	.8617	.0934	.1084	10.703	9.222	4.377	40.365	10
11	1.178	.8489	.0843	.0993	11.863	10.071	4.851	48.855	11
12	1.196	.8364	.0767	.0917	13.041	10.907	5.322	58.054	12
13	1.214	.8240	.0702	.0852	14.237	11.731	5.791	67.943	13
14	1.232	.8118	.0647	.0797	15.450	12.543	6.258	78.496	14
15	1.250	.7999	.0599	.0749	16.682	13.343	6.722	89.694	15
16	1.269	.7880	.0558	.0708	17.932	14.131	7.184	101.514	16
17	1.288	.7764	.0521	.0671	19.201	14.908	7.643	113.937	17
18	1.307	.7649	.0488	.0638	20.489	15.673	8.100	126.940	18
19	1.327	.7536	.0459	.0609	21.797	16.426	8.554	140.505	19
20	1.347	.7425	.0432	.0582	23.124	17.169	9.005	154.611	20
21	1.367	.7315	.0409	.0559	24.470	17.900	9.455	169.241	21
22	1.388	.7207	.0387	.0537	25.837	18.621	9.902	184.375	22
23	1.408	.7100	.0367	.0517	27.225	19.331	10.346	199.996	23
24	1.430	.6995	.0349	.0499	28.633	20.030	10.788	216.085	24
25	1.451	.6892	.0333	.0483	30.063	20.720	11.227	232.626	25
26	1.473	.6790	.0317	.0467	31.514	21.399	11.664	249.601	26
27	1.495	.6690	.0303	.0453	32.987	22.068	12.099	266.995	27
28	1.517	.6591	.0290	.0440	34.481	22.727	12.531	284.790	28
29	1.540	.6494	.0278	.0428	35.999	23.376	12.961	302.972	29
30	1.563	.6398	.0266	.0416	37.539	24.016	13.388	321.525	30
36	1.709	.5851	.0212	.0362	47.276	27.661	15.901	439.823	36
40	1.814	.5513	.0184	.0334	54.268	29.916	17.528	524.349	40
48	2.043	.4894	.0144	.0294	69.565	34.042	20.666	703.537	48
50	2.105	.4750	.0136	.0286	73.682	35.000	21.428	749.955	50
52	2.169	.4611	.0128	.0278	77.925	35.929	22.179	796.868	52
60	2.443	.4093	.0104	.0254	96.214	39.380	25.093	988.157	60
70	2.835	.3527	.00817	.0232	122.363	43.155	28.529	1231.15	70
72	2.921	.3423	.00781	.0228	128.076	43.845	29.189	1279.78	72
80	3.291	.3039	.00655	.0215	152.710	46.407	31.742	1473.06	80
84	3.493	.2863	.00602	.0210	166.172	47.579	32.967	1568.50	84
90	3.819	.2619	.00532	.0203	187.929	49.210	34.740	1709.53	90
96	4.176	.2395	.00472	.0197	211.719	50.702	36.438	1847.46	96
100	4.432	.2256	.00437	.0194	228.802	51.625	37.529	1937.43	100
104	4.704	.2126	.00405	.0190	246.932	52.494	38.589	2025.69	104
120	5.969	.1675	.00302	.0180	331.286	55.498	42.518	2359.69	120
240	35.632	.0281	.00043	.0154	2308.8	64.796	59.737	3870.68	240
360	212.700	.00470	.00007	.0151	14113.3	66.353	64.966	4310.71	360
480	1269.7	.00079	.00001	.0150	84577.8	66.614	66.288	4415.74	480

Compound interest factors

	Single Payment		Uniform Payment Series				Uniform Gradient		
	Compound Amount Factor	Present Worth Factor	Sinking Fund Factor	Capital Recovery Factor	Compound Amount Factor	Present Worth Factor	Gradient Uniform Series	Gradient Present Worth	
	Find F Given P	Find P Given F	Find A Given F	Find A Given P	Find F Given A	Find P Given A	Find A Given G	Find P Given G	
n	F/P	P/F	A/F	A/P	F/A	P/A	A/G	P/G	n
1	1.020	.9804	1.0000	1.0200	1.000	0.980	0	0	1
2	1.040	.9612	.4951	.5151	2.020	1.942	0.495	0.961	2
3	1.061	.9423	.3268	.3468	3.060	2.884	0.987	2.846	3
4	1.082	.9238	.2426	.2626	4.122	3.808	1.475	5.617	4
5	1.104	.9057	.1922	.2122	5.204	4.713	1.960	9.240	5
6	1.126	.8880	.1585	.1785	6.308	5.601	2.442	13.679	6
7	1.149	.8706	.1345	.1545	7.434	6.472	2.921	18.903	7
8	1.172	.8535	.1165	.1365	8.583	7.325	3.396	24.877	8
9	1.195	.8368	.1025	.1225	9.755	8.162	3.868	31.571	9
10	1.219	.8203	.0913	.1113	10.950	8.983	4.337	38.954	10
11	1.243	.8043	.0822	.1022	12.169	9.787	4.802	46.996	11
12	1.268	.7885	.0746	.0946	13.412	10.575	5.264	55.669	12
13	1.294	.7730	.0681	.0881	14.680	11.348	5.723	64.946	13
14	1.319	.7579	.0626	.0826	15.974	12.106	6.178	74.798	14
15	1.346	.7430	.0578	.0778	17.293	12.849	6.631	85.200	15
16	1.373	.7284	.0537	.0737	18.639	13.578	7.080	96.127	16
17	1.400	.7142	.0500	.0700	20.012	14.292	7.526	107.553	17
18	1.428	.7002	.0467	.0667	21.412	14.992	7.968	119.456	18
19	1.457	.6864	.0438	.0638	22.840	15.678	8.407	131.812	19
20	1.486	.6730	.0412	.0612	24.297	16.351	8.843	144.598	20
21	1.516	.6598	.0388	.0588	25.783	17.011	9.276	157.793	21
22	1.546	.6468	.0366	.0566	27.299	17.658	9.705	171.377	22
23	1.577	.6342	.0347	.0547	28.845	18.292	10.132	185.328	23
24	1.608	.6217	.0329	.0529	30.422	18.914	10.555	199.628	24
25	1.641	.6095	.0312	.0512	32.030	19.523	10.974	214.256	25
26	1.673	.5976	.0297	.0497	33.671	20.121	11.391	229.169	26
27	1.707	.5859	.0283	.0483	35.344	20.707	11.804	244.428	27
28	1.741	.5744	.0270	.0470	37.051	21.281	12.214	259.936	28
29	1.776	.5631	.0258	.0458	38.792	21.844	12.621	275.703	29
30	1.811	.5521	.0247	.0447	40.568	22.396	13.025	291.713	30
36	2.040	.4902	.0192	.0392	51.994	25.489	15.381	392.036	36
40	2.208	.4529	.0166	.0366	60.402	27.355	16.888	461.989	40
48	2.587	.3865	.0126	.0326	79.353	30.673	19.755	605.961	48
50	2.692	.3715	.0118	.0318	84.579	31.424	20.442	642.355	50
52	2.800	.3571	.0111	.0311	90.016	32.145	21.116	678.779	52
60	3.281	.3048	.00877	.0288	114.051	34.761	23.696	823.692	60
70	4.000	.2500	.00667	.0267	149.977	37.499	26.663	999.829	70
72	4.161	.2403	.00633	.0263	158.056	37.984	27.223	1034.050	72
80	4.875	.2051	.00516	.0252	193.771	39.744	29.357	1166.781	80
84	5.277	.1895	.00468	.0247	213.865	40.525	30.361	1230.413	84
90	5.943	.1683	.00405	.0240	247.155	41.587	31.793	1322.164	90
96	6.693	.1494	.00351	.0235	284.645	42.529	33.137	1409.291	96
100	7.245	.1380	.00320	.0232	312.230	43.098	33.986	1464.747	100
104	7.842	.1275	.00292	.0229	342.090	43.624	34.799	1518.082	104
120	10.765	.0929	.00205	.0220	488.255	45.355	37.711	1710.411	120
240	115.887	.00863	.00017	.0202	5744.4	49.569	47.911	2374.878	240
360	1247.5	.00080	.00002	.0200	62326.8	49.960	49.711	2483.567	360
480	13429.8	.00007		.0200	671442.0	49.996	49.964	2498.027	480

Compound interest factors

	Single Payment		Uniform Payment Series				Uniform Gradient		
	Compound Amount Factor	Present Worth Factor	Sinking Fund Factor	Capital Recovery Factor	Compound Amount Factor	Present Worth Factor	Gradient Uniform Series	Gradient Present Worth	
	Find F Given P	Find P Given F	Find A Given F	Find A Given P	Find F Given A	Find P Given A	Find A Given G	Find P Given G	
n	F/P	P/F	A/F	A/P	F/A	P/A	A/G	P/G	n
1	1.040	.9615	1.0000	1.0400	1.000	0.962	0	0	1
2	1.082	.9246	.4902	.5302	2.040	1.886	0.490	0.925	2
3	1.125	.8890	.3203	.3603	3.122	2.775	0.974	2.702	3
4	1.170	.8548	.2355	.2755	4.246	3.630	1.451	5.267	4
5	1.217	.8219	.1846	.2246	5.416	4.452	1.922	8.555	5
6	1.265	.7903	.1508	.1908	6.633	5.242	2.386	12.506	6
7	1.316	.7599	.1266	.1666	7.898	6.002	2.843	17.066	7
8	1.369	.7307	.1085	.1485	9.214	6.733	3.294	22.180	8
9	1.423	.7026	.0945	.1345	10.583	7.435	3.739	27.801	9
10	1.480	.6756	.0833	.1233	12.006	8.111	4.177	33.881	10
11	1.539	.6496	.0741	.1141	13.486	8.760	4.609	40.377	11
12	1.601	.6246	.0666	.1066	15.026	9.385	5.034	47.248	12
13	1.665	.6006	.0601	.1001	16.627	9.986	5.453	54.454	13
14	1.732	.5775	.0547	.0947	18.292	10.563	5.866	61.962	14
15	1.801	.5553	.0499	.0899	20.024	11.118	6.272	69.735	15
16	1.873	.5339	.0458	.0858	21.825	11.652	6.672	77.744	16
17	1.948	.5134	.0422	.0822	23.697	12.166	7.066	85.958	17
18	2.029	.4936	.0390	.0790	25.645	12.659	7.453	94.350	18
19	2.107	.4746	.0361	.0761	27.671	13.134	7.834	102.893	19
20	2.191	.4564	.0336	.0736	29.778	13.590	8.209	111.564	20
21	2.279	.4388	.0313	.0713	31.969	14.029	8.578	120.341	21
22	2.370	.4220	.0292	.0692	34.248	14.451	8.941	129.202	22
23	2.465	.4057	.0273	.0673	36.618	14.857	9.297	138.128	23
24	2.563	.3901	.0256	.0656	39.083	15.247	9.648	147.101	24
25	2.666	.3751	.0240	.0640	41.646	15.622	9.993	156.104	25
26	2.772	.3607	.0226	.0626	44.312	15.983	10.331	165.121	26
27	2.883	.3468	.0212	.0612	47.084	16.330	10.664	174.138	27
28	2.999	.3335	.0200	.0600	49.968	16.663	10.991	183.142	28
29	3.119	.3207	.0189	.0589	52.966	16.984	11.312	192.120	29
30	3.243	.3083	.0178	.0578	56.085	17.292	11.627	201.062	30
31	3.373	.2965	.0169	.0569	59.328	17.588	11.937	209.955	31
32	3.508	.2851	.0159	.0559	62.701	17.874	12.241	218.792	32
33	3.648	.2741	.0151	.0551	66.209	18.148	12.540	227.563	33
34	3.794	.2636	.0143	.0543	69.858	18.411	12.832	236.260	34
35	3.946	.2534	.0136	.0536	73.652	18.665	13.120	244.876	35
40	4.801	.2083	.0105	.0505	95.025	19.793	14.476	286.530	40
45	5.841	.1712	.00826	.0483	121.029	20.720	15.705	325.402	45
50	7.107	.1407	.00655	.0466	152.667	21.482	16.812	361.163	50
55	8.646	.1157	.00523	.0452	191.159	22.109	17.807	393.689	55
60	10.520	.0951	.00420	.0442	237.990	22.623	18.697	422.996	60
65	12.799	.0781	.00339	.0434	294.968	23.047	19.491	449.201	65
70	15.572	.0642	.00275	.0427	364.290	23.395	20.196	472.479	70
75	18.945	.0528	.00223	.0422	448.630	23.680	20.821	493.041	75
80	23.050	.0434	.00181	.0418	551.244	23.915	21.372	511.116	80
85	28.044	.0357	.00148	.0415	676.089	24.109	21.857	526.938	85
90	34.119	.0293	.00121	.0412	827.981	24.267	22.283	540.737	90
95	41.511	.0241	.00099	.0410	1012.8	24.398	22.655	552.730	95
100	50.505	.0198	.00081	.0408	1237.6	24.505	22.980	563.125	100

Compound interest factors

	Single Payment		Uniform Payment Series				Uniform Gradient		
	Compound Amount Factor	Present Worth Factor	Sinking Fund Factor	Capital Recovery Factor	Compound Amount Factor	Present Worth Factor	Gradient Uniform Series	Gradient Present Worth	
	Find F Given P	Find P Given F	Find A Given F	Find A Given P	Find F Given A	Find P Given A	Find A Given G	Find P Given G	
n	F/P	P/F	A/F	A/P	F/A	P/A	A/G	P/G	n
1	1.060	.943	1.0000	1.0600	1.000	0.943	0	0	1
2	1.124	.8900	.4854	.5454	2.060	1.833	0.485	0.890	2
3	1.191	.8396	.3141	.3741	3.184	2.673	0.961	2.569	3
4	1.262	.7921	.2286	.2886	4.375	3.465	1.427	4.945	4
5	1.338	.7473	.1774	.2374	5.637	4.212	1.884	7.934	5
6	1.419	.7050	.1434	.2034	6.975	4.917	2.330	11.459	6
7	1.504	.6651	.1191	.1791	8.394	5.582	2.768	15.450	7
8	1.594	.6274	.1010	.1610	9.897	6.210	3.195	19.841	8
9	1.689	.5919	.0870	.1470	11.491	6.802	3.613	24.577	9
10	1.791	.5584	.0759	.1359	13.181	7.360	4.022	29.602	10
11	1.898	.5268	.0668	.1268	14.972	7.887	4.421	34.870	11
12	2.012	.4970	.0593	.1193	16.870	8.384	4.811	40.337	12
13	2.133	.4688	.0530	.1130	18.882	8.853	5.192	45.963	13
14	2.261	.4423	.0476	.1076	21.015	9.295	5.564	51.713	14
15	2.397	.4173	.0430	.1030	23.276	9.712	5.926	57.554	15
16	2.540	.3936	.0390	.0990	25.672	10.106	6.279	63.459	16
17	2.693	.3714	.0354	.0954	28.213	10.477	6.624	69.401	17
18	2.854	.3503	.0324	.0924	30.906	10.828	6.960	75.357	18
19	3.026	.3305	.0296	.0896	33.760	11.158	7.287	81.306	19
20	3.207	.3118	.0272	.0872	36.786	11.470	7.605	87.230	20
21	3.400	.2942	.0250	.0850	39.993	11.764	7.915	93.113	21
22	3.604	.2775	.0230	.0830	43.392	12.042	8.217	98.941	22
23	3.820	.2618	.0213	.0813	46.996	12.303	8.510	104.700	23
24	4.049	.2470	.0197	.0797	50.815	12.550	8.795	110.381	24
25	4.292	.2330	.0182	.0782	54.864	12.783	9.072	115.973	25
26	4.549	.2198	.0169	.0769	59.156	13.003	9.341	121.468	26
27	4.822	.2074	.0157	.0757	63.706	13.211	9.603	126.860	27
28	5.112	.1956	.0146	.0746	68.528	13.406	9.857	132.142	28
29	5.418	.1846	.0136	.0736	73.640	13.591	10.103	137.309	29
30	5.743	.1741	.0126	.0726	79.058	13.765	10.342	142.359	30
31	6.088	.1643	.0118	.0718	84.801	13.929	10.574	147.286	31
32	6.453	.1550	.0110	.0710	90.890	14.084	10.799	152.090	32
33	6.841	.1462	.0103	.0703	97.343	14.230	11.017	156.768	33
34	7.251	.1379	.00960	.0696	104.184	14.368	11.228	161.319	34
35	7.686	.1301	.00897	.0690	111.435	11.498	11.432	165.743	35
40	10.286	.0972	.00646	.0665	154.762	15.046	12.359	185.957	40
45	13.765	.0727	.00470	.0647	212.743	15.456	13.141	203.109	45
50	18.420	.0543	.00344	.0634	290.335	15.762	13.796	217.457	50
55	24.650	.0406	.00254	.0625	394.171	15.991	14.341	229.322	55
60	32.988	.0303	.00188	.0619	533.126	16.161	14.791	239.043	60
65	44.145	.0227	.00139	.0614	719.080	16.289	15.160	246.945	65
70	59.076	.0169	.00103	.0610	967.928	16.385	15.461	253.327	70
75	79.057	.0126	.00077	.0608	1300.9	16.456	15.706	258.453	75
80	105.796	.00945	.00057	.0606	1746.6	16.509	15.903	262.549	80
85	141.578	.00706	.00043	.0604	2343.0	16.549	16.062	265.810	85
90	189.464	.00528	.00032	.0603	3141.1	16.579	16.189	268.395	90
95	253.545	.00394	.00024	.0602	4209.1	16.601	16.290	270.437	95
100	339.300	.00295	.00018	.0602	5638.3	16.618	16.371	272.047	100

Compound interest factors

8%   8%

	Single Payment		Uniform Payment Series				Uniform Gradient		
	Compound Amount Factor	Present Worth Factor	Sinking Fund Factor	Capital Recovery Factor	Compound Amount Factor	Present Worth Factor	Gradient Uniform Series	Gradient Present Worth	
	Find F Given P	Find P Given F	Find A Given F	Find A Given P	Find F Given A	Find P Given A	Find A Given G	Find P Given G	
n	F/P	P/F	A/F	A/P	F/A	P/A	A/G	P/G	n
2	1.166	.8573	.4808	.5608	2.080	1.783	0.481	0.857	2
3	1.260	.7938	.3080	.3880	3.246	2.577	0.949	2.445	3
4	1.360	.7350	.2219	.3019	4.506	3.312	1.404	4.650	4
5	1.469	.6806	.1705	.2505	5.867	3.993	1.846	7.372	5
6	1.587	.6302	.1363	.2163	7.336	4.623	2.276	10.523	6
7	1.714	.5835	.1121	.1921	8.923	5.206	2.694	14.024	7
8	1.851	.5403	.0940	.1740	10.637	5.747	3.099	17.806	8
9	1.999	.5002	.0801	.1601	12.488	6.247	3.491	21.808	9
10	2.159	.4632	.0690	.1490	14.487	6.710	3.871	25.977	10
11	2.332	.4289	.0601	.1401	16.645	7.139	4.240	30.266	11
12	2.518	.3971	.0527	.1327	18.977	7.536	4.596	34.634	12
13	2.720	.3677	.0465	.1265	21.495	7.904	4.940	39.046	13
14	2.937	.3405	.0413	.1213	24.215	8.244	5.273	43.472	14
15	3.172	.3152	.0368	.1168	27.152	8.559	5.594	47.886	15
16	3.426	.2919	.0330	.1130	30.324	8.851	5.905	52.264	16
17	3.700	.2703	.0296	.1096	33.750	9.122	6.204	56.588	17
18	3.996	.2502	.0267	.1067	37.450	9.372	6.492	60.843	18
19	4.316	.2317	.0241	.1041	41.446	9.604	6.770	65.013	19
20	4.661	.2145	.0219	.1019	45.762	9.818	7.037	69.090	20
21	5.034	.1987	.0198	.0998	50.423	10.017	7.294	73.063	21
22	5.437	.1839	.0180	.0980	55.457	10.201	7.541	76.926	22
23	5.871	.1703	.0164	.0964	60.893	10.371	7.779	80.673	24
24	6.341	.1577	.0150	.0950	66.765	10.529	8.007	84.300	24
25	6.848	.1460	.0137	.0937	73.106	10.675	8.225	87.804	25
26	7.396	.1352	.0125	.0925	79.954	10.810	8.435	91.184	26
27	7.988	.1252	.0114	.0914	87.351	10.935	8.636	94.439	27
28	8.627	.1159	.0105	.0905	95.339	11.051	8.829	97.569	28
29	9.317	.1073	.00962	.0896	103.966	11.158	9.013	100.574	29
30	10.063	.0994	.00883	.0888	113.283	11.258	9.190	103.456	30
31	10.868	.0920	.00811	.0881	123.346	11.350	9.358	106.216	31
32	11.737	.0852	.00745	.0875	134.214	11.435	9.520	108.858	32
33	12.676	.0789	.00685	.0869	145.951	11.514	9.674	111.382	33
34	13.690	.0730	.00630	.0863	158.627	11.587	9.821	113.792	34
35	14.785	.0676	.00580	.0858	172.317	11.655	9.961	116.092	35
40	21.725	.0460	.00386	.0839	259.057	11.925	10.570	126.042	40
45	31.920	.0313	.00259	.0826	386.506	12.108	11.045	133.733	45
50	46.902	.0213	.00174	.0817	573.771	12.233	11.411	139.593	50
55	68.914	.0145	.00118	.0812	848.925	12.319	11.690	144.006	55
60	101.257	.00988	.00080	.0808	1253.2	12.377	11.902	147.300	60
65	148.780	.00672	.00054	.0805	1847.3	12.416	12.060	149.739	65
70	218.607	.00457	.00037	.0804	2720.1	12.443	12.178	151.533	70
75	321.205	.00311	.00025	.0802	4002.6	12.461	12.266	152.845	75
80	471.956	.00212	.00017	.0802	5887.0	12.474	12.330	153.800	80
85	693.458	.00144	.00012	.0801	8655.7	12.482	12.377	154.492	85
90	1018.9	.00098	.00008	.0801	12724.0	12.488	12.412	154.993	90
95	1497.1	.00067	.00005	.0801	18701.6	12.492	12.437	155.352	95
100	2199.8	.00045	.00004	.0800	27484.6	12.494	12.455	155.611	100

Compound interest factors

	Single Payment		Uniform Payment Series				Uniform Gradient		
	Compound Amount Factor	Present Worth Factor	Sinking Fund Factor	Capital Recovery Factor	Compound Amount Factor	Present Worth Factor	Gradient Uniform Series	Gradient Present Worth	
	Find F Given P	Find P Given F	Find A Given F	Find A Given P	Find F Given A	Find P Given A	Find A Given G	Find P Given G	
n	F/P	P/F	A/F	A/P	F/A	P/A	A/G	P/G	n
1	1.100	.9091	1.0000	1.1000	1.000	0.909	0	0	1
2	1.210	.8264	.4762	.5762	2.100	1.736	0.476	0.826	2
3	1.331	.7513	.3021	.4021	3.310	2.487	0.937	2.329	3
4	1.464	.6830	.2155	.3155	4.641	3.170	1.381	4.378	4
5	1.611	.6209	.1638	.2638	6.105	3.791	1.810	6.862	5
6	1.772	.5645	.1296	.2296	7.716	4.355	2.224	9.684	6
7	1.949	.5132	.1054	.2054	9.487	4.868	2.622	12.763	7
8	2.144	.4665	.0874	.1874	11.436	5.335	3.004	16.029	8
9	2.358	.4241	.0736	.1736	13.579	5.759	3.372	19.421	9
10	2.594	.3855	.0627	.1627	15.937	6.145	3.725	22.891	10
11	2.853	.3505	.0540	.1540	18.531	6.495	4.064	26.396	11
12	3.138	.3186	.0468	.1468	21.384	6.814	4.388	29.901	12
13	3.452	.2897	.0408	.1408	24.523	7.103	4.699	33.377	13
14	3.797	.2633	.0357	.1357	27.975	7.367	4.996	36.801	14
15	4.177	.2394	.0315	.1315	31.772	7.606	5.279	40.152	15
16	4.595	.2176	.0278	.1278	35.950	7.824	5.549	43.416	16
17	5.054	.1978	.0247	.1247	40.545	8.022	5.807	46.582	17
18	5.560	.1799	.0219	.1219	45.599	8.201	6.053	49.640	18
19	6.116	.1635	.0195	.1195	51.159	8.365	6.286	52.583	19
20	6.728	.1486	.0175	.1175	57.275	8.514	6.508	55.407	20
21	7.400	.1351	.0156	.1156	64.003	8.649	6.719	58.110	21
22	8.140	.1228	.0140	.1140	71.403	8.772	6.919	60.689	22
23	8.954	.1117	.0126	.1126	79.543	8.883	7.108	63.146	24
24	9.850	.1015	.0113	.1113	88.497	8.985	7.288	65.481	24
25	10.835	.0923	.0102	.1102	98.347	9.077	7.458	67.696	25
26	11.918	.0839	.00916	.1092	109.182	9.161	7.619	69.794	26
27	13.110	.0763	.00826	.1083	121.100	9.237	7.770	71.777	27
28	14.421	.0693	.00745	.1075	134.210	9.307	7.914	73.650	28
29	15.863	.0630	.00673	.1067	148.631	9.370	8.049	75.415	29
30	17.449	.0573	.00608	.1061	164.494	9.427	8.176	77.077	30
31	19.194	.0521	.00550	.1055	181.944	9.479	8.296	78.640	31
32	21.114	.0474	.00497	.1050	201.138	9.526	8.409	80.108	32
33	23.225	.0431	.00450	.1045	222.252	9.569	8.515	81.486	33
34	25.548	.0391	.00407	.1041	245.477	9.609	8.615	82.777	34
35	28.102	.0356	.00369	.1037	271.025	9.644	8.709	83.987	35
40	45.259	.0221	.00226	.1023	442.593	9.779	9.096	88.953	40
45	72.891	.0137	.00139	.1014	718.905	9.863	9.374	92.454	45
50	117.391	.00852	.00086	.1009	1163.9	9.915	9.570	94.889	50
55	189.059	.00529	.00053	.1005	1880.6	9.947	9.708	96.562	55
60	304.482	.00328	.00033	.1003	3034.8	9.967	9.802	97.701	60
65	490.371	.00204	.00020	.1002	4893.7	9.980	9.867	98.471	65
70	789.748	.00127	.00013	.1001	7887.5	9.987	9.911	98.987	70
75	1271.9	.00079	.00008	.1001	12709.0	9.992	9.941	99.332	75
80	2048.4	.00049	.00005	.1000	20474.0	9.995	9.961	99.561	80
85	3229.0	.00030	.00003	.1000	32979.7	9.997	9.974	99.712	85
90	5313.0	.00019	.00002	.1000	53120.3	9.998	9.983	99.812	90
95	8556.7	.00012	.00001	.1000	85556.9	9.999	9.989	99.877	95
100	13780.6	.00007	.00001	.1000	137796.3	9.999	9.993	99.920	100

Compound interest factors

	Single Payment		Uniform Payment Series				Uniform Gradient		
	Compound Amount Factor	Present Worth Factor	Sinking Fund Factor	Capital Recovery Factor	Compound Amount Factor	Present Worth Factor	Gradient Uniform Series	Gradient Present Worth	
	Find F Given P	Find P Given F	Find A Given F	Find A Given P	Find F Given A	Find P Given A	Find A Given G	Find P Given G	
n	F/P	P/F	A/F	A/P	F/A	P/A	A/G	P/G	n
1	1.120	.8929	1.0000	1.1200	1.000	0.893	0	0	1
2	1.254	.7972	.4717	.5917	2.120	1.690	0.472	0.797	2
3	1.405	.7118	.2963	.4163	3.374	2.402	0.925	2.221	3
4	1.574	.6355	.2092	.3292	4.779	3.037	1.359	4.127	4
5	1.762	.5674	.1574	.2774	6.353	3.605	1.775	6.397	5
6	1.974	.5066	.1232	.2432	8.115	4.111	2.172	8.930	6
7	2.211	.4523	.0991	.2191	10.089	4.564	2.551	11.644	7
8	2.476	.4039	.0813	.2013	12.300	4.968	2.913	14.471	8
9	2.773	.3606	.0677	.1877	14.776	5.328	3.257	17.356	9
10	3.106	.3220	.0570	.1770	17.549	5.650	3.585	20.254	10
11	3.479	.2875	.0484	.1684	20.655	5.938	3.895	23.129	11
12	3.896	.2567	.0414	.1614	24.133	6.194	4.190	25.952	12
13	4.363	.2292	.0357	.1557	28.029	6.424	4.468	28.702	13
14	4.887	.2046	.0309	.1509	32.393	6.628	4.732	31.362	14
15	5.474	.1827	.0268	.1468	37.280	6.811	4.980	33.920	15
16	6.130	.1631	.0234	.1434	42.753	6.974	5.215	36.367	16
17	6.866	.1456	.0205	.1405	48.884	7.120	5.435	38.697	17
18	7.690	.1300	.0179	.1379	55.750	7.250	5.643	40.908	18
19	8.613	.1161	.0158	.1358	63.440	7.366	5.838	42.998	19
20	9.646	.1037	.0139	.1339	72.052	7.469	6.020	44.968	20
21	10.804	.0926	.0122	.1322	81.699	7.562	6.191	46.819	21
22	12.100	.0826	.0108	.1308	92.503	7.645	6.351	48.554	22
23	13.552	.0738	.00956	.1296	104.603	7.718	6.501	50.178	24
24	15.179	.0659	.00846	.1285	118.155	7.784	6.641	51.693	24
25	17.000	.0588	.00750	.1275	133.334	7.843	6.771	53.105	25
26	19.040	.0525	.00665	.1267	150.334	7.896	6.892	54.418	26
27	21.325	.0469	.00590	.1259	169.374	7.943	7.005	55.637	27
28	23.884	.0419	.00524	.1252	190.699	7.984	7.110	56.767	28
29	26.750	.0374	.00466	.1247	214.583	8.022	7.207	57.814	29
30	29.960	.0334	.00414	.1241	241.333	8.055	7.297	58.782	30
31	33.555	.0298	.00369	.1237	271.293	8.085	7.381	59.676	31
32	37.582	.0266	.00328	.1233	304.848	8.112	7.459	60.501	32
33	42.092	.0238	.00292	.1229	342.429	8.135	7.530	61.261	33
34	47.143	.0212	.00260	.1226	384.521	8.157	7.596	61.961	34
35	52.800	.0189	.00232	.1223	431.663	8.176	7.658	62.605	35
40	93.051	.0107	.00130	.1213	767.091	8.244	7.899	65.116	40
45	163.988	.00610	.00074	.1207	1358.2	8.283	8.057	66.734	45
50	289.002	.00346	.00042	.1204	2400.0	8.304	8.160	67.762	50
55	509.321	.00196	.00024	.1202	4236.0	8.317	8.225	68.408	55
60	897.597	.00111	.00013	.1201	7471.6	8.324	8.266	68.810	60
65	1581.9	.00063	.00008	.1201	13173.9	8.328	8.292	69.058	65
70	2787.8	.00036	.00004	.1200	23223.3	8.330	8.308	69.210	70
75	4913.1	.00020	.00002	.1200	40933.8	8.332	8.318	69.303	75
80	8658.5	.00012	.00001	.1200	72145.7	8.332	8.324	69.359	80
85	15259.2	.00007	.00001	.1200	127151.7	8.333	8.328	69.393	85
90	26891.9	.00004		.1200	224091.1	8.333	8.330	69.414	90
95	47392.8	.00002		.1200	394931.4	8.333	8.331	69.426	95
100	83522.3	.00001		.1200	696010.5	8.333	8.332	69.434	100

Compound interest factors

	Single Payment		Uniform Payment Series				Uniform Gradient		
	Compound Amount Factor	Present Worth Factor	Sinking Fund Factor	Capital Recovery Factor	Compound Amount Factor	Present Worth Factor	Gradient Uniform Series	Gradient Present Worth	
	Find F Given P	Find P Given F	Find A Given F	Find A Given P	Find F Given A	Find P Given A	Find A Given G	Find P Given G	
n	F/P	P/F	A/F	A/P	F/A	P/A	A/G	P/G	n
1	1.180	.8475	1.0000	1.1800	1.000	0.847	0	0	1
2	1.392	.7182	.4587	.6387	2.180	1.566	0.459	0.718	2
3	1.643	.6086	.2799	.4599	3.572	2.174	0.890	1.935	3
4	1.939	.5158	.1917	.3717	5.215	2.690	1.295	3.483	4
5	2.288	.4371	.1398	.3198	7.154	3.127	1.673	5.231	5
6	2.700	.3704	.1059	.2859	9.442	3.498	2.025	7.083	6
7	3.185	.3139	.0824	.2624	12.142	3.812	2.353	8.967	7
8	3.759	.2660	.0652	.2452	15.327	4.078	2.656	10.829	8
9	4.435	.2255	.0524	.2324	19.086	4.303	2.936	12.633	9
10	5.234	.1911	.0425	.2225	23.521	4.494	3.194	14.352	10
11	6.176	.1619	.0348	.2148	28.755	4.656	3.430	15.972	11
12	7.288	.1372	.0286	.2086	34.931	4.793	3.647	17.481	12
13	8.599	.1163	.0237	.2037	42.219	4.910	3.845	18.877	13
14	10.147	.0985	.0197	.1997	50.818	5.008	4.025	20.158	14
15	11.974	.0835	.0164	.1964	60.965	5.092	4.189	21.327	15
16	14.129	.0708	.0137	.1937	72.939	5.162	4.337	22.389	16
17	16.672	.0600	.0115	.1915	87.068	5.222	4.471	23.348	17
18	19.673	.0508	.00964	.1896	103.740	5.273	4.592	24.212	18
19	23.214	.0431	.00810	.1881	123.413	5.316	4.700	24.988	19
20	27.393	.0365	.00682	.1868	146.628	5.353	4.798	25.681	20
21	32.324	.0309	.00575	.1857	174.021	5.384	4.885	26.330	21
22	38.142	.0262	.00485	.1848	206.345	5.410	4.963	26.851	22
23	45.008	.0222	.00409	.1841	244.487	5.432	5.033	27.339	24
24	53.109	.0188	.00345	.1835	289.494	5.451	5.095	27.772	24
25	62.669	.0160	.00292	.1829	342.603	5.467	5.150	28.155	25
26	73.949	.0135	.00247	.1825	405.272	5.480	5.199	28.494	26
27	87.260	.0115	.00209	.1821	479.221	5.492	5.243	28.791	27
28	102.966	.00971	.00177	.1818	566.480	5.502	5.281	29.054	28
29	121.500	.00823	.00149	.1815	669.447	5.510	5.315	29.284	29
30	143.370	.00697	.00126	.1813	790.947	5.517	5.345	29.486	30
31	169.177	.00591	.00107	.1811	934.317	5.523	5.371	29.664	31
32	199.629	.00501	.00091	.1809	1103.5	5.528	5.394	29.819	32
33	235.562	.00425	.00077	.1808	1303.1	5.532	5.415	29.955	33
34	277.963	.00360	.00065	.1806	1538.7	5.536	5.433	30.074	34
35	327.997	.00305	.00055	.1806	1816.6	5.539	5.449	30.177	35
40	750.377	.00133	.00024	.1802	4163.2	5.548	5.502	30.527	40
45	1716.7	.00058	.00010	.1801	9531.6	5.552	5.529	30.701	45
50	3927.3	.00025	.00005	.1800	21813.0	5.554	5.543	30.786	50
55	8984.8	.00011	.00002	.1800	49910.1	5.555	5.549	30.827	55
60	20555.1	.00005	.00001	.1800	114189.4	5.555	5.553	30.846	60
65	47025.1	.00002		.1800	261244.7	5.556	5.554	30.856	65
70	107581.9	.00001		.1800	597671.7	5.556	5.555	30.860	70
75	46122.1				1367339.2	5.556	5.555	30.862	75
100	15424131.9				85689616.2	5.556	5.555	30.864	100

Statistics and Sampling

OUTLINE

As stated in Standard Methods for the Examination of Water and Wastewater[1]: "Data quality objectives are systematic planning tools based on the scientific method. They are used to develop data collection designs and to establish specific criteria for the quality of data to be collected." The obvious purpose of any environmental sampling event is to gather information as to the extent of contamination to the air, water, and/or soil. This information is used to determine if a need exists for environmental restoration, to develop an appropriate action plan to address the identified target species, and to monitor the effectiveness of the implemented effort. As with all sampling plans, the primary intent is to determine the number, spatial distribution, and frequency of sampling events that provides a statistically significant picture of the extent of contamination. The primary focus of Appendix B is to use a statistical approach to determine the number of samples required to ensure that a set of measurements are significant with a desired level of confidence.

DESCRIPTIVE STATISTICS

Descriptive statistics is concerned with recording and summarizing data. One way of summarizing data is to compute a value about which the data are centered. Three

---

1   APHA, AWWA, WEF (1998), p. 1-18.

commonly used *measures of central tendency* are the mean, the mode, and the median. The *mean* is computed as

$$\bar{x} = (1/n)\sum x_i \tag{B.1}$$

where n = sample size (number of observations) and x_i = the individual observation of x. The *mode* is simply the most frequently occurring value in the data set, and the *median* is the middle number (50th percentile) in an ordered set of numbers.

In addition to summarizing a data set in terms of a representative value such as its mean, median, or mode, we should also be concerned about the amount of variability (or spread) in the data. The dispersion of a set of observations is most often measured in terms of the deviations of the observations from their mean, as calculated from the following equation:

$$s^2 = (1/n-1)\sum(x_i - \bar{x})^2 \tag{B.2}$$

where s^2 = sample variance of x. The square root of the variance is called the standard deviation. To facilitate calculator computations, Equation B.2 can be manipulated into the following form:

$$s^2 = (1/n-1)\sum(x_i)^2 - (n/n-1)(\bar{x})^2 \tag{B.3}$$

The mean is also a random variable with a standard deviation that can be calculated from s according to the following formula:

$$s_{\bar{x}} = s/(n)^{1/2} \tag{B.4}$$

where $s_{\bar{x}}$ = the standard deviation of the mean (sometimes called the *standard error of the mean*).

At this point, a few comments on notation and terminology are in order. In the vocabulary of the statistician, the terms *population, sample, parameter,* and *statistic* have very important distinctions. *Population* refers to the largest collection of persons, places, or things—including measurements—in which we have an interest; for example, the ages of all environmental engineers in this country. A *sample* is a part or subset of a population, such as the ages of environmental engineers in a particular city or firm. A *parameter* is a measure that describes a population, such as population mean or population variance. Population parameters are generally unknown. A *statistic* is a measure computed from sample data, such as a sample mean. The population variance and mean are denoted by σ^2 and μ, respectively. The corresponding measures of a sample are s^2 and $\bar{x}$. Sample data (statistics) are used to estimate population parameters.

PROBABILITY

Probability is a measure that describes the likelihood that an event will occur. Probability is a dimensionless number that ranges from 0 (meaning the event cannot occur) to 1 (meaning the event is certain to occur). The probability of occurrence of an event, A, is simply the number of ways A may occur divided by the total number of possible outcomes of the experiment under consideration:

$$P(A) = n_A/N \tag{B.5}$$

where $P(A)$ = probability event A will occur, n_A = number of ways A can occur, and N = total number of possible outcomes. For example, the probability of being dealt an ace from a newly shuffled deck of 52 cards is 4 (number of aces in the deck) divided by 52 (total number of possible outcomes), or 0.08.

In situations involving the probability of the occurrence of multiple events, the following rules apply:

Rule 1. Addition rule for mutually exclusive events (*A* or *B* can occur, but not simultaneously):

$$P(A \text{ or } B) = P(A) + P(B) \tag{B.6}$$

Rule 2. Multiplication rule for independent events (two events could both occur, but the occurrence of one does not influence the occurrence of the other):

$$P(A \text{ and } B) = P(A)P(B) \tag{B.7}$$

Rule 3. General addition rule (*A* or *B* can occur, but outcomes are not necessarily mutually exclusive):

$$P(A \text{ or } B) = P(A) + P(B) - P(A \text{ and } B) \tag{B.8}$$

Rule 4. General multiplication rule (Probability of two events occurring when they are not independent; that is, the probability of one event occurring is different depending on whether the other has occurred):

$$P(A \text{ and } B) = P(A)P(B \setminus A) \tag{B.9}$$

where $P(B \setminus A)$ is read as "the probability of *B* given the occurrence of *A*."

The basic rules of probability outlines above assume that the distribution of the values of the variables in question is known. This is not usually the case, and a procedure for computing probabilities when dealing with unknown populations is needed. That is to say, it would be very useful to be able to write mathematical equations, graphs, or tables to describe the population in question. A number of *theoretical distributions* are often used to describe the possible values of a variable and the probabilities that each value will occur, including the normal distribution, the Poisson distribution, the exponential distribution, and the chi-square distribution. Here, we will focus on the normal and Poisson distributions.

The Normal Distribution

The theoretical distribution most frequently used by engineers and scientists is the normal distribution. Many variables such as highest test scores and linear dimensions in general conform to a normal (or approximately normal) distribution. Many non-normal distributions can be transformed to induce normality (for example, by taking the square root or logarithm of the variable). Even if the distribution in the subject population is far from normal, the distribution of sample means tends to become normal in a sufficiently large number of repeated random samples. This is the single most important reason for the use of the normal distribution.

The normal distribution is a symmetric, continuous distribution. The normal "density function" is written as

$$f(x) = \left[1 / \left(\sigma\sqrt{2\pi} \right) \exp - \left(x - \mu \right)^2 / \left(2\sigma^2 \right) \right] \tag{B.10}$$

where σ = standard deviation of *x*, π = 3.1416 …, and μ = the mean of *x*. The probability distribution functions for continuous distribution are often referred to as "density functions." Probability functions for discrete distributions are referred to as "mass functions."

The probability of an outcome falling in a specified range is given by the corresponding area under the density function, integrating the density function between the desired limits. As can be seen from Equation B.10, the normal distribution is actually a family of distributions, each with unique values of μ and σ. To facilitate tabulation of areas under the normal curve (that is, probabilities), the "standard" normal curve is used. The standard normal curve has a mean = 0, a standard deviation = 1.0, and a total area = 1.0. Any normal distribution can be transformed into the standard normal distribution of the random variable Z by the following equation:

$$Z = (x - \mu) / \sigma \qquad \text{(B.11)}$$

where Z = the transformed value of x (Z is a measure of the deviation from the mean in units of standard deviations).

Tabulations of the area under the cumulative standard normal distribution can be found in standard statistics textbooks. Table B.1 summarizes some important probabilities in the standard normal distribution. For example, approximately 95% of the area under the curve lies within ±2.0 standard deviations from the mean. Thus, we could expect a normally distributed variable to be within this range about 95% of the time.

Table B.1   Area under the standard normal curve

Boundaries	Area between Boundaries
$\mu \pm 0.5\sigma$	0.383
$\mu \pm 1.0\sigma$	0.683
$\mu \pm 1.5\sigma$	0.866
$\mu \pm 2.0\sigma$	0.954
$\mu \pm 2.5\sigma$	0.977
$\mu \pm 3.0\sigma$	0.988
$\mu \pm 3.5\sigma$	0.999

The Poisson Distribution

The second theoretical distribution that has many useful applications in environmental engineering is the Poisson distribution. The Poisson distribution is useful in describing the occurrence of discrete random events. The probability of x events occurring in time t is given by the following expression:

$$P(x) = (e^{-m} m^x) / x! \qquad \text{(B.12)}$$

where e = base of the natural logarithms, m = mean frequency of occurrence, or number of time periods, and ! = factorial operator (for example, $3! = 3 \times 2 \times 1 = 6$).

The terms of the Poisson distribution may be summed to give the probability of fewer than or more than x events per time periods. If the engineer is interested in the probability of $\leq x$ events occurring, it may be expressed as

$$P(\leq x) = \sum [(m^i e^{-m}) / i!], \text{ for } i = 0, 1 \ldots, x \qquad \text{(B.13)}$$

For the case of fewer than x,

$$P(< x) = \sum [(m^i e^{-m}) / i!], \text{ for } i = 0, 1 \ldots, x - 1 \qquad \text{(B.14)}$$

For the case of more than x,

$$P(> x) = 1 - \sum [(m^i e^{-m})/i!], \text{ for } i = 0, 1 \ldots, x \quad \textbf{(B.15)}$$

For the case of x or more,

$$P(\geq x) = 1 - \sum [(m^i e^{-m})/i!], \text{ for } i = 0, 1 \ldots, x - 1 \quad \textbf{(B.16)}$$

or

$$P(\geq x) = \sum [(m^i e^{-m})/i!], \text{ for } i = x, x+1 \ldots, \infty \quad \textbf{(B.17)}$$

For the case of at least x but not more than y,

$$P(x \leq i \leq y) = \sum [(m^i e^{-m})/i!], \text{ for } i = x, \ldots, y \quad \textbf{(B.18)}$$

It should be noted that for the Poisson distribution, the mean variances are equal. Therefore, when the ratio of the variance to the mean is markedly different from 1.0, this is an indication that the observed data do not follow a Poisson distribution.

Tabulated values of the Poisson distribution can be found in most standard statistics textbooks.

CONFIDENCE BOUNDS

Referring to Table B.1, it can be seen that in a normal distribution there is a probability of approximately 0.95 that an outcome will fall within about two standard deviations of the mean. More precisely, there is a probability of 0.95 that an outcome will lie within 1.96 standard deviations of the mean. That is

$$P(-1.96 \leq Z \leq +1.96) = 0.95 \quad \textbf{(B.19)}$$

Recall that any variable x can be expressed in "standard form" by changing from x to $(x - \mu)/\sigma$. For $\bar{x}$ (the sample mean), the corresponding expression is

$$Z = (\bar{x} \pm \mu) / (\sigma \sqrt{n}) \quad \textbf{(B.20)}$$

where $\sigma / \sqrt{n}$ is the standard error of the mean. From Equation B.18 it follows that

$$P\left[\bar{x} - (1.96\sigma / \sqrt{n}) \leq \mu \leq \bar{x} + (1.96\sigma / \sqrt{n})\right] = 0.95 \quad \textbf{(B.21)}$$

Equation B.21 can be read as "the probability that the true (population) mean lies in the interval $\bar{x} - (1.96\sigma / \sqrt{n})$ and $\bar{x} + (1.96\sigma / \sqrt{n})$ is 0.95." This is referred to as a *confidence bound*, or confidence interval, on the estimate of the true mean. Other confidence intervals can be constructed by replacing 1.96 with the Z-score corresponding to the desired confidence probability. The general expression for confidence limits on $\bar{x}$ is

$$CL = \bar{x} \pm Z(s / \sqrt{n}) \quad \textbf{(B.22)}$$

The general expression for the probability of a value of x greater or less than the mean is

$$x = \bar{x} \pm Zs \quad \textbf{(B.23)}$$

In the preceding discussion, the Z values of the standard normal distribution are, strictly speaking, applicable only when the sample size is "large" ($n > 30$). For small samples ($n < 30$) values from the t-distribution should be used in place of

Z. Tabulated t values can be found in most standard statistics textbooks; see also Table B.2.

If the required confidence bound (or tolerable error) is specified, the sample size required to estimate the mean with this confidence can be estimated by solving the following expression for n:

$$1.96\left(\sigma / \sqrt{n}\right) \le e \qquad \text{(B.24)}$$

and

$$n = (1.96s)^2 / e^2 \qquad \text{(B.25)}$$

where e = the desired tolerance (error).

t-DISTRIBUTION

The t-distribution is often used to test an assumption about a population mean when the parent population is known to be normally distributed but its standard deviation is unknown. In this case, the inferences made about the parent mean will depend upon the size of the samples being taken.

It is customary to describe the t-distribution in terms of the standard variable t and the number of degrees of freedom v. The degrees of freedom is a measure of the number of independent observations in a sample that can be used to estimate the standard deviation of the parent population; the number of degrees of freedom v is one less than the sample size ($v = n - 1$).

The density function of the t-distribution is given by

$$f(t) = \frac{\Gamma\left(\frac{v+1}{2}\right)}{\sqrt{v\pi}\,\Gamma\left(\frac{v}{2}\right)\left(1 + t^2/v\right)^{(v+1)/2}} \qquad \text{(B.26)}$$

and is provided in Table B.2. The mean is $m = 0$ and the standard deviation is

$$\sigma = \sqrt{\frac{v}{v-2}} \qquad \text{(B.27)}$$

Probability questions involving the t-distribution can be answered by using the distribution function $t_{\alpha,v}$ shown in Figure B.1. Table B.2 gives the value of t as a function of the degrees of freedom v down the column and the area (α) of the tail across the top. The t-distribution is symmetric. As an example, the probability of t falling within ± 3.0 when a sample size of $8(v = 7)$ is selected is one minus twice the tail ($\alpha = 0.01$):

$$P\{-3.0 < t < 3.0\} = 1 - (2 \times 0.01) = 0.98$$

The t-distribution is a family of distributions which approaches the Gaussian distribution for large n.

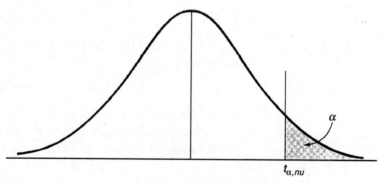

Figure B.1

Table B.2. *t*-Distribution; values of $t_{\alpha,v}$

Degrees of Freedom, *v*	Area of the Tail				
	$\alpha = 0.10$	$\alpha = 0.05$	$\alpha = 0.025$	$\alpha = 0.01$	$\alpha = 0.005$
1	3.078	6.314	12.706	31.821	63.657
2	1.886	2.920	4.303	6.965	9.925
3	1.638	2.353	3.182	4.541	5.841
4	1.533	2.132	2.776	3.747	4.604
5	1.476	2.015	2.571	3.365	4.032
6	1.440	1.943	2.447	3.143	3.707
7	1.415	1.895	2.365	2.998	3.499
8	1.397	1.860	2.306	2.896	3.355
9	1.383	1.833	2.262	2.821	3.250
10	1.372	1.812	2.228	2.764	3.169
11	1.363	1.796	2.201	2.718	3.106
12	1.356	1.782	2.179	2.681	3.055
13	1.350	1.771	2.160	2.650	3.012
14	1.345	1.761	2.145	2.624	2.977
15	1.341	1.753	2.131	2.602	2.947
16	1.337	1.746	2.120	2.583	2.921
17	1.333	1.740	2.110	2.567	2.898
18	1.330	1.734	2.101	2.552	2.878
19	1.328	1.729	2.093	2.539	2.861
20	1.325	1.725	2.086	2.528	2.845
21	1.323	1.721	2.080	2.518	2.831
22	1.321	1.717	2.074	2.508	2.819
23	1.319	1.714	2.069	2.500	2.807
24	1.318	1.711	2.064	2.492	2.797
25	1.316	1.708	2.060	2.485	1.787
26	1.315	1.706	2.056	2.479	2.779
27	1.314	1.703	2.052	2.473	2.771
28	1.313	1.701	2.048	2.467	2.763
29	1.311	1.699	2.045	2.462	2.756
inf.	1.282	1.645	1.960	2.326	2.576

HYPOTHESIS TESTING

Hypothesis testing belongs to that branch of statistics known as inferential statistics and is the basis for statistical decision making. To test a hypothesis is to make a decision regarding the reasonableness of the results obtained from statistical analyses.

In statistics, there are two hypotheses: the *null hypothesis* (H_0), and the *alternative hypothesis* (H_1). The test procedure uses sample data to make one of two statistical decisions: (1) reject the null hypothesis (as false) or (2) decide *not* to reject the null hypothesis. The test is performed on the null hypothesis. When we reject the null hypothesis, we accept the alternative hypothesis as being true. In making this decision, we incur two possible types of errors: (1) Type I error (concluding the hypothesis is true when it is really false) and (2) Type II error (concluding the hypothesis is false when it is really true). These errors are commonly referred to as α and β, respectively. The basic steps in hypothesis testing are outlined as follows:

Step 1: *State the statistical hypotheses.* If the parameters of interest are the means of two populations, the following hypotheses could be considered:

$$H_0 : \mu_1 = \mu_2, \qquad H_1 : \mu_1 \neq \mu_2$$
$$H_0 : \mu_1 \leq \mu_2, \qquad H_1 : \mu_1 > \mu_2$$
$$H_0 : \mu_1 \geq \mu_2, \qquad H_1 : \mu_1 < \mu_2$$

The first case is an example of a "two-sided" or "two-tailed" hypothesis. In this case, we are asking, "Can we conclude that the two populations have different means?" If the issue is which population has the larger mean, then the second or third statements would be appropriate. These represent "one-sided" or "one-tailed" hypotheses. In hypothesis testing, the alternative hypothesis is the statement of what we expect to be able to conclude. If the questions is whether Population 1 has a larger mean than Population 2, the $H_0 : \mu_1 > \mu_2$ would be tested. If this H_0 can be rejected, we accepted the $H_1 : \mu_1 > \mu_2$ as true.

Step 2: *Calculate the test statistics.* To test the hypothesis, the analyst selects an appropriate test statistics and specifies its distribution when H_0 is true; that is, the test procedure is based on the underlying distribution of the statistics used to estimate the parameters in question. For example, *testing the significance of the difference between means from two independent samples* may be based on the t statistics

$$t = (\bar{x}_1 - \bar{x}_2) / [s_p^2 (1/n_1 + 1/n_2)]^{1/2} \qquad \textbf{(B.28)}$$

where $\bar{x}_1$, $\bar{x}_2$ and n_1, n_2 refer to the means and sample sizes of the two groups, and s_p^2 is obtained by pooling the two sample variances s_1^2 and s_2^2:

$$s_p^2 = [(n_1 - 1)s_1^2 + (n_2 - 1)s_2^2] / (n_1 + n_2 - 2) \qquad \textbf{(B.29)}$$

Step 3: *State the "decision rule."* The issue here is to determine whether the magnitude of the test statistic computed from sample data in Step 2 is sufficiently extreme (either too large or too small) to justify rejecting H_0. Two basic approaches are commonly used to formulate the decision rule. In the first, the analyst rejects H_0 if the probability of obtaining a value of the test statistic of a given or more extreme value is equal to or less than some small number α (referred to as the level of significance). Commonly used values for α are 0.10, 0.05, or 0.01. The second approach involves stating the decision rule in terms of critical values of the test statistic. Because critical values are a function of the level of significance, the two approaches are equivalent. Tabulated values for α and the corresponding critical values for commonly used probability distribution can be found in many statistics textbooks. (The decision rule is usually formulated prior to even stating the statistical hypotheses. Its location in the sequence of steps presented here is largely illustrative.)

Step 4: Apply the decision rule. If the probability of obtaining the computed or larger value of the test statistic is $\leq \alpha$, reject H_0 and conclude that H_1 is true. Alternatively, if the computed value of the test statistic is greater than the critical value for the stated level of significance, reject H_0 and conclude that H_1 is true.

A summary of test statistics for use in several other important hypothesis testing situations is as follows:

Testing the difference between an observed and a hypothesized mean

$$Z = \left(\bar{x} - \mu_0 \right) / \left(\sigma / \sqrt{n} \right) \tag{B.30}$$

$$t = \left(\bar{x} - \mu_0 \right) / \left(s / \sqrt{n} \right) \tag{B.31}$$

where $\bar{x}$ is the observed mean and μ_0 is the value of the hypothesized mean, such as a know population mean. The test statistic in this case is compared to values in the standard normal distribution or the t-distribution, depending upon the size of n.

Tests concerning population variances

$$\chi^2 = (n-1)s^2 / \sigma_0^2 \tag{B.32}$$

$$F = s_1^2 / s_2^2 \tag{B.33}$$

Equation B.32 tests the difference between a hypothesized population variance σ_0^2 and a sample variance (s^2) using the chi-square distribution (χ^2). Equation B.33 uses values from an F distribution to determine whether two populations have equal variances. Values of the chi-square and F distribution can be found in standard statistics textbooks.

SAMPLING

The *confidence level* is the probability that a measurement is within a set interval and can be defined as $(1 - \alpha)$, or (1 minus the probability of a type I error), or equivalently, the probability of not making a type I error. A type I error is defined as *the probability of deciding a constituent is present when it is actually absent*, often called a false positive. It is also necessary to define the *power* as $(1 - \beta)$, or (1 minus the probability of a type II error), or equivalently, the probability of not making a type II error. A type II error is defined as *the probability of not detecting a constituent when it actually is present*, often called a false negative. Finally, the *minimum detectable relative difference* (*MDRD*) is defined as the *relative increase over the background measurement that is detectable with a probability of* $(1 - \beta)$. Mathematically, this may be expressed as

$$MDRD = \frac{\mu_s - \mu_b}{\mu_b} \times 100 \qquad \text{(B.34)}$$

where μ_s and μ_b are the sample and background mean values, respectively. The coefficient of variation (CV) is a measure of the sample variance relative to the sample mean and may be calculated as

$$CV = \frac{s^2}{\bar{x}} \qquad \text{(B.35)}$$

Once values for CV, power, confidence level, and MDRD are calculated or established by protocol, the number of samples required to achieve that MDRD is determined from Table B.3.

Table B.3 Number of samples required to achieve MDRD at $(1 - \alpha)$ and $(1 - \beta)$

Coefficient of Variation [%]	Power [%]	Confidence Level [%]	Minimum Detectable Relative Difference [%]				
			5	10	20	30	40
15	95	99	145	39	12	7	5
		95	99	26	8	5	3
		90	78	21	6	3	3
		80	57	15	4	2	2
	90	99	120	32	11	6	5
		95	79	21	7	4	3
		90	60	16	5	3	2
		80	41	11	3	2	1
	80	99	94	26	9	6	5
		95	58	16	5	3	3
		90	42	11	4	2	2
		80	26	7	2	2	1
25	95	99	397	102	28	14	9
		95	272	69	19	9	6
		90	216	55	15	7	5
		80	155	40	11	5	3
	90	99	329	85	24	12	8
		95	272	70	19	9	6
		90	166	42	12	6	4
		80	114	29	8	4	3

(continued)

Coefficient of Variation [%]	Power [%]	Confidence Level [%]	Minimum Detectable Relative Difference [%]				
			5	10	20	30	40
	80	99	254	66	19	10	7
		95	156	41	12	6	4
		90	114	30	8	4	3
		80	72	19	5	3	2
35	95	99	775	196	42	25	15
		95	532	134	35	17	10
		90	421	106	28	13	8
		80	304	77	20	9	6
	90	99	641	163	43	21	13
		95	421	107	28	14	8
		90	323	82	21	10	6
		80	222	56	15	7	4
	80	99	495	126	34	17	11
		95	305	78	21	10	7
		90	222	57	15	7	5
		80	140	36	10	5	3

Example **B.1**

MDRD

A set of samples from an abandoned industrial site detected a suspected groundwater contaminant at a mean concentration of 0.045 µg/L with $s^2 = 0.007$ µg/L. How many samples are required to be collected at a confidence level of 95% and power of 90% if the mean background concentration of the contaminant is 0.035 µg/L?

Solution

The minimum detectable relative difference (MDRD) is calculated from the mean concentration (C_{avg}) and the background concentration (C_{bg}) as follows:

$$\text{MDRD} = \frac{C_{avg} - C_{bg}}{C_{bg}} \times 100 = \frac{0.045 - 0.035}{0.035} \times 100 = 28.6\%$$

Also, the coefficient of variation (CV) may be calculated as

$$\text{CV} = \frac{s^2}{\bar{x}} = \frac{0.007}{0.045} = 15.6\%$$

Therefore, from Table B.3 at MRDD = 30% and CV = 15% with $(1 - \alpha) = 95\%$ and $(1 - \beta) = 90\%$,

$$\Rightarrow n = 4$$

Periodic Table of the Elements

Periodic Table of the Elements

	6	Atomic Number
	C	Symbol
	Carbon	Name
	12.0107	Average Atomic Mass

Legend

Hydrogen
Semiconductors
(also known as metalloids)

Metals
- Alkali metals
- Alkaline-earth metals
- Transition metals
- Other metals

Nonmetals
- Halogens
- Noble gases
- Other nonmetals

* The systematic names and symbols for elements greater than 111 will be used until the approval of trivial names by the IUPAC.

The discoveries of elements with atomic numbers 112, 114, and 116 have been reported but not fully confirmed.

The atomic masses listed in this table reflect the precision of current measurements. (Each value listed in parentheses is the mass number of that radioactive element's most stable or most common isotope.)

Group 1	Group 2	Group 3	Group 4	Group 5	Group 6	Group 7	Group 8	Group 9	Group 10	Group 11	Group 12	Group 13	Group 14	Group 15	Group 16	Group 17	Group 18
1 **H** Hydrogen 1.007 94																	2 **He** Helium 4.002 60
3 **Li** Lithium 6.941	4 **Be** Beryllium 9.012 182											5 **B** Boron 10.811	6 **C** Carbon 12.0107	7 **N** Nitrogen 14.0067	8 **O** Oxygen 15.9994	9 **F** Fluorine 18.998 4032	10 **Ne** Neon 20.1797
11 **Na** Sodium 22.989 769 28	12 **Mg** Magnesium 24.3050											13 **Al** Aluminum 26.981 5386	14 **Si** Silicon 28.0855	15 **P** Phosphorus 30.973 762	16 **S** Sulfur 32.065	17 **Cl** Chlorine 35.453	18 **Ar** Argon 39.948
19 **K** Potassium 39.0983	20 **Ca** Calcium 40.078	21 **Sc** Scandium 44.955 912	22 **Ti** Titanium 47.867	23 **V** Vanadium 50.9415	24 **Cr** Chromium 51.9961	25 **Mn** Manganese 54.938 045	26 **Fe** Iron 55.845	27 **Co** Cobalt 58.933 195	28 **Ni** Nickel 58.6934	29 **Cu** Copper 63.546	30 **Zn** Zinc 65.409	31 **Ga** Gallium 69.723	32 **Ge** Germanium 72.64	33 **As** Arsenic 74.921 60	34 **Se** Selenium 78.96	35 **Br** Bromine 79.904	36 **Kr** Krypton 83.798
37 **Rb** Rubidium 85.4678	38 **Sr** Strontium 87.62	39 **Y** Yttrium 88.905 85	40 **Zr** Zirconium 91.224	41 **Nb** Niobium 92.906 38	42 **Mo** Molybdenum 95.94	43 **Tc** Technetium (98)	44 **Ru** Ruthenium 101.07	45 **Rh** Rhodium 102.905 50	46 **Pd** Palladium 106.42	47 **Ag** Silver 107.8682	48 **Cd** Cadmium 112.411	49 **In** Indium 114.818	50 **Sn** Tin 118.710	51 **Sb** Antimony 121.760	52 **Te** Tellurium 127.60	53 **I** Iodine 126.904 47	54 **Xe** Xenon 131.293
55 **Cs** Cesium 132.905 4519	56 **Ba** Barium 137.327	57 **La** Lanthanum 138.905 47	72 **Hf** Hafnium 178.49	73 **Ta** Tantalum 180.947 88	74 **W** Tungsten 183.84	75 **Re** Rhenium 186.207	76 **Os** Osmium 190.23	77 **Ir** Iridium 192.217	78 **Pt** Platinum 195.084	79 **Au** Gold 196.966 569	80 **Hg** Mercury 200.59	81 **Tl** Thallium 204.3833	82 **Pb** Lead 207.2	83 **Bi** Bismuth 208.980 40	84 **Po** Polonium (209)	85 **At** Astatine (210)	86 **Rn** Radon (222)
87 **Fr** Francium (223)	88 **Ra** Radium (226)	89 **Ac** Actinium (227)	104 **Rf** Rutherfordium (261)	105 **Db** Dubnium (262)	106 **Sg** Seaborgium (266)	107 **Bh** Bohrium (264)	108 **Hs** Hassium (277)	109 **Mt** Meitnerium (268)	110 **Ds** Darmstadtium (271)	111 **Rg** Roentgenium (272)	112 **Uub*** Ununbium (285)	113 **Uut*** Ununtrium (284)	114 **Uuq*** Ununquadium (289)		116 **Uuh*** Ununhexium (292)		

58 **Ce** Cerium 140.116	59 **Pr** Praseodymium 140.907 65	60 **Nd** Neodymium 144.242	61 **Pm** Promethium (145)	62 **Sm** Samarium 150.36	63 **Eu** Europium 151.964	64 **Gd** Gadolinium 157.25	65 **Tb** Terbium 158.925 35	66 **Dy** Dysprosium 162.500	67 **Ho** Holmium 164.930 32	68 **Er** Erbium 167.259	69 **Tm** Thulium 168.934 21	70 **Yb** Ytterbium 173.04	71 **Lu** Lutetium 174.967
90 **Th** Thorium 232.038 06	91 **Pa** Protactinium 231.035 88	92 **U** Uranium 238.028 91	93 **Np** Neptunium (237)	94 **Pu** Plutonium (244)	95 **Am** Americium (243)	96 **Cm** Curium (247)	97 **Bk** Berkelium (247)	98 **Cf** Californium (251)	99 **Es** Einsteinium (252)	100 **Fm** Fermium (257)	101 **Md** Mendelevium (258)	102 **No** Nobelium (259)	103 **Lr** Lawrencium (262)

I N D E X